TREATISE ON ANALYTICAL CHEMISTRY

A comprehensive account in three parts

PART I

THEORY AND PRACTICE

PART II

ANALYTICAL CHEMISTRY OF
INORGANIC AND ORGANIC COMPOUNDS

PART III

ANALYTICAL CHEMISTRY IN INDUSTRY

TREATISE ON ANALYTICAL CHEMISTRY

Edited by I. M. KOLTHOFF
School of Chemistry, University of Minnesota

and PHILIP J. ELVING
Department of Chemistry, University of Michigan

with the assistance of ERNEST B. SANDELL
School of Chemistry, University of Minnesota

PART II

ANALYTICAL CHEMISTRY OF INORGANIC AND ORGANIC COMPOUNDS

VOLUME 10

AN INTERSCIENCE ® PUBLICATION

JOHN WILEY & SONS, New York—Chichester—Brisbane—Toronto

An Interscience® Publication

Copyright © 1978 by John Wiley & Sons, Inc.

Library of Congress Catalogue Card Number: 59–12439

ISBN 0-471-49998-6

Printed in the United States of America

10 9 8 7 6 5 4 3 2 1

TREATISE ON ANALYTICAL CHEMISTRY

PART II

ANALYTICAL CHEMISTRY OF INORGANIC AND ORGANIC COMPOUNDS

SECTION A

Systematic Analytical Chemistry of the Elements

VOLUME 10: **ANTIMONY • ARSENIC • BORON CARBON • MOLYBDENUM TUNGSTEN**

AUTHORS OF VOLUME 10

Robert S. Braman

W. E. Chambers

Paul D. Coulter

Ronald A. Greinke

Richard B. Hahn

Gordon A. Parker

Richard F. Skonieczny

Authors of Volume 10

R. S. Braman

University of South Florida Department of Chemistry, Tampa, Florida

W. E. Chambers

Union Carbide Corporation, Carbon Products Division, 270 Park Avenue, New York, New York

P. D. Coulter

Union Carbide Corporation, Parma Technical Center, P.O. Box 6116, Cleveland, Ohio

Ronald A. Greinke

Union Carbide Corporation, Parma Technical Center, P.O. Box 6116, Cleveland, Ohio

Richard B. Hahn

Wayne State University, Department of Chemistry, Detroit, Michigan

Gordon A. Parker

Department of Chemistry, University of Toledo, 2801 W. Bancroft Street, Toledo, Ohio

Richard F. Skonieczny

Department of Health, Herman Kiefer Health Complex, 1151 Taylor, Detroit, Michigan

PART II. ANALYTICAL CHEMISTRY OF INORGANIC AND ORGANIC COMPOUNDS

CONTENTS—VOLUME 10

SECTION A. Systematic Analytical Chemistry of the Elements

Boron. By *Robert S. Braman*

Carbon. By *W. E. Chambers, Paul D. Coulter, and Ronald A. Greinke*

Antimony. By *Richard B. Hahn*

Molybdenum.   By *Gordon A. Parker*

Tungsten. By *Gordon A. Parker*

TREATISE ON ANALYTICAL CHEMISTRY

A comprehensive account in three parts

PART I

THEORY AND PRACTICE

PART II

ANALYTICAL CHEMISTRY OF INORGANIC AND ORGANIC COMPOUNDS

PART III

ANALYTICAL CHEMISTRY IN INDUSTRY

BORON

By Robert S. Braman, *Department of Chemistry,*
University of South Florida, Tampa, Florida

Contents

I. INTRODUCTION

The discussion of the analytical chemistry of boron presented in this chapter is confined largely to inorganic compounds of boron and to methods for total boron assay in the case of organoboron compounds. The analytical chemistry or organoboron compounds and the boranes is presented elsewhere (general reference 1).

Applications of methods of analysis for boron are extensive in the literature. As a consequence an attempt has been made to be selective rather than exhaustive in referencing applications.

A number of reviews and general references are available on the analytical chemistry of boron: A selection of these appear under General References.

Most of the analytical methods for boron reviewed are total boron assay methods. Biological, soil, fertilizer, alloy, and food samples are invariably chemically treated so as to oxidize organic matter, remove interfering ions, and generally prepare boron in the boric acid or borate anion chemical form. Separations, where used, generally involve treatment of borate anion or boric acid. Finishing determinative procedures generally involve boric acid or boric anion chemistry.

Sample treatment and separation steps are critical parts of any approach to boron analysis, particularly where microgram amounts of boron are to be determined. Metal ions, weak acid anions, and oxidizing agents can be troublesome interfering materials. Borosilicate glassware, useful for just about any other wet analytical chemistry except trace boron analysis, can definitely lead to high boron results. Dry ashing samples in ovens can cause gains in boron—or losses.

In adapting methods for boron reviewed here to their own needs, analytical chemists must consider interferences, sample size, and convenience of the analytical method. Extraction of boric acid from aqueous media into chlorinated solvents with 2-ethylhexane-1,3-diol, originating with work by DeFord and George (154), is becoming more popular. Colorimetric determinitive methods involving curcumin, carminic acid, or 1,1-dianthrimide continue to be the most popular choices for small amounts of boron. The mannitol–borate titration procedure has not been challenged by any other approach for amounts of boron above (very) approximately 1 mmole, with perhaps the exception of neutron absorptiometry.

Naturally, essentially all instrumental analytical methods have been used for boron analysis or for determination of selected boron compounds. Judgment of applicability of these to specific problems is left to the reader.

A. OCCURRENCE

It has been estimated that boron constitutes approximately 0.001% of the Earth's crust. It is a comparatively mobile element geochemically and is found in sea water, which contains 4.6 ppm boron by weight. Boron is one of the conserved elements in sea water; it remains in comparatively constant ratio to other major sea water constituents. Chemically, boron is found in nature almost entirely as borate or derivative anions. Boron-rich minerals or deposits are usually found where they were laid down as the result of evaporation of inland lakes of volcanic or marine origin. Boric acid is present in the waters of certain volcanic springs in Italy and Sicily. These sources probably are a consequence of the steam volatility of boric acid and the solubility of alkali metal borates.

The most economical, extractable boron minerals (see also Table 1) are the borax series: borax, $Na_2B_4O_7 \cdot 10H_2O$; kernite, $Na_2B_4O_7 \cdot 4H_2O$; and tincalconite $Na_2B_4O_7 \cdot 5H_2O$. A lens-shaped deposit containing these minerals was found in Kern County, California in 1925. This deposit probably formed from the evaporation of a lake of volcanic origin containing dissolved sodium borate and boric acid. It is almost pure borax minerals. Searles Lake, California brine is a second major source of boron and together with the Kern county deposit supplies most of the world's boron. Less economical deposits of calcium and magnesium borates such as colemanite, $Ca_2B_6O_{11} \cdot 5H_2O$; ulexite, $NaCaB_5O_9 \cdot 8H_2O$; and kurnakovite, $Mg_2B_6O_{11} \cdot 15H_2O$ are commonly found throughout the world. These deposits were utilized largely before the discovery of the major California deposits. Major boron mineral deposits in the Lake Inder region of Kazakhstan, USSR, are of the alkaline earth borate type. Boron also occurs as a complex borosilicate, tourmaline, in many silicate rocks.

As a consequence of its chemical nature, boron is found associated with sodium, calcium, magnesium, aluminum, silicon, manganese, and iron. It is the need to separate boron from these elements, transition metals in general, fluoride ion, and weak acids that complicates analytical procedures.

TABLE 1[a]

A Selection of Boron-Bearing Minerals

Mineral	Composition
Borax	$Na_2O \cdot 2B_2O_3 \cdot 10H_2O$
Kernite	$Na_2O \cdot 2B_2O_3 \cdot 4H_2O$
Tincalconite	$Na_2O \cdot 2B_2O_3 \cdot 5H_2O$
Ulexite	$Na_2O \cdot 2CaO \cdot 5B_2O_3 \cdot 16H_2O$
Colemanite	$2CaO \cdot 3B_2O_3 \cdot 5H_2O$
Hydroboracite	$CaO \cdot MgO \cdot 3B_2O_3 \cdot 6H_2O$
Kurnakovite	$2MgO \cdot 3B_2O_3 \cdot 15H_2O$
Ezcurrite	$2Na_2O \cdot 5B_2O_3 \cdot 7H_2O$
Priceite(Pandermite)	$4CaO \cdot 5B_2O_3 \cdot 7H_2O$
Boracite	$5MgO \cdot MgCl_2 \cdot 7B_2O_3$
Tourmaline	$XY_3Al_6(BO_3)_3(Si_6O_{18})(OH)_4$ (X = Na, Ca Y = Al, Fe^{3+}, Li, Mg)
Datolite	$2CaO \cdot B_2O_3 2SiO_2 \cdot H_2O$
Danburite	$CaO \cdot 2SiO_2 \cdot B_2O_3$
Sassolite	H_3BO_3
Ferruccite	$NaBF_4$

[a] General reference 2.

B. INDUSTRIAL USES OF BORON

Borate chemicals and boric acid have a wide variety of industrial applications. Borax is used in the ceramics industry, principally for the manufacture of borosilicate and similar glasses. Boric oxide in these glasses influences both the refractive index and the coefficient of expansion.

Borax has long been used as a washing compound, a disinfectant, and a mouthwash. These applications make use of its mild alkaline buffering and disinfectant properties.

Boric acid and borates are used in treating textiles, lumber, plywood, and other combustible materials to impart fire-retardant properties. Borax and certain organoboron compounds have fungicidal properties that make them good preservatives or additive agents.

Borax is an excellent solvent for metal oxides and has long been used as a welding flux and as a cleansing agent in nonferrous metallurgy. Borax glass is a common fusion material used in analytical spectroscopy.

Elemental boron is widely used in the metal industry to increase the hardenability of steel. Added as the ferroboron alloy, it is usually present in steels in the range, 0.001–0.005%. Boron is used as a deoxidizer for nonferrous alloys and as a degasifier in high conductance copper and copper base alloys, and it is used to refine the grain of aluminum castings. A small amount of elemental boron is used in the semiconductor industry as a doping agent in silicon and germanium.

Because of its high thermal neutron absorption cross-section, boron and its compounds have been employed alone or dispersed in hydrogenous neutron moderators as neutron shields.

With the discovery some 50 years ago that boron is one of the dozen or so known mineral elements essential for plant growth, a considerable interest has developed in boron in plant tissues, soil, fertilizers, and mineral waters. Although necessary for plant growth, borates when present in large quantities act as unselective herbicides.

Boron trifluoride and boron trichloride are used as catalysts in organic synthesis and analytical esterification procedures.

The most recent interest in the field of boron chemistry rests with the boron hydrides and organoboron compounds. Boron hydrides, in particular, have been produced, at least in pilot plant quantities, for study as high-energy fuels. An excellent review of the industrial uses of boron and boron compounds is available in the Kirth-Othmer Encyclopedia of Chemical Technology, second edition (220).

C. TOXICITY

Finely divided elemental boron constitutes the same type of hazard as other metals. It is a strong reducing agent. Boric acid and borate type compounds are not highly toxic and are not considered industrial poisons. Boric acid has been the cause of fatal poisoning in children by accidental ingestion. The fatal dose for adults is estimated to be above 15–20 grams.

Boron halides, by virtue of their reactivity with moisture in air, produce HF, HCl, and HBr mists, which are themselves toxic and corrosive. Threshold limit values for these compounds are in the 3–5 ppm range.

Boron hydrides are by far the most hazardous of the boron compounds. Some are pyrophoric in addition to being toxic.

II. PHYSICAL AND CHEMICAL PROPERTIES

Physical and chemical properties of boron and some selected boron compounds are given in Tables 2 and 3. Elemental boron, rarely encountered commercially, is usually found as the amorphous powder. It is comparatively unreactive at ordinary temperatures but reacts to form binary compounds with halides, oxygen, sulfur, and many metals at elevated temperatures. Fluorine reacts to form boron trifluoride at room temperature. Crystalline boron is less reactive than amorphous boron but undergoes the same reactions at the same or higher temperatures. It reacts with ammonia to form boron nitride at elevated temperatures.

The borides are largely high melting, unreactive, hard solids. Most are not attacked by water but are dissolved by acids or oxidizing agents.

Boric acid, metaboric acid, and boric oxide are three members of a series of compounds that can be considered to be various degrees of dehydration of boric acid. In metaboric acid, boric oxide, and intermediate compounds, oxygen is shared between two boron atoms, thus leading to polymeric structures of various degrees of polymerization. The boric acid series of compounds forms esters with alcohols, diols, or polyols under nonaqueous, dehydrating conditions. This reaction is basic to the formation of boron complexes in colorimetric determinations and for the formation of methyl borate, which is distilled in separation procedures. Fluoride ion is the only complexing agent capable of severing the boron–oxygen bond in aqueous solutions. The oxygen compounds of boron can be reduced, either electrolytically or by

TABLE 2[a]

Physical Properties of Boron

Atomic number: 5
Atomic weight: 10.811 + 0.003
Atomic electron configuration: $1s^2\ 2s^2\ 2p^1$

Isotopes	Half life	Natural occurrence	Thermal neutron absorption cross-section
B^8	0.78 sec	—	—
B^{10}	Stable	19.8%	3898 barns
B^{11}	Stable	80.2%	0.05 barns
B^{12}	0.19 sec	—	—
B^{13}	0.035 sec	—	—

Melting point: 2100°C approx.
Boiling point: 2600°C approx.
Density (grams/ml):   2.3 (amorphous, 20°C)
                      2.31–2.5 (crystalline, 20°C)
Crystalline modifications: tetragonal, α, β rhombohedral
Ionic radius (crystal)   3+ charge; 0.23 Å
Oxidative states in boron compounds:

Compound type	Oxidation state
H_3BO_3	3+
$RB(OH)_2$	1+
R_2BOH	1−
B_2H_6	3−

[a] General reference 3.

the thermite reaction and other high-temperature processes, to produce intermediates in the preparation of the hydrides, halides, or crystalline boron. The alkoxy compounds may also be reduced by alkali hydrides along another path to hydride or boron alkyl preparations.

Borax, sodium borate, and other borate-type compounds are salt-like and have weak acid or base properties. They are otherwise comparatively unreactive.

Sodium borohydride and other alkali metal borohydride compounds are likewise salt-like compounds, water soluble, and stable in aqueous solution above pH 7. The borohydride ion is rapidly hydrolyzed in acidic medium, but sodium borohydride in slightly alkaline solution can even be used as a standard titrimetric reducing agent. It must be standardized, nevertheless, and restandardized each day. Sodium and potassium borohydrides are used as reducing agents for ketone groups in organic synthesis and analysis.

The alkali-metal tetraphenylborates, which may be considered to be derivatives of alkali-metal borohydrides, have achieved considerable

TABLE 3[a]

Properties of Selected Boron Compounds

	mp (°C)	bp (°C)	Density (g/ml)	Remarks
BCl_3	−107	12.5	1.432 (200°C)	Corrosive, hydrolyzable
BF_3	−127	−101	1.87 (−130°C)	Corrosive, hydrolyzable
B_2O_3	325	1500	1.844 (20°C)	Poorly soluble, unreactive
H_3BO_3	185d.		1.435 (15°C)	Poorly soluble, unreactive
$Ha_2B_4O_7 \cdot 10H_2O$	75°C	−10H$_2$O 200°C	1.73	Salt-like
B_4C	2400	3500	2.52	Unreactive, dissolves in alkalies
$B(OCH_3)_3$	−29	68.7	0.932	Readily hydrolyzed
$NaBH_4$	500d.		1.04 (20°C)	Water stable above pH 7, reducing agent
$NaB(C_6H_5)_4$	d.	d.	—	Analytical reagent for K^+, Rb^+, Cs^+, NH_4^+
KBF_4	530		2.5	Unreactive, insoluble
$B_3H_6N_3$	−58 to −56	53	0.8618 (0°C)	Similar to benzene
MgB_2	2235	—	2.1–2.3 (15°C)	Reacts with aq. acids, forms boron hydrides
TiB_2	2900	—	—	Unreactive
$Ca(BO_2)_2 \cdot 2H_2O$	−H$_2$O (200–300)	—	—	Salt-like
BN	2500 subl.	—	2.29	Cubic crystalline form is extremely hard

[a] From several sources (general references 1–3).

analytical interest. Sodium tetraphenylborate has become a widely used gravimetric reagent for the determination of potassium, rubidium, cesium, ammonium ion, basic nitrogen compounds, and certain heavy metal ions. The solubility of the potassium, rubidium, cesium, and ammonium tetraphenylborates ranges from 1 to 5 $\mu g/100$ ml water (153). Both gravimetric (340) and titrimetric (391) methods have been reported. A series of reviews have appeared on analytical applications (22,23).

Borazine and its derivatives resemble benzene and its derivatives in both physical and chemical properties. Esters or boric acid are generally easily hydrolyzable. Comparatively stable compounds are formed with triethanolamine and its derivatives. Triethanolamine borate is only partially hydrolyzed in neutral aqueous solutions (257).

Boron–oxygen compounds are not electroactive. Nevertheless, some polarographic work has been done with borate complexes and is described Section VII.J. Boron hydrides and the borinic and boronic acids are reducing agents and are consequently electroactive. For a general presentation on electrochemical properties of boron, *Oxidation Potentials* by Latimer (245) is recommended.

III. SAMPLE HANDLING (Loss of Boron by Volatilization)

Considerable care must be exercised in the preparation of boron-containing samples in order to avoid loss of boron by volatilization. Precautions are obviously required in the handling of volatile boron compounds such as methyl borate, boron trifluoride, or the boranes. Nevertheless, the loss of boron from borate-ion-containing samples in such operations as boiling of acid solutions, evaporation, drying, ignition, and fusion is a real if more subtle problem.

Tchijewski in 1884 was apparently the first of numerous authors to point out that boric acid is volatilized from aqueous solutions when the latter are boiled. Studies since then have indicated that the loss of boron from boiled or evaporated aqueous solutions is dependent upon the acidity, temperature, and the other chemicals present.

In a study of solutions that were from 0.03 to 5 M in boric acid and were boiled at atmospheric pressure, it was found that the ratio of the weight percentage of boric acid in the gas phase to the weight percentage in the liquid phase is 0.0036 ± 0.0003 (200). Volatility was found to be independent of the boiling rate.

It has been reported (20) that boric acid volatility is unaffected by the presence of strong acids such as HCl, HBr, or HNO_3 in solution. A recent careful study (131) has shown that hydrochloric acid solutions of boric acid exhibit substantially greater losses of boron upon evaporation to dryness on a water bath (75–80°C) than solutions of other mineral acids. In general, salts that lower the solubility of boric acid in water increase its volatility, while those that increase its solubility decrease its volatility. The volatility of boric acid is effectively reduced by compounds such as tartrates, citrates, mannitol, and sugars that form nonvolatile complexes with boric acid (248).

Based on this data, it is evident that the loss of boric acid should be negligibly small on brief boiling to effect sample dissolution or removal of carbon dioxide (upon acidification) prior to the standard titration method for boron. For example, less than 0.1% of the boric acid in solution should be lost from a solution boiled for a period sufficient to evaporate 10% of the original volume. This is supported in the literature (439) and is also in accord with the experience of the author. Consequently, extra precautions such as boiling in a special apparatus at low temperature under reduced pressures to remove carbon dioxide are usually unnecessary.

Although the amount of boron in the vapor space above boiling dilute solutions containing boric acid is small, as solution temperatures rise the concentration of boron in the vapor increases rapidly. For example, the saturated boric acid vapor in equilibrium with a saturated solution of boric acid at 104.4°C contains 28.8 mmol boric acid/kg water vapor, 93.5 mmol at 119.9°C, and 372 mmol at 140.4°C (201). The vapor above a saturated aqueous solution of boric acid at its boiling point (103.2°C) contains 27.4 mmol/kg vapor (440).

The volatility of boric acid in the presence of steam at elevated temperatures (1100–1300°C) has actually been made the basis of a separation technique in a pyrohydrolysis method.

Additional data on the volatility of boric acid can be found in the work of Bezzi (41) and Tazaki (415). These authors indicate that boric acid is volatilized as H_3BO_3 and not as a hydrate or as metaboric acid and substantiate the contention that boron concentration in vapor increases with temperature.

The partial pressures of undissociated boric acid and of water above solid H_3BO_3 at various temperatures are given in Table 4 (41).

Apparently even fused boric oxide, which is obtained from boric acid upon heating to 500°C, loses 5–6% of its weight on heating for 3–5 hr (262).

There is considerable doubt concerning the volatility of borax. In an old (1898) but apparently carefully done study borax was found to be lost as such when heated in a blast lamp. Kolthoff (223) was unable to

TABLE 4
Partial Pressure of Water and Boric Acid Over Solid Boric Acid

Temperature (°C)	Partial pressure of boric acid (mm Hg)	Partial pressure of water (mm Hg)
20	—	1.1
40	—	3.3
60	—	13.3
80	—	41.6
90	0.17	70.3
100	0.317	118
105	0.41	—
110	0.59	—
119		290

detect any loss of borax when heated for 2 hr at 800°C in an electric furnace. On the other hand, in other work (62,63) borax lost sodium oxide when heated for several hours at 700–800°C.

Cole and Taylor (88) studied the vapor pressure of $Na_2O \cdot B_2O_3$ and $Na_2O \cdot 2B_2O_3$. Assuming vapor densities correspond to these formulas, the vapor pressures for the two compounds are calculated to be 2.55 and 2.1 mm Hg at 1150°C, 7.2 and 5.5 mm Hg at 1200°C, and 62 and 25.5 mm Hg at 1300°C, respectively. Both Na_2O and B_2O_3 volatilized in each case in an approximate 1:1 molar ratio. Metal borates apparently decompose and volatilize in the range, 1200–1400°C (397), but there is no indication of boron volatilization at temperatures below 1000°C for alkaline-boron residues.

Some authors have reported the loss of boron from solutions made alkaline with ammonia. This is not surprising because the solutions become more acidic upon evaporation:

$$BO_2^- + NH_4^+ + H_2O \rightleftharpoons H_3BO_3 + NH_3$$

The loss of boron upon digestion or evaporation to dryness of sample solutions or mixtures has been frequently noted. This problem is merely an extension to more severe conditions or the loss of boron by volatilization from solutions. Serious losses occur in the evaporation of mineral acid solutions of boric acid with the exception that phosphoric acid apparently completely inhibits boron loss (24). This is probably due to the formation of boron phosphate. Volatilization of boron from solutions of boric acid in water, nitric acid, sulfuric acid, and perchloric acid at 75°C on a water bath is less than 3% until low solution volumes (80% of original evaporated) are reached. Hydrochloric acid causes serious losses of boron earlier in the evaporation than do most other acids. Mannitol in a 10:1 mole excess over the boron effectively prevents the loss of boron upon evaporating of acidified solutions. Boron is also volatilized from concentrated sulfuric acid at 228°C and from fuming perchloric acid (131). Precautions should be taken to prevent evaporation to dryness of localized splatters on the side of glassware when heating acidified solutions.

Methanol enhances the volatilization of boric acid from basic residues. For example, it was reported (374) that 41% of the boron was lost when a solution containing 12 mg of boric acid, 112 mg of potassium hydroxide, and 100 ml of methanol was evaporated. The addition of 100 ml of water decreased boron loss to 2%. The methanol concentration should be kept below 50% to avoid loss of boron, even in the presence of a 25:1 excess of strong base over the boron present.

Boron is easily lost from solutions containing compounds or ions that can react to form volatile substances. In addition to the loss of boron as methyl borate indicated previously, boron is also lost as boron trifluoride or fluoboric acid from solutions containing hydrofluoric acid.

Selection of the best treatment of natural materials by wet and dry ashing is controversial. Wet ashing of plant materials leads to significant losses of boron (177). On the other hand, boron contamination was observed during dry ashing in an electric furnace at 550°C (451). The latter authors indicated that tall-form porcelain crucibles apparently eliminated the furnace contamination. Dry ashing in the presence of calcium hydroxide in tall-form crucibles appears at this time to be preferable to wet-ashing procedures unless modifications are made to minimize boron loss by volatilization. Calcium hydroxide is needed in the ashing of plant vegetative tissue to avoid boron losses (159).

Boron values are reported to be high on the first use of new platinum crucibles (91).

Wet-ashing procedures that have proven to be acceptable generally are of the closed-reaction-chamber type. For example, oxidation of organoboron compounds in a Paar bomb (19), Carius oxidation (384), and the Schöninger flask technique (461) is generally acceptable.

IV. SEPARATION METHODS

A. INTRODUCTION

Nearly all of the chemical methods for the determination of boron are subject to interference by a large number of compounds or ions. Particularly troublesome are weak acids and certain transition metal ions. Experience has indicated that in nearly every situation in which boron is to be determined, a separation from other constituents is required.

Of the variety of methods proposed for the isolation of boron, especially from inorganic materials, the distillation method developed by Chapin (445) in 1908 was long considered the most reliable, most versatile, and perhaps the most exact method available. Its main disadvantages are the time required, the comparative complexity of the procedure, and the need for boron-free glassware. Many methods for precipitating interfering substances while leaving the boron in solution have been proposed, but nearly all are of limited applicability or are inexact due to the tendency of boron to be lost by coprecipitation.

Progress is being made in the development of extraction techniques. Of particular interest is the 1,3-diol extraction method originally discovered by DeFord and George (154). Only one earlier method based upon the extraction of the boron by an immiscible solvent has been described; the applicability of this method is restricted to the analysis of glasses and similar products. A significant number of methods of general applicability including 1,3-diol extraction have been appearing, and further work is definitely indicated.

Some promising work has been done with ion-exchange methods for removing interfering ions. Recent gas-chromatographic methods have dealt almost exclusively with volatile boron compounds. Electrolytic separations are reported but have limited application.

B.   DISTILLATION METHODS

1.   Distillation of Boron as the Methyl Ester

The separation of boron by distillation of the methyl ester was first proposed independently by Gooch and by Rosenbladt in 1887. Many investigators, particularly Chapin (445), have since modified and improved the original methods. The distillation method is still in use today and remains a versatile, general method for the separation of boron. Separation is based upon the volatilization of methyl borate, $B(OCH_3)_3$, when acid solutions are boiled with methyl alcohol. The rate of formation and the rate of hydrolysis of this ester are very fast, in fact essentially instantaneous. The ester itself boils at 68.7°C. With methanol the ester forms an azeotrope containing 75.5% by weight of the ester and boiling at 54°C at 750 mm (372). The method serves equally well for the isolation of boron prior to its determination and for the removal of boron from solutions. Boron also forms esters with higher alcohols, but only the methyl ester has been extensively used in analytical separations. In brief, the solubilized sample is mixed with a dehydrating agent such as anhydrous calcium chloride, methyl alcohol and an acid (HCl) are added, and the distillation is started. The distillate is hydrolyzed and analyzed for boron by titration or by a suitable colormetric procedure.

2. Modifications of the Distillation Method

A modification of the Chapin distillation method especially designed for the determination of boron in natural waters and in plant leaves has been reported (449). Copper flasks and beakers or boron-free glassware were used to avoid extraction of boron from the glassware. This precaution may not have been necessary, since no evidence for extraction of boron from ordinary glassware has been found by numerous later investigators. The precaution of using boron-free glassware depends upon the amount of boron being distilled and the opportunity for glassware to be dissolved in sample treatment steps. Blank runs are advised in all cases. If less than 1 mg of boron is being sought, contamination from glassware can become important.

Research at the National Bureau of Standards (186, p. 757) has shown that the residue remaining in the distillation flask at the end of the distillation must be subjected to further treatment if all of the boron is to be recovered. This treatment is particularly necessary if high silicon and boron contents are encountered. The residue is filtered, washed, ignited, fused, and again carried through the distillation procedure.

Various investigators have suggested the use of drying agents other than calcium chloride. Sulfuric acid has been employed frequently in preparative work; this acid is undesirable even for this purpose, however, because of side reactions, particularly the formation of dimethyl ether (372).

Zinc chloride (375) and phosphoric acid (350) have been used to effect the dehydration. The evaporation of the boron-containing solution to a syrupy state before distillation and without addition of a dehydrating agent has been reported (411). Three distillations were required to remove all of the boron. The success of the technique probably results from dehydration of the sample by distillation of the methanol-water azeotrope.

Difficulty is experienced in determining trace quantities of boron (micrograms) in the presence of large amounts of fluoride when using the distillation technique. Addition of a methanol solution of aluminum chloride is effective in complexing the fluoride, and recovery of boron is better than experienced in the presence of calcium chloride alone (146). Large excesses of aluminum are apparently to be avoided, however, since large amounts of aluminum render complete distillation of the boron difficult (143).

Calcium chloride effectively retains fluoride in ordinary distillations. For example, in a sample containing 0.1500 gram of boric oxide and 0.2 gram of sodium fluoride, 0.1517 gram of boric oxide was found after separation by distillation in the presence of calcium chloride. Even under these adverse conditions the error amounts to only 1%. With low concentrations of fluoride the error is negligible.

Interest in distillation continues, a recent example being a study in which trimethylborate was found to be quantitatively distilled by isothermal distillation (244). Coal analysis has been done by titration after methylborate distillation (7). Nickel alloys have been analyzed for boron after methylborate distillation (458). BF_3-borazine complexes have been decomposed and analyzed by titration after distillation of methylborate (128).

3. Distillation Separation of Boron as BCl_3 or as BF_3

Successful methods based upon the volatilization of boron trichloride have been described (276,456). This technique, although little used now, has been found suitable for the separation of boron from iron alloys and is claimed to be more rapid and convenient than the Chapin distillation method. Finely divided metal samples are treated with chlorine gas in a tube heated to 450–500°C. The volatile boron trichloride is removed and allowed to react with water, producing boric acid and hydrochloric acid. Phosphorous interferes. The method is suitable only for the analysis of samples in which boron can be converted entirely to BCl_3. Boron-oxygen compounds are not volatilized as BCl_3.

A method involving separation of boron as the trifluoride has been reported for the determination of boron in soils. Samples were first heated with calcium peroxide, placed in the distillation flask, treated with a minimum of fluoride and excess perchloric acid, and distilled. The quinalizarin colorimetric method is used to finish the analysis. Fluoride ion does not interfere in the quinalizarin method, but it must be remembered that silicon will accompany the boron. The distillation of boron as the trifluoride has often been used to remove boron prior to the determination of other constituents.

4. Pyrohydrolysis

The volatility of boric oxide at melting-glass temperatures coupled with the facility of the steam distillation of boric acid suggested that the pyrohydrolysis technique might be applied to the separation of boron from inorganic materials. Williams et al. (452) developed and applied the technique to the analysis of glass. Samples were mixed with uranium oxide and sodium metasilicate monohydrate, placed in a platinum tube, and heated to 1300–1350°C while passing steam over the mixture. Boric acid in the distillate was determined by titration. Lead and zinc above 4–6% in samples interfered as did 1% phosphorus pentoxide. Application of this technique appears best for samples containing 1% or more B_2O_3. The distillation is time consuming, but up to 5% fluorine in samples apparently did not interfere. Application of the same technique to the analysis of alloys has been reported (448). The method has also been studied for the assay of elemental boron (114) and boron alloys (232).

C. PRECIPITATION METHODS

The borates of several heavy metals are only slightly soluble. Nevertheless, no method for the quantitative precipitation of metal borate has been yet perfected. The precipitates are of indefinite and variable composition and tend to carry by coprecipitation large amounts of other ions in the solution. References to the early literature describing attempts to precipitate boron quantitatively may be found in the Gmelin Handbook (156). The precipitation of boron as nitron tetrafluoborate is quantitative and is also used as a gravimetric procedure (256). It has not been applied purely as a separation technique.

Two new gravimetric reagents have been reported, 1,1-diantipyrinyl-butane and α,α-diantipyrinyltoluene (4). These presumably could also be used for separation alone.

Most of the precipitation separation methods have been developed for removing interfering ions from solutions prior to boron analysis. Although most are effective, again there remains the danger of coprecipitation and subsequent loss of boron. Precipitation separations offer a comparatively rapid means of effecting removal of interfering ions in routine analyses, but few, if any, of the methods that have been described can be recommended for exact work.

Most of the early reported precipitation methods involve treatment of the sample solution with an alkali or alkaline earth hydroxide or carbonate. Interfering metal ions are precipitated as the hydrous oxides or carbonates. Barium, strontium, or calcium hydroxides or carbonates are employed in most of the procedures in order to effect precipitation of phosphate, chromate, fluoride anions, and metal cations.

Barium carbonate, together with barium hydroxide or sodium hydroxide, has been used to make solutions slightly alkaline for the precipitation of interfering ions in the determination of boron in metal borides

(45,46). Chromate ion, if present, is precipitated as barium chromate. A similar method has been used for the precipitation of interfering ions in the analysis of ferrous alloys (64). When aluminum and fluoride ions are present, calcium carbonate is used to insure quantitative precipitation of both. Iron and aluminum salts, phosphate, and ammonia are removed by the addition of barium chloride and barium hydroxide prior to the determination of boron in fertilizers (353). This precipitation method was included in an official method of the A.O.A.C. (14) for the determination of water-soluble boron in fertilizers.

The addition of strontium chloride has been recommended for the precipitation of silicate, soaps, carbonate, and orthophosphate prior to the determination of boron in soaps and detergents (43,44).

Boron is heavily coprecipitated in the precipitation of metal ions as hydrous oxides (362). Hydrous oxides absorb boron, usually causing interference when the concentration of the latter is high, but the absorption is negligible when the boron concentration is low (224, p. 117).

A method was developed employing a single precipitation with 8-quinolinol for the removal of zinc, nickel, iron, lead, and aluminum (362). Undoubtedly other metals that are precipitated by this reagent are also removed, or could be with modification. The precipitation was effected in alkaline medium at 60°C. Magnesium chloride was added to precipitate most of the excess reagent. Activated charcoal was added to remove final traces of reagent. After acidification and boiling to remove CO_2, the analysis was finished by acidimetric titration of the boron. The use of silver ion was recommended for removal of phosphate and arsenate ions. Silicate was removed by adding ammonium chloride to the alkaline solution followed by neutralization with HCl and digestion.

The quinolinol precipitation method appears to be sufficiently exact and rapid to be of considerable use. Nearly all of the other precipitation methods either to isolate boron or to remove interferences appear unsuitable because of lengthy reprecipitations necessary to recover all of the boron. Nevertheless, in many quality-control applications the less exact methods may find use.

D.   EXTRACTION SEPARATIONS

1.   Liquid Extractions

Despite the comparatively low solubility of boric acid in water, it is not readily extracted with water-immiscible solvents. Far better success has been achieved with the extraction of boron as a 1,3-diol complex or as a BF_3 complex.

A variety of water-immiscible solvents have been investigated for the possible extraction of boric acid (132). Distribution coefficients of boric acid are all unfavorable, ranging from approximately 0.02 for chloroform, ether, benzene, benzaldehyde, and carbon tetrachloride to 0.22 and 0.18 for n-pentanol and octanol, respectively. The calculated distribution coefficients for water-immiscible solvents containing possible chelating agents of boric acid are likewise poor; boric acid distribution coefficients are less than 0.4 even in favorable cases. For example, triethanolamine in chloroform exhibited a distribution coefficient for boron of 0.056; ethyleneglycol in ether gave 0.029. The low values obtained with these two good complexing agents for boric acid is attributed to the anionic complexes that they form with boron in aqueous media.

The distribution of boric acid between amyl alcohol and water has also been studied (1,295). The distribution coefficient was found to be approximately 0.3. It should be possible to extract the boric acid quantitatively by multiple or continuous extraction with the immiscible alcohols.

Distribution of the boric acid–catechol complex between aqueous phase (0.2 M catechol) and dichloromethane is 95–98% complete at pH 4.6 (47). This is superior to other 1,2-diol type complex extractions and deserves further attention.

DeFord and George (154) have recently discovered that although 1,2-diol complexes of boric acid are not readily extractable from aqueous media, complexes of certain 1,3-diols are readily extractable into immiscible solvents. Extraction of the 1,3-diols is attributed to their formation of a neutral, planar complex with boric acid. Distribution coefficients of boric acid between aqueous, acidic solutions and chloroform containing 2-ethyl-1,3-hexanediol, 2-methyl-2,3-pentanediol, or 2,2-diethyl-1,3-propanediol ranged from 6 to 10. Of these, 2-ethyl-1,3-hexanediol, the least water soluble complexing agent, exhibited the most favorable distribution coefficient. Quantitative extraction of boric acid from acidified aqueous media was achieved by four extractions utilizing chloroform solutions in the 1,3-diol. Chloroform was found to be superior to toluene, ether, carbon tetrachloride, and n-butanol as an immiscible solvent for the extraction. Increasing ionic strength of aqueous solutions improved boric acid distribution coefficients. None of the common metal ions and acids tested interfered in the extraction procedure, but copper, zinc, lead, fluoride, phosphate, and acetate ions are extracted to a slight extent and interfere in the subsequent titrimetrimetric determination of boric acid in the extract. To eliminate interferences from fluoride ion and from metal ions, an EDTA-phosphate buffer technique was developed. A phosphate buffer of pH 6.0–6.5 was prepared and EDTA added. At this pH the tetrafluoborate anion does not form, and EDTA complexes many cations. The EDTA–phosphate buffer eliminates the interferences mentioned previously. Boric acid could even be quantitatively extracted from samples of fluoboric acid if the latter was first hydrolyzed on a steam bath for 3 or 4 hr in the presence of the EDTA–phosphate buffer.

Other similar studies include boron extraction by 3-methylbutane-1,3-diol (388) and by several diols, including 2,2-dimethylpropane-1,3-diol and 2-ethylpropane-1,3-diol (111).

This extraction method is the most promising to date. As little as 1 mg of boron can be extracted with 1–2% accuracy. Although the lower limit of the extraction was not established, there is not apparent reason to prevent its use with microgram quantities of boron upon proper adaptation.

The extraction procedure has been used prior to a flame photometric determination (3). Applications of the 1,3-diol extractions have been made to analysis of fertilizers (322) with a carmine colorimetric reagent and to analysis of boron in plants (283) with an atomic absorption finishing step. Titration of boron is also possible after 3-methylbutane-1,3-diol extraction into chloroform (387).

A liquid–liquid extraction method involving the extraction of boric acid with an ether–alcohol mixture has been reported (155). The technique was developed for the determination of boron in glass and involves a special application of the partition law. Boric acid is extracted into ethyl ether from a strongly acidic aqueous solution containing 50% alcohol by volume. The fused glass sample of fixed weight (0.5 gram) is dissolved, neutralized, and excess sulfuric acid is added. The sample volume is adjusted to 25.0 ml, 25.0 ml of absolute ethanol is added, and the solution is extracted with 50.00 ml of ether. An aliquot of the ether layer is then titrated with standard base.

The percentage of boric oxide in the glass sample is calculated from the equation:

$$\%B_2O_3 = 4(B_2O_3)_{Et_2O}\left[V_{Et_2O} + \frac{V_{H_2O}}{k}\right]$$

where $(B_2O_3)_{Et_2O}$ represents the weight of boric oxide found in the 50 ml ether aliquot, the Vs represent the volumes of the two phases, and k represents the distribution coefficient obtained from the following equation:

$$k = (0.417 - 0.00232T)$$

For temperatures between 20 and 30°C results agreed to within ±0.1% absolute with the known percentage of B_2O_3 in glass samples. Several other modifications of the technique have been reported (341,442). The technique is faster than the Chapin distillation technique and appears satisfactory for routine work where high accuracy is not necessary. Si, Ca, Ba, Mg, Al, Na, Ti, Fe, Zn, Pb, and As do not interfere; F^- gives low values. It does not, however, appear to be as suitable for accurate work as the 1,3-diol extraction method.

Tetrafluoborate anion complexes with a number of other organic compounds are quantitatively extractable into organic solvents. This approach has received considerable interest. Table 5 gives data on a selected group of references. A number of different finishing, analysis approaches have been used. The extraction procedures generally are applied to small amounts of boron.

2. Solid–Liquid Extractions

Several examples of the extraction of boric acid directly from solids have been reported. For example, boric acid is separated from talcum powder (also containing calcium and/or magnesium carbonates, zinc stearate, and perfume) by extraction with boiling water (25). Boiling water has also been used to extract boric acid from soils (379). These techniques are limited of course because all soluble compounds are extracted with the boric acid. Pyrohydrolysis (see Section IV.B.4) could be considered a special case of solid–liquid extraction, but it is really a distillation method.

Solid extractions employing ethanol have also been reported in very old work (293) for the analysis of glass samples; acetone has been used for the extraction of boric acid from the ash remaining after the ignition of organic materials (74) and ether for the extraction of boric acid from evaporates of mineral waters (33).

E. ION EXCHANGE METHODS

Several investigators have studied the applicability of ion-exchange resins for the separation of boron from various cations. All work to date has been confined to aqueous solutions in which the boron exists as borate ion, boric acid, or a borate-polyalcohol complex. Samuelson (361) has recommended the use of cation-exchange resins for the separation of borate ion, along with other interfering anions, from cations prior to the usual qualitative tests for cations. No studies on the separation of borate ion from other anions have been reported.

TABLE 5

Extraction of BF_4^- Complexes

BF_4^- Complex with	Solvent	Remarks	References
Methyl violet	benzene	Colorimetric Steel analysis	329
Methyl green	dichloroethane	Mg range, colorimetric	320
	Trichloroethylene	0.1–5 μg B/ml	413
Tetrabutylammonium cation	methylisobutyl ketone	Several micrograms, flame photometric	260
Rhodamine B	Butylacetate	Microgram range, colorimetric	314
Tris(1,10-phenanthroline) Iron(II) sulfate	nitrobenzene	At ppm range, colorimetric	308
	butyronitrile	colorimetric, 0.3–8 μmol B	13
Nile blue A	chlorobenzene	Fertilizer analysis, 0.002–2.0 μg B/ml, colorimetric	147
Janus green		2–5 μg in 5 ml	166
Phenazone dyes	$CHCl_3$-CCl_4 (1:1)	Method study up to 5 μg B	69

Numerous separations of borate ion or boric acid from ions that interfere in the titration of boric acid have been reported. Table 6 gives some selected references for separations.

Among the advantages claimed (270) for the ion-exchange methods are freedom from occlusion or coprecipitation of boric acid, freedom from a blank correction (in most cases), no requirement of boron-free glassware since only acidic solutions of boric acid need to be boiled, no lengthy evaporations, simple equipment, more rapid determinations, and greater accuracy.

In the separation of microgram quantities of boron, a mixed-bed resin column of strong cation-exchange resin (Nalcite HCR) and a weak-base anion-exchange resin (Amberlite IR 45) has been used (459). Passage of the sample through the resin resulted in a nearly deionized aqueous solution containing boric acid and other weak acids such a silicic acid and carbonic acid. The method was not recommended as a separation technique for microgram quantities of boron prior to other finishing analysis techniques because of errors in the 3–5 μg range. In later work the source of error was found to be due to boron in the anion-exchange resins (73). Anion resins completely regenerated with sodium hydroxide have lost all residual boron. Unfortunately, the freshly regenerated resins were found to pick up considerable boron from subsequent samples. A rinsing procedure for anion resins prior to use in separations involving microgram quantities of boron was recommended. The boron content of rinse water from a weak-base resin in the regenerated form as received from the manufacturer decreased to less than 1 μg in 250 ml after 1 liter of rinsing. Rinsing must be repeated because the boron content of the aqueous phase in the resin bed increases upon standing. The method appears to be excellent for the separation of milligram quantities of boron from most interfering substances except fluoride ion. The separation of microgram quantities of boron requires careful conditioning of the column to avoid contamination or loss of significant amounts of boron. It is apparent that further investigation of the mixed-bed column technique using other weak-base anion resins now available is warranted.

The sugar-borate complexes are retained on anion-exchange resins of the strongly basic type (217). This behavior has been made the basis of methods for the separation of sugars and of glycerol, but applications to the separation of boron have not been studied.

An anion-exchange column technique has been used to simultaneously remove interfering substances and convert boron to the BF_4^- anion form before detection by an ion-specific electrode (75). Boron in solution is passed through a column of Amberlite XE-243 boron-specific resin.

TABLE 6

Ion Exchange Separation of Interfering Ions from Boron

Sample	Ions removed	Ion-exchange material	Reference
$Pb(BF_4)_2$, HBF_4, electrolyte	Pb^{2+}	Cation resin	98
Ni^{2+}, H_3BO_3, electrolyte	Ni^{2+}	Cation resin	2
Ti^{4+}, H_3BO_3, HCl	Ti^{4+}	Cation resin	305
Zn, Ni, H_3BO_3, plating bath	Ni^{2+}, Zn^{2+}	Cation resin (Amberlite IR 120)	145
Zn, Al, etc. in deodorants	Zn^{2+}, Al^{3+}, etc.	Cation resin (Amberlite IR 120)	84
Silicates	Al^{3+}, Fe^{3+}, Si	Cation resin (Ion-X)	236
Ti-B alloys	Ti^{4+}	Cation resin	310
Natural waters	Fe^{3+}, Al^{3+}, Zn^{2+}, NH_4^+, etc.	Cation resin	297
Fertilizers	Phosphate	Anion resin (Amberlite IRA 400)	377
Solutions	Ca^{2+}	Cation resin	142
Solutions	CrO_4^{2-}, VO_3^-	Cation resin	270
Solutions	NH_4^+, Be^{2+}, Mg^{2+}, Ca^{2+}, Zn^{2+}, Cd^{2+}, Fe^{2+}, Ni^{2+}, Co^{2+}, Ca^{2+}, Hg^{2+}, Al^{3+}, Fe^{3+}, Th, Sn^{4+}, Ti^{4+}, U, and Zr	Cation resin	66, 297

Interfering nitrate and iodide ions are removed from the column by washing it with dilute aqueous ammonia. A solution of 10% hydrofluoric acid is then added to the column to produce BF_4^- anion. Excess hydrofluoric acid is removed by water, and the BF_4^- anion is eluted by adding 0.3 M NaOH. The eluate is passed directly through a Dowex 5W-X8 cation exchange resin in the acid form, and the resulting eluate is measured for fluoroboric acid content using a BF_4^- ion-specific liquid ion-exchange electrode. The concentration limit of detection is approximately 0.01 mmol or 0.1 ppm as boron.

F. CHROMATOGRAPHIC METHODS

A few chromatographic separations of inorganic boron compounds have been reported. They have been performed largely on paper and have dealt with microgram quantities. Boron, silicon, and molybdenum have been separated by paper chromatography (240,241). Acetone acidified with HCl was used as developer; the separation was performed in an atmosphere saturated with 2-butanone plus 5% concentrated HCl. A chromatographic method has been used for the isolation of boron prior to its determination in mineral waters (338).

Gas-chromatographic methods have been employed widely for the separation and determination of the volatile organoboron compounds and boranes in mixtures. These are outside the intended scope of coverage of this chapter.

Gas chromatography of metal halides including boron trichloride has been reported (57). Separation of BHF_2 and BF_3 can be obtained on a 50% Kel-F oil or Chromosorb-W column (94).

Boranes are less difficult than the halides but still require careful handling because of their pyrophoric behavior. The series, $B_2H_6-B_{10}H_{14}$, was separated on a 6% OV-17 on 80/100 mesh Chromosorb-W column using a programmed temperature gas chromatograph and a flame emission type detector (399). A response study of the flame emission type detector shows that the lower limit of detection for boranes is approximately 0.7 ng (398).

G. SEPARATIONS BY ELECTROLYSIS

Electrolytic deposition into a mercury cathode is a convenient and rapid means for removing many metals from solution. Nearly all metals below zinc in the electrolytic series can be removed in this way. Boron is not removed from solution by this technique. Unfortunately, several other metals such as aluminum, beryllium, magnesium, the alkaline earths, vanadium, zirconium, and the rare earths are also left in the aqueous solution. Since many of these metals interfere in the determination of boron, the electrolytic method is of limited value.

The mercury cathode has been used for the removal of iron, chromium, manganese, etc. prior to the determination of boron in steel (210) and alloys (157). Interfering substances were removed prior to colorimetric determinations.

V.   DETECTION AND IDENTIFICATION

A.   INTRODUCTION

A great variety of methods have been described that are suitable for the detection of boron. Limits of detection range from as little as a few nanograms to several milligrams of boron. Some are applicable to fractional part per million concentrations of boron; others require concentrations of the order of several parts per thousand. Some are almost completely specific for boron, while others are subject to many interferences. Methods have been developed for the detection of boron in nearly every type of natural and commercial product.

Although it is not feasible to describe all qualitative tests that have been proposed, an attempt has been made in this section to cite a selection of significant papers dealing with each of the more important methods. Critical comments have been included where possible. Details of a few typical procedures are included.

Obviously, methods for detection and identification of boron can nearly all be used for quantitative analysis purposes and vice versa. Consequently, some overlap of information exist between this section and that on quantitative methods of analysis (Section VII). Included here are all referenced organic complexes of boron. Only those in comparatively wide use are treated in the quantitative analysis section. Most instrumental methods for boron analysis are treated only in the sections on quantitative methods.

The more widely used methods for the detection of boron may be grouped into three major classes: (1) color and fluorescence tests, (2) flame tests, and (3) tests based upon changes in pH.

Boric acid reacts with several types of organic reagents to produce colored or fluorescent products. In most cases the colored compounds are chelate complexes, although in a few instances other types of colored products are formed.

Volatile compounds of boron, when injected into a flame, impart a green coloration to the flame. This phenomenon forms the basis of several procedures for the detection of boron.

Boron acid reacts with many polyols to form a tetraalkoxy borate ion and hydronium ion. This behavior serves as the basis for the detection of boric acid, since the pH of the reaction solution containing boric acid decreases when a polyalcohol such as mannitol is added.

B.   COLOR AND FLUORESCENCE TESTS EMPLOYING ORGANIC REAGENTS

No inorganic compounds of boron are suitable for its colorimetric or fluorometric detection. Consequently, all the detection methods described in this section are based upon a reaction between boric acid and an organic reagent. Boric acid forms many colored complexes with organic reagents (mostly bidentate ligands) possessing hydroxyl groups. The hydroxyl groups may be alcoholic, phenolic, or enolic in character. Some dehydrating solvent such as concentrated sulfuric acid is invariably used as a solvent for color reactions.

Korenman (229) reviewed compounds that develop color or fluorescence with boric acid and concluded that the reaction is a six-membered ring closed by boron. Korenman grouped color reagents into the first five of the following groups. To Korenman's original groups have been added compounds that form five-membered rings, derivatives of catechol and haematoxylin, amines, and miscellaneous compounds. Thus the color and fluorescence reagents are grouped for discussion into the following classifications.

1.   Derivatives of 1-hydroxyanthraquinone
2.   Derivatives of flavone
3.   Derivatives of 1,8-dihydroxynaphthalene
4.   Derivatives of salicylic acid
5.   Derivatives of acetylacetone
6.   Derivatives of catechol and haematoxylin
7.   Amines
8.   Miscellaneous materials

1.   Derivatives of 1-hydroxyanthraquinone

Hydroxyanthraquinones and similar compounds are all phenolic in nature and form esters with acids. Furthermore, the hydroxyanthraquinones are soluble in sulfuric acid, and esterification can occur in this solution because the solvent acts as a dehydrating agent driving the reaction to completion. When boric acid solutions of the reagents in concentrated sulfuric acid are heated, ring formation occurs (107–109,130). An example with quinalizarin can be represented as follows:

The color reactions of various compounds in the hydroxyanthraquinone class are summarized in Table 7. The most widely used reagents are discussed in more detail in the following.

a.   Quinalizarin

Quinalizarin (alizarin 3R(By), alizarin bordeaux) is 1,2,5,8-tetrahydroxyanthraquinone. It dissolves readily in sulfuric acid with the formation of a bluish-violet color. The reagent forms a blue color in sulfuric acid in the presence of boric acid. Nitrates, oxidizing agents, and fluorides interfere with color formation. Quinalizarin is more sensitive

TABLE 7

Hydroxyquinone Complexes of Boric Acid in H_2SO_4

	Color of	
Complexing agent	Reagent	Complex
1-Hydroxyanthraquinone	Yellow	Orange
2-Hydroxyanthraquinone	Red-yellow	Red-yellow
1,2-Dihydroxyanthraquinone	Wine-red	Violet
1,3-Dihydroxyanthraquinone	Yellow	Yellow
1,4-Dihydroxyanthraquinone	Red	Yellow-fluorescent
1,5-Dihydroxyanthraquinone	Red	Red-violet
1,8-Dihydroxyanthraquinone	Red-yellow	Red
2,3-Dihydroxyanthraquinone	Brown	Brown
2,6-Dihydroxyanthraquinone	Red	Red
2,7-Dihydroxyanthraquinone	Red	Red
Alizarinsulfonic acid	Red-yellow	Red
1,2,3-Trihydroxyanthraquinine	Brown	Violet-brown
1,2,4-Trihydroxyanthraquinine	Orange	Wine-red
1,4,5,8-Tetrahydroxyanthraquinone	Blue	Blue
1,2,5,8-Tetrahydroxyanthraquinone	Violet	Blue

than purpurin (369) and is almost specific for boron. Of materials similar to boric acid, only germanic acid produces a color with quinalizarin (410). Approximately 0.001 mg of boron in a concentration of 1 ppm can be detected by the color formation in sulfuric acid.

The effect of concentration of sulfuric acid has been studied (393) and found to be optimum at 93%. Quinalizarin and many of the other color reagents for detecting boron are also used in quantitative procedures discussed in Section VII.

b. Alizarin-S

Alizarin-S is the sodium salt of alizarinsulfonic acid (1,2-dihydroxyan-thraquinone-3-sulfonic acid). When dissolved in sulfuric acid it becomes red-yellow in color and changes to red upon the addition of boric acid (130). As little as 1 μg of boron may be detected.

When observed under an ultraviolet light, as little as 0.02 μg of boron can be detected (410). Interfering ions include ferric, iodide, chlorate, and antimony (III). These ions interfere in both the color and fluorescence test when present at levels of 1 part per thousand. Ferrous, antimony(V), carbonate, sulfide, sulfate, phosphate, chloride, ammonium, and many other common mono- and divalent cations do not interfere. It was found that silicate, bromide, nitrate, cobalt, and chromium interfered in the color test but not in the fluorescence test.

c. Purpurin

Purpurin is 1,2,4-trihydroxyanthraquinone. In concentrated sulfuric acid purpurin gives an orange color that changes to a wine-red upon addition of boric acid. A 0.5% solution of purpurin in sulfuric acid may be substituted for quinalizarin or alizarin-S in the test procedures to yield a method that will detect as little as 6 μg of boron (130).

d. Cochineal, Carmine, Carminic Acid, and Carmine Red

Cochineal is the dried female insect, *Dactylopius cocus* (218). Carmine is the aluminum lake of cochineal; hence neither are pure compounds. Carminic acid comprises about 10% of cochineal. Carmine red is the substance obtained by boiling a solution of carminic acid with a few drops of hydrochloric acid. Carminic acid in sulfuric acid changes from red to blue in the presence of boric acid. As little as 0.1 μg of boron can be detected in 0.03 ml of solution (464). Qualitative test procedures parallel those of color reagents already mentioned.

e. 1-Amino-1-hydroxyanthraquinone

A solution of this reagent in concentrated sulfuric acid will produce an intense orange-brown fluorescence in the presence of boric acid. As little as 0.1 ppm borax will give a positive test (339). The reagent was the most suitable of 60 compounds tested as reagents for the fluorescent detection of boric acid (120).

f.   1-Hydroxy-4-*p*-toluidinoanthraquinone

This derivative of 1-hydroxyanthraquinone called CI Solvent Violet 13 has been used for analysis of boron in steel. The color is developed in 85% H_2SO_4 and is suitable for boron down to the 10 μg sample size (32).

2.   Derivatives of Flavone

Early work by Wilson (454) on the color formation of flavones with boric acid pointed out the necessity of the following type of structure for color formation:

$$\underset{R-C_v-C_w-C_x}{\overset{\overset{a}{|}}{}}\overset{\overset{O}{\|}}{-C_y\!\!=\!\!C_z-R'}$$

in which a is an auxochrome group such as $=$O, —OH, or —OCH$_3$; R, C_v, and C_w may form a benzene ring; and C_x, C_y, and C_z may form a portion of a pyran ring. R' is always a hydroxylated or methoxylated benzene ring (454). Morin and pentamethylquercetin are the two main color indicators of the flavone class.

Morin is 3,5,7,2',4'-pentahydroxyflavone. In acetone solution containing citric acid Morin gives a yellow color that becomes deeper yellow upon the addition of boric acid. This change in intensity of color can be better used in quantitative analysis. Morin with oxalic acid also complex with boric acid. Used as a fluorescent method the limit of detection is 0.4 ng B/ml (336).

Quercetin is 3,5,7,3',4'-penthydroxyflavone. Pentamethylquercetin is the pentamethyl ether of quercetin and has been used for boron detection (342). When the reagent is dissolved in acetone containing citric acid, it forms a colorless solution. Upon addition of boric acid a yellow color develops. By means of this color change, approximately 30 ppm of boric acid is detectible. The test is sensitive to water, which discharges the color.

Other naturally occurring hydroxyflavones that give a color with boric acid are chrysin, luteolin, kaempferol, quercetin, quercetagetin, quercetagitin, quercimeritrin, cannabiscetin, cannabiscitrin, herbacetin, gossypetin, gossypitrin, hibiscetin, hibiscitrin, and patuletin. Methyl ethers of some of these also give color tests. Chalkone derivatives such as 2,4-dihydroxy-4'-methoxychalkone and other chalkones also give color reactions. It is probable that sensitivity is similar to the pentamethylquercetin test for boric acid.

Complexes of polyhydroxyflavones and oxalic acid with boron have been recently studied (337). A sensitivity of 0.02 μg B has been obtained using Morin and oxalic acid (92).

3. Derivatives of 1,8-Dihydroxynaphthalene

The main compounds of interest in this group is p-nitrobenzeneazo-chromotropic acid, also called Chromotope-2B, or its sodium salt. The acid, 3-(p-nitrobenzeneazo)-4,5-dihydroxy-2,7-naphthalenesulfonic acid, in concentrated sulfuric acid gives a blue color that changes to greenish-blue upon addition of boric acid. A color change to bluish-violet or greenish-blue occurs when as little as 0.08 μg of boron is present in 0.04 ml of sample (227,330). Fluorides and oxidizing agents interfere. A study of chromotropic acid complexes has been reported (26).

Azomethine H, which is 4-hydroxy-5-(salicylideneamino)naphthalene-2,7-disulfonic acid, can also be placed in this group (385). It has been used in an autoanalyzer method for plant tissue analysis (27).

4. Derivatives of Salicyclic Acid

Reagents in this class have received little attention. Although salicylic acid itself does not give a color reaction with boric acid, its derivatives containing chromophores do. One such derivative is aluminon. Korenman described two procedures utilizing this reagent (229). The concentration limit of detection for boron was as low as 1 ppm. Tungstic, molybdic, hydrofluoric, phosphoric, chromic, and other acids did not interfere.

Acetylsalicylic acid has been studied as a fluorometric reagent for boron (328). The complex has an emission maximum near 410 nm compared to 430 nm for the reagent. Down to 0.01 μg B/ml may be detected.

Boron complexes with salicylic acid interact with dyes such as Rhodamine-B. Possible use of this in analysis has been studied (434).

5. Derivatives of Acetylacetone

The only reagent in this class that has been studied extensively is curcumin or turmeric. Although the terms curcumin and turmeric are often used synonymously, they do not refer to the identical reagent. Turmeric is the powder obtained from grinding the rhizomes of various species of Curcuma; hence it is a mixture. Turmeric contains cellulose, gum, starch, mineral matter, volatile oil, brown coloring matter, and a characteristic yellow coloring matter that is called curcumin (418).

Curcumin is 1,7-bis(4-hydroxy-3-methoxyphenyl)-1,6-heptadiene-3,5-dione (198). Curcumin is one of the oldest reagents for boron, having been used in early work as curcuma or tumeric paper. When the yellow aqueous acidified solution of curcumin is heated with boric acid, a redish-brown isomer, called rosocyanine, is formed. When rosocyanine is treated with base it is converted to a bluish compound that can be converted back to a rosocyanine by treatment with acid. The color reaction test for boron may be carried out either on a glass plate, in solution, or on paper or other fibers impregnated with the reagent. Numerous old references can be found concerning or utilizing the paper test procedure for boron.

Bertrand and Agulhon (38,39) used the curcuma paper technique for the detection of as little as 0.5 μg B/ml in the presence of organic matter. These authors developed their paper test into their well-known quantitative procedure. The A.O.A.C. included methods based upon tumeric paper for the detection of boron in water, brine, salt, and preservatives.

In 1951 Hegedus (179) critically reviewed the methods based upon the color reaction of yellow curcumin and boric acid to form rosocyanine and reported that the following substances interfere with the reaction: fluoride ion, titanium, zirconium, molybdenum, beryllium, tungsten, and oxidizing agents such as nitrate, chlorate, bromate, iodate, and nitrite ions. To prevent interference it was suggested that boron be distilled as the methyl borate before the test and that oxidizing agents be reduced before the test.

The usual procedure for the curcumin test involves treatment of the sample with HCl and an alcoholic solution of 0.1% of curcumin and evaporation to dryness. A red or red-brown residue indicates the presence of boron. The limit of detection of the curcumin test has been variously reported and is approximately 0.01–0.02 μg at a concentration of 0.4 ppm (286).

6.   Derivatives of Catechol and Hematoxylin

Korenman and Sheyanova (230) are apparently the first authors reporting the use of azo dye derivatives of catechol and hemotoxylin as color agents. These types give five-membered rings with boron. Limiting detectible concentrations of the several reagents investigated ranged from 2 to 0.05 parts per thousand.

Nitropyrocatechols were studied for use as color reagents (167). The best was 4-nitropyrocatechol, which formed 2:1 and 1:1 complexes with borates in aqueous solution. A sensitivity of 15 ppm per absorbance limit was observed at pH 7.5–8.5.

7. Amine Derivatives

The use of triethanolamine as a reagent for boron has been studied by Jaffe (199) and Lucchesi (255). Copper sulfate, when added to a solution of boron in triethanolamine, is green. The test is not sensitive; the detection limit is in the 10–40 μg/ml region.

Possibly the most important amines in this class are 1,1'-dianthrimide (120) and 4,4'-diamino-1,1'-dianthraquinoyl amine (30). Of these 1,1'-dianthrimide is the most widely used. It has the structure

which is related to the hydroxyanthraquinones. These reagents have been used in quantitative methods for boron and presumably can be used in suitably devised spot tests.

Mimosa (also called tincture of mimosa, titan yellow, clayton yellow, cotton yellow, thiazole yellow, and color index 813) in sulfuric acid yields a brownish-yellow solution. When a solution containing boric acid is made weakly alkaline, boiled, treated with mimosa and hydrochloric acid until the yellow color disappears, and then evaporated to dryness, a yellow residue remains. If boron is present, a red color results when the yellow residue is made alkaline with sodium carbonate. The limit of detection is 0.07 μg. Tartaric acid, citric, oxalic, and acetic acids interfere and must be removed by ignition before applying the test. Most of the work with mimosa has been done by Robin (349).

Diaminoanthrarufin (diamino-9,10-anthraquinone) also can be placed in this class of reagents (460).

8. Miscellaneous Color Reagents

A few reagents not falling in the previous categories have been used in color tests for boron. Notable are congo red (403), α-nitroso-β-napthol (343), resacetophenone (301), benzoin (327,446), dibenzoylmethane (265), and opium alkaloids (344). Some are based upon pH effects, and none appear to be superior to previously mentioned indicators.

Boron has been determined as the 4'-chloro-2-hydroxy-4-methoxybenzophenone complex. The color is developed in sulfuric acid and has a fluorescence maximum at 490 nm with a limit of detection at 10 ng B in water samples of solution (250,290).

Some sixteen different 9-substituted 2,6,7-trihydroxy-xanthen-3-ones have been studied as fluorescent reagents for boron. Phenylfluorene was considered best; down to 10 μg B could be determined (300).

Azure A and Azure C dyes have been studied for boron analyses (443).

C. FLAME TESTS

The flame test is probably the oldest qualitative test for boron and was known as early as 1732. It is certain that in that year Claude Geoffroy knew that boric acid imparted a characteristic green color to an alcohol flame. Volatile boron compounds impart a green color to flames. The color is attributed to the B—O band system in the 500–600 nm region. The flame test for boron can be made of general use so long as nonvolatile boron-containing compounds can be converted to volatile boron compounds. Flame tests may be used directly on solids or solutions, but judgment must be exercised to avoid interference from matrix materials that may obscure the boron flame color or impart their own green coloration to the test flame. The performance of the flame test with volatile boron compounds appears to be superior chiefly because of the elimination of interference in the volatilization process.

Sensitivity of the flame tests for boron depend upon the luminosity and color of the flame source, and luminosity and color caused by the matrix materials introduced into the flame with the sample or boron compound sought. It was found, for example, that hydrocarbon flames of butane gas exhibited much less sensitivity for the detection of volatile boron compounds than did hydrogen–air flames (55). Hydrocarbon–air flames have a high luminosity in the visible region due to C_2 and C—H bands and black body radiation from carbon particles. This explains in part the higher sensitivity obtained by several early authors who employed alcohol or hydrogen–air flames in the flame tests for boron. Alcohol–air flames have a lower luminosity in the visible region than hydrocarbon flames.

The flame test for boron has, of course, been widely used for the determination of boron by flame photometry. The limit of detection for boron compounds can be quite low. Braman and Gordon (55) report the detection of 10–20 parts per billion pentaborane in air. Detection limits for boron will be lower if volatile boron compounds are detected in gas phase than if aqueous solutions are employed. The limit of detection for boron in solutions by flame photometry is approximately 10 ppm. The difference between this sensitivity and that obtained with gas samples may be attributed to the volatilization of the water sample and the cooling effect of water on the flame.

Early work on the detection of boron involved the direct introduction of the boron-containing sample into the flame with little or no treatment. This is of course the same procedure employed in the conventional flame-photometric determination of boron. As mentioned before, sensitivity will be influenced by matrix elements copresent and will be difficult at best unless the test is performed by an instrumental analytical technique with a monochrometer to isolate the boron wavelengths.

Volatile boron trifluoride was used by Turner to detect boron as early as 1827. Several modifications of his method have been reported. The method basically consists of converting borates or boric acid to boron trifluoride in a tube or some apparatus that facilitates injection of vapors from the reaction into a flame. Bertrand and Agulhon (38) treated samples with hydrofluoric and sulfuric acid and forced the evolved gases into a capillary with hydrogen. Geilman and Bode (152) treated 5–10 mg of sample with an equal weight of calcium fluoride and moistened with a drop of concentrated sulfuric acid. The mixture taken up on a platinum coiled loop and held 2–3 mm from the flame imparted a green color to it if 2–3 μg of boron were present. In a modification of this technique samples prepared as before were placed in a boron-free tube and heated while a gas stream was passed through the heated tube. The effluent gas was ignited, and the sample was warmed carefully until evolution of fumes ceased. As little as 0.25 μg of boron gives an easily discernible green coloration. If boron-containing samples are heated with $(NH_4)_2SiF_6$, most of the boron is converted to the volatile NH_4BF_4, which produces a green color in flames. Copper, phosphoric acid, and molybdic acid do not interfere. Ammonium tetrafluoroborate is a sublimate that can be dissolved and introduced into a flame or vaporized directly into a flame by heating.

The oldest and most widely used flame test employing volatile boron compounds is with the methyl borate ester. Boron is converted from whatever form it is in to methyl borate. This usually requires a dehydrating agent, an acid, and methanol. Tests are usually performed in an apparatus that introduces the volatile ester directly into the test flame.

One of the best series of articles on the methyl borate flame test is by Stahl (402). Stahl employed an Erlenmeyer flask fitted with a capillary exit tube for injection of the sample into the air inlet port of a burner. The sample was treated with sulfuric acid and methanol. Air was passed through the warmed sample solution. The best ratio of methanol to sulfuric acid was found to be 5:1; up to 3% water in the samples did not interfere. Use of a micro apparatus permitted detection of from 0.8 to 5 μg of boron. Methanol was found superior to ethanol and isopropanol.

D.   METHODS DEPENDING UPON CHANGE IN pH

Several qualitative tests depend upon the increased acidity of boric acid when mannitol or other suitable diols are added to solutions. Either colored indicators or a pH meter can be used to determine the pH changes. Sensitivity of the method depends upon the buffer capacity of the solution in which the test is being performed and upon the chemical form of the boron. Presumably suitable pretreatment can be performed to avoid interferences from organic acids or bases and carbonate ion. Interferences may be expected from transition-metal ions and fluoride ion.

The procedure of Dodd (110) is typical. Sodium hydroxide is added to the test solution until it is alkaline to methyl red. The solution is boiled and filtered, cooled, and then acidified with sulfuric acid until the solution is barely neutralized to the methyl red color change. Addition of mannitol will produce an acidic reaction if boron is present.

Another method for the analysis of samples containing from 0 to 200 μg B/liter is based upon the color change of bromothymol blue indicator (135). The solution is adjusted to pH 7.3 with the aid of a pH meter, mannitol is added, and the change in the bromothymol blue color is measured. The method is claimed to be superior to the carmine and quinalizarin methods.

E.   MISCELLANEOUS TESTS

A few methods based upon the microscopic character of crystals of specific boron compounds have been reported. For example, KBF_4 forms characteristic orthorhombic crystals (80). The reaction of yohimbine with borax solutions yields characteristic long, fan-shaped bundles of crystals (272). The limit of detection is 2 μg at a concentration of 1:5000. As little as 0.1 μg of borax can be detected microscopically when precipitated as $Ba(BO_2)_2$ (273). Boron can also be detected by the characteristic crystals formed in the barium borotartrate precipitation $(Ba_5B_2C_{12}H_8O_{24} \cdot 4H_2O)$. As little as 5 μg of boron at a concentration of 10 ppm can be detected (151). A neutron radiographic approach is also available for detection (see Section VII.K on nuclear techniques).

F.   DETECTION OF SPECIFIC BORON COMPOUNDS

Boron hydrides are strong reducing agents and are, in addition, strong electron-pair acceptors in complexation reactions. The reducing power of the hydrides is used by detection with diethanolamine silver nitrate solutions. Transition elements do not interfere because of their complexation by the excess amine.

The complexation properties of the hydrides are used in the detection of decaborane and alkylated decaboranes in the reaction of these hydride with amines, such as pyridine or quinoline and their derivatives (56). The higher hydrides and polymeric hydrides form colored (usually red) complexes with the amines. Pentaborane and lower hydrides do not form highly colored complexes.

Borinic acids, diborane, and alkylated diborane have been detected by reaction with 8-hydroxyquinoline in nonaqueous media to form highly fluorescent reaction products. The compounds with ethyldiboranes is more suitable than the one with diborane, which is the simple amine borane complex (54). For the detection of specific compounds, instrumental analytical techniques such as gas chromatography, infrared spectrophotometry, and mass spectrometry must be depended upon to a great extent.

VI. ALKALIMETRIC TITRATION OF BORIC ACID

The most precise, reliable, and generally useful method for the determination of boron is the titration of boric acid with standard base. Although boric acid is itself too weak to permit direct titration, it forms strongly acidic complexes with certain polyalcohols, and it is these acidic complexes that can be titrated with high precision. Titration methods are subject to interferences from other acidic or basic constituents in the sample and from certain neutral salts as well. Under certain circumstances the method is applicable to the determination of very small amounts (of the order of 10 μg or more) of boron as well as to the determination of macro amounts (10–50 mg).

A. THE ACID STRENGTH OF BORIC ACID

Of the several acids that exist in the boric oxide–water systems, only orthoboric acid, H_3BO_3, is capable of existence in significant concentration in dilute aqueous solution under ordinary conditions of temperature and pressure. A study of the boric oxide–water systems has been made (235).

Orthoboric acid is a very weak acid in aqueous solution. The most reliable value of its ionization constant is probably 5.84×10^{-10} (pK = 9.234) at 25°C (264). Most workers report values above pK 9.00; one recent value (195) was pK_a = 8.98. Boric acid is so very weak in its second and third ionizations that for all practical purposes it can be treated as a monofunctional acid in aqueous solutions. Measurements (228) gave values of 5×10^{-13} and 5×10^{-14}, respectively, for the second and third equilibrium constants. The ionization constant decreases slightly as the temperature is lowered ($pK = 9.508$ at $0°C$) and increases slightly at higher temperatures ($pK = 9.031$ at $60°C$).

When the concentration of boric acid in aqueous solution exceeds about $0.1 \ M$, the apparent ionization constant increases with increasing concentration. When the concentration has reached $1.0 \ M$, the pH of the solution is about 3.4, corresponding to an apparent ionization constant of 1.6×10^{-7} (408). This anomalous behavior can be most satisfactorily explained by assuming that boric acid polymerizes in concentrated solutions to form condensed acids having larger ionization constants than boric acid itself. The concentration of any polymeric species present is low since the presence of such species can hardly be detected from measurements of colligative properties such as vapor pressure or freezing-point depression. Evidently the polymerization proceeds to a significant extent only above $0.025 \ M$ boric acid. Nevertheless, the polymers formed are fairly strong acids and have a considerable influence on the pH of the solution. Several different investigators have attempted to elucidate the nature of the polymeric species and to formulate the equilibria involved (78,118,408,419). The predominant polymeric species is considered a cyclic trimer consisting of two boric acid and one borate moieties (12,117,196).

An examination of the experimental studies on the self-complexation of boric acid indicates that in nearly all cases the investigators have been forced to make certain assumptions before their data could be treated mathematically. It has been difficult to reach firm conclusions as to the formulas, formation constants, and ionization constants of the complexes that are present. Nevertheless, it seems quite certain that such polymeric species are present and that equilibrium between monomeric and condensed species is reached very quickly.

Studies (116) on the Raman spectra of borate solutions indicated that the borate ion in aqueous solution is $B(OH)_4^-$. There is no evidence that the borate ion is polymeric, even in very concentrated solutions.

B. THE EFFECT OF NEUTRAL SALTS ON THE ACID STRENGTH OF BORIC ACID

Relatively low concentrations (of the order of a few tenths molar or less) of neutral salts have no more than a slight effect on the acid strength of boric acid, but high concentrations of some salts cause an increase in the apparent acid ionization constant. For example, in solutions that were $2 \ M$ in the chlorides of potassium, sodium, lithium, barium, strontium, and calcium, the apparent ionization constant, as determined from the pH of half-titrated solutions of 0.1 M boric acid, is 0.9, 2.2, 5.0, 58, 110, and 295 × 10^{-9}, respectively (363). The effectiveness of any particular salt in increasing the apparent acidity of boric acid seems to be related to the charge to radius ratio of the cation; small, highly charged ions are the most effective.

The reasons for this salt effect on the acidity are not completely clear. The ions that produce the most pronounced effect are all highly hydrated (386), and perhaps such ions effectively increase the concentration of polymeric boric acid species by decreasing the activity of the water molecules in the solution. It is also possible that weak complexes between the metal ion and borate ion may exist; they would, of course, increase the apparent ionization constant of the acid.

C.   THE EFFECT OF POLYALCOHOLS ON THE ACID STRENGTH OF BORIC ACID

As early as 1842 Biot reported that boric acid became distinctly acid to litmus upon the addition of certain sugars. This and later observations formed the basis for the titrimetric methods for boric acid developed by Thomson (417), Barthe, Jörgensen, and Honig and Spitz in the period, 1893–1896, and for the studies on the structures of sugars and other polyhydroxy compounds. Since that time these studies have been continued by a large number of investigators. Of particular note is the work of Boeseken, whose classic studies extended over a period of many years. A concise summary of work in this field has been published (48).

The work of the many investigators who have studied the polyolborate reaction leaves little doubt that the marked changes in properties that result are due to the formation of polyolborate complexes. Nevertheless, the exact nature of the polyol-boric acid complexes and attendant equilibria reactions has often been a matter of some controversy. For example, two of the more recent studies (306,307) reports that only 1:1 boric acid:mannitol complexes exist. Later work showed that under the experimental conditions employed therein, 1:1 complexes only would be formed. There has also been considerable disagreement on values of certain equilibrium constants. The mannitol-boric acid system has been the most extensively studied and seems likely to adequately serve as a model system for all polyols. From the excellent recent work of Kankare (209) and that of Knoeck and Taylor (221) on the mannitol-boric acid system, the following reactions and complexes must be considered.

$$B(OH)_3 + H_2O \rightleftharpoons B(OH)_4^- + H^+ \tag{1a}$$
$$B(OH)_3 + OH^- \rightleftharpoons B(OH)_4^- \tag{1b}$$
$$B(OH)_4^- + LH_2 \rightleftharpoons LB(OH)_2^- + 2H_2O \tag{2}$$
$$B(OH)_4^- + 2LH_2 \rightleftharpoons L_2B^- + 4H_2O \tag{3}$$
$$2B(OH)_4^- + LH_2 \rightleftharpoons LB_2(OH)_4^{2-} + 4H_2O \tag{4}$$
$$B(OH)_3 + LH_2 \rightleftharpoons LBOH + 2H_2O \tag{5}$$
$$B(OH)_3 + 2LH_2 \rightleftharpoons L_2BH + 3H_2O \tag{6}$$
$$2B(OH)_3 + LH_2 \rightleftharpoons LB_2(OH)_2 + 4H_2O \tag{7}$$
$$LB(OH)_2^- + LH_2 \rightleftharpoons L_2B^- + 2H_2O \tag{8}$$
$$LBOH + LH_2 \rightleftharpoons L_2BH + H_2O \tag{9}$$

Structures for some of these complexes are as follows:

$$
\begin{array}{c}
\mid \\
-\text{C}-\text{O} \\
\mid \qquad\quad \diagdown \\
\qquad\qquad \text{B}-\text{OH} \\
\mid \qquad\quad \diagup \\
-\text{C}-\text{O} \\
\mid
\end{array}
$$

LB$_2$(OH)$_2$

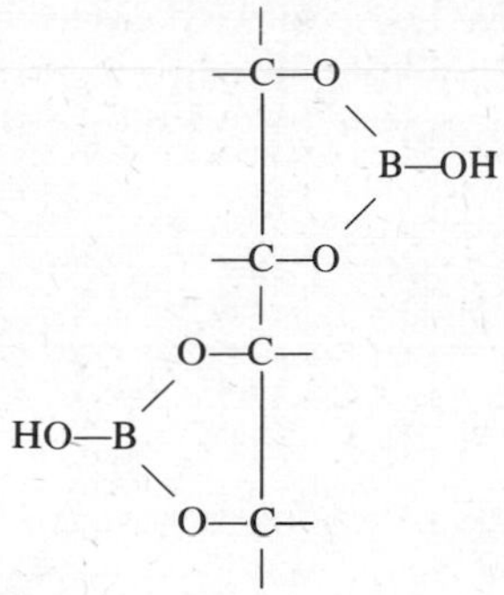

$$
\begin{array}{c}
\text{O}-\text{C}- \\
\diagup \qquad\quad \mid \\
\text{HO}-\text{B} \\
\diagdown \qquad\quad \mid \\
\text{O}-\text{C}- \\
\mid
\end{array}
$$

L$_2$BH   Undissociated form of L$_2$B$^-$

The existence of these complexes is obviously dependent upon concentration, pH, and nature of the polyol involved. It is instructive to review some of the experimental work.

The first compound of type LBOH isolated was the boric acid complex of 2,4-dimethylpentane-2-4-diol (182):

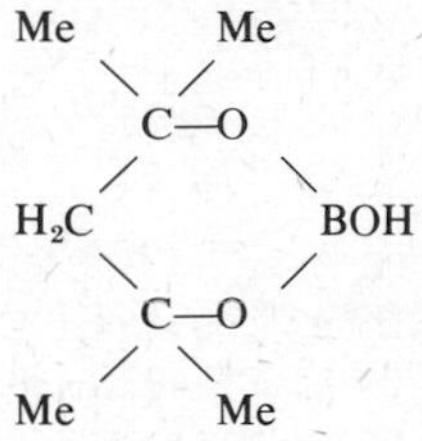

$$
\begin{array}{c}
\text{Me} \qquad \text{Me} \\
\diagdown \quad \diagup \\
\text{C}-\text{O} \\
\diagup \qquad\qquad \diagdown \\
\text{H}_2\text{C} \qquad\qquad \text{BOH} \\
\diagdown \qquad\qquad \diagup \\
\text{C}-\text{O} \\
\diagup \quad \diagdown \\
\text{Me} \qquad \text{Me}
\end{array}
$$

Analogous acids of still other diols have been prepared (319,348,364). As was indicated in the section on separations, 1,3-diols proved to be excellent extraction reagents. The preparation of the sodium salts of the mannitol, d-galactose, and d-glucose complexes of boric acid has been reported (287). All of the acids of the LBOH type so far studied have proven to be very weak acids.

A number of compounds that appear to contain the LB(OH)$_2^-$ type of ion have been prepared. The boric acid complex of mannitol (139) and a similar complex of cis-cycloheptane-1,2-diol, have been reported (101). It seems quite possible that these compounds should be formulated as LB(OH)$_2^-$-type complexes rather than as hydrates of LBOH-type complexes (284) as suggested. Several salts of the mannitol-boric acid complex were also prepared (139). All of these salts appear to be examples of the LB(OH)$_2^-$ type of complex. Several of the salts of glycol complexes of boric acid are hydrates and are undoubtedly LB(OH)$_2^-$-type complexes.

The existence of complexes of the L_2B^- type has been demonstrated. A large number of salts of the pyrocatechol complex have been prepared (182,284,364,365). There seems to be no doubt that these salts have the structure:

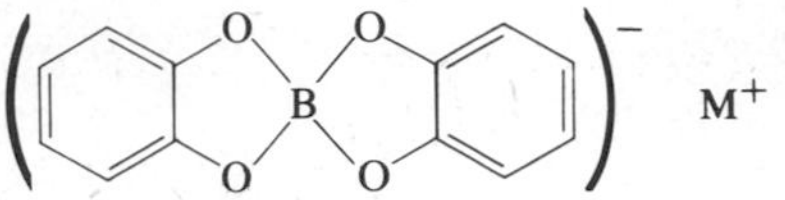

M^+

The corresponding acid, as well as salts of the 3- and 4-nitro derivatives of pyrocatechol, have also been prepared (284). Mannitol, xylose, and fructose complexes containing two sugar molecules per molecule of boric acid have been prepared (425). The characterization of these complexes was not adequate to show whether or not a L_2B^--type structure is involved. Nevertheless, the salicylic acid complex

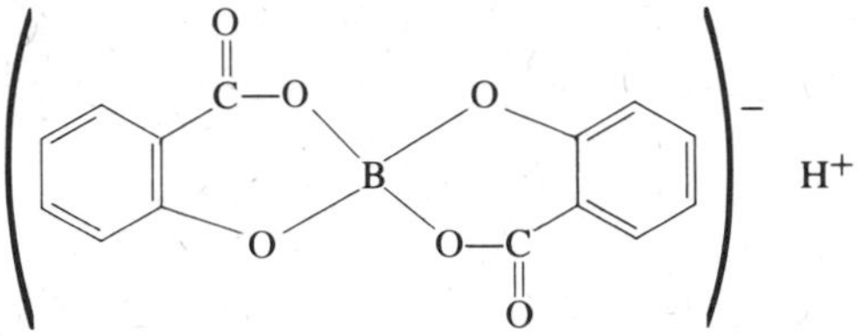

H^+

has been prepared. The acid was resolved into optical isomers through the use of its alkaloid salts (285). The fact that the acid can be resolved indicated that the complex is not planar, and hence that the boron valencies are probably tetrahedrally disposed. X-ray diffraction measurements indicate a tetrahedral configuration with 109° bond angles for the boron atom in complexes of the $LB(OH)_2^-$ and L_2B^- types (183). On the other hand, the configuration around the boron atom in LBOH-type complexes is probably planar with 120° bond angles.

Since the boric acid is a weak acid its conductivity is low; the conductivity of the polyols is also generally negligible. Any increase in conductivity observed when polyols are added to boric acid solutions must then result from reactions such as

$$B(OH)_3 + 2LH_2 \rightleftharpoons H^+ + L_2B^- + 3H_2O \qquad (10)$$

in which borate-diol complexes are formed. This method will not detect the formation of undissociated or unchanged complexes.

In the aliphatic series only glycols or polyols with adjacent hydroxyl groups show any appreciable tendency to form complexes of the $LB(OH)_2^-$ and L_2B^- types. The simple 1,2-glycols show an almost negligible tendency to form complexes of this type. Glycerol, on the other hand, shows a small but readily detectable tendency to form such complexes, and mannitol has a pronounced effect in increasing the acidity. Boeseken has explained this behavior by suggesting that the hydroxyl groups, due to mutual repulsion, tend to be disposed *trans* to one another in the simple 1,2-glycols and hence are too far apart to permit formation of the cyclic system involved in the $LB(OH)_2^-$ and L_2B^- complexes. In the case of compounds having more than two adjacent hydroxyl groups, it is no longer possible for the hydroxyl groups to be 180° apart, and hence the tendency toward complexation is enhanced. In general, the greater the number of adjacent hydroxyl groups, the greater is the tendency toward complexation. The configuration around each of the asymmetric carbon atoms likewise has an effect on the proximity of neighboring hydroxyl groups, and hence an effect on the tendency toward complexation. In the case of the alicyclic 1,2-diols, the hydroxyl groups are rigidly held in a very favorable configuration if the hydroxyl groups are *cis*; such compounds are particularly effective complexing agents. The *trans* diols, as might be expected, show no tendency to form complexes. Similarly the cyclic anhydrides of the sugar alcohols are usually more effective than the alcohols themselves provided two adjacent *cis* hydroxyl groups are present in the cyclic system. For example, the equilibrium constant for reaction (10) is approximately tenfold greater for mannitan than for mannitol (437); the formation constant of the erythritan complex is about the same as that of the mannitan complex (237).

In the aromatic series, *ortho* dihydroxy compounds are effective complexing reagents, as might be expected, since the two hydroxyl groups are rigidly held in positions especially favorable for the formation of the necessary cyclic system. *Meta* and *para* dihydroxy compounds show no tendency to form complexes. A variety of techniques have been employed to gain a further insight into the nature of the complexes formed in aqueous solution and into the stability of these complexes. The results of many investigators (93,422,423) who have made polarimetric studies on polyol-borate and polyolboric acid systems are uniformly in accord in the conclusion that boric acid has either a very small effect or no effect at all on the rotation of optically active polyols, but that the borate ion produces marked changes in rotation. Polyols that have been studied in this way include mannitol, ducitol, mannose, glycose, galatose, fructose, xylose, arabinose, and diethyl tartrate; all have given essentially the same results. The fact that boric acid produces no appreciable change in rotation seems to indicate beyond reasonable doubt that complexation of boric acid by polyols, at least by polyols of the types studied, is very feeble. In other words, the equilibrium constants for reactions (5) and (10) must be very small indeed. On the other hand, the fact that very pronounced changes in rotation occur when borates are added to any of the polyols listed previously indicates that extensive complexation occurs. It seems likely that the equilibrium constants for one or both of the reactions

$$B(OH)_4^- + LH_2 \rightleftharpoons LD(OH)_2^- + 2H_2O$$
$$B(OH)_4^- + 2LH_2 \rightleftharpoons L_2B^- + 4H_2O$$

must be large.

The fact that complexes of the LBOH type can be readily prepared by crystallization from glycol-boric acid solutions seems to be in disagreement with the statement that formation constants of these complexes are small. Nevertheless, compounds of this type that have been prepared are relatively sparingly soluble in water and can be prepared in good yield because of their insolubility rather than because of a favorable equilibrium. Complexes of the LBOH type involving polyols such as the sugars and sugar alcohols are relatively soluble; hence the formation of complexes of these polyols proceeds only to the point at which an equilibrium is reached in the solution. Furthermore, complexes of the $LB(OH)_2^-$ and L_2B^- types may be favored because of structural factors. Groups may be too close together to permit the formation of strain-free planar complexes of the LBOH type, which involve oxygen-boron-oxygen bond angles of 120° (197).

In addition to the complexes already mentioned it is possible to prepare other types of complexes under special circumstances. By dehydrating mixture of mannitol and boric acid it is possible to prepare (58) not only a compound of the LBOH type

$$CH_2OH \cdot CH \cdot CH \cdot CHOH \cdot CHOH \cdot CH_2OH$$

but also a compound containing two boric acids per mannitol

There is substantial evidence that compounds of the last type exist in appreciable concentrations in aqueous solutions. Some evidence has been presented that complexes made up of three polyol molecules per borate ion may exist in small concentrations in aqueous solutions containing very high concentrations of mannitol or fructose.

Evidence from polarimetric studies has been obtained for the existence in mannitol-borate solutions of polymeric complexes of the L_2B^- type

$$
\left[
\begin{array}{l}
CH_2OH \\
| \\
-OCH \\
| \\
-OCH \\
| \\
HCO \\
| \quad\diagdown \quad - \\
\quad\quad B \\
HCO\diagup \quad - \\
| \\
CH_2OH
\end{array}
\right]_n^{-n}
$$

involving four hydroxyl groups of the mannitol, that is, complexes in which mannitol acts as a di-diol. Similarly, the $LB(OH)_2$ type of complex in the mannitol system apparently involves four hydroxyl groups of the mannitol (197):

$$
\begin{array}{c}
CH_2OH \\
| \\
HO \quad\quad OCH \\
\diagdown \diagup \quad\quad | \\
B \\
\diagup \diagdown \quad\quad | \\
HO \quad\quad OCH \\
| \\
HCO \quad\quad OH \\
\diagdown \diagup \\
B \\
\diagup \diagdown \\
HCO \quad\quad OH \\
| \\
CH_2OH
\end{array}
$$

Recent work by Kankare (209) using a potentiostatic titration method, and a thorough mathematical treatment of data including a computer method of curve fitting appears to be definitive. In addition to $LB(OH)_2^-$ and L_2B^- complexes found previously, the existence of $LB_2(OH)_4^{2-}$, L_2BH, and $LB_2(OH)_2$ must be assumed to obtain a satisfactory fit of experimental data to theory. From the calculated stability constants reported by Kankare and given in Table 8, it appears as if the unchaged L_2BH is the least important in defining the composition of a mannitol-boric acid system.

A number of measurements of the equilibrium constants for the formation of the various polyol-borate complexes have been reported. Some of the data that are available are tabulated in Table 9. Although the constants reported by various investigators seem to agree within one order of magnitude or better, the agreement is far from satisfactory. In view of the probability that mannitol acts as a di-diol, the values of the constants reported for the mannitol complexes must be regarded as questionable since all were calculated on the assumption that mannitol acts as a mono-diol.

An attempt has been made to evaluate the various reported values of formation constants, and the following conclusions seem to be warranted:

1. When polyols are present in relatively large excess in relatively dilute solutions of boric acid or borates (as in titration conditions), the most significant complexes formed are $LB(OH)_2^-$ and L_2B^- types. Under normal titration conditions the polyol concentration is approximately 0.5 M. In this case the concentration of the $LB(OH)_2^-$ complex is small relative to that of the L_2B^- complex. When glycerol is used to enhance the acidity of boric acid it is normally added in concentrations greatly exceeding 0.5 M; under these conditions the concentration of the L_2B^- type of complex is considerably larger than that of the $LB(OH)_2^-$ type if

TABLE 8
Stability Constants of Mannitol-Boric Acid
Complexes[a]

	pK
$\dfrac{[L_2BH]}{[B(OH)_3][LH_2]^2}$	-0.21 ± 0.06
$\dfrac{[LB_2(OH)\cdot]}{[B(OH)_3]^2[LH_2]}$	2.01 ± 0.04
$\dfrac{[LB(OH)_2^-]}{[B(OH)_4^-[LH_2]}$	2.98 ± 0.01
$\dfrac{[L_2B^-][H^+]}{[B(OH)_3][LH_2]^2}$	-3.913 ± 0.008
$\dfrac{[LB_2(OH)_4^{2-}]}{[B(OH)_4^-]^2[LH_2]}$	4.41 ± 0.05

[a] From Kankare (209).

TABLE 9

Formation Constants of Borate-Diol Complexes[a]

Diol	$K_2 = \dfrac{[LB(OH)_2^-]}{[B(OH)_4^-][LH_2]}$	$K = \dfrac{[L_2B^-]}{[LB(OH)_2^-][LH_2]}$	$K_3 = \dfrac{[L_2B^-]}{[B(OH)_4^-][LH_2]^2}$	$K = \dfrac{[H^+][LB(OH)_2]}{[B(OH)_3][LH_2]}$	$K = \dfrac{[H^+][L_2B^-]}{[B(OH)_3][LH_2]^2}$	Ref.
Mannitol	4×10^3	50	2.9×10^5		1.7×10^{-4}	49
	4.3×10^3	44	1.9×10^5	$2\text{--}3 \times 10^{-6}$	1.1×10^{-4}	437
	3.4×10^3	44	1.9×10^5	2.5×10^{-6}	1.10×10^{-4}	50
	3.0×10^2	170	1.5×10^5	2.0×10^{-6}	8.8×10^{-5}	368
	—	—	5×10^4	1.75×10^{-7}	3×10^{-5}	102
			1.7×10^5		1.0×10^{-5}	354
Sorbitol	5.6×10^2	207	1.16×10^5	3.3×10^{-7}	6.8×10^{-5}	420
	8×10^3	40	3.6×10^5	$4\text{--}5 \times 10^{-6}$	2.1×10^{-4}	437
	6.3×10^3	57	3.6×10^5	3.7×10^{-6}	2.11×10^{-4}	50
Glycerol	5×10^2	—	—	3×10^{-7}	—	226
Fructose	1.5×10^2	3	4.3×10^2	$8\text{--}10 \times 10^{-8}$	2.5×10^{-7}	437
	—	—	1.9×10^5	$2\text{--}4 \times 10^{-6}$	1.1×10^{-4}	49
	3×10^3	50	1.5×10^5	4.0×10^{-6}	9.0×10^{-5}	437
	6.9×10^2	120	8.3×10^4	$4\text{--}5 \times 10^{-6}$	4.8×10^{-5}	420
Mannitan	8×10^3	300	2.7×10^6		1.6×10^{-3}	437
Diethyl tartrate	6.6×10^{-2}	1.3×10^{-3}	8.5×10^{-5}	3.9×10^{-11}	5.0×10^{-14}	424
Pyrocatechol	1.1×10^4	2.3	2.5×10^4	6×10^{-6}		368
					1.5×10^{-5}	364

[a] K_2 and K_3 refer to equations 2 and 3. Others refer to appropriate equilibrium equations not given in the text.

the data in Table 9 are correct. In most cases, then, only the L_2B^- complex needs to be taken into account in considering the titration of boric acid in polyol systems.

2. Significant amounts of LBOH-type complexes may be formed in boric acid solutions containing high concentrations of polyols. These complexes appear to be present in relatively low concentrations, are very weak acids, and have a negligible influence on pH.

3. Mannitol presents a special case in that it can apparently function as a di-diol, forming $LB_2(OH)_4^{2-}$ complexes.

4. From the values of the equilibrium given in Table 9 it is possible to calculate the effective acid dissociation constant of boric acid in the presence of several polyols. The effective dissociation constant may be defined as:

$$K_a' = \frac{[H^+][L_2B^-]}{[B(OH)_3]}$$

A few values of K_a' for several polyols commonly used in boric acid titrations are given in Table 10.

5. From the work of many investigators it can be concluded that the following sugars and sugar alcohols have a relatively pronounced effect in increasing the acidity of boric acid: mannitol, dulcitol, sorbitol, xylitol, fructose (or invert sugar, which is an equimolar mixture of fructose and glucose), erythritan, and mannitan. Of this group, only mannitol and fructose (as invert sugar) have been used extensively in the titration of boric acid, largely because they are readily available. It appears that sorbitol is slightly more effective than mannitol. Compounds such as pyrocatechol and pyrogallol are effective in increasing the acidity of boric acid but cannot be used in the titration of boric acid since they are themselves weak acids and hence interfere in the titration (368). Polyols that have a relatively small effect in increasing the acidity of boric acid include sucrose, maltose, lactose, glucose, rhamnose, mannose galactose, arabinose, xylose, erythritol, glycerol, propylene glycol, trimethylene glycol, and ethylene glycol. Although some of this group of polyols, particularly glycerol, have been employed in the titrimetric determination of boric acid, they must be used in very high concentration to obtain an adequate increase in the acidity of boric acid.

6. The stability of the borate-polyol complexes increases significantly with decreasing temperatures (310), and hence the effectiveness of a polyol in increasing the acidity of boric acid increases as the temperature is lowered.

7. In general, the rates of formation and dissociation of the polyol-borate complexes are fast, and hence equilibrium is reached quickly. In

TABLE 10

Effective Acid Dissociation Constant of Boric Acid, K'_a, in the Presence of Representative Polyols

Polyol	K_7	Polyol concentration (M)				
		0.1	0.2	0.5	2.0	4.0
Mannitol	1×10^{-4}	1×10^{-6}	4×10^{-6}	3×10^{-5}	—	—
Sorbitol	2×10^{-4}	2×10^{-6}	8×10^{-6}	5×10^{-5}	—	—
Glycerol	3×10^{-7}	—	—	—	1×10^{-6}	5×10^{-6}
Fructose	1×10^{-4}	1×10^{-6}	4×10^{-6}	3×10^{-5}	—	—
Mannitan	2×10^{-3}	2×10^{-5}	1×10^{-4}	5×10^{-4}	—	—

those cases in which a sugar must undergo mutarotation or pyranose-furanose interconversion before complexation is possible, the system naturally reaches equilibrium slowly since the rates of mutarotation and interconversion are slow.

D. DIRECT TITRATION OF BORIC ACID

Since boric acid is a very weak acid, it is difficult to determine it precisely by means of a direct titration. In the absence of "activators" to increase acid strength, the titration curve is comparatively flat in the vicinity of the equivalence point, and hence there is no sharp end point. Approximate values of the pH at the equivalence point together with the titration errors resulting from failure to locate the equivalence point precisely are given in Table 11. Even with a pH meter or with a comparison solution it is seldom possible to locate the equivalence point to within better than ±0.1 pH unit. Thus under the most favorable circumstances the precision that may be expected is of the order of ±1–2% in concentration ranges normally encountered. If ordinary acid–base indicators are used without the aid of a comparison solution, the errors that may be expected are of the order of ±4–10%. Errors resulting from a shift of the equivalence point pH due to such factors as temperature change or change in the ionic strength of the solution would generally lead to greater overall errors.

Despite the imprecision of a direct titration of boric acid, several authors have reported methods involving such a titration. Nitramine and tropeolin-O have been suggested as indicators. Both have transition intervals of pH 11.0–13.0. It is apparent from the data in Table 11 that the color changes occur after the equivalence point; hence neither of these indicators is ideally suited for the titration.

TABLE 11

Equivalence Point pH and Titration Errors in the Direct Titration of Boric Acid

Borate concentration at equivalence point	Equivalence point pH	Error (%)	
		0.1 pH unit	0.5 pH unit
0.1	11.11	0.7	3.5
0.05	10.96	1.0	5
0.02	10.76	1.4	7
0.01	10.61	2	10

Titrations of boric acid solutions in the concentration range, 0.01–0.1 M, can be performed conductometrically with accuracies of the order of ±0.3–0.5% (225). Although the conductometric method is more time consuming than indicator methods and requires careful control of the temperature, it would appear to be the method of choice for locating the equivalence point in any direct titration of boric acid. It is possible also to obtain fairly good results (of the order of ±2–3%) in the conductometric titration of boric acid in the presence of other weak acids such as acetic acid. Nevertheless the titration of boric acid in the presence of neutral salts such as 2 M sodium chloride or 1 M calcium chloride or in the presence of polyols is much more precise and accurate than the direct titration. There seems to be no reason to prefer to direct titration method in most cases.

E.   TITRATION OF BORIC ACID IN THE PRESENCE OF POLYOLS

1.   Titration Curves of Boric Acid in the Presence of Polyols

Thomson in 1893 (417) was the first to propose that the titration of boric acid be carried out in the presence of a polyol in order to realize the greater precision that results from an increase in the strength of the acid titrated.

The effect of polyols in increasing the strength of boric acid and in increasing the sharpness of the break at the equivalence point can be seen readily in Figs. 1, 2, and 3, which show the titration curves of boric acid solutions in the presence of various concentrations of three of the polyols frequently employed in this titration. The curves shown are taken from Schafer (366) and are similar to the curves reported by many other authors. All curves in the figures represent titrations of 20.0 ml aliquots of a 0.0925 M solution of boric acid with 2.09 M sodium hydroxide at 18°C using a hydrogen electrode.

Several conclusions can be drawn from these titration curves:

In no case is a satisfactory titration curve obtained unless considerably more than 2 mmol polyol is added for each millimole of boric acid present in the sample. In fact, one would predict from equations 2 and 3 (Section VI.C) that the acid strength of boric acid depends not only upon the ratio of polyol to boric acid but rather upon some power of the concentration of polyol in excess of that required to form the complex species, $LB(OH)_2^-$ and L_2B^- with the borate present. This prediction is borne out by the titration curves and also by the data of Hollander and Rieman (187), who found that the equivalence point pH depended primarily upon the concentration of the polyol and only slightly upon the polyol–boric acid ratio.

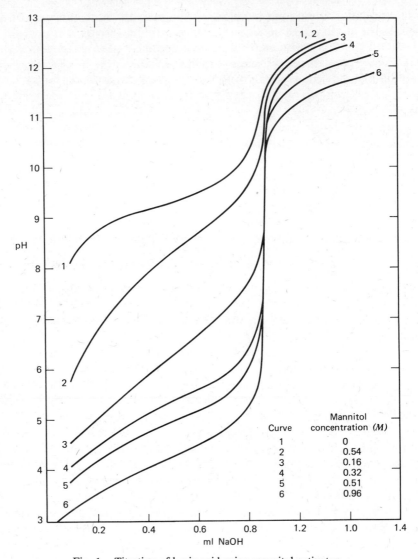

Fig. 1. Titration of boric acid using mannitol activator.

The optimum amount of mannitol or fructose is approximately 1.0 gram/10 ml of sample solution. The optimum amount of either of these "activators" is independent of the concentration of boric acid for all concentrations normally encountered in titrimetric analysis (up to about 0.1 M) (376). Increasing the amount of either of these activators to more than 1 gram/10 ml produces no significant increase in the sharpness of the end-point break; in fact, the magnitude of the end-point break decreases at very high concentrations of fructose (Fig. 2, Curve 7).

Although frequently used as an "activator" in boric acid titrations, glycerol is much less efficient than either mannitol of fructose. The addition of 4.5 grams glycerol/10 ml of solution gives a titration curve

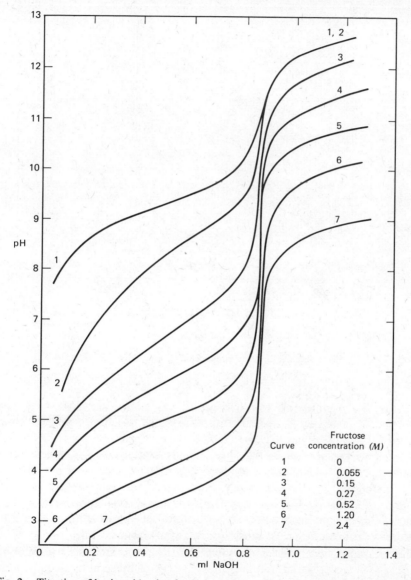

Curve	Fructose concentration (M)
1	0
2	0.055
3	0.15
4	0.27
5	0.52
6	1.20
7	2.4

Fig. 2. Titration of boric acid using fructose activator (fructose concentration shown).

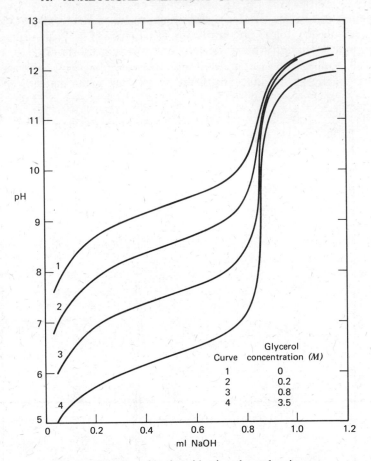

Curve	Glycerol concentration (M)
1	0
2	0.2
3	0.8
4	3.5

Fig. 3.   Titration of boric acid using glycerol activator.

that, in the vicinity of the end-point, is nearly identical with that
obtained by using only 1.0 gram fructose/10 ml of solution. A comparison of the mannitol and fructose titration curves shows that mannitol is
very distinctly superior to fructose. The magnitude of the break in the
titration curve when the optimum amount of mannitol is used is
approximately 1.5 pH units greater than that observed with the optimum
amount of fructose. The superiority of mannitol results not because it is
more effective in increasing the strength of boric acid, but rather
primarily because fructose, although a very weak acid ($K_a = 9 \times 10^{-13}$)
(420) is a stronger one than mannitol. Because of this, a solution

containing excess fructose is buffered at a lower pH following the boric acid equivalence point, and hence the height of the break is diminished.

The pH at the equivalence point depends upon the polyol used and upon its concentration. This fact must be kept in mind in selecting a suitable indicator for the titration. A recent comparison of sorbitol to mannitol has shown that sorbitol is slightly better than mannitol in the potentiometric titration procedure (31). Mannitol would be the more readily available reagent.

2.   Titration Procedure

The following procedure is typical for the titration of boric acid in macro amounts (1 mg or more) and in the absence of interfering substances except carbohydrates.

To the sample solution in a 250 ml beaker is added sufficient 10% H_2SO_4 to render the solution slightly acidic to methyl red indicator. The solution is then boiled for 3–5 min to expel CO_2.

After cooling to room temperature the excess acid is titrated with standard 0.1 M NaOH (carbonate free) to the methyl red end point. To the solution is added phenolphthalein indicator and 1 gram of mannitol for each 10 ml of solution. The solution is then titrated to the phenolphthalein end point. The volume of standard titrant used from the methyl red end point to the phenolphthalein end point is used to compute the amount of boric acid present.

The titration should be carried out with solutions of boric acid from 0.01 to 0.1 M in order to achieve a sharp end point. The first equivalence point occurs at pH 5.1 for 0.1 M boric acid. The presence of neutral salts in relatively high concentration such as 1 M calcium chloride or 2 M sodium chloride will shift the equivalence point to lower values. Neutral salts also may cause significant changes in the transition intervals of acid–base indicators.

The second equivalence point occurs at pH 7.5–8.6 depending upon boric acid concentration.

Temperature, the presence of neutral salts, boric acid concentration, and type of activators all influence the equivalence point pH values. For example, the acid dissociation constant of boric acid increases with rising temperature (pK_a equal to 9.508 and 9.031 at 0 and 60°C, respectively) (264). The ion product of water also changes quite markedly with changes in temperature (pK_w equals 14.93, 13.98, and 12.24 at 0, 25, and 100°C, respectively). The use of a pH meter and the plotting of complete titration curves eliminate many of these errors and is preferred for precise work.

3. Interfering Agents

Principal interfering agents in the mannitol-borate titration procedures are weak acids and bases of all types. Transition-metal cations precipitate upon the addition of base and thus act as weak acids in the conventional sense. Carbonate or bicarbonate ion cause the most often encountered interference. They can be removed by boiling the acidified sample solution. Fluoride ion is perhaps the most difficult interfering agent to remove. It may be precipitated as calcium fluoride followed by filtration or distillation of the boron as methyl borate. See also the section on separations for the removal of interfering agents.

4. The Identical pH Method

An interesting variation of the conventional titration methods, proposed by Foote (137), has come to be known as the identical pH method (358,367,449). In this method the solution is first adjusted to a somewhat arbitrary pH value, usually in the least-buffered region, pH 6.5–7.5. Mannitol or some other activator is added, and the solution is titrated to the identical original pH value.

It is at once apparent from titration curves that it is not possible to choose any one pH value that corresponds to the equivalence point pH of both the strong acid and the mannitol-borate complex. It is essential, therefore, that the method be carefully standardized if accurate results are desired. Salt concentrations, amount of activator added, sample volumes, and temperature should be controlled, and standardization titrations should be run with several different concentrations of boric acid in the range of interest (188).

Because of its relative simplicity the technique has gained considerable popularity. It has the distinct advantage that it permits the titration of boric acid in the presence of small amounts of many metal salts, phosphate and fluoride ion, weak acids, and weak bases that interfere with the conventional method. Solutions with higher buffer capacity exhibit indistinct end points.

In many cases the elimination of separation procedures outweighs the small loss in accuracy inherent in the identical pH method. EDTA has been used as a complexing agent in this type of titration to reduce metal-ion interferences (326). Identical pH titration to pH 6.9 using a citrate-mannitol mixture avoids interference from precipitation of metal hydroxides (303). Samples containing 5 mg B in the presence of 35 mg of Ti, Fe, Mg, Al, Zr, Cu, Pb, Sn, Zn, Co, Ni, Ca, N, or Si may be determined ±1% relative.

F. SPECIAL TITRATION TECHNIQUES

Very little work has been reported on the titration of boric acid in nonaqueous solutions. End points have been poor in all cases reported to date (51,294).

Satisfactory results have been obtained in the thermometric titration of boric acid with strong base (172,253).

An iodometric method for the determination of boric acid has been described. Strong acids or bases in the samples are first neutralized, mannitol is added, and the sample is then treated with an excess of a potassium iodate-potassium iodide solution. One equivalent of iodine is released for each equivalent of boric acid present. The iodine released may be titrated with standard thiosulfate. A more precise modification of this method is to add excess thiosulfate and back titrate after the first addition of iodate-iodide solution.

An unusual redox titration method for boron involves the precipitation of barium borotartrate, filtration, treatment with excess standard ceric reagent, and back titration of excess ceric ion with standard oxalic acid. Errors are reported to be less than 2.3% relative in the 0.1–10 mg B range (271).

A coulometric titration method has been reported (266).

G. TITRATION OF BORATES

The borate ion is a reasonably strong base, comparable in strength to ammonia. It is possible, therefore, to determine borate in a sample by titration with strong acid (334). The equivalence point in the titration occurs at about pH 5.1–5.5.

H. TITRATION OF BORIC ACID IN THE PRESENCE OF NEUTRAL SALTS

Neutral salts have been used in the place of polyols to activate boric acid (82). Calcium chloride is more effective than lithium chloride and is apparently the best reported to date. Unfortunately, calcium chloride often contains small amounts of basic impurities or traces of boron. A careful check of this reagent should be made before use.

A procedure for the titration of boric acid containing solutions utilizing calcium chloride as an activator is as follows, assuming interference is absent:

Titrate no more than a 40 ml sample solution to a yellow-orange end point of methyl red indicator to neutralize the strong acids or bases present. Add sufficient calcium chloride to saturate the solution, and titrate with standard base to the same yellow-orange end point. Add more calcium chloride to resaturate the solution, and continue the titration to the end point dropwise.

The method was reported (363) to be good within ± 0.1 mg of B_2O_3 for samples containing from 16 to 65 mg of B_2O_3. Titrations have also been carried out with a pH meter.

VII.   OTHER QUANTITATIVE METHODS FOR THE DETERMINATION OF BORON

A.   GRAVIMETRIC METHODS

Most of the early approaches to the determination of boron were based upon gravimetric methods. Boron has been weighed as the borate of various metals such as barium, calcium, cadmium, and silver. None of the early methods are as reliable as the currently used and much more convenient titrimetric methods.

Despite the early and continued interest, the search for suitable sparingly soluble, stoichiometric compounds that can serve as a basis of gravimetric determination of boron has not been very successful. Only two methods merit consideration, the barium borotartrate method (150,-151) and the nitron tetrafluoborate method (256).

In the barium borotartrate method the analyte solution is treated with 8–10 times its volume of a reagent solution prepared by adding 1 vol. of concentrated ammonium hydroxide to 10 vol. of a solution made of 13 grams barium chloride dihydrate, 14 grams dextrotartaric acid, 240 grams ammonium chloride, and 1000 ml water. The ammonium hydroxide must be added to the acid barium tartrate solution immediately before use. The precipitate, which has the composition $4BaC_4H_4O_6 \cdot Ba(BO_2)_2 \cdot 4H_2O$, is washed with ammonium hydroxide solution, with 1:1 water–acetone, and finally with 95% alcohol before drying at 110°C prior to weighing. Aluminum ions interfere (347) by preventing the formation of the precipitate. Germanium interferes (373) because it too forms an insoluble barium tartrate complex. A similar method based upon the precipitation of a barium borosaccharate has been reported.

The nitron tetrafluoborate method is based upon its insolubility in excess hydrofluoric acid. Aqueous solutions of the sample should contain from 125 to 250 mg of boric acid. The solution is diluted to 60 ml with water, and to this is added 15 ml of a solution of 37.5 grams of nitron in 250 ml of 5% acetic acid in water, and 1–1.25 grams of 48% hydrofluoric acid. The solution is allowed to stand for 10–20 hr to permit formation of the fluoborate ion and then the complex. The precipitate is cooled, filtered, and washed with a saturated solution of the complex in the wash water. The precipitate may be dried at 105–110°C for 2 hr and weighed. Nitrate, perchlorate, iodide, thiocyanate, chromate, chlorate, nitrite, and bromide ion form insoluble salts with nitron and so would interference with the analysis. The method is not subject to interference from fluoride ion, or from most weak acids or weak bases. This is the chief advantage of the method, which is capable of ±1% precision.

B. SPECTROPHOTOMETRIC AND COLORIMETRIC METHODS

Dozens of color-forming reagents are suitable for the determination of boron. All reagents useful for detection can in theory be employed in quantitative procedures. It is not feasible or necessary in this section to cover all reagents that have been used. Instead, some of the more widely used or easily applied reagents have been selected for more detailed coverage. Several investigators have compared many color reagents for boron (120,126,162,462). Of all the color reagents reported, 1,1'-dianthrimide in sulfuric acid solution appears to be the most generally useful.

The main interfering agents in all of the colorimetric methods are fluoride ion, nitrates, oxidizing agents, and materials in the sample that react with concentrated sulfuric acid to give colored materials. Sample pretreatment such as ignition with calcium hydroxide, distillation of the boron as methyl borate, or extraction have usually been employed to avoid interference. It is recommended that the maximum concentration of possible interfering substances be determined or estimated before including a sample pretreatment step when developing a routine analytical procedure. Some reagents can tolerate considerable amounts of interfering ions before accuracy is affected.

1. Quinalizarin

The quinalizarin reagent has been widely used for determination of trace quantities of boron in various materials. Numerous procedures have been reported: a selection are in Table 12. A typical procedure is the following.

Place 5 ml of the unknown or standard solution containing up to 30 μg of boron in a 125 ml flask; add 50 ml of 98.5% sulfuric acid and 5 ml of 20 N sulfuric acid from a burette. Stopper, stir, and cool to 80°F in a water bath. Add 1.0 ml of the quinalizarin reagent (0.20 mg dried quinalizarin in 1 liter of 98.5% sulfuric acid), place in an absorption cell, and measure the absorption at 600 nm. Very little change in the absorption of the reagent or the complex occurs on standing for 24 hr.

Several authors have noted that both temperature and sulfuric acid concentration in the final solution are critical. The greatest sensitivity has been variously reported to occur in 93–98% sulfuric acid. Specific uses of the reagent should include a check of the effect of sulfuric acid

TABLE 12
Selected References for Quinalizarin Methods

Material analyzed	Remarks, sensitivity, range	Reference
Alloys	Maximum sensitivity at 93%, H_2SO_4, 0.005–0.25 mg B	393
	Substituted glacial acetic acid for some of the H_2SO_4	268·
	Nickel alloys	458
Plants and soils	0.0001 mg detection limit	35,36
	Studied effect of H_2SO_4 concentration, 0–30 μg in 2 ml sample	203
Light metal alloys (Al)	High Si not an interference	371
Wines	Microgram range of B	42
Plant, soils, fertilizers	0–5 μg and 0–40 μg B	370
	A.O.A.C. method 0–2 μg	15
	A.O.A.C. method 0–8 μg B	259

concentration in the development of a satisfactory procedure. A limit of detection of 0.1 μg/ml of boron is typical for quinalizarin. The use of a spectrophotometer is superior to visual observation of a standard series or the use of a filter photometer, mainly because of the deep color of most reagent blanks and the small shift in the wavelength of maximum absorption when the boron complex is present.

Fluoride ion above 500 ppm interferes (35) but can be removed by precipitation with thorium chloride. Nitrate ion also interferes but can be removed by ignition. Germanium is at least 100 times less sensitively detected than boron. Transition metal ions may interfere by complex formation with the colorimetric reagent.

2. Alizarin-S, Alizarin Blue-S, and Solway Purple

Alizarin blue-S in concentrated sulfuric acid changes from purple through brown to green in the presence of boric acid. The difference in the wavelength of maximum absorption in the two colored forms of the reagent is greater than that of the two forms of quinalizarin and of alizarin red-S (421). Alizarin-S has approximately the same limit of detection as quinalizarin, 0.1 μg/ml.

The unsulfonated base of solway purple is twice as sensitive as alizarin blue-S. Waxolone Purple AS is similar to this reagent and gives similar reactions (184). This latter method does not require as high a concentration of sulfuric acid as the others. The procedure for the use of these reagents is the same as for quinalizarin except that different color filters or wavelengths are used. Alizarin blue-S is subject to interference

TABLE 13
References for Alizarin-S, Alizarin blue-S, and Solway Purple

Dye	Remarks, range, sensitivity	Reference
Alizarin-S	0–5 ± 0.2 μg B; compared color with a standard series prepared from methyl orange	104,105
Alizarin-S	Soils analyzed	100
Alizarin-S (Na salt)	Concentration, lower limit 0.1 ppm; 560 nm recommended for maximum sensitivity	79
Alizarin blue-S	0–10 nm range; filter photometer method	421
Solway Purple (unsulfonated dye base)	0–10 nm B range; twice as sensitive as Alizarin blue-S	421
Waxoline Purple AS (similar to Solway Purple)	1–8 nm B range; lower H_2SO_4 concentration required	184

from fluoride and nitrate ions as are all hydroxyanthraquinone derivatives.

The unsulfonated base of solway purple slowly sulfonates to a blue compound in concentrated sulfuric acid solutions.

3. Carmine, Carminic Acid, Carmine Red, and Cohineal

The use of carmine as a quantitative reagent for boron was first reported in 1947 (126). The main work in the development of the technique was reported by Hatcher and Wilcox (178). Carmine in concentrated sulfuric acid changes from a bright red to a stable purple or blue in the presence of boron. The color change is greatest at approximately 610, nm and it follows Beer's Law in the 0–20 μg of boron region. A general procedure employed is as follows:

TABLE 14
References for Carmine and Similar Dyes

Dye	Remarks	Reference
Carmine	—	126
	1–10 μg B; in boron hydrides; spectrophotometric method	185
	0–40 μg B; studied method	332
	Give absorption curves maximum sensitivity at approx. 600 nm	242
	0–20 μg B; sensitivity is ±0.2 μg	178
Carminic acid	0–40 μg B	73
Carmine red	—	215

Pipet 2 ml of sample into an Erlenmeyer flask, add 2 drops of concentrated hydrochloric acid, add 10 ml of concentrated sulfuric acid, mix, cool, and finally add 10 ml of carmine reagent solution (0.05% carmine in sulfuric acid). The color is permitted to develop for 45 min and remains stable for at least 24 hr.

Color development is a function of temperature (331). At 30°C the color develops in approximately 1 hr, but at 90°C it develops rapidly. Heating the test mixture in a water bath should speed color development. At higher temperatures, however, the color fades more rapidly. The limit of detection of the method is approximately 0.1 $\mu g/ml$. Interfering agents include fluoride, nitrate, arsenous, citrate, iodate, iodide, and vanadate ions.

4. Turmeric (Curcumin)

Turmeric is one of the oldest colorimetric reagents, the first method having appeared in 1902. The most widely used procedure is attributed to Bertrand and Aghulon (38). The color-forming step depends upon the specific reaction of boric acid and curcumin. A typical procedure (103) follows:

TABLE 15
References to Curcumin Methods

Material analyzed	Remarks	Reference
Solutions	0–20 μg B, on filter paper 0.2 ppm	103, 426
Solutions	Studied method, 0–3 μg B; 530 nm	463
Plant ash; soil	0.5–8.0 μg B	298
Solutions	0–5 μg B; studied method	323, 324, 428
Solutions	0–20 μg B; acetone solvent; studied method	390
Solutions	Studied effect of Ti, Mo, and Zr	457
Natural waters	—	138, 68
Solutions	Stability studied	453
Urine	After dry ashing	115
Copper alloys	—	316
Food (caviar)	After wet ashing	26

A 1.00 ml aliquot of an aqueous solution of the sample (0–2.0 μg B) is placed in a 250 ml beaker of boron-free glass. Add 4 ml of a curcumin-oxalic acid reagent (0.04 gram of finely ground curcumin and 5 grams of oxalic acid in 100 ml of 95% ethanol) and mix. Evaporate on a water bath at 55 ± 3°C, and bake at the same temperature for at least 15 min more. Cool and treat the reaction products with 25 ml of 95% ethanol. Filter directly into an absorption cell and read at 540 nm.

The method has numerous interferences. Ferric iron, molybdenum, titanium, niobium, tantalum, and zirconium form red-brown colors with curcumin (129). Oxidizing agents such as peroxides, CrO_4^{2-}, MnO_4^-, NO_2^-, and ClO_3^- prevent or retard the formation of the color.

Prior extraction of boron by 2-ethyl-1,3-hexanediol before analysis by curcumin likely avoids some of these interferences (261).

There has been considerable speculation concerning the colored agent formed when curcumin is treated with boric acid. Earlier investigators reported that the complex contained no boron. It has been established (400) that curcumin is

and the colored complex, rosocyanin, is

Rosocyanin is apparently formed only in the absence of oxalic acid. When oxalic acid is present, a 2:2:2 compound of boric acid, oxalic acid, and curcumin is formed.

The proposed structure is similar to that of rosocyanin except that one oxalic acid bridges between two boric acid molecules complexed with curcumin. The other oxalic acid molecule bridges between two phenol groups.

In sulfuric acid–acetic acid solution the curcumin–boron ratio is reported to be 2:1 (112). The structure of 1:1 chelates of curcumin with boric acid and benzeneboronic acid in acetic acid, benzene, or dioxane has been studied (427).

5. 1,1′-Dianthrimide

Several investigators have selected 1,1′-dianthrimide as the best colorimetric reagent for boron (60,120,213,333). This substance gives a

TABLE 16
References for the 1,1'-Dianthrimide Method

Material analyzed	Remarks	Reference
—	Study of 60 compounds	120
Aluminum alloys	Precision: ±0.0016% B for 0.022–0.22% B samples	60
Ti alloys	Method found better than carmine and quinalizarin method	85
Plant materials	Sensitivity 2 μg B	158
Fertilizers	1 μg B ± 5% Ca(OH)₂ ignition removes all interferences	355
	Studied procedure	24
Biological materials	Improved method	205
	Optimum conditions studied	165

yellowish-green color when dissolved in sulfuric acid, but in the presence of boric acid the color changes to blue; maximum sensitivity is achieved at 620 nm. The complex follows the Beer–Lambert law over a wide concentration range, the reaction is more sensitive than the quinalizarin method, the reagent is easily purified, and reproducibility of results is excellent. Fluoride ion in high concentrations interferes. Nitrate and nitrite ions interfere but can easily be removed during sample preparation. Oxidizing agents such as chromate, periodate, and perchlorate also interfere.

The method is often applicable in cases in which the carmine, curcumin, or quinalizarin methods are not readily adaptable, such as, for example, the determination of boron in aluminum alloys (60). The color of the complex is a function of temperature; the complex is destroyed at 90°C.

6. Miscellaneous Reagents

References to a number of quantitative procedures utilizing a variety of miscellaneous reagents are given in Table 17.

C. ULTRAVIOLET AND INFRARED SPECTROPHOTOMETRIC METHODS

The application of uv and ir techniques has largely been to organoboron compounds and boranes. Little use of either technique has been made to the determination of inorganic boron compounds. Table 18 gives references to some work that has appeared.

D. FLUOROMETRIC METHODS

Many compounds have been reported to give fluorescence in the presence of boric acid, and a number of methods based upon this effect

TABLE 17
Reference to Miscellaneous Reagents

Color-forming reagent	Remarks	Reference
Chromotrope-2 B	0–4 μg B in acetic acid	269
	620–630 nm	406, 407
	In H_2SO_4	11
	Study of the method	26
Diaminochrysazin	2–7 μg B in H_2SO_4	87
Tetrabromochrysazin	1–10 μg B in H_2SO_4; 540 nm	113, 212, 462
Pentamethyl quercetin	0.3–1.4 mg B in 25 ml acetone solvent	429, 342, 302
Morin	10–200 μg B in 30 ml acetone	202
Morin and oxalic acid	0.2 μg sensitivity, fluorometric	92
4;4''amino-1; 1'-dianthroquinoly amine	Sensitivity 0.1 μg B/ml	30
Barium chloranilate	530 nm	401
Victoria Violet	540 nm	345
Azored N-Resorcinol	1–66 μg B range in 50 ml	164
Monomethylthionine	658 nm, 0.1–200 μg B	321
Brilliant green	Analyzed beryllium oxide samples	211

have been reported. Fluorometric methods are more sensitive but are subject to more interferences than photometric methods.

The most suitable fluorometric reagents of sixty compounds tested was found to be 1-amino-4-hydroxyanthraquinone (120). A linear response to boron concentration was found with a 546 nm Hg excitation

TABLE 18
References to ir and uv Spectra of Inorganic Boron Compounds

Compounds	Technique	Reference
B_2O_3 in glasses	ir	206
B_2O_3 fused	ir	174
H_3BO_3	ir 5–15 nm	383
Borates of K, Ca, Mn, Cu, Pb	ir 25–50 nm	317
BCL_3	ir 3–15 nm	381
Fluoborates of NH_4^+, K, Na	ir 2–25 nm	90
B_4Cl_4	ir, Raman	252
BCl_3, B_2H_6	ir	242
Various borates of Na, Mg, Pb, K, Mn, and BN	ir	288
La, Sc, In borates	ir 7–20 nm	404
$B(OD)_3$	ir	40
Borax ores	Raman	238
B_4C, BN	ir	208, 277
Organo-borate esters	ir	382

line. The chemical procedure based upon this reagent was essentially identical with that based upon 1,1'-dianthrimide.

Benzoin, when added to a slightly alkaline solution of boric acid in 85% ethanol, gives a greenish-white fluorescence. A method using this reagent is applicable in the 0–10 μg B range with a precision of ±0.1 μg B (327,394,446,447).

Curcumin, resacetaphenone, and benzoin have been compared as fluorometric reagents (396). Benzoin was found most suitable.

E. EMISSION SPECTROGRAPHIC METHODS

Boron is present in trace amounts in many samples of interest. The spectrographic method is especially well suited for the determination of small quantities of boron, and consequently many spectrographic methods for boron have been reported. The method is generally much more rapid and interference free than the colorimetric methods. Precision and accuracy of the spectrographic procedures are not as high as for the colorimetric methods but are quite sufficient for many applications.

Boron has one of the simplest arc spectra of any element and has only two particularly intense lines, the doublet B 2496.778 and B 2497.733. The latter line is more sensitive and is freer from interference. Spectrographic methods have been used in the analysis of steels, soil, glass, and other materials. A few specific applications are listed in Table 19.

Generally the sensitivity of the method is in the 0.1 ppm range in the more favorable cases. A typical range in which boron can be determined is 0.001–0.005%. Precision and accuracy are approximately $\pm10\%$ for quantitative procedures. Iron has a line at 2497.82 Å. This line has an excitation potential of 16.1 eV (compared to 4.9 eV for the B 2497.73 line), and difference in excitation energy helps to eliminate the interference.

The plasma jet has been used for analysis of steel after dissolution and extractive removal of iron (160). Boron in the 0–10 ppm range in solution is determined. A hollow cathode source has also been used for boron analysis (175). Boron in white cast iron by hollow cathode makes use of a MgF_2 additive to volatilize the boron (356).

Boron isotope analyses can be made by observing the difference in emission band wavelengths of B^{10}—O and B^{11}—O bands (81).

F. FLAME PHOTOMETRY

Boron is sensitively detected by flame photometry by virtue of a series of boron oxygen bands with peak wavelengths of approximately 546, 518, and 492 nm in order of decreasing sensitivity (95). Sensitivity is

TABLE 19
References to Spectrographic Procedures

Material analyzed	Remarks	Reference
Solutions	Paper pellets; used dc arc; Fe 2483 Å reference; 1–10 μg B ±12%	28
Solutions; 0–250 ppm B	20 μl samples used; sensitivity, 1 ppm ±10%	77
Steels	Studied the method; ac, dc arc, methods of eliminating Fe interference	89
Graphite	Sensitivity ±0.1 ppm B	99
Urine	Used Si 2435.1 A line as internal standard	119
Solutions; tourmaline	Solution technique; precision ±4%; used Zn internal standard	296
Uranium-base materials	Carrier distillation method	380
Rocks	Used Sb; internal standard; precision ±10%	243
	—	291
Fertilizers, plants, ash	Porous cup method 0.4–10 ppm B range	380
Steels	0.0005–0.003% range dc arc	405
Glass	—	171
Plants, biological materials	—	189, 279, 455, 431
Pu and U nitrate solutions	After ion-exchange separation	114

in the 5–10 ppm range for aqueous solutions. Solutions containing high percentages of methanol exhibit up to 20 times higher sensitivity. The use of photomultiplier detector units with the spectrophotometer is preferred for the highest sensitivity. This permits use of narrow slits for discrimination against interference.

As in other flame-photometric methods, variations in sample solution composition influence flame intensities and must be studied in developing techniques for specific analyses (61). Sodium has been found to be an interference in analysis of biological materials (189).

The flame-photometric method has the advantage that it does not suffer from interference from most organic materials and from many of the inorganic materials that interfere in other methods.

A flame-photometric detection method has been reported for the determination of boron hydrides in air (55). No solution is used; sample air is passed continuously through a chamber in which a small hydrogen flame is burning. Volatile boron compounds in the 0.02–5 ppm range in air are detected. The increase in sensitivity over that of solution methods is attributed to the lack of water or solvent in the sample and low background luminosity of the hydrogen-air flame.

G.   ATOMIC ABSORPTION METHODS

Borates are not easily reduced to atomized boron in flames. Thus atomic absorption methods should not be expected to be easily devised or to have a high sensitivity. Nevertheless, boron has been determined by atomic absorption utilizing a nitrous oxide–acetylene flame (173,263,-313). Using the 249.7 nm line, limits of detection have been found to be 15 ppm (18). Somewhat better limits of detection (3 ppm) are obtained if boron extracted as the 2-ethylhexane-1,3-diol complex is used. Fluoride ion interferes in this (283) and probably the other referenced atomic absorption methods. Until an improved technique for atomizing boron compounds is found, the method is not recommended for boron analysis.

H.   POLARIMETRIC METHODS

In 1887 Biot first observed that the optical activity of tartaric acid is increased when boric acid is added to the solution. Since that time many investigators have studied the effect of boric acid and borates on the optical activity of a large number of optically active hydroxy-acids, sugars, sugar alcohols, and other related compounds. Despite the many studies that have been reported, very little effort has been directed toward the development of polarimetric methods for the determination of boric acid or of borates.

An early method specifically intended for the polarimetric determination of boron was based upon the change in optical rotation of d-tartaric acid when boric acid is added (352). Briefly the method consisted in adding the boric acid sample to 10.0 ml of 1.0 M tartaric acid, diluting to 20.0 ml, and measuring the rotation in a 2.2 dm tube with sodium light. Under these conditions the calibration curve of optical rotation versus milligrams of boric acid exhibited a slight curvature; the average slope of the curve was approximately 0.04°/mg boric acid. Optical rotations could be measured to within about ±0.01°, thus permitting boric acid in the range from 0 to 30 mg to be determined with a precision of about ±0.25 mg.

The procedure suffers from the disadvantage that the optical rotation is significantly altered by neutral salts such as sodium chloride or sodium sulfate. Divalent cations were not studied, but one would expect many of these ions to interfere seriously because of the formation of tartrate complexes.

Any polarimetric method for boron in which tartaric acid is used as the reagent suffers from the handicap that the specific rotation of tartaric acid and its salts are themselves high. The analysis depends, therefore, upon the measurement of a small change in a large rotation.

The nature of the complexation reactions between the borate ion and polyhydroxy compounds, such as the sugars and sugar alcohols, has already been discussed in the preceding section on acid-base titrimetry. In many cases the specific rotations of the polyol-borate complexes are very different from the rotations of the polyols themselves and offer a possible means for the polarimetric determination of the borate ion. The specific rotations of several sugars and sugar alcohols, as well as their borate complexes, are listed in Table 20. All of the rotations given Table 20 were measured with a Bates saccharimeter at 20°C (197). All specific rotations are based on the weight of polyol in the solution rather than on the weight of the complex.

Only the rotations of the L_2B^- complexes are given, since they are the major form present in solutions containing an excess of a diol reagent. Since mannitol also acts as a di-diol, some of the $LB_2(OH)_4^{2-}$ complex is also present.

The change in the rotations of both fructose and sorbose upon the addition of borates is very large, but the use of either as a reagent for the polarimetric determination of borate has two serious drawbacks; the rotation of the reagent itself is large, and small errors in the measurement of the amount of reagent used may lead to large percentage errors in the analysis. Furthermore, both of these reagents undergo mutarotation, which may lead to large errors unless care is taken to be sure the equilibrium has been established. Although mannitol does not exhibit as large a change in rotation upon the addition of a given amount of borate as either fructose or sorbose, it does not suffer from either of these disadvantages and would seem to be the preferred reagent for the polarimetric determination of borate.

A method for the polarimetric determination of borate, using mannitol as the reagent, has been described (96). The method consists in adding 25.0 ml of mannitol-buffer reagent (mannitol, 100 grams/liter; ammonium hydroxide, 1.6 M; ammonium chloride, 2.0 M) to the sample, diluting to 50.0 ml, and measuring the rotation in a 2 dm tube with a sodium lamp.

TABLE 20
Specific Rotation of Polyols and Their Borate Complexes

Polyol	Specific rotation (degrees)	
	Polyol	Complex
d-Glucose	+53	+40
d-Fructose	−92	−45
l-Sorbose	−43	−5
Mannitol	−0.5	+22
Sorbitol	−2.2	+3

This method permits the determination of boric oxide with a precision of ±2 mg. The method is relatively free from interference from many of the ions, such as fluoride and phosphate, that interfere in the alkalimetric titration of boric acid. However, if the fluoroborate ion is present in samples containing boric acid and fluorides, this ion must be destroyed by hydrolysis prior to the determination of the total boron content of the sample. The method is likewise free from interference from acids or bases in the sample. Since the specific rotation of the reagent itself is low, the amount of reagent added need not be extremely careful controlled. The calibration curve is linear up to borate–mannitol ratios of 1:2 on a molar basis.

If an alkaline mannitol solution (mannitol, 100 grams/liter; sodium hydroxide, 0.6 M) is substituted for the mannitol-buffer solution in the preceding procedure, the method is applicable to the determination of borate or other forms of "oxidized" boron in borohydride samples. Nevertheless, mannitol is partially converted to its rather strongly levorotatory basic form in the presence of sodium hydroxide; hence both the slope and the intercept of the calibration curve depend upon the concentration of mannitol and of sodium hydroxide. Both of these variables must be controlled within close limits to obtain accurate analyses with the modified procedure.

Although the methods employing mannitol are less subject to interference from other constituents in the sample than is that employing tartaric acid, the sensitivity of the mannitol method if only about one-fifth that of the tartaric acid method (222).

I. MASS SPECTROMETRY

Mass spectrometry has been widely used for the study and analysis of organoboron compounds and the boron hydrides. In addition to the use for isotope abundance determinations on boron trifluoride (282), an application (170) for the determination of boron in semiconductors has been reported. The latter method employed spark excitation. Oil-soot particles selected by microscope from filtered air have been analyzed by mass spectrometry (281). Particles were vaporized by an ion microprobe.

J. POLAROGRAPHIC METHODS

Neither boric acid nor borates are themselves electrochemically active, and no direct polarographic methods for their determination are possible. However, several indirect polarographic methods have been investigated.

A kinetic wave, with a half-wave potential of about -1.9 V versus the saturated calomel electrode, is obtained when boric acid is electrolyzed in aqueous tetramethylammonium iodide solutions with the dropping mercury electrode (216,234). The height of this wave is not a linear function of boric acid concentration and is only a small fraction of the height expected from a diffusion-controlled process. The experimental evidence indicates that the wave is due to the reduction of hydrogen ion produced by a slow ionization of the boric acid. The rate constant for this dissociation was calculated to be 1.3×10^{-3}/sec. In the presence of an excess of sorbitol, the dissociation apparently becomes almost complete.

Borate ion decreases the height of the fructose reduction wave. Both the height and the half-wave potential of the fructose wave are dependent upon pH. The wave height increases with increasing pH up to a pH of about 12, where it reaches a maximum, and then decreases with further increase in pH. The half-wave potential at pH 12 is -1.85 V versus the saturated calomel electrode (420). As the pH increases, the wave height increases, because hydroxide ion catalyzes the conversion of the ring structure into the open keto structure. The reduction in the height of the wave that results from the addition of borate ion is apparently due to the formation of an unreducible fructose-borate complex.

The effect of boric acid on the fructose reduction wave was investigated with a view toward developing an analytical method for boron (83). Solutions were 0.1 M in lithium chloride, 0.01 M in lithium hydroxide, and 1.0×10^{-3} M in fructose. Both the fructose wave height and the diminution of the wave height in the presence of borate depended not only upon pH but also upon the time of standing. The temperature coefficient of the wave height was unusually large, it doubles when the temperature is raised from 20°C to 40°C. For low concentrations of borate, the fructose wave height decrease is approximately linear with borate concentration. The wave height–concentration relationship departs markedly from linearity as the borate concentration approaches that of the fructose. The behavior of the sorbose–borate system is similar to that of the fructose–borate system.

The wave height for the polarographic reduction of benzil in sodium hydroxide solution is also diminished when borate ion is added to the solution (318). The data indicate that a complex consisting of one benzil and two borate ions is formed. The formation constant of the complex is approximately 1.9×10^3.

The effect of borate ion on the diffusion current of 2,2′-dihydroxybenzoin solutions can be used for the polarographic estimation of boron in the concentration range from 0.7 to 1.6 mM (133).

One of the reduction waves of thenoyltrifluoroacetone is suppressed in borate buffers, presumably because of the formation of a thenoyltrifluoroacetone-borate complex (121). It is probable, however, that this complex is too weak to permit a polarographic determination of boron to be based upon the diminution of the thenoyltrifluoroacetone wave height.

In a similar, indirect manner, the suppression of the reduction current of 3-nitrocatechol has been used as a boron analysis method at the d.m.e. at pH 8. The limit of detection is near 1.5 ppm (168).

Boric acid is capable of forming bisulfite ion when added to an unbuffered aqueous sulfite solution containing polyhydroxy compounds such as mannitol (249). Although sulfite ion is not polarographically active, the bisulfite ion is reduced in two distinct steps with half-wave potentials of -0.5 and -1.0 V versus a mercury pool anode. The height of the bisulfite waves produced upon the addition of boric acid to a sulfite–polyol solution may be used as a quantitative measure of the quantity of boric acid added, although the wave height–concentration relationship is not linear. Obviously many other acids or bases would interfere in this method.

K. NUCLEAR TECHNIQUES

1. Neutron Absorptiometry

The thermal neutron absorption cross-section of boron is markedly higher than that of most other elements. Table 21 contrasts this property

TABLE 21
Cross-Sections of Selected Elements

Element	Cross-section	
	Barns	cm²/gram
Boron	750	49.0
Hydrogen	0.33	0.20
Carbon	0.0045	0.00023
Nitrogen	1.78	0.0765
Oxygen	0.0002	0.000008
Fluorine	0.009	0.0003
Sodium	0.49	0.13
Magnesium	0.059	0.0015
Aluminum	0.22	0.0049
Silicon	0.13	0.0028
Chlorine	32	0.54
Calcium	0.42	0.0063

for several elements. Only four of the elements listed (hydrogen, nitrogen, sodium, and chlorine) would be expected to interfere significantly. The interference from these elements should be insignificant unless the weights present exceed that of the boron. The thermal neutron absorption is a comparatively unique property of boron, and several authors have studied methods based upon this property for determining boron. The first methods were for the determination of boron in glass and similar materials (161,267,414). The technique employed by these investigators was simply a neutron transparency measurement. Samples were placed between a neutron source and a detector, and the measurement resembled the in-line geometrical arrangement of conventional absorption spectroscopy. Sensitivity and precision obtained by the early techniques were poor. For example, 5% boron in samples could be determined to ±0.1% absolute. The method, even though imprecise, was used for the analysis of boron in ores, for which it was apparently well suited. The hydrogen content of samples was found to be troublesome and caused variation in counting rates.

DeFord and Braman (52,97) were the first to extensively study the thermal neutron absorption technique and develop it into a practical, useful method. The greatest sensitivity in using the neutron absorption method is obtained when the optimum geometrical arrangement of source, sample, and detector is employed. This was found to be one in which the sample is in an annular form around the detector as shown in Fig. 4. The equipment consists of a paraffin wax neutron moderator, an annular sample cell, a boron trifluoride (boron-10 enriched) thermal neutron-detector tube, and a conventional high-voltage supply and scaler. A 20–40 mCi radium–beryllium neutron source is recommended. Source size is limited by the response time of the boron–trifluoride counter tube. Larger sources permit more rapid analyses or give improved precision. Sample cells may be made from any material that has a low thermal neutron-absorption cross-section. Aluminum and stainless steel are best. Pyrex or even soft glass cannot be used because of boron content.

The source of neutrons is most effective when placed near the wall of the sample cell. Analyses are performed by dissolving the sample in water or other suitable solvent, filling the cell to above the moderator level and counting thermal neutrons for the desired period of time. A standard series of known boron concentration is used to establish a calibration curve.

In all experiments in which an apparatus of the general type shown in Fig. 4 was used, it was found that the concentration of the neutron-absorbing element in the sample solution was related to the counting rate

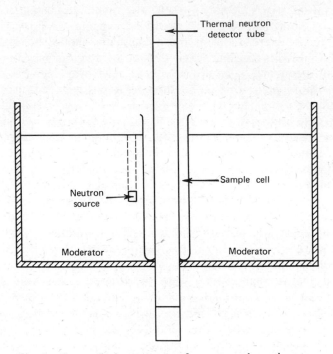

Fig. 4.   Geometrical arrangement for neutron absorptiometry.

by the expression

$$C = C_0 \frac{N_0 - n}{N - n} - C_0 \tag{1}$$

where C is the concentration of boron or other absorber present, N is the number of counts recorded in a fixed counting period, N_0 is the number of counts recorded in the counting period of the same duration with pure solvent in the sample cell, and C_0 and n are arbitrary constants depending upon the size and shape of the moderator, the counter tube characteristics, and other similar factors. The constant n apparently is related to background neutron counting rate or to neutrons that do not pass through the sample cell before being counted. It was noted that as n increases, precision and sensitivity decrease rapidly. If proper materials of construction and proper arrangement of the apparatus are used, n can be rendered insignificant and equation 1 becomes

$$C = \frac{1}{N} C_0 N_0 - C_0 \tag{2}$$

It is apparent that counting rate is inversely proportional to boron concentration.

Precision of analyses is related to both the number of events counted and to the amount of boron in the sample cell. The optimum amount of boron per sample for maximum precision for most apparatus arrangements will be from 0.5 to 1 gram. Samples containing down to 100 mg of boron may be readily analyzed. Precision in the range ±0.3% relative may be routinely obtained with 20–30 mCi sources and optimum boron concentrations. With longer counting periods the precision is improved to ±0.1% relative or better.

The method has been used for the routine analysis of organoboron compounds. Routine analyses require approximately a 5 min counting time for ±0.2–0.3% precision. The isotope abundance of samples analyzed must be the same as for the boron used to calibrate the method. Pharmaceutical preparations have been analyzed by neutron absorption (106).

The neutron-absorption technique has been used for the determination of boron isotope abundances (53,169). It is not as accurate nor as precise as the mass-spectrometric method for this purpose.

2. Neutron Autoradiography

Boron on the surface of solids or in a very thin film of sample may be detected and determined making use of the $^{10}B(n,\alpha)^7Li$ reaction. Cellulose acetate butyrate, cellulose triacetate, or cellulose nitrate films are placed directly in contact with the material to be analyzed; both are irradiated with a neutron flux from a nuclear reactor. The cellulose film is removed and etched to bring out the alpha particle tracks, which are then observed under a microscope. Tracks in the film are on the order of 2 μm in diameter. Limits of detection depend upon the neutron flux. Lelental (247) found that in a neutron dose of 5.5×10^{11} neutrons/cm^2, one alpha track corresponds to 3×10^{-14} g of boron when using cellulose nitrate, the best of the three films. Nevertheless, because of a 3.2×10^3 track/cm^2 background at this dose rate, the limit of detection is in the nanogram/cm^2 range. Quantitative results may be obtained, since a linear relation exists between the number of alpha tracks and boron content of samples. The method has little interference and has had considerable recent use. Boron has been determined in metals and solids (191), iron and steel (148), water (136), and glass (76).

The preceding are only a selection of references. Earlier work includes that of Bean et al. (29) and Hughes et al. (190).

3. Proton Reactions

Boron undergoes a $^{11}B(p,n)^{12}C$ reaction upon irradiation with 15 MeV protons. This has been used for analysis of crystaline materials (125) with a limit of detection near 0.13–1.3 ppb (124). Proton activation has been used for the analysis of water after preconcentration (21) and nickel alloys with a limit of detection near 0.0014% B (207). Lower energy protons (900 KeV) give the $^{11}B(p,\gamma)^{12}C$ reaction in which the prompt gamma rays are detected (432). Activation of boron by deuterons, charged particles, photons, and alpha particles has been studied (122,-123,378).

4. Neutron Reactions

Boron analysis by detection of the prompt gamma ray in the $^{10}B(n,\alpha)^{7}Li$ reaction has been tried (181) with limits of detection near 0.6 μg of boron. Steel has been analyzed by prompt gamma detection (204). Measurement of short-lived beta or gamma emit less after rapid (0.6 sec) sample transfer through a reactor (258) gave a detection limit of 1.1 μg of boron. Finally, the helium produced in the $^{10}B(n,\alpha)^{7}Li$ reaction has been detected by heating neutron irradiated steel to remove the helium (444). Approximately 0.1 ppm B may be detected.

L. ELECTRON SPECTROSCOPY (ESCA)

ESCA methods are, of course, primarily of use in studying the molecular bonds in compounds rather than in quantitative analysis. A number of boron hydrides, halides, and borane adducts have been studied. Allison et al. (8) and Finn and Jolly (134) have measured the $1s$ binding energies of selected boron halides, hydrides, and some borane adducts in the vapor state. Similar studies on some 25 boron compounds in the solid state have also been reported (180). Boron hydrides and derivatives make particularly interesting study materials by ESCA because of electron deficiencies these compounds possess. Chemical shifts observed are generally linearly related to boron atom charges estimated by Pauling, CNDO, and extended Hückel methods.

VIII. PRIMARY STANDARDS

The best boron-containing primary standard is probably borax, $Na_2B_4O_7 \cdot 10H_2O$. It has been tested by many investigators and has been recommended by several of these authors (193,194,246,430). Borax is stable under proper storage conditions, does not absorb water during weighing, and has a high equivalent weight. Its main disadvantage is that thorough tests of its purity are difficult to perform.

A very pure borax may be obtained by recrystallizing a good commercial product three times from water. The crystals are quickly filtered and dried to constant weight in an atmosphere of 70% relative humidity. This is conveniently maintained in a large dessicator containing a solution saturated with both sodium chloride and sucrose. Exposure of borax to the air should be avoided because it takes up a little carbon dioxide and thus becomes contaminated.

Boric acid has been proposed as a primary standard (251). It must be dried below 55°C, however, to avoid loss of water through decomposition. Above this temperature boric acid is converted to metaboric acid and other polymeric species.

Boric oxide has been used as a primary standard material by many investigators, but no systematic study of its suitability for this purpose has been made. Boric acid, when heated above 443°C, is converted completely to boric oxide. Upon cooling the fused oxide forms a glass that must be broken up and stored in a completely dry atmosphere to avoid rapid absorption of water. Boric oxide is difficult to weigh and store without absorption of water.

This author prefers to use the mannitol–borate titration procedure for precise definition of boron assay.

IX. DETERMINATION OF BORON IN SPECIFIC MATERIALS

A. BORON IN PLANTS, BIOLOGICAL MATERIALS, AND SOILS

The boron requirement of plants confines its concentration to a narrow range to avoid toxicity on the one hand and deficiency on the other. The amounts present are normally in the part per million range, and consequently the colorimetric and spectrographic procedures are best suited for analyses. A few of the other methods have been applied. The main difficulty in the analysis of biological materials and plants is sample treatment. Separation from interfering materials is generally a lesser problem. Because of the small amounts of boron generally present in biological samples, sample treatment methods must be very carefully carried out to avoid loss of boron and to avoid contamination from glassware that contains large amounts of boron.

Almost all the methods proposed for the determination of boron in biological materials involve the ignition of the sample. In this step, boron is almost certainly lost in all cases. Contamination of materials with boron from the oven in which they are being dry-ashed may also be experienced and could be much more serious than boron losses. It should be noted that boric oxide is often present in the materials of construction of ovens and in high-temperature cements. The oxide is somewhat volatile at elevated temperatures. Dry-ashing with calcium hydroxide in tall form beakers appears to be the best of all reported techniques. There are so many sources of loss or contamination, however, that a careful check on the ashing step is imperative in the development of any procedure. Extraction of samples with various solutions and solvents has been employed. Extraction efficiency should be established before using this technique. Organic materials are always extracted along with the boron and these must often be removed prior to use of many finishing analytical procedures.

Table 22 lists references to several reported methods for the analysis of biological materials. Spectrographic and flame-photometric analytical procedures appear to be more convenient than ones requiring extensive sample treatment after ashing.

An autoanalyzer method utilizing Azomethine-H for plant tissue has been developed (27).

Soils have often been analyzed by extraction of boron as boric acid or soluble borate. Due to the low boron concentration, colorimetric procedures have usually been employed after separation of boron from interferences.

TABLE 22

Selected References to Methods for the Analysis of Biological Materials

Material analyzed	Remarks	Reference
Plant material	Studied sample treatment	457, 103, 24
Plant material	West ashed and dry ashed	120, 16, 17
	Spectrographic method	274, 431, 392, 312
	Flame-photometric procedures	280, 71
Urine	Spectrographic method	119
Plants and soils	Discussion of boron in plants including analysis procedures	324
Biological materials	Ashing procedures studied	315, 394
Tissues	Wet ashed, 1,1'-dianthamide method	205
	Radio frequency plasma ashed, 1,3-diol extraction, curcumin method	261
Blood	Fluorescent complex	290
Soils, rocks, and minerals	Colorimetric	275

B. ANALYSIS OF ORES, ROCKS, AND MINERALS

With the exception of a few boron mineral ores, boron is present in rocks and minerals only in small quantities. Consequently, spectrographic techniques have often been used (86,243,296,325). Samples containing more than 0.1% boron, especially the borax ores, can more readily be analyzed by titrimetric procedures after dissolution of the sample and removal of interference (7,176,236,438).

The neutron-absorption technique has been used for the routine analysis of minerals containing above 1% boron (239). It is particularly well suited unless chlorine content of the ore exceeds that of boron. The sample must be solubilized, but the method is almost interference free and capable of high precision and accuracy. The neutron absorption apparatus and procedures of DeFord and Braman (97) are preferred to neutron-transmition-type measurements.

C. ANALYSIS OF CERAMICS AND GLASSES

Glasses and ceramic materials usually contain sufficient boron to permit the use of titrimetric determinations after dissolution and removal of interference. Samples are usually fused in sodium carbonate and dissolved in acid. Interfering substances are then removed by precipitation, ion exchange, or distillation of boron as methyl borate (10,188,367,-395,441). Spectrographic (171) and neutron-absorption (52,267) procedures have also been used.

A nuclear-magnetic resonance method has been reported for the determination of boron in glass (311). The shift in the F-19 resonance line is determined upon formation of BF_4^- ions in the sample treatment. More conventional quantitative methods are superior.

See also other sections in this treatise on the analysis of glass and ceramics.

D. DETERMINATION OF BORON IN STEEL AND ALLOYS

With the exception of borides, boron is usually present in steels and alloys in low percentages (0.0001–0.05%). Consequently the colorimetric and spectrographic procedures are most frequently employed. Separation of metallic ions is required for the colorimetric techniques. Table 23 gives references to some selected methods for steel or alloy analysis. See also other sections in this treatise on the analysis of steel and alloys. Rodden has reviewed the analysis of borides, elemental boron, alloys, and boron carbides (351).

TABLE 23
Selected References to the Analysis of Steel and Alloys

Material analyzed	Remarks	Reference
Steel	Spectrographic method; 0.0001–0.02% B	89
	High resolution spectrographic method; low B	219
	Spectrographic method, 0.005–0.003% B range	495
	Quinalizarin method	357
	Spectrographic method	359
Titanium alloys	1,1'-dianthrimide method	85
Aluminum alloys	Titrimetric method	347
Uranium alloys	Spectrographic method	380
Titanium and Ti alloys	Carminic acid method	72
Iron	Spectrographic standards	389
Boron silicides	Potentiometric titration	140
Boron carbides	Free boron in carbides determined	231
Boron nitride, graphite rods, and Ti borides	Mannitol–borate titration	416
Crude grade boron	Special dissolution method; titrimetric	433
Rare-earth borides	Free boron determined	233
Nickel alloys	Quinalizarin	458

E. DETERMINATION OF BORON IN WATER

Boron in fresh water is ordinarily present in concentrations less than 1 ppm. Colorimetric (163,304) and spectrographic (296,360) procedures have been employed. One of the more useful methods, however, is the identical pH method (137,214,450). A sensitivity of ±0.05 ppm B can be obtained with this method if a 500 ml sample is used. The method has the very attractive feature that no separation or special sample treatment are required. Use of a pH meter is recommended instead of colorimetric detection (34,412).

For the analysis of water samples containing more than 1 ppm boron the standard mannitol borate procedure may be employed after removal of interfering substances. Bicarbonate ion must usually be removed by boiling acidified samples. Ion exchange removal of dissolved interfering metal ions can also be employed.

Sea water represents a special case because high precision and accuracy are generally sought. The ratio of boron to total salt content is almost constant throughout the world, and high precision is required to detect differences. Only removal of bicarbonate ion is required for direct application of the mannitol–borate titration procedure. All other major ions present in sea water do not interfere. Titration interferences such as the transition-metal ions and fluoride ion are too low in concentration to be troublesome. Polyhydroxy organic matter in sea water is claimed to cause low results (149,309) by complexing the boron. Oxidation of sea water with permanganate ion or other agents is recommended in these references. Nevertheless, small quantities of polyhydroxy organic compounds should not constitute an interference in titration, as can be seen from the small complex formation constants (Table 8). Recent work by Byrne and Kester (70) should be consulted for information on the inorganic forms of boron in sea water. Organic complexes in sea water should be very weak and constitute only a small fraction of the total boron present.

The use of permanganate ion or other oxidizing agents can lead to high results for boron through side reactions of the reduced form of the oxidizing agent. For example, manganous ion can react with dissolved air to give the dioxide and hydrogen ions during the mannitol borate titration, thus causing a systematic positive error. In the development of procedures for the analysis of highly polluted water requiring the removal of organic matter, care must be exercised to avoid introduction of systematic positive errors and boron losses upon wet digestion.

An autoanalyzer method has been developed for boron in sea water, presumably applicable to other natural waters. Based upon a colorimetric method, boron at the 3 mg/liter concentration may be determined with a ±1.5% relative standard deviation (192). An autoanalyzer method using carminic acid for water, sewage, and sewage effluents has been developed (254). A fluorescent method using 4'-chloro-2-hydroxy-4-methoxybenzophenone has a limit of detection near 10 ppb (250).

F. BORON IN AIR

Little reliable data is available on boron in air. Boric acid is not sufficiently volatile at ordinary ambient temperatures to be present in any vapor form. Steam distillation effects may occur in isolated geothermal areas such as volcanoes. Boron in air is associated with the particulate phase and likely also with sea water aerosolization. Analyses of air for boron have been done by a colorimetric method (299) results being in the 0.1–0.2 ng B/liter range.

G. BORON IN COMMERCIAL AND NATURAL PRODUCTS

Numerous methods have been reported for the analysis of various commercial and natural products. The main differences in the methods

TABLE 24

Analysis of Selected Commercial and Natural Products

Materials analyzed	Remarks	Reference
Soaps	Not applicable to synthetic detergents	43, 44, 37
Deodorants	Sample ashed	84
Borax	ASTM method	9
Talcum powder	Extraction method	25
H_3BO_3 Salves	—	59
Salves	Titration method	127
Pharmaceutical preparations	Various methods of sample treatment	376
	Curcumin method	292
Wines		5, 409
Foods	Titration after ashing and distillation	6
	Curcumin method after wet ashing	67
Butter	Ashed, titration	346
Oranges	—	144
Fruits	Titration	65
Meat products	Spectrographic	289
Plating solutions	Separated cations by ion exchange	270
	Titration procedures	435, 436
Borides	Titration procedure	335

lie in sample treatment, which varies with the product as necessary to eliminate interference. Table 24 lists specific publications on the analysis of several commercial and natural products. These are only a representative few of the many methods published.

X. RECOMMENDED LABORATORY PROCEDURES

The following procedures for sample treatment, separations, detection, and determination of boron will be generally useful for most analytical problems. Nevertheless, modifications of each method may be required for application to specific determinative problems. For more description of specific procedures the reader is referred to appropriate previous sections and references.

A. SAMPLE TREATMENT

1. Wet-Ashing Procedures

No completely general wet-ashing procedure is recommended for the treatment of plant, animal, and other organic matter before analysis. Each analysis sample type is different but some recommendations can be made. Wet ashing is considered less desirable than dry ashing and should not be used unless no loss of boron can be demonstrated in developmental work. Highly acidic infusions of material can lead to serious losses of boron. The wet chemical oxidation of organoboron compounds in a Paar bomb, by Carius oxidation, or in a Schoeninger flask is acceptable since no loss of boron is encountered in these methods.

2. Dry-Ashing Procedures

The preferred method for treating plant material is dry ashing in an oven in the presence of boron-free calcium hydroxide in platinum, quartz, or porcelain dishes. Covered, tall form dishes are recommended to minimize loss of boron by volatilization and gain of boron from electric furnace atmospheres.

Procedure

Weigh samples from 0.5 to 2 grams and place in a tall form crucible, mix with 5 grams of calcium or barium hydroxide, and heat on an electric furnace at a temperature not in excess of 550°C for 4 hr. Dissolve the ash in acid, and then determine the boron by a suitable spectrographic, colorimetric, or flame photometric method.

3. Sample Fusion

Glass or ceramic materials are fused with sodium carbonate in platinum crucibles.

Procedure

Weigh a 0.5 gram ground sample into a platinum crucible, and mix with 3 grams of sodium carbonate. Heat the covered crucible slowly until a clear fusion is obtained. Cool and decompose the fused sample with hydrochloric acid. Carefully break up the fused mass, and continue slow addition of acid until evolution of gases ceases. The sample may now be treated by any of several methods for analysis.

Alloys and slag samples may be analyzed after fusion of the sample with sodium peroxide in zirconium metal crucibles.

Procedure

Weigh a 0.5 gram sample into a zirconium crucible and mix with 5 grams of sodium peroxide. Heat the covered crucible to red heat slowly, and continue heating for 15 min until a fusion is obtained. Cool and decompose the melt with hydrochloric acid. Break up the fused mass, filter, and wash with hot water. The sample solution may now be analyzed by any of several techniques.

B. SEPARATIONS

1. Distillation as Methyl Borate

The procedure outlined by Chapin is probably the most generally useful of the several procedures that have been proposed. The Chapin method is designed for the quantitative recovery of boron only; the calcium chloride that is added to the sample for dehydration purposes usually makes analysis of the nonvolatile residue impractical. If the object of the distillation is the separation of boron prior to the determination of other constituents in the sample, the use of drying agents should be avoided. In this case, the sample solution should be evaporated to dryness, the residue treated with absolute methanol saturated with hydrogen chloride, and the evaporation repeated. Usually from two to four such treatments are required for quantitative removal of the boric acid. It is virtually impossible to achieve complete volatilization of the boric acid from aqueous solutions, but extraction with a 1,3-diol (see Section X.B.2) will probably suffice.

Reagents

A good grade of *methanol* should be distilled from lime after it has been heated under reflux for several hours in contact with the lime. The more nearly anhydrous the alcohol, the better. If volatile acids are left in the alcohol, high results for boron will be experienced. *Calcium chloride* should be granular, anhydrous, and free from boron.

Procedure

The major distillation apparatus is shown in Fig. 5.

If the mineral is soluble in hydrochloric acid, transfer 1 gram of it to flask B without letting any adhere to the neck, and treat with not more than 5 ml of hydrochloric acid (1:1). Heat gently on a water bath until dissolution is complete.

If the mineral is not soluble, add to it exactly six times its weight of sodium carbonate or of an equimolar mixture of sodium and potassium carbonates, mix, and fuse. Without removing from the crucible, decompose the melt with 1:1 hydrochloric acid in a calculated amount, added

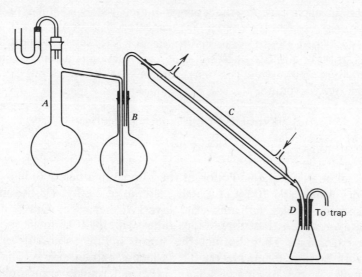

Fig. 5. Distillation apparatus for boron. *A,* flask, 500 ml with U-tube trap, heated by heating mantle; *B,* flask, 250 ml, heated by water bath or heating mantle; *C,* glass condenser; *D,* receiving flask, 250 ml, one of several with attached water trap, cooled.

in small portions. While this is being done, the crucible should rest in a casserole, and the lid should be kept in place as much as possible. Toward the end it may be necessary to heat a little, but care should be taken not to boil it, because boric acid would be lost with the steam. Pour the solution into flask B, and rinse the crucible with very little water.

Add pure anhydrous calcium chloride, using about 1 gram for each milliliter of solution and running it through a paper funnel to keep the neck of the flask clean. Swirl the flask to allow the chloride to take up the water, connect the flask to the remainder of the apparatus, and raise the water bath until the flask rests in the water but does not touch the bottom. Begin the distillation from flask A, taking care that the open end of the capillary "boiling tube" is free from alcohol and that the U-tube attached to the received is trapped with water.

Do not heat the decomposition flask B until about 25 ml of alcohol have condensed in it. After that, heat the water bath with a small flame, thus keeping the flask hot enough to prevent further condensation of alcohol. The distillation should not be so rapid that condensation of methyl borate in the glass condenser is inefficient.

When a distillate of about 100 ml has collected, exchange the receiver for a fresh one, and collect a second distillate. Add the contents of the trap tube in each case to the corresponding receiver. Two or more distillates are used only when it is suspected that boron is not completely contained in the first distillate. For the separation of small amounts of boron prior to analysis, only one fraction is usually required. After distillation the analysis may be finished by titration or other determinative technique.

2. Extraction from Aqueous Solutions

Procedure

To 10 ml of an acidified solution of the sample (pH below 6) in a 25 ml separatory funnel add 10 ml of a 10% solution of 2-ethyl-1,3-hexanediol in chloroform. Shake and allow the layers to separate. Draw off the organic layer, and extract three more times with fresh 10 ml portions of organic extractant. Then extract the boron in the combined organic layers with 20 ml of 0.5 M NaOH. The aqueous extract may be analyzed by titrimetry or by any one of several other determinative methods. To eliminate fluoride or metal interferences, add an EDTA–phosphate buffer (pH 6–6.5) to the sample solution prior to extraction.

C. QUANTITATIVE DETERMINATION OF BORON

1. Spectrophotometric Methods

a. QUINALIZARIN

Place 5 ml of the unknown or standard boron solution containing up to 0.03 mg of boron in a 125 ml flask, and add 5 ml of 10 M sulfuric acid and 50 ml of 98% sulfuric acid from burets. Stopper, stir, and cool to 80°F in a water bath. Add 1.0 ml of the quinalizarin reagent (0.20 mg dried quinalizarin in 1 liter 98.5% sulfuric acid), place in an absorption cell, and measure the intensity versus a reagent blank carried through the same procedure. Maximum sensitivity is obtained with a spectrophotometric measurement at 600 nm (or organge filter).

b. CARMINIC ACID

Pipet 2 ml of the sample into an Erlenmeyer flask, and add 2 drops of concentrated HCl. Add 10 ml of the concentrated sulfuric acid, mix, and cool. Add 10 ml of carminic acid solution (0.05% carminic acid in concentrated sulfuric acid), mix, allow to stand for at least 45 min to permit color development, and read transmittance at 585 nm against a blank of 2 ml of distilled water carried through the same procedure.

2. Mannitol–Borate Titration Procedure

This procedure is recommended for the titration of samples containing more than 1 mg boric acid/100 ml of sample solution when interfering substances are absent.

Methods for the removal of cation and fluoride ion interference are recommended in Sections IX.B.4 and IX.B.5 and extraction isolation of boron in Section X.B.2.

Reagents

The reagents used in this procedure are sodium hydroxide 0.1 M standard, carbonate-free; mannitol, solid powder free from weak acids and bases; methyl red indicator solution, 0.1% in water; phenophthalein indicator; 10% Acid (Use boron-free HCl or H_2SO_4); and 10% Base (Use boron- and carbonate-free NaOH).

Procedure

To the sample solution in a 250 ml beaker containing 5 drops of methyl red indicator add 10% H_2SO_4 of sodium hydroxide until the solution is slightly acidic to methyl red. Boil for 3–5 min to expel carbon dioxide. Cool to room temperature, and titrate excess acid with standard 0.1 M carbonate-free base to the methyl red end point. Add 5 drops of phenolphthalein indicator. Add 1 gram of mannitol for each 10 ml of solution, and titrate to the phenolphthalein end point. The volume of titrant used from the methyl red to the phenolphthalein end point is used to compute the amount of boric acid present.

The titration can also be accomplished electrometrically with a pH meter and glass electrodes. For small amounts of boron 0.01 M base may be used.

ACKNOWLEDGMENT

The original review on the analytical chemistry of boron from which this chapter eventually has evolved was initiated many years ago by Professor Donald D. DeFord and two of his (then) students, Dr. Claude A. Lucchesi and Dr. James M. Thoburn, at Northwestern University—the impetus largely stemming from "Project Zip" at Calley Chemical Co. and the intense interest at that time in boron hydride high-energy fuels. A general increase in both fundamental and analytical chemistry knowledge of boron originated with these projects.

GENERAL REFERENCES

1. Braman, R. S., "Boron Compounds," in F. D. Snell and C. L. Hilton, Eds., *Encyclopedia of Industrial Chemical Analysis,* Wiley-Interscience, New York, 1968, pp. 329–358.

2. Gale, W. A., "History and Technology of the Borax Industry" in R. M. Adams, Ed., *Boron, Metallo-Boron Compounds and Boranes,* Wiley, New York, 1964, pp.

3. Newkirk, A. E., "Elemental Boron" in R. M. Adams, Ed., *Boron, Metallo-Boron Compounds and Boranes,* Wiley, New York, 1964, pp.

4. Boon, J. L., "Refractory Boron Compounds," under "Boron Compounds" in A. Standen, Ed., *Kirk-Othmer Encyclopedia of Chemical Technology,* Vol. 3, 2nd edit., Wiley-Interscience, New York, 1964, pp. 673–680.

5. Campbell, Jr., G. W., "Boron Hydrides," under "Boron Compounds" in A. Standen, Ed., Kirk-Othmer Encyclopedia of Chemical Technology, Vol. 3, 2nd ed., Wiley-Interscience, New York, pp. 684–707.

6. Newson, H. C., "Boric Acid Esters," under "Boron Compounds" in A. Standen, Ed., Kirk-Othmer Encyclopedia of Chemical Technology, Vol. 3, 2nd ed., Wiley-Interscience, New York, pp. 652–673.

7. Newson, H. C., "Organic Boron Compounds (Boron Nitrogen)," under "Boron Compounds" in A. Standen, Ed., Kirk-Othmer Encyclopedia of Chemical Technology, Vol. 3, 2nd ed., Wiley-Interscience, New York, pp. 728–737.

8. Nies, N. P., "Boron Oxides, Boric Acid and Borates," under "Boron Compounds" in A. Standen, Ed., Kirk-Othmer Encyclopedia of Chemical Technology, Vol. 3, 2nd ed., Wiley-Interscience, New York, pp. 608–652.

9. Woods, W. G., "Organic Boron Compounds (Boron Carbon)," under "Boron Compounds" in A. Standen, Ed., Kirk-Othmer Encyclopedia of Chemical Technology, Vol. 3, 2nd ed., Wiley-Interscience, New York, pp. 707–727.

10. Gerrard, W., The Organic Chemistry of Boron, Academic, New York, 1961.

REFERENCES

1. Abegg, R., C. J. J. Foxx, and W. Herz, *Z. Anorg. Chem.,* **35,** 129–147 (1903).

2. Abesgauz, D. M., and Z. I. Kheifets, *Tr. Leningrad Tekhnol. Inst.* im *Lensoveta,* No. 35, 183–185 (1956).

3. Agazzi, E. J., *Anal. Chem.,* **39,** 233–235 (1967).

4. Akimov, V. K., A. I. Vusev, and P. I. Andzhaparidge, *Zh. Anal. Khim.,* **26,** 2434–2436 (1971).

5. Alberti, C., *Ann. Chim. Appl.,* **28,** 483–487 (1938).

6. Alcock, R. S., *Analyst (London)* **62,** 522–526 (1937).

7. Allan, F. J., F. S. Dundas, and D. A. Lambie, *J. Inst. Fuel,* **42,** 29–30 (1969).

8. Allison, D. A., G. Johansson, C. J. Allan, U. Gelius, H. Siegbahn, J. Allison, and K. Siegbahn, *J. Electron Spectrosc.,* **1**(3), 269–283 (1973).

9. American Society for Testing Materials, ASTM Standards, Part 22, 1973, Designation D 501 and D 929-50.

10. American Society for Testing Materials, ASTM Standards, Part 13, 1973, Supplement, Designation C 169-69.

11. Andreeva, Z. F., and Z. E. Leikova, *Dokl. Mosk. Selskokoz Akad. K.A.T.*, No. 149, 293–296 (1969).

12. Antikainen, P. J., *Suomen Kemistilehti*, B, **30**, 74 (1957).

13. Archer, V. S., F. G. Doolittle, and L. M. Young, *Talanta*, **15**, 864–866 (1968).

14. Association of Official Agricultural Chemists, *Official and Tentative Methods of Analysis*, Sixth Ed., 1945.

15. Association of Official Agricultural Chemists, *Official Methods of Analysis*, 9th ed., 1960. Published by Association of Official Agricultural Chemists, Washington D.C.

16. Austin, C. M., and J. S. McHargue, *J. Ass. Offic. Agr. Chem.* **31**, 284–285 (1948).

17. Austin, C. M., and J. S. McHargue, *J. Ass. Offic. Agr. Chem.*, **31**, 427–431 (1948).

18. Bader, H., and H. Bradenberger, *At. Absorp. Newsl.*, **7**, 1–3 (1968).

19. Bailey, J. J., and D. G. Gehring, *Anal. Chem.*, **33**, 1970–1762 (1961).

20. Banchi, G., and M. Giannotti, *Ann. Chim. Appl.* **20**, 271–301 (1930).

21. Bankert, S. F., S. D. Bloom, and G. D. Sauter, *Anal. Chem.*, **45**, 692–697 (1973).

22. Barnard, A. J., Jr., *Chem. Anal.*, **44**, (1955); *ibid.*, **45**, 110 (1956).

23. Barnard, A. J., Jr., and H. Buchl, *Chem. Anal.*, **46**, 16 (1957); *ibid.*, **47**, 46 (1958); *ibid.*, **48**, 44 (1958).

24. Baron, H., *Z. Anal. Chem.*, **143**, 339–349 (1954).

25. Barry, H. C., *J. Ass. Offic. Agr. Chem.*, **33**, 359–362 (1950).

26. Bartusek, M., and L. Havelkova, *Collect. Czech. Chem. Commun.*, **32**, 3853–3862 (1967).

27. Basson, W. D., R. G. Boehmer, and D. A. Stanton, *Analyst (London)*, **94**, 1135–1141 (1969).

28. Bayliss, N. S., and D. J. David, *J. Soc. Chem. Ind.*, **67**, 357–358 (1948).

29. Bean, C. P., R. L. Feischer, P. S. Swartz, and H. R. Hart, *J. Appl. Phys.*, **37**, 2218 (1966).

30. Beckett, E. G., and M. R. H. Webster, *Analyst (London)*, **68**, 306 (1943).

31. Belcher, R., G. W. Tully, and G. Svehla, *Anal. Chim. Acta*, **50**, 261–267 (1970).

32. Bell, D., and K. McArthur, *Analyst (London)*, **93**, 298–305 (1968).

33. Bellocq, A., *Chem. Zentralbl.*, **II**, 563 (1896).

34. Bengolea, D. J., and J. A. Raggio, *Bol. Obras. Sanit. Nacion (Buenos Aires)*, **8**, 110–114 (1944).

35. Berger, K. C., and E. Truog, *Ind. Eng. Chem. Anal. Ed.*, **11**, 540–545 (1939).

36. Berger, K. C., and E. Truog, *Soil Sci.*, **57**, 25–36 (1944).

37. Bernstein, R., and M. Haftel, *J. Amer. Oil Chem. Soc.*, **27**, 45–47 (1950).

38. Bertrand, G., and H. Agulhon, *Bull. Soc. Chim.*, **8**, 90–99 (1910).

39. Bertrand, G., and H. Agulhon, *C.R. H. Acad. Sci.*, **157**, 1433–1436 (1913).

40. Bethell, D. E., and N. Sheppard, *Trans. Faraday Soc.*, **51**, 9–15 (1955).

41. Bezzi, S., *Gazz. Chim. Ital.*, **65**, 766–772 (1935); *Ann. Chim. Appl.* **22**, 713–725 (1932).

42. Bionda, G., and E. Bruno, *Z. Anal. Chem.*, **155**, 183–186 (1957).

43. Blank, E. W., and M. R. Griffin, *J. Amer. Oil Chem. Soc.*, **25**, 327–328 (1948).

44. Blank, E. W., and A. Troy, *Oil Soap*, **23**, 50–55 (1946).

45. Blumenthal, H., *Anal. Chem.*, **23**, 992–994 (1951).

46. Blumenthal, H., *Powder Met. Bull.*, **6**, 80–82 (1951).

47. Bock, R., and M. Vrchlabsky, *Z. Anal. Chem.*, **246**, 228–230 (1969).

48. Boeseken, J., in W. W. Pigman and M. L. Wolfram, Eds., *Advances in Carbohydrate Chemistry*, Vol. 4, Academic, New York, 1949, pp. 189–210.

49. Boeseken, J., N. Vermaas, and A. T. Kuchlin, *Rec. Trav. Chim.*, **49**, 711–716 (1930).

50. Boeseken, J., N. Vermaas, W. H. Zaayer, and J. L. Leefers, *Rec. Trav. Chim.* **54**, 853–860 (1935).

51. Bork, V. A., and Sar'nikova, *Z. Anal. Khim.*, **23**, 901–907 (1968).

52. Braman, R. S., Doctoral Thesis, *Dissertation Abstr.* **17**, 27 (1957).

53. Braman, R. S., *Talanta,* **10**, 991–996 (1963).

54. Braman, R. S., Unpublished work.

55. Braman, R. S., and E. S. Gordon, *Inst. Elec. Electr. Eng., Transact. Instr. Measurements,* March 1965.

56. Braman, R. S., and T. N. Johnston, *Talanta,* **10**, 810–813 (1963).

57. Brazhnikov, V. V., and K. I. Sakodynskii, *Zh. Prikl. Khim., Leningrad,* **43**, 2247–2253 (1970).

58. Bremer, C., U.S. Pat. 2,223,948 (1940); U.S. Pat. 2,224,011 (1940).

59. Brandrup, W., *Apoth. Ztg.,* **44**, 149 (1929).

60. Brewster, D. A., *Anal. Chem.*, **23**, 1809 (1951).

61. Bricker, C. E., W. A., Dippel, and N. H. Furman, *Nuclear Sci. Abstr.*, **6**, 212 (1952).

62. Briscoe, H. V. A., and P. L. Robinson, *Nature,* **118**, 374 (1926).

63. Briscoe, H. V. A., P. L. Robinson, and G. E. Stephenson, *J. Chem. Soc.*, **127**, 150–162 (1925).

64. Brown, E. S., Paper presented before the Analytical Chemistry Group, Pittsburgh Section, Amer. Chem. Soc., Symposium on Boron, Nov. 14, 1952.

65. Brown, W. B., *Analyst*, **61**, 671–680 (1936).

66. Brunisholz, G., and J. Bonnet, *Helv. Chim. Acta,* **34**, 2074–2075 (1951).

67. Brunstad, J. W., *J. Ass. Offic. Anal. Chem.*, **51**, 987–991 (1968).

68. Bunton, N. G., and B. H. Tait, *J. Amer. Water Works Ass.*, **61**, 357–359 (1969).

69. Busev, A. I., P. Ya. Yakovlev, and G. V. Kozina, *Zh. Anal. Khim.*, **22**, (8) 1227–1233 (1967).

70. Byrne, R. H. Jr., and D. R. Kester, *J. Marine Res.*, **32**, 119–127 (1974).

71. Calfee, R. K., and J. S. McHargue, *Ind. Eng. Chem. Anal. Ed.*, **9**, 288–290 (1937).

72. Calkins, R. C., and V. A. Stenger, *Anal. Chem.*, **28**, 399–402 (1956).

73. Callicoat, D. L., and J. D. Wolszon, *Anal. Chem.*, **31**, 1434 (1959).

74. Camus, N., *An. Soc. Quim. Argentina*, **2**, 123–130 (1914).

75. Carlson, R. M., and J. L. Paul, *Anal. Chem.*, **40**, 1292–1295 (1968).

76. Carpenter, B. S., *Anal. Chem.*, **44**, 600–602 (1972).

77. Carpenter, P. G., T. D. Morgan, and E. D., Parsona, *World Oil,* **136**, 214, 216, 218, 220 (1953).

78. Carpini, G., *Bull. Soc. Chim. Fr.*, **1952**, 1010–1014.

79. Cellini, R. F., and F. A. Gonzalez, *An. Real Soc. Espan. Fis. Quim., Ser. A*, **50**, 59–70 (1954).

80. Chamot, E. M., and C. W. Mason, *Handbook of Chemical Microscopy*, Vol. II, Wiley, New York, 1940.

81. Chaney, C. L., *Appl. Spectrosc.*, **22**, 584–586 (1968).

82. Cikritova, M., and K. Sandra, *Chem. Listy*, **19**, 179–182 (1925).

83. Clayton, J. C., Doctoral Thesis, U. of Pennsylvania, Aug. 1954.

84. Clements, J. E., *J. Ass. Offic. Agr. Chem.*, **36**, 791–793 (1953).

85. Codell, M., and G. Norwitz, *Anal. Chem.*, **25**, 1446–1449 (1953).

86. Cody, R. D., *Appl. Spectrosc.*, **22**, 272–274 (1968).

87. Cogbill, E. C., and J. H. Yoe, *Anal. Chem.*, **29**, 1251 (1957).

88. Cole, S. S., and N. W. Taylor, *J. Amer. Ceram. Soc.*, **18**, 82–85 (1935).

89. Corliss, C. H., and B. F. Scribner, *J. Res. Nat. Bur. Stand.*, **36**, 351–364 (1946).

90. Cote, G. L., and H. W. Thompson, *Proc. Roy. Soc., Ser. A.*, **210**, 217–223 (1951).

91. Couch, E. L., *Clays Clay Mines*, **17**, 38–41 (1969).

92. Dabrowski, J., and L. Pszonicki, *Chem. Anal.*, **16**, 51–57 (1971).

93. Darmois, E., and R. Peyroux, *C.R. H. Acad. Sci.*, **193**, 1182–1185 (1931).

94. Dazord, J., and H. Mongeot, *Bull. Soc. Chim. Fr.*, **1971** (1) 51–53.

95. Dean, J. A., and C. Thompson, *Anal. Chem.*, **27**, 42–46 (1955).

96. DeFord, D. D., A. S. Blonder, and R. S. Braman, *Anal. Chem.*, **33**, 471 (1961).

97. DeFord, D. D., and R. S. Braman, *Anal. Chem.*, **30**, 1765 (1958).

98. Degtyarenko, Ya. A., *Ukr. Khim. Zh.*, **22**, 813–815 (1956).

99. DeKeyser, W. L., and R. Cypres, *Bull. Centre Phys. Nucleaire Univ. Bruxelles*, **No. 35**, 28 (1952).

100. Dermott, W., and N. Trinder, *J. Agr. Sci.*, **37**, 152–155 (1947).

101. Derx, H. G., *Rec. Trav. Chim.*, **41**, 312–342 (1922).

102. Deutsch, A., and S. Osoling, *J. Am. Chem. Soc.*, **71**, 1637–1640 (1949).

103. Dible, W. T., E. Truog, and K. C. Berger, *Anal. Chem.*, **26**, 418–421 (1954).

104. Dickinson, D., *Analyst (London)*, 106–109 (1943).

105. Dickinson, D., *Analyst (London)*, **73**, 395–396 (1948).

106. Dillinger, P., J. Tolgyessy, and M. Sarsunova, *Cesk. Farm. 1969*, *18* (9) 433–437.

107. Dimroth, O., *Ann. Chem.*, **446**, 97–108 (1925).

108. Dimroth, O., and T. Faust, *Berichte*, **54**, 3020–3034 (1921).

109. Dimroth, O., and F. Ruck, *Ann. Chem.*, **446**, 123–131 (1925).

110. Dodd, A. S., *Analyst*, **54**, 282–285 (1929).

111. Dyrssen, D. W., L. R. Uppstroem, and M. Zangen, *Anal. Chim. Acta*, **46**, 55–61 (1969).

112. Dyrssen, D. W., P. Yuri, and L. R. Uppstroem, *Anal. Chim. Acta*, **60**, 139–151 (1972).

113. Eberle, A. R., and M. W. Lerner, *Anal. Chem.*, **32**, 146 (1960).

114. Eberle, A. R., L. J. Pinto, and M. W. Lerner, *Anal. Chem.*, **36**, 1282 (1964).

115. Edmond, C. R., *AMDEL Bull.*, No. 10, 1–11 (1970).

116. Edwards, J. O., *J. Am. Chem. Soc.*, **75**, 6151–6153 (1953).

117. Edwards, J. O., *J. Am. Chem. Soc.*, **75**, 6154 (1953).

118. Edwards, J. O., G. C. Morrison, V. F. Ross, and J. W. Schultz, *J. Am. Chem. Soc.*, **77**, 266–268 (1955).

119. Eichhoff, H. J., and H. Geil, *Biochem. Z.*, **322**, 494–496 (1952).

120. Ellis, G. H., E. B. Zook, and O. Baudisch, *Anal. Chem.*, **21**, 1345–1348 (1949).

121. Elving, P. J., and C. M. Callahan, *J. Am. Chem. Soc.*, **77**, 2077 (1955).

122. Engelmann, C., *Isotop. Radiat. Technol.*, **8**, 118–121 (1970).

123. Engelmann, C., *J. Radioanal. Chem.*, **7**, 281–298 (1971).

124. Engelmann, C., *J. Radioanal. Chem.*, **7**, 89–101 (1971).

125. Engelmann, C., J. Gosset, and J. M. Rigaud, *Radiochem. Radioanal. Lett.*, **5**, 319–329 (1970).

126. Evans, C. A., and J. S. McHargue, *J. Ass. Offic. Agr. Chem.*, **30**, 308–310 (1947).

127. Faber, T., *Pharm. Ztg.*, **59**, 163–164 (1900).

128. Fedotova, L. A., and M. G. Voronkov, *Zh. Anal. Khim.*, **22**, 1431–1433 (1967).

129. Feigl, F., *Spot Tests in Inorganic Analysis*, 5th edit., Elsevier, Amsterdam, 1958.

130. No volume number is given in this C.A. reference; the year should be sufficient.

131. Feldman, C., *Anal. Chem.*, **33**, 1916 (1961).

132. Fernelius, W. C., and C. M. Callahan, Report No. CCC-1024-TR-57, Pennsylvania State University, State College, Pa., October 26, 1954.

133. Fernelius, W. C., J. M. Schempf, and C. L. Fili, Jr., *Nuclear Sci. Abstr.*, **9**, 628 (1955).

134. Finn, P. and W. L. Jolly, *J. Am. Chem. Soc.*, **95**, 1540–1542 (1972).

135. Fizzotti, C., and L. Selmi, *Chim. Ind. (Milan)*, **34**, 265–266 (1952).

136. Fleischer, R. L., and D. B. Lovett, *Geochim. Cosmochim. Acta* **32**, 1126–1128 (1968).

137. Foote, F. J., *Ind. Eng. Chem. Anal. Ed.*, **4**, 39–42 (1932).

138. Foster, M. D., *Ind. Eng. Chem. Anal. Ed.*, **1**, 27–28 (1929).

139. Fox, J. J., and A. J. H. Gauge, *J. Chem. Soc.*, **99**, 1075–1079 (1911).

140. Frank, A. J., *Anal. Chem.*, **35**, 830 (1963).

141. Friedlander, G., J. W. Kennedy, and J. M. Miller, *Nuclear and Radiochemistry*, 2nd edit., Wiley, New York, 1964.

142. Frizzell, L. D., *Ind. Eng. Chem. Anal. Ed.*, **16**, 615–616 (1944).

143. Funk, H., and H. Winter, *Z. Anorg. Allg. Chem.*, **142**, 257–268 (1925).

144. Furlong, C. R., *Analyst (London)*, **73**, 498–500 (1948).

145. Gabrielson, G., *Plating*, **41**, 47–54 (1954).

146. Gaestel, C., and J. Hure, *Bull. Soc. Chim. Fr.*, **16**, 830–831 (1949).

147. Gagliard, E., and E. Wolf, *Mikrochim. Acta, 1968* (1) 140–147.

148. Garnish, J. D., H. D. H. Hughes, *J. Inst. Metals*, **101**, 108–111 (1973).

149. Gast, J. A., and T. G. Thompson, *Anal. Chem.*, **30**, 549 (1958).

150. Gautier, J. A., and P. Pignard, *C. R. H. Acad. Sci.*, **235**, 242–244 (1952).

151. Gautier, J. A., and P. Pignard, *Mikrochem. Mikrochim. Acta*, **36/37**, 793–800 (1951).

152. Geilmann, W., and H. Bode, *Z. Anal. Chem.*, **129**, 3–5 (1949).

153. Geilmann, W., and W. Gebauhr, *Z. Anal. Chem.*, **139**, 161–181 (1953).

154. George, R. S., *Dissertation Abstr.*, **24**, 1814 (1963).

155. Glaze, F. W., and A. N. Finn, *J. Res. Nat. Bur. Stand.*, **27**, 33–37 (1941).

156. Gmelin, L., *Handbuch der anorganischen Chemie*, Part 8, Boron, Verlag Chemie, G.M.B.H., Leipzig-Berlin, 1926.

157. Golubtsova, R. B., *Zh. Anal. Khim.*, **15**, 481 (1960).

158. Garfinkiel, E., and A. G. Pollard, *J. Sci. Food Agr.*, **3**, 622–624 (1952).

159. Gopal, N. H., *J. Agr. Food Chem.*, **17**, 1146–1147 (1969).

160. Goto, H., and I. Atsuya, *Z. Anal. Chem.*, **240**, 102–110 (1968).

161. Govaerts, J., *Experientia*, **6**, 459–461 (1950).

162. Goward, G. W., and V. R. Wiederkehr, *Anal. Chem.*, **35**, 1542 (1963).

163. Graells de Kempny, R. S., *Rev. Obras Sanit. Nacion (Buenos Aires)*, **16**, 100–110 (1952).

164. Grezo, V. A., and E. N. Poluektova, *Zh. Anal. Khim.*, **13**, 434 (1958).

165. Gupta, H. K. L., and D. F. Boltz, *Anal. Lett.*, **4**, 161–167 (1971).

166. Guseinov, I. K., I. L. Bagbanly, and A. K. Posadovskaya, *Azerb. Khim. Zh.*, *1970* (3) 122–125.

167. Hakoila, E. J., J. J. Kankare, and T. Skarp, *Anal. Chem.*, **44**, 1857–1860 (1972).

168. Hakoila, E. J., J. J. Kankare, and T. Skarp, *Acta Chem. Fenn.*, **45**, 176–178 (1972).

169. Hamlen, R. P., and W. S. Koski, *Anal. Chem.*, **28**, 1631 (1956).

170. Hannay, N. B., and A. J. Ahern, *Anal. Chem.*, **26**, 1056–1058 (1954).

171. Harai, K., *J. Chem. Soc. Japan*, **62**, 463–466, 933–934, 1183–1184 (1941).

172. Harries, R. J. N., *Talanta*, **15**, 1345–1352 (1968).

173. Harris, R., *At. Absorp. Newsl.*, **8**, 42–43 (1969).

174. Harrison, A. J., *J. Amer. Ceramic Soc.*, **30**, 362–366 (1947).

175. Harrison, W. W., and N. J. Prakash, *Anal. Chim. Acta*, **49**, 141–159 (1970).

176. Harvey, C. O., *Analyst (London)*, **68**, 211–212 (1943).

177. Hatcher, J. T., *Anal. Chem.*, **32**, 726 (1960).

178. Hatcher, J. T., and L. V. Wilcox, *Anal. Chem.*, **22**, 567 (1950).

179. Hagedus, A., *Magyar Kem. Fol.*, **57**, 112–116 (1951).

180. Hendrickson, D. N., J. M. Hollander, and W. L. Jolly, *Inorg. Chem.*, **9**, 612–615 (1970).

181. Henkelmann, R., *Radiochim. Acta*, **15**, 169–180 (1971).

182. Hermans, P. H., *Z. Anorg. Chem.*, **142**, 83–110 (1925).

183. Hermans, P. H., *Rec. Trav. Chim.*, **47**, 123, 423 (1938).

184. Higgins, D. J., *J. Sci. Food Agr.*, **2**, 498–503 (1951).

185. Hill, W. H., J. M. Merrill, and B. J. Palm, *Nucl. Sci. Abstr.* **11**(9), 4820, p. 518 (1957).

186. Hillebrand, W. F., G. E. F. Lundell, H. A. Bright, and J. I. Hoffman, *Applied Inorganic Analysis*, 2nd edit., Wiley, New York, 1953, pp. 749–765.

187. Hollander, M., and W. Rieman, III, *Ind. Eng. Chem. Anal. Ed.*, **17**, 602–603 (1945).

188. Hollander, M., and W. Rieman, III, *Ind. Eng. Chem. Anal. Ed.*, **18**, 788–789 (1946).

189. Hoogland, P. L., *Anal. Chim. Acta*, **2**, 831–838 (1948).

190. Hughes, J. D. H., M. A. P. Dewey, and G. W. Briers, *Nature*, **223**, 498–499 (1969).

191. Hughes, J. D. H., and G. T. Rogers, *J. Inst. Metals*, **95**, 299–302 (1967).

192. Hulthe, P., L. Uppstroem, and G. Oestling, *Anal. Chim. Acta*, **51**, 31–37 (1970).

193. Hurley, F. H., Jr., *Ind. Eng. Chem. Anal. Ed.*, **8**, 220–221 (1936).

194. Hurley, F. H., Jr., *Ind. Eng. Chem. Anal. Ed.*, **9**, 237–238 (1937).

195. Ingri, N., *Acta Chem. Scand.*, **16**, 439 (1962).

196. Ingri, N., G. Lagerstrom, M. Frydman, and L. G. Sillen, *Acta Chem. Scand.*, **11**, 1034 (1952).

197. Isbell, H. S., J. F. Brewster, N. B. Holt, and H. L. Frush, *J. Res. Nat. Bur. Stand.*, **40**, 129–149 (1948).

198. Jackson, C. L., and L. Clarke, *Amer. Chem. J.*, **45**, 48–58 (1911).

199. Jaffe, E., *Ind. Chim.*, **9**, 750–752 (1934).

200. Jaulmes, P., and A. Gontard, *Bull. Soc. Chim.*, **4**, 139–148 (1937).

201. Jaulmes, P., and E. Galhac, *Bull. Soc. Chim.*, **4**, 149–157 (1937).

202. Jewsbury, A., and G. H. Osborn, *Anal. Chim. Acta*, **3**, 481–488 (1949).

203. Johnson, E. A., and M. J. Toogood, *Analyst* (*London*), **79**, 493–496 (1954).

204. Juna, J., K. Konecny, and M. Vobecky, *Collect. Czech. Chem. Commun.*, **34**, 1605–1610 (1969).

205. Kaczmarczyk, A., J. R. Messer, and C. E. Peirce, *Anal. Chem.*, **43**, 271–272 (1971).

206. Kaiser, K., *Glastech. Ber.* **12**, 198–202 (1934).

207. Kamada, H., R. Inoue, M. Terasawa, Y. Gohshi, H. Kamei, and I. Fujii, *Anal. Chim. Acta*, **46**, 107–112 (1969).

208. Kammori, O., K. Sato, and F. Kurosawa, *Jap. Anal.*, **17**, 1270–1273 (1968).

209. Kankare, J. J., *Anal. Chem.*, **45**, 2050–2056 (1973).

210. Kar, H. A., *Metals Alloys*, **9**, 175–177 (1938).

211. Karalove, Z. K., and A. A. Nemodruk, *Zh. Anal. Khim.*, **17**, 985 (1962).

212. Karpen, W. L., *Anal. Chem.*, **33**, 738 (1961).

213. Kasiura, K., *Chem. Anal.*, **16**, 219–223 (1971).

214. Kawaguchi, H., *Jap. Anal.*, **4**, 307–310 (1955).

215. Kazarinova-Okina, V. A., *Zavod. Lab.*, **14**, 263–265 (1948).

216. Kemula, W., and J. Witwicki, *Roczniki Chem.*, **28**, 305 (1954).

217. Khym, J. X., and L. P. Zill, *J. Am. Chem. Soc.*, **73**, 2399 (1951).

218. Kingzett, C. T., *Kingzett's Chemical Encyclopedia*, 9th edit., Van Nostrand, New York, 1966, p. 238.

219. Kirchgessner, W. G., and N. A. Finkelstein, *Anal. Chem.*, **25**, 1034–1038 (1953).

220. *Kirth-Othmer Encyclopedia of Chemical Technology*, Vol. 3, Wiley, New York, 1964.

221. Knoeck, J., and J. K. Taylor, *Anal. Chem.*, **41**, 1730 (1969).

222. Kodama, K., and H. Shiio, *Anal. Chem.*, **34**, 106 (1962).

223. Kolthoff, I. M., *J. Am. Chem. Soc.*, **48**, 1447 (1926).

224. Kolthoff, I. M., and V. A. Stenger, *Volumetric Analysis*, Vol. II, Wiley-Interscience, New York, 1942.

225. Kolthoff, I. M., *Z. Anorg. Chem.*, **111**, 1, 28, 97 (1920).

226. Kolthoff, I. M., *Rec. Trav. Chim.* **44,** 975–982 (1925).

227. Komarovskii, A. G., and N. S. Poluektov, *Mikrochemie,* **14,** 317–320 (1934).

228. Konopik, N., and O. Leberl, *Monatsh. Chem.,* **80,** 655–669 (1949).

229. Korenman, I. M., *Zh. Anal. Khim.,* **2,** 153–158 (1947).

230. Korenman, I. M., and F. R. Sheyanova, *Zh. Anal. Khim.,* **7,** 128–130 (1952).

231. Kotlyar, E. E., and T. N. Nazarchuk, *Zh. Anal. Khim.,* **15,** 207–210 (1960).

232. Kovalenko, O. A., V. D. Zhalybina, and I. K. Maiboroda, *Zavod. Lab.,* **36,** 920–922 (1970).

233. Kugai, L. N., and T. N. Nazarchuk, *Zh. Anal. Khim.,* **16,** 205 (1961).

234. Kuta, J., *Chem. Listy,* **48,** 1493 (1954); *Chem. Listy,* **50,** 884 (1956).

235. Kracek, F. C., G. W. Morey, and H. E. Merwin, *Amer. J. Sci.,* **35A,** 143–171 (1938).

236. Kramer, H., *Anal. Chem.,* **27,** 144–145, 1024 (1955).

237. Krantz, J. C., Jr., C. J. Carr, and F. F. Beck, *J. Phys. Chem.,* **40,** 927–931 (1936).

238. Krishnamurti, D., *Proc. Indian Acad. Sci., Sect. A.,* **41,** 7–11 (1955).

239. Kristyanov, V. K., and G. I. Panov, *Zh. Anal. Khim.,* **12,** 362–366 (1957).

240. LaCourt, A., G. Sommereyns, and G. Wantier, *Mikrochem. Mikrochim. Acta,* **39,** 396–403 (1952).

241. LaCourt, A., G. Sommereyns, and G. Wantier, *C. R. H. Acad. Sci.,* **232,** 2426–2428 (1951).

242. LaCourt, A., G. Sommereyns, and M. Claret, *Mikrochem. Mikrochim. Acta,* **38,** 444–455 (1951).

243. Landergren, S., *Ark. Kemi Mineral Geol.,* **19A,** 7 (1945).

244. Landry, J. C., M. F. Landry, and D. Monnier, *Anal. Chim. Acta,* **62,** 117–186 (1972).

245. Latimer, W. M., *Oxidation Potentials,* 2nd edit., Prentice-Hall, New York, 1952.

246. Lazarkevich, N. A., *Ukr. Khim. Zh.,* **4,** Sci. Pt. 405–428 (1929).

247. Lelental, M., *Anal. Chem.,* **44,** 1270–1272 (1972).

248. Levi, G. R., and R. Curti, *Gass. Chim. Ital.,* **68,** 376–380 (1938).

249. Lewis, D. T., *Analyst,* **81,** 531–536 (1956).

250. Liebich, B., D. Monnier, and M. Marcantonatos, *Anal. Chim. Acta,* **52,** 305–312 (1970).

251. Liem, H. T., *Pharm. Tijdschr. Nederland. Indie,* **13,** 291–296 (1936).

252. Linevsky, M. J., E. R. Shull, D. E. Mann, and T. Wartik, *J. Am. Chem. Soc.,* **75,** 3287–3288 (1953).

253. Linde, H. W., L. B. Rogers, and D. N. Hume, *Anal. Chem.,* **25,** 404–407 (1953).

254. Lionnel, L. J., *Analyst (London),* **95,** 194–199 (1970).

255. Lucchesi, C. A., *Dissert. Abstr.,* **15,** 2007 (1955).

256. Lucchesi, C. A., and D. D. DeFord, *Anal. Chem.,* **29,** 1169 (1957).

257. Lucchesi, C. A., and D. D. DeFord, *J. Inorg. Nucl. Chem.,* **14,** 290 (1960).

258. Lukens, H. R., *J. Radioanal. Chem.,* **1,** 349–354 (1968).

259. MacDougall, D., and D. A. Biggs, *Anal. Chem.,* **24,** 566 (1952).

260. Maeck, W. J., M. E. Kussy, B. E. Ginther, G. V. Wheeler, and J. E. Rein, *Anal. Chem.,* **35,** 62 (1963).

261. Mair, Jr., J. W., and H. G. Day, *Anal. Chem.*, **44**, 2015–2017 (1972).

262. Maksimenko, M. S., V. N. Krylov, and M. A. Lipinskii, *J. Appl. Chem. (USSR)*, **19**, 154–159 (1946).

263. Manning, D. C., *At. Absorp. Newsl.*, **6**, 35–37 (1967).

264. Manov, G. G., N. J. DeLollis, and S. F., Acree, *J. Res. Nat. Bur. Stand.*, **33**, 287–306 (1944).

265. Marcantonatos, M., G. Gamba, and D. Monnier, *Helv. Chim. Acta*, **52**, 538–543 (1969).

266. Marinenko, G., and C. E. Champion, *J. Res. Nat. Bur. Stand.*, *A*, **75**, 421–428 (1971).

267. Martelly, J., and P. Sue, *Bull. Soc. Chim.*, **13**, 103–106 (1946).

268. Martin, G., and M. Maes, *Bull. Soc. Chim. Biol.*, **34**, 1178–1182 (1952).

269. Martin, G., *Bull. Soc. Chim. Biol.*, **36**, 719–729 (1954).

270. Martin, J. R., and J. R. Hayes, *Anal. Chem.*, **24**, 182–185 (1952).

271. Martin, L. R., M. A. Henry, and J. R. Hayes, *Microchem. J.*, **13**, 529–533 (1968).

272. Martini, A., *Mikrochemie*, **26**, 227–232 (1939).

273. Martini, A., *Pub. Inst. Invest. Microquim., Univ. Nac. Litoral (Argentina)*, **6**, 87–93 (1942).

274. Mathis, W. T., *Anal. Chem.*, **25**, 943–947 (1953).

275. Maurice, J., *Ann. Agron.*, **19**, 699–704 (1968).

276. Mazzeti, C., and F. DeCarli, *Atti Congr. Naz. Chim. Pura Appl.* **1923**, 444–447 (1923).

277. McCarthy, D. E., *Appl. Opt.* **7**, 1997–2000 (1968).

278. McHargue, J. S., and W. S. Hodgkiss, *J. Ass. Offic. Agr. Chem.*, **24**, 250–252 (1941).

279. McHargue, J. S., E. B. Offutt, and W. S. Hodgkiss, *Soil Sci. Soc. Amer. Proc.*, **4**, 308–309 (1939).

280. McHargue, J. S., and R. K. Calfee, *Ind. Eng. Chem. Anal. Ed.*, **4**, 385–388 (1932).

281. McHugh, J. A., and J. F. Stevens, *Anal. Chem.*, **44**, 2187–2192 (1972).

282. Melton, C. E., L. O. Gilpatrick, R. Baldock, and R. M. Healy, *Anal. Chem.*, **28**, 1049–1051 (1956).

283. Melton, J. R., W. L. Hoover, P. A. Howard, and J. L. Ayers, *J. Ass. Offic. Anal. Chem.*, **53**, 682–685 (1970).

284. Meulenhoff, J., *Rec. Trav. Chim.*, **44**, 150–160 (1925).

285. Meulenhoff, J., *Z. Anorg. Allg. Chem.*, **142**, 373–382 (1925).

286. Michel, F., *Mikrochem. Mikrochim. Acta*, **29**, 63–72 (1941).

287. Michl, H., *Monatsh. Chem.*, **83**, 737–747 (1952).

288. Miller, F. A., and C. H. Wilkins, *Anal. Chem.*, **24**, 1253–1294 (1952).

289. Mitteldorf, A. J., *Appl. Spectrosc.*, **6**, 21–24 (1951).

290. Monnier, D., C. A. Menzinger, and M. Marcantonatos, *Anal. Chim. Acta*, **60**, 233–237 (1972).

291. Moore, D. M., *Appl. Spectrosc.*, **23**, 278–279 (1969).

292. Moore, R. A., *Proc. Soc. Anal. Chem.*, **9**, 35–36 (1972).

293. Morse, H. N., and W. M. Burton, *Amer. Chem. J.*, **10**, 154–158 (1888); *Z. Anal. Chem.*, **28**, 240 (1889).

294. Moss, M. L., J. H. Elliott, and R. T. Hall, *Anal. Chem.*, **20**, 784–788 (1948).

295. Mueller, P., and R. Abegg, *Z. Phys. Chem.*, **57**, 513–532 (1906).

296. Musha, S., *Sci. Rep. Res. Inst., Tohoku Univ., Ser. A*, **2**, 437–442 (1950).

297. Muto, S., *J. Chem. Soc. Japan, Pure Chem. Sect.*, **72**, 976–979 (1951).

298. Naftel, J. A., *Ind. Eng. Chem. Anal. Ed.*, **11**, 407–409 (1939).

299. Nakaga, S., and M. Nishimura, *Jap. Anal.*, **20**, 967–970 (1971).

300. Nazarenko, V. A., and S. Ya. Vinkovetskaya, *J. Anal. Chem. (USSR)*, **26**, 700–703 (1971).

301. Neelakantam, K., and L. Ramachandra Row, *Proc. Indian Acad. Sci., Sect. A*, **15**, 81–88 (1942).

302. Neelakantam, K., and S. Rangaswami, *Proc. Indian Acad. Sci., Sect. A*, **18**, 171–178 (1943).

303. Negina, V. R., E. A. Kozyreva, A. V. Balakshina, and L. S. Chikisheva, *Zavod. Lab.*, **34**, 278–279 (1968).

304. Nemejc, J., *Chem. Listy*, **32**, 340–345, 361–364 (1938).

305. Newstead, E. G., and J. E. Gulbierz, *Anal. Chem.*, **29**, 1673–1674 (1957).

306. Nickerson, R. F., *J. Inorg. Nucl. Chem.*, **30**, 1447 (1968).

307. Nickerson, R. F., *J. Inorg. Nucl. Chem.*, **32**, 1400 (1970).

308. Nishimura, M., and S. Nakaya, *Jap. Anal.*, **18**, 148–154 (1969).

309. Noakes, J. E., and D. W. Hood, *Deep-Sea Res.*, **8**, 120 (1961).

310. Norwitz, G., and M. Codell, *Anal. Chim. Acta*, **11**, 233–238 (1954).

311. Noshiro, M., and Y. Jitsugiri, *Jap. Anal.*, **18**, 1200–1203 (1969).

312. Obolenskaya, L. I., *Issledovaniya Priklad. Khim., Akad. Nauk SSSR, Otdel. Khim. Nauk, Sbornik Rabot*, **1955**, 337–340 (1956).

313. Olivier, M., *Z. Anal. Chem.*, **248**, 145–148 (1969).

314. Onishi, H., and H. Nagai, *Jap. Anal.*, **17**, 345–348 (1968).

315. Otting, W., *Angew. Chem.*, **64**, 670–679 (1952).

316. Pakalns, P., *Analyst (London)*, **94**, 1130–1134 (1969).

317. Parodi, M., *C. R. H. Acad. Sci.*, **204**, 1111–1112 (1937).

318. Pasternak, R., *Helv. Chim. Acta*, **30**, 1894–1899 (1947).

319. Pastureau, P., and M. Veiler, *C. R. H. Acad. Sci.*, **202**, 1683–1685 (1936).

320. Pasztor, L. C., J. D. Bode, and Q. Fernando, *Anal. Chem.*, **32**, 277 (1960).

321. Pasztor, L. C., and J. D. Bode, *Anal. Chem.*, **32**, 1530 (1960).

322. Peterson, H. P., and D. W. Zoromski, *Anal. Chem.*, **44**, 1291 (1972).

323. Philipson, T., *Lantbruks-Hogskol. Ann.*, **12**, 251–258 (1944–1945).

324. Philipson, T., *Acta Agr. Scand.*, **3**, 121–242 (1953).

325. Pieruccini, R., *Periodica Mineral. (Rome)*, **19**, 209–238 (1950).

326. Piryutko, M. M., and N. V. Benediktova-Lodochnikova, *Zh. Anal. Khim.*, **25**, 136–141 (1970).

327. Podchainova, V. N., and L. V. Skornyyakova, *Tr. Ural. Politekh. Inst. 1967*, 60–64.

328. Podchainova, V. N., L. V. Skornyakova, and B. L. Dvinyaninov, *Izv. Vyssh. Ucheb. Zaved., Khim. Khim. Tekhnol.*, **11**, 241–243 (1968).

329. Poluektov, N. S., L. I. Kononenko, and R. S. Pauer, *Zh. Anal. Khim.*, **13**, 396 (1958).

330. Poluektov, N. S., and M. P. Nikonova, *Tr. Komissii Anal. Khim., Akad. Nauk SSSR*, **3**, 188–199 (1951).

331. Post, B., F. W. Glaser, and D. M. Moskowitz, *Acta Met.*, **2**, 20–25 (1954).

332. Powell, W. A., and E. H. Poindexter, Anal. Abstr. 5, Abstr. No. 1139(1958).

333. Powell, W. A., and E. H. Poindexter, Anal. Abstr. 5, Abstr. No. 1140(1958).

334. Primbsch, E. O., *Emailwaren-Industrie*, **17**, 125–126 (1940).

335. Privalova, M. M., and G. P. Makhova, *Zavod. Lab.*, **35**, 159–160 (1969).

336. Pszonicki, L., and W. Tkacz, *Chem. Anal. (Warsaw)*, **15**, 809–816 (1970).

337. Pszonicki, L., and W. Tkacz, *Chem. Anal. (Warsaw)*, **15**, 1097–1104 (1970).

338. Quentin, K. E., *Z. Lebensm.-Unters. Forsch.*, **95**, 305–313 (1952).

339. Radley, J. A., *Analyst (London)*, **69**, 47–48 (1944).

340. Raff, P., and W. Brotz, *Z. Anal. Chem.*, **133**, 241 (1951).

341. Ramos, M. L., *Bol. Ass. Quim. Brasil*, **9**, 44–48 (1951).

342. Rangaswami, S., and T. R. Seshadri, *Proc. Indian Acad. Sci., Sect. A*, **16**, 129–134 (1942).

343. Reichard, C., *Pharm. Ztg.*, **51**, 298–299 (1906); *Chem. Zentralbl.*, **77**, 1714 (1906).

344. Reichard, C., *Pharm. Ztg.*, **51**, 817–819 (1906); *Chem. Zentralbl.* **77**, 1714 (1906).

345. Reynolds, C. A., *Anal. Chem.*, **31**, 1102 (1959).

346. Richmond, H. D., and J. B. P. Harrison, *Analyst (London)*, **27**, 179–182 (1902).

347. Ripeltauer, E., and G. Jangg, *Z. Anal. Chem.*, **138**, 18–29 (1953).

348. Rippere, R. E., and V. K. LaMer, *J. Phys. Chem.*, **47**, 204–234 (1943).

349. Robin, L., *Bull. Soc. Chim.*, **13**, 602–606 (1913).

350. Robinson, K. L., *Analyst (London)*, **64**, 324–328 (1939).

351. Rodden, C. J., "Analysis of Essential Nuclear Reactor Materials," Division of Technical Information, U.S. Atomic Energy Commission; in *Annu. Rev. Nucl. Sci.*, **1**, 343–362 (1952).

352. Rosenheim, A., and F. Lyser, *Z. Anorg. Allg. Chem.*, **119**, 1–38 (1921).

353. Ross, W. H., and R. B. Deemer, *Amer. Fertilizer*, **52**, 62–65 (1920).

354. Ross, S. D., and A. J. Catotti, *J. Amer. Chem. Soc.*, **671**, 3563 (1949).

355. Roth, H., and W. Beck, *Z. Anal. Chem.*, **141**, 404–414 (1954).

356. Rudnevskii, N. K., A. N. Tumanova, L. V. Kutergina, and N. A. Pozdnyakova, *Zh. Prikl. Spektrosk.*, **8**, 571–573 (1968).

357. Rudolph, G. A., and L. C. Flickinger, *Steel*, **112**, 114, 131–139, 149 (1943).

358. Ruehle, A. E., and D. A. Shock, *Ind. Eng. Chem. Anal. Ed.*, **17**, 453–454 (1945).

359. Runge, E. F., L. S. Brooks, and F. R. Bryan, *Anal. Chem.*, **27**, 1543–1546 (1955).

360. Russell, R. G., *Anal. Chem.*, **22**, 904–907 (1950).

361. Samuelson, O., *Ion Exchangers in Analytical Chemistry*, Wiley, New York, 1953.

362. Schafer, H., and A. Sieverts, *Z. Anal. Chem.*, **121**, 170–183 (1941).

363. Schafer, H., and A. Sieverts, *Z. Anorg. Allg. Chem.*, **246**, 149–157 (1941).

364. Schafer, H., Z. Anorg. Allg. Chem., **250**, 127–144 (1942).

365. Schafer, H., Z. Anorg. Allg. Chem., **259**, 255–264 (1949).

366. Schafer, H., Z. Anorg. Allg. Chem., **247**, 96–112 (1941).

367. Schafer, H., and A. Sieverts, Z. Anal. Chem., **121**, 161–169 (1941).

368. Schafer, H., FIAT Rev. Inorg. Chem., **III**, 225–233 (1946).

369. Scharrer, K., and R. Gottschall, Z. Pflanzenernahr., Dungung Bodenk., **39**, 178–197 (1935).

370. Scharrer, K., and H. Kuhn, Ber. Ges. Natur. Heilk. Giessen, Naturw. Abt., **27**, 72–84 (1954).

371. Scharrnbeck, C., Chem. Tech., **9**, 416–418 (1957).

372. Schlesinger, H. I., H. C. Brown, D. L. Mayfield, and J. R. Gilbreath, J. Amer. Chem. Soc., **75**, 213–215 (1953).

373. Schrauzer, G. N., Mikrochim. Acta, **1953**, 124–130 (1953).

374. Schulek, E., and O. Szakacs, Z. Anal. Chem., **137**, 5–7 (1952).

375. Schulek, E., and P. Rozsa, Hidrol. Kozlony, **27**, 69–79 (1947).

376. Schulek, E., and G. Vastagh, Z. Anal. Chem., **84**, 167–184 (1931).

377. Schultz, E., Mitt. Lebensm. Hyg., **44**, 213–227 (1953).

378. Schuster, E., and K. Wohlleben, Z. Anal. Chem., **245**, 239–244 (1969).

379. Scott, W. W., and S. K. Webb, Ind. Eng. Chem. Anal. Ed., **4**, 180–181 (1932).

380. Scribner, B. R., and H. R. Mullin, J. Res. Nat. Bur. Stand., **37**, 379–389 (1946).

381. Scruby, R. E., J. R. Lacher, and J. D. Park, J. Chem. Phys., **19**, 386–387 (1951).

382. Sefidvash, F., Anal. Chem., **40**, 1165–1166 (1968).

383. Sen, M. K., Indian J. Phys., **11**, 9–11 (1937).

384. Shaheen, D. G., and R. S. Braman, Anal. Chem., **33**, 893–894 (1961).

385. Shanina, T. M., N. E. Gel'man, and V. S. Mikhailovskaya, Zh. Anal. Khim., **22**, 782–787 (1967).

386. Shishido, S., Bull. Chem. Soc. Japan, **25**, 199–202 (1952).

387. Shvarts, E. M., A. E. Dzene, and A. F. Ievin'sh, Izv. Akad. Nauk Latv. SSR, Ser. Khim., **1968**, 749 (1968).

388. Shvarts, E. M., A. E. Dzene, and A. F. Ievin'sh, Zavod Lab., **35**, 787 (1969).

389. Shyne, J. C., and E. R. Morgan, Anal. Chem., **27**, 1542–1543 (1955).

390. Silverman, L., and K. Trego, Anal. Chem., **25**, 1264–1267 (1953); Anal. Chem., **25**, 1639 (1953).

391. Smith, D. L., D. R. Jamieson, and P. J. Elving, Anal. Chem., **32**, 1253 (1960).

392. Smith, F. M., W. G. Schrenk, and H. H. King, Anal. Chem., **20**, 941–943 (1948).

393. Smith, G. S., Analyst (London), **60**, 735–739 (1935).

394. Smith, W. C., Jr., A. J. Goudie, and J. N. Sivertson, Anal. Chem., **27**, 295–297 (1955).

395. Society of Glass Technology, Technical Committee, No. 1, J. Soc. Glass Technol., **34**, 305T–309T (1950).

396. Sommer, L., Chem. Listy, **51**, 2032–2036 (1957).

397. Solomin, N. V., and L. V. Potemkina, Dokl. Akad. Nauk SSSR, **96**, 91–93 (1954).

398. Sowinski, E. J., and I. H. Suffet, J. Chromatogr. Sci., **9**, 632–634 (1971).

399. Sowinski, E. J., and I. H. Suffet, *Anal. Chem.*, **44**, 2237–2239 (1972).

400. Spicer, G. S., and J. D. H. Strickland, *J. Chem. Soc.*, **1952**, 4644–4650 (1952); *J. Chem. Soc.* **1952**, 4650–4653 (1952).

401. Srivastava, R. D., P. R. Van Buren, and H. Gesser, *Anal. Chem.*, **34**, 209 (1962).

402. Stahl, W., *Z. Anal. Chem.*, **101**, 342–347 (1935).

403. Stamm, J., *Pharmacia*, **1924**, 18, 25 (1924).

404. Steele, W. C., and J. C. Decius, *J. Chem. Phys.*, **25**, 1184–1188 (1956).

405. Steinberg, R. H., *Appl. Spectrosc.*, **7**, 176–178 (1953).

406. Stettbacher, A., *Mitt. Lebensm. Hyg.*, **29**, 201–217 (1938).

407. Stettbacher, A., *Mitt. Lebensm. Hyg.*, **34**, 90–97 (1943).

408. Stetten, D., Jr., *Anal. Chem.*, **23**, 1177–1179 (1951).

409. Sumuleanu, C., and G. Ghimicescu, *Ann. Sci. Univ. Jassy*, **21**, 361–368 (1935).

410. Szebelledy, L., and S. Tanay, *Z. Anal. Chem.*, **107**, 26–30 (1936).

411. Tabata, K., and S. Moriyasu, *Res. Electrotech. Lab. Japan*, **152**, 32 (1925).

412. Tagaya, T., *Bull. Inst. Phys. Chem. Res. (Tokyo)*, **21**, 165–180 (1942).

413. Tarayan, V. M., E. N. Ovsepyan, and S. R. Barkhudaryan, *Dokl. Akad. Nauk Armyan SSR*, **48**, 52–53 (1969).

414. Taylor, T. I., and W. W. Havens, Jr., *Nucleonics*, **6**, 54–66 (1950).

415. Tazaki, H., *J. Sci. Hiroshima Univ. A*, **10**, 109–112, 113–116 (1940).

416. Tereshko, J. W., *Anal. Chem.*, **35**, 157 (1963).

417. Thomson, R. T., *J. Soc. Chem. Ind.*, **12**, 432 (1893).

418. Thorpe, T. E., *Dictionary of Applied Chemistry*, 4th edit. Vol. XI, Longman's Green, New York, 1954, p. 757.

419. Thygesen, J. E., *Z. Anorg. Allg. Chem.*, **237**, 101–112 (1938).

420. Torssell, K., *Ark. Kemi. Mineral Geol.*, **3**, 571 (1952).

421. Trinder, N., *Analyst (London)*, **73**, 494–497 (1948).

422. Tuzuki, Y., and Y. Kimura, *Bull. Chem. Soc. Japan*, **15**, 27–31 (1940).

423. Tuzuki, Y., *Bull. Chem. Soc. Japan*, **16**, 23–31 (1941).

424. Tuzuki, Y., *Bull. Chem. Soc. Japan*, **13**, 337–349 (1938).

425. Tung, B., and H. Chang, *J. Chinese Chem. Soc.*, **9**, 125–133 (1942).

426. Umland, F., and A. Janssen, *Z. Anal. Chem.*, **249**, 186–188 (1970).

427. Umland, F., and F. Pottkamp, *Z. Anal. Chem.*, **241**, 223–234 (1968).

428. Uppstrom, L. R., *Anal. Chim. Acta*, **43**, 475–486 (1968).

429. Urs, M. K., and K. Neelakantam, *J. Sci. Ind. Res. (India)* **11B**, 259–260 (1952).

430. Vandaveer, R. L., *J. Ass. Offic. Agr. Chem.*, **22**, 563–567 (1939).

431. Vanselow, A. P., *Proc. Amer. Soc. Hort. Sci.*, **46**, 15–20 (1945).

432. Vasilev, S. S., G. I. Mikhailov, L. P. Starchik, and L. V. Konanykin, *Zavod. Lab.*, **35**, 299–300 (1969).

433. Vasileva, M. G., and A. L. Sokolova, *Zh. Anal. Khim.*, **17**, 530 (1962).

434. Vasilevskaya, A. E., *Nauch. Tr. Vses. Inst. Min. Resursov*, *1971* 5, 22–31.

435. Verma, M. R., and K. C. Agrawal, *Electroplat. Metal Finish.*, **7**, 171–172 (1954).

436. Verma, M. R., and K. C. Agrawal, *Electroplat. Metal Finish.*, **7**, 403–404, 458–459 (1954).

437. Vermaas, N., *Rec. Trav. Chim.*, **51**, 67–92 (1932).

438. Volodchenkova, A. I., and B. N. Melentiev, *C. R. Acad. Sci. USSR*, **30**, 140–143 (1941).

439. Von Polheim, P., *Z. Anal. Chem.*, **137**, 8–15 (1952).

440. Von Stackelberg, M., F. Quatram, and J. Dressel, *Z. Elektrochem.*, **43**, 14–28 (1937).

441. Webster, P. A., *J. Amer. Ceram. Soc.*, **23**, 235–241 (1940).

442. Webster, P. A., and A. K. Lyle, *J. Amer. Ceram. Soc.*, **23**, 235–241 (1940).

443. Weir, C. C., *J. Sci. Food Agric.*, **21**, 545–547 (1970).

444. Weitman, J., N. Daverhoeg, and S. Farrolden, *Nucl. Appl. Technol.*, **9**, 408–415 (1970).

445. Wherry, E. T., and W. H. Chapin, *J. Amer. Chem. Soc.*, **30**, 1687–1701 (1908).

446. White, C. E., *J. Chem. Educ.*, **28**, 369–372 (1951).

447. White, C. E., A. Weissler, and D. Busker, *Anal. Chem.*, **19**, 802–805 (1947).

448. Wiederkehr, V. R., and G. W. Goward, *Anal. Chem.*, **31**, 2102 (1959).

449. Wilcox, L. V., *Ind. Eng. Chem. Anal. Ed.*, **4**, 38–39 (1932); *ibid.*, **2**, 358–361 (1930).

450. Wilcox, L. V., U.S.D.A. Tech. Bull. No. 962, 1948.

451. Williams, D. E., and J. Vlamis, *Anal. Chem.*, **33**, 967 (1961).

452. Williams, J. P., D. E. Campbell, and T. S. Magliocca, *Anal. Chem.*, **31**, 1560, 1766 (1959).

453. Williams, D. E., and J. Vlamis, *Anal. Chem.*, **33**, 1098 (1961).

454. Wilson, C. W., *J. Amer. Chem. Soc.*, **61**, 2303 (1939).

455. Wilson, S. H., and M. Fieldes, *New Zealand J. Sci. Tech.*, **24B**, 98–102 (1942).

456. Winslow, E. H., and H. A. Liebhafsky, *J. Amer. Chem. Soc.*, **64**, 2725–2726 (1942).

457. Winsor, H. W., *Anal. Chem.*, **20**, 176–181 (1948).

458. Wojtowicz, M., and M. Kubica, *Chem. Anal. (Warsaw)*, **13**, 65–72 (1968).

459. Wolszon, J. D., and J. R. Hayes, and W. H. Hill, *Anal. Chem.*, **29**, 829 (1957).

460. Wuensch, G., *Z. Anal. Chem.*, **258**, 30–31 (1972).

461. Yasuda, S. K., and R. N. Rogers, *Microchem. J.*, **4**, 155 (1960).

462. Yoe, J. H., and R. L. Grob, *Anal. Chem.*, **26**, 1465–1468 (1954).

463. Zaletel, B. V., *Rec. Trav. Inst. Recherches Structure Matiere (Belgrade)*, **2**, 31–59 (1953).

464. Zorkin, F. P., *J. Appl. Chem. (USSR)*, **9**, 1505–1506 (1936).

CARBON

By W. E. Chambers, Paul D. Coulter, and Ronald A. Greinke, *Carbon Products Division, Parma Technical Center, Union Carbide Corporation, Parma, Ohio*

Contents

I. INTRODUCTION

Graphite and diamond have been known from ancient times. Graphite (Greek "to write") was sometimes confused with other substances, and "plumbago" (like lead) was a synonym for graphite. For example, the "lead" in a lead pencil is actually a graphite mixture. Diamond was mentioned in the "Vedas," ca. 1100 B.C. Although individual modifications of carbon were well known, the identification of diamond and graphite as different forms of elemental carbon was not made until the early 1800s. Amorphous carbon, a form that resembles finely divided graphite, yet having properties of its own, has been established by industrial usage.

In its ability to form a large number of compounds, carbon is unique among the elements. Over one million compounds have been identified. The total number of compounds possible is virtually unlimited. The vast majority of these are organic compounds. Only the analytical chemistry of elemental carbon and of inorganic compounds of carbon is discussed in this chapter. The determination of carbon in organic compounds is discussed in Volume II, pp. 297–404.

A.   OCCURRENCE

1.   In Nature

Carbon, one of the more abundant elements, is found in a large variety of forms. Its two crystallized forms, diamond and graphite, have industrial value and are obtained by mining. Diamond is sparsely distributed in different parts of the world. The principal sources are South Africa and South America. Although graphite is widely distributed, most deposits are very low grade; hence only a few deposits are economical to mine. The principal sources are Ceylon, Austria, Bavaria, Siberia, the United States, and Canada.

Carbon is the principal constituent in coal, peat, and crude oil. Compounds of carbon are found in natural gas, carbon dioxide in the air, carbonic acid and carbonates in natural bodies of water, and carbonates of calcium, magnesium, and iron in rock.

2.   In Industrial Processes and Products

Large quantities of carbon for industrial uses are prepared from petroleum products, coal, gaseous hydrocarbons, wood, and other carbon-containing substances. Methods of preparation vary, and the type of carbon obtained (amorphous or crystalline) depends on processing techniques and the choice of raw materials. Some carbonaceous deposits such as coal, bitumen, and peat are widely used as fuel.

3.   As an Impurity

Carbon is encountered industrially as an impurity usually in the production of high-purity metals, metal carbides, and in some alloys.

B.   INDUSTRIAL PROCESSES

Carbon is processed and used in all three of its common forms (diamond, graphite, and amorphous). Industrial-quality diamond is manufactured by subjecting carbon to high pressure (about 1.8 million psi) and very high temperature (about 2700°C). Diamond finds use as dies for drawing wire, abrasive on grinding wheels, the cutting edge on drill bits used for rock drilling, and wherever the very hard nature of diamond can usefully be employed. Good-quality natural diamond is a valuable gem.

Graphite is manufactured in large quantities by passing an electrical current through a mixture of petroleum coke with a pitch binder. Graphite is mixed with clay to form refractory crucibles used in the melting of metals and in other metallurgical operations. Graphite electrodes are used in electrochemical industries and in the manufacturing of metal alloys. Very-high-purity graphite rods are used for analysis in emission-spectrographic procedures. High-purity or nuclear-grade graphite is used in reactors. Other uses for graphite are lubricants, stove polish, batteries, brushes for electric motors, and when mixed with clay, as "lead" pencils. Graphite is highly resistant to weathering and to attack by most chemical reagents and is used when these properties are needed.

Amorphous carbon is manufactured by a number of techniques and in a variety of types that can be grouped according to their general use: namely, coloring agents or pigment, specific adsorption agents due to high surface activity, and the fabrication of certain consumer products where porosity control is essential. Carbon black (the product of incomplete combustion of a gas with the flame in contact with a metallic surface), acetylene black (the product of thermatomic cracking of acetylene), and lampblack (the smoke particles from an unobstructed hydrocarbon flame) are used in the plastic and rubber industry and as coloring agents and pigments. Bone black (the carbonaceous residue obtained by destructive distillation of bones) and charcoals (obtained by heating materials in a closed container) are employed as decolorizing agents and gas adsorbents. The action of the char is very dependent on its precursor. Some chars are very specific in their adsorbent qualities. All the blacks, partially graphitized blacks, and graphite are used to fabricate electrodes, brushes, arc carbons, batteries, refractory materials, and other related products.

Metallurgical coke is manufactured in large quantities by heating bituminous coal in ovens having oxygen-deficient atmosphere. Coke is used for reducing ores or iron and other metals and for fuel. Cokes, pitches, and coal tars are used in the fabrication of many carbon or graphite materials.

Pyrolytic graphite is formed by high-temperature (2000°C) decomposition of a hydrocarbon gas. It is dense (about 2.2 grams/cc), nonporous (will hold a very high vacuum), and resistant to chemical attack. The physical properties of pyrolytic graphite are very anisotropic. For example, the thermal conductivity along the plane of pyrolytic graphite is several times that of copper but parallel to the c axis it is a good insulator. It is used in aerospace applications such as a liner for rocket nozzles. Oxidation resistant crucibles and other apparatus are made for metallurgical and laboratory use from pyrolytic graphite.

Carbon yarn and carbon cloth have been made by carbonizing and graphitizing cellulose yarn and cloth. The yarn and cloth have been used in aerospace applications and as packing for furnaces. The yarn has potential as a fabricating material, since it has one of the highest strength-to-weight ratios and a very high modulus of elasticity.

C. TOXICOLOGY

Elemental carbon is considered a physiologically inert substance. As industrial contaminant, it is regarded as a nuisance dust. The A.C.G.I.H. (The American Conference of Governmental Industrial Hygienists) has adopted a threshold limit of 5.0 mg/m^3 of air.

Carbon dioxide, although not toxic in the usual sense, can be hazardous both by suffocation and by stimulating the respiration rate. Daily 1 hr exposures of 8–10% carbon dioxide do not have any marked, deleterious effect. The A.C.G.I.H. threshold for an 8 hr exposure is 5000 ppm.

Carbon monoxide due to its colorless, odorless, and very toxic nature is a very dangerous industrial contaminant. The physiological response is asphyxia. Carbon monoxide combines with the hemoglobin and interferes with the transportation of oxygen to the cells. This reaction is reversible, and, in general, mild doses of carbon monoxide do no permanent damage; however, damage done by severe asphyxia such as brain and nervous system damage is not reversible. The symptoms of carbon monoxide poisoning are listed in Table 1.

Elimination of carbon monoxide is solely through the lungs. Elimination can be accelerated by inhaling pure oxygen or oxygen with 5–7% carbon dioxide added. The A.C.G.I.H. upper limit of exposure for 8 hr is 50 ppm.

Cyanides are among the most toxic of all industrial chemicals. The physiological response to inhalation of higher concentrations of hydrogen cyanide vapor or ingestion of 50–100 mg of sodium or potassium cyanide is almost instantaneous collapse and cessation of respiration. At lower dosages early symptoms may be weakness, headaches, confusion, and occasionally nausea and vomiting. For milder cases of cyanide poisoning the use of artificial respiration and inhalation of amyl nitrite

TABLE 1

Symptoms Caused by Various Amounts of Carbon Monoxide Hemoglobin in the Blood
(327)

Blood saturation (% CO hemoglobin)	Symptoms
0–10	No symptoms
10–20	Tightness across forehead, possibly slight headache, dilation of cutaneous blood vessels
20–30	Headache and throbbing in temples
30–40	Severe headache, weakness, dizziness, dimness of vision, nausea, vomiting, and collapse
40–50	Same as previous item with more possibility of collapse and syncope, and increased respiration and pulse
50–60	Syncope, increased respiration and pulse, coma with intermittent convulsions, and Chenye-Stokes respiration
60–70	Coma with intermittent convulsions, depressed heart action and respiration, and possibly death
70–80	Weak pulse and slow respiration, respiratory failure, and death

vapor may be sufficient for recovery. The A.C.G.I.H. limit for hydrogen cyanide is 10 ppm, and for cyanide dusts is 5 mg/m^3 expressed as CN.

Metal carbonyls are very toxic. Nickel carbonyl is the most common form, being manufactured in tonnage quantities in the Mond process. Immediately after exposure, giddiness and headache occur, sometimes accompanied with dyspnea and vomiting. Fresh air brings relief of symptoms. Dyspnea returns 12–36 hr later, cyanosis and leucocytosis appear, the temperature begins to rise, coughing with more or less bloodstained sputum occurs, pulse rate increases, and delirium and other signs of disturbances of the central nervous system appear. Death occurs in fatal cases between 4 and 11 days.

There is some possibility that nickel carbonyl is a carcinogenic agent. Mond process workers experience cancer of the lung and nose with a rate of five times normal. Kincaid and co-workers (207) have suggested an air concentration of nickel carbonyl vapor of 0.04 ppm by volume as the maximal limit for avoidance of acute effects in man. The A.C.G.I.H. has set an upper limit of 0.001 ppm nickel carbonyl for repeated 8 hr daily exposures.

Coal-tar-pitch volatiles, generated from heating bituminous coal in coke ovens are carcinogenic (302,304). The A.C.G.I.H. has set the upper limit for coal-tar-pitch volatiles in working place air at 0.15 mg/m^3 of benzene-soluble material.

II. PROPERTIES OF CARBON

A. PHYSICAL PROPERTIES

The appearance of elemental carbon varies from the dull gray black of carbon to the sparkling beauty of clear gem diamond. The softness of graphite (less than talc) and the hardness of diamond (the hardest naturally occurring substance) illustrate the great difference between these two allotropic modifications of carbon. Amorphous carbon particles are believed to have the graphite structure. The physical properties of carbon differ for the several forms and are summarized in Table 2. The physical properties for some carbon compounds are given in Table 3. The nuclear properties are given in Table 4, the optical properties in Table 5.

Although carbon is generally inert in electrochemical cells, sometimes being used as an indicating electrode, its electrochemical behavior is of prime importance in the production of dry cells. Some of its electrochemical properties are given in Table 6.

B. CHEMICAL PROPERTIES

Of the common reagents only chromic, nitric, and perchloric acids attack carbon. Even by these acids, graphite and diamond are only very slowly oxidized. Graphite gives a number of unusual oxidation products that have been attributed to its layer structure. Oxidation by potassium chlorate in a mixture of nitric and sulfuric acids results in the swelling of the graphite with the insertion of oxygen between the carbon layers. The product is known as "graphitic oxide." Oxidation of graphite in the presence of a strong acid, such as nitric, perchloric, sulfuric, or phosphoric acid, is accompanied by separation of the layers. Salt-like substances called "graphitic salts" are formed. Potassium, rubidium, cesium, fluorine, bromine, and ferric chloride have been shown to be able to expand the graphite structure.

Three oxides of carbon are known: carbon suboxide, C_3O_2; carbon monoxide, CO; and carbon dioxide, CO_2. Carbon suboxide is prepared by the dehydration *in vacuo* of malonic acid by phosphoric anhydride at temperatures of 140–150°C. The oxide, a vile smelling gas, is extremely soluble in water with the formation of the parent acid.

Carbon monoxide can be formed almost quantitatively by the combustion of carbon in an insufficient supply of air at 1000°C. Carbon monoxide is a mildly reactive substance. The monoxide burns in air with a pale blue flame, forming the dioxide, and combines directly with halogens, sulfur, selenium, and some metals to give carbonyl com-

TABLE 2

Physical Properties of Carbon

Property	Diamond	Carbon (all forms)	Graphite
Atomic number	—	6	—
Atomic wt. (C^{12} = 12.000)	—	12.0115	—
Atomic volume (cc/gram at 1 atm)	3.42	—	—
Density (gram/cc at 20°C)	3.15–3.53 (3.513 gem)	—	1.9–2.26
Mass number of isotopes	—	10, 11, 12, 13, 14, 15, 16	—
Relative frequency of naturally occurring isotopes	—	98.892 (C^{12}) 1.108 (C^{13})	—
Atomic electron arrangement	—	$1S^2\, 2S^2\, 2P^2$	—
Valence	—	2, 3, 4	—
Melting Point (°C)	—	3,570 (?)	—
Boiling Point (°C)	—	3,470 (sub)	—
Specific heat (cal/gram)			
0°C	0.1044	—	0.170
20°C	0.120	—	0.254
140°C	0.222	—	0.254
Heat of combustion (kg·cal/g) ($C + O_2 \rightarrow CO_2$)	7.860	—	7.900
Coefficient of linear thermal expansion (cm/cm·C)	1.18×10^{-6}	—	$0.7–4.0 \times 10^{-6}$
Electrical resistivity (ohm·cm at 0°C)	—	—	$0.45–1.0 \times 10^{-3}$
Thermal neutron absorption cross section (mbarn)	—	3.73	—
Hardness (Mohs' scale)	10	—	0.5–1
Magnetic susceptibility (20°C, 10^{-6} cgs)	−0.49	—	−3.5

TABLE 3

Properties of Some Carbon Compounds

Properties		Compounds				
	CO	CO_2	CS_2	CCl_4	HCN	CH_4
Melting point (°C)	−207	79° sub	−111.6	−22.9	−13.4	−182.5
Boiling point (°C)	−192	—	46.3	76.8	25.6	−161.5
Specific gravity (grams/liter)	1.250 (0°C)	1.977 (0°C)	1.262 (20°C)	1.595 (20°C)	0.699 (20°C)	0.7168 (0°C)
Water Solubility (g or cc/100 ml)	3.5 cc (0°C) 20 cc (20°C)	0.348g (0°C) 0.058g (60°C)	0.22g (22°C) 0.014g (50°C)	very slight(0°C)	—	9 cc (20°C)
Molar magnetic susceptibility (10^6 cgs)	−9.8	−21	−42.2 (liq.)	−66.8	—	12.2
Heat of formation (kcal/mol at 25°C)	+26.428 (gas)	+94.385 (gas)	−22.0 (gas)	33.190 (liq.)	−30.108	19.1 (gas)
Dielectric constant (20°C)	1.00070 (0°C)	1.00985 (0°C)	2.647	2.24	116	1.000944 (0°C)
Ionization constant						
K_1	—	4.2×10^{-7} (H_2CO_3)	—	—	4.9×10^{-10}	—
K_2	—	4.8×10^{-11}	—	—	—	—

TABLE 4
Nuclear Properties of Carbon

Isotope	Atomic mass ($0^{16} = 16.0000$)	Half life ($T_{1/2}$)	Modes of decay	Nuclear spin (I)	Thermal neutron cross section (barn)
$^{10}_6C$	10.02084	19.0 sec	β_+	0	—
$^{11}_6C$	11.01499	20.5 min.	β_+	3/2	—
$^{12}_6C$	12.00386	—	—	0	0.0037
$^{13}_6C$	13.00756	—	—	1/2	0.0009
$^{14}_6C$	14.00770	5770 yr	β_-	0	—
$^{15}_6C$	—	2.25 sec	β_-	1/2	—
$^{16}_6C$	—	0.74 sec	β_-	0	—

pounds. At elevated temperatures, the monoxide reduces metal oxides and water.

Carbon dioxide is formed as the product of complete oxidation of free or combined carbon. It is thermally stable up to about 1000°C. Carbon dioxide dissolves only to a limited extent in water, forming carbonic acid, H_2CO_3. Two types of salts are formed by carbonic acid: carbonates, CO_3^{2-}, and bicarbonates, HCO_3^-. In general, only the alkali-metal carbonates (excluding lithium) and ammonium carbonate are soluble. Bicarbonates, on the other hand, are usually much more soluble than other common salts.

A number of binary compounds are formed by carbon. The metal carbides can be divided into three classes: (1) salt-like carbides such as calcium carbide; (2) interstitial carbides formed by transition metals and

TABLE 5
Optical Properties of Carbon

Property	Diamond	Carbon (all forms)	Graphite
Crystal structure	Cubic (fcc)	—	Hexagonal
Lattice parameter (A)			
A_0	3.5667	—	2.464
C_0	—	—	6.736
Axial ratio	—	—	2.734
Space group	$O_H^7 - FD3M$	—	$P6_3/MMC$
Principal diffraction lines (A) (relative intensity)	2.06 (100)	—	3.37 (100)
	1.26 (27)	—	1.68 (8)
	1.08 (16)	—	1.23 (6)
K X-ray absorption edge (A)	—	43.648	—
K Critical X-ray absorption energy (eV)	—	0.284	—

TABLE 6

Electrochemical Properties of Carbon

Property	Value
First ionization potential (V)	11.217
Second ionization potential (V)	24.27
Third ionization potential (V)	47.65
Fourth ionization potential (V)	64.22
Electrochemical equivalents (valence 4)	
mg/C	0.03111
g/amp·hr	0.11201
Contact potential[a] against copper metal of (V)	
$C + NH_3$	+0.079
$C + H_2$	0.096
$C + N_2$	0.129
$C + CO_2$	0.130
$C + NO$	0.136
$C + O_2$	0.142
$C + O_3$	0.155

[a] Coconut charcoal saturated with gas named.

by molybdenum and tungsten; and (3) those carbides formed by chromium, manganese, iron, cobalt, and nickel.

The salt-like carbides are further classified as to whether they apparently contain carbon as C^{4-}, C_2^{2-}, or C_3^{4-}. The salt-like carbides are colorless or only faintly colored, transparent substances that do not conduct electricity at ordinary temperatures and are decomposed by water with hydrocarbons being liberated. Be_2C and Al_4C_3 are examples of the first class of salt-like carbides and yield methane when decomposed by water. Na_2C_2, Cu_2C_2, and CaC_2 are examples of the second class of salt-like carbides and yield acetylene when decomposed by water. Only Mg_2C_3 is known for the third class, and it forms chiefly allylene when decomposed by water.

Interstitial carbides are formed by the entry of carbon into the interstices of the crystal lattices of the transition metals. Two types are formed: MC, a cubic close-packed lattice; and M_2C, a hexagonal close-packed lattice. They are metal-like, being opaque, having luster, and a high electrical conductivity. They are hard and nearly inert to all chemical agents with the exception of strong oxidants.

Of the remaining group of carbides, Cr_3C_2 and M_3C (M = Mn, Fe, Co, and Ni) have been prepared. These carbides react readily with dilute acids. Some react with water to give a variety of products. For example, MnC_3 gives a mixture of methane and hydrogen when decomposed by water. Hydrogen, various hydrocarbons, and free carbon are the products formed when Fe_3C and NiC_3 are decomposed by hydrochloric acid.

Boron and silicon carbides have been formed. A number of boron carbides have been reported, but only B_4C has been characterized. Silicon carbide is prepared commercially by reduction of silica with carbon at about 3500°C.

The four tetrahalides of carbon have been prepared. Carbon tetrabromide and tetraiodide are unstable. The chloride is manufactured by passing dry chlorine through carbon disulfide containing a little iodine or antimony trichloride as a catalytic agent. Carbon tetrafluoride is made by direct union of the elements or by the reaction of fluorine and another halogen with carbon or by the action of an interhalogen or carbon.

Carbon disulfide can be formed by the direct union of the elements, but is now generally manufactured by the reaction of natural gas (methane) with sulfur vapor. Carbon disulfide burns in air to give carbon dioxide and sulfur dioxide. Carbon disulfide can be partially converted to the subsulfide, C_3S_2, by heating to high temperature.

There is a large family of compounds and complexes containing the CN^- group. In terms of composition, the simplest compound possessing the cyanide group is cyanogen, $(CN)_2$. A cyanide compound is known for nearly every element. Hydrogen cyanide is formed when a metal cyanide is treated with dilute acid. Discussion of the cyanide complexes is beyond the scope of this chapter.

Carbon monoxide and metals unite to form metal carbonyls. The carbonyls are prepared by a variety of methods. For example, nickel tetracarbonyl and iron pentacarbonyl are formed by direct combination of the metal and carbon monoxide. Detailed information on the metal carbonyls and complexes has been given (100).

III. SAMPLING OF CARBON-CONTAINING MATERIALS

Obtaining a representative sample from carbon or carbon-containing materials can be a difficult task. The tendency for impurities to form inclusions is well known. Even high-purity graphite has local segregation of impurities. Heat-treated carbons and graphites sometimes show a different impurity level in the center of a block from that of the edges or sides. Bulk materials such as cokes, pitches, and coal show different impurity levels according to size of the sampled pieces. The theory of coal sampling has been studied (51). The sampling of metals for carbon analysis is complicated by carbon's tendency to form inclusions in the metals. It is apparent, therefore, that no one sampling technique will suffice.

For bulk carbonaceous materials and unfabricated carbon and graphite materials American Society for Testing Materials (ASTM) procedures

(11) have been established. The designated methods are: for coal, D-271-70; for coke, D-346-35; for bituminous materials, D-140-55; for petroleum products, D-270-65; for carbon blacks, D-1900-63 and D-1799-65; and for graphite, C-560-69. Charcoal and related chars are sampled by the same techniques as coal and coke.

For most metals ASTM methods for sampling are available. More details on particular metals can be obtained by consulting the appropriate chapter in this treatise.

Special precautions must be observed in the handling of high-purity carbon and nuclear grade graphite. Samples and stock should be handled with cotton gloves that have been laundered with boron-free soap. The hands should also be washed with boron-free soap. Sampling should be done with saws, drills, or machine tools of known composition. Tools with high vanadium, titanium, molybdenum, and so forth should be avoided. Some evaluation of the amount of contamination of the material during sampling operations can be accomplished by weighing of the sampling equipment before and after use. The amount of contamination can be confirmed by analyzing some of the material in chunk form without tool contact, for the specific elements of which the tools are composed. These results can be compared with the analytical results of the same materials after accomplishing sampling and homogenization procedures.

The sampling procedures for industrial fabricated materials of carbon and graphite have not yet been well defined. Due to large variations in the size of production lots and to the degree of purity, no one sampling procedure has found wide use. The ASTM procedure, C-560-69 (12), for graphite is to crush and grind the entire sample to pass a No. 60 (250 μm) sieve in a roll crusher. The sample may have been reduced in size initially by drilling the test bar with silicon carbide-tipped drills. Some sampling techniques (56) currently in use for particular industrial materials are:

1. *Massive carbon or graphite electrodes for smelting or steel refining furnaces.* (The cation impurity is of importance only where it effects the oxidation rate.) The piece is sampled (2–5% of total) by core plug drilling the ends of the electrodes.

2. *Graphite for electronic applications.* (The typical limits for calcium and total ash are 60 and 300 ppm, respectively.) The piece is sampled (5–25%) by sawing $\frac{1}{4}$ in. from the ends of the cylindrical electrodes, and if possible, a similar disk is taken 24 in. from the end of each electrode. The disks are homogenized by passing through a jaw crusher and then through laboratory pulverizers. The saw, jaw crusher,

and pulverizer must be cleaned and carefully protected from sources of contamination.

3. *Graphite for use in chlor-alkali production cells*. (A low concentration of vanadium and similar metals is of prime importance for mercury cells. Typical limits of 2–50 ppm are found.) The graphite block (20–50 lb. each) is sampled (5–25% of total) by the collection of drillings from $\frac{1}{4}$ in. holes made in each block in several locations. (The locations chosen are usually those where larger holes will subsequently be located in the final machined product.) The drillings from each block are then either composited or analyzed individually for contaminants as dictated by experience or specifications.

4. *Graphite for use in nuclear reactors*. (It is essential that elements with high-neutron cross-section be absent or in low concentration. Typical boron concentration is 0.1–1.0 ppm, and ash is 5–1000 ppm.) Two adjacent slabs are sawed from 5 to 20% of the pieces at a distance from the end equivalent to $\frac{1}{3}$ the length of the stock. One slab in cross-section equal to that of the stock pieces is $\frac{1}{4}$ in. thick; the adjacent sample slab is cut to be $\frac{1}{8}$ in. thick. The saw used must be clean and should be used only for sawing this type of material.

Both the $\frac{1}{4}$ in. slabs used for the determination of contaminating elements and the $\frac{1}{8}$ in. thick slabs taken for ash determination must have at least $\frac{3}{16}$ in. trimmed from the original surface for each of the four sides of the cross-section. The amount removed from the original surface of the sample slabs is such that the final cross-section of each test slab is equivalent to the maximum cross-section that will be used in the field.

The purpose of this trimming is to remove impurities and contamination that might have been incorporated into the outer surfaces of the bar stock during graphitization procedures and subsequent storage and handling of the stock. It is important that the handling and sampling procedures be performed so that the chemist will measure impurities in the stock as it will be used in the field. The trimmed $\frac{1}{4}$ in. slabs are homogenized by jaw-crushing and milling into one homogeneous sample as described before. Each $\frac{1}{8}$ in. thick test slab is broken by hand in such a manner that portions representing the outside and the inside sections of each slab make up the final 50 g sample for determination of ash.

IV. SEPARATION AND ISOLATION OF CARBON

In analysis of materials for carbon without regard to its chemical or physical state (graphite, amorphous, free, or combined), prior separation of carbon is seldom made. To a certain extent, however, all techniques in which carbon dioxide is formed could be classified as separation techniques in that the carbon is separated from the bulk material as carbon dioxide. For the purposes of this section, we consider only those techniques of separation that do not involve the formation of carbon dioxide as the means of separation. Two general types of separation are considered; the separation of free carbon from materials and the separation of graphitic carbon from other forms of carbon.

A. SEPARATION OF "FREE" CARBON

The term "free carbon" as applied to tars, pitches, rubbers, synthetic elastomers, and similar material has come to mean the material that is not soluble in the solvent. The insoluble material includes high-molecular-weight aromatics, polymeric material, and other carbonaceous substances in addition to uncombined carbon. The trend today is to call the undissolved material, insolubles (i.e., benzene insolubles). Numerous extraction solvents have been suggested. Benzene, quinoline, a mixture of glacial acetic acid and toluene, toluene, aniline, chloroform, carbon disulfide, pyridine, and xylene have been employed as solvents. Extraction by fatty oils such as olive oil has been suggested. The nature of the material restricts the choice; however, benzene, quinoline, carbon disulfide, and xylene are preferred, benzene and quinoline being the more common choices.

Various methods for separating free carbon from rubber and synthetic elastomers have been suggested. Most are based on digestion of the sample with concentrated nitric acid followed by extraction with organic solvents. A general procedure involving a nitric acid digestion has been described for use with all vulcanized elastomers (232, 212).

Selective oxidation (244,246) and electrochemical (123,163) methods have been applied to the separation of free carbon from metals and metal carbides. In the selective oxidation methods, the choice of temperature and atmosphere allows the oxidation of carbon compounds but not of free carbon. Electrochemical methods remove the metal matrix and metal carbides by a combination of electrochemical and chemical action, leaving the uncombined carbon as a residue. A more detailed discussion of the determination of free carbon in metal carbides is presented in Section VIII.D.

B. SEPARATION OF GRAPHITIC CARBON FROM NONGRAPHITIC CARBON

The classical technique for separation of graphitic carbon from other forms of carbon is the formation of graphitic acid. The carbon mixture is mixed with concentrated sulfuric acid, and then small portions of potassium chlorate are added. A precipitate of golden-yellow to light green flakes of graphitic acid are obtained. The formula for graphitic acid is arbitrarily taken as $C_{28}H_{10}O_{15}$. Although a tedious and somewhat tricky procedure, reproducible results can be obtained with careful work. A modified version of this procedure (sometimes called Berthelot's method) has been described (1).

Most methods are based on the different density of graphitic carbon and amorphous carbon (338). ASTM Committee D-2 proposed the separation of nongraphitic carbon from graphite by use of ethylene bromide (10). The sample is ground and mixed with ethylene bromide and allowed to stand. The nongraphitic portion floats and can be removed from the upper portion of the liquid. This procedure is very dependent on the physical nature of the sample. Quite often erroneous results are obtained. To date (1977) this procedure has not received tentative approval.

V.   DETECTION AND IDENTIFICATION

A.   DETECTION OF CARBON

Carbon is commonly identified by chemical, instrumental, and petrographic techniques. Emission spectroscopy is the most common instrumental technique used to identify carbon; however, mass spectrometry and activation analysis have been successfully employed for the identification of carbon. Fusion (Lassaigne) with metallic sodium or potassium (138,114,276,277), fusion with potassium (110) or sodium azide (113), and ignition with magnesium powder (196,197) are commonly used chemical tests for detection of carbon in any form. Carbon is converted to cyanide ion by fusion with either sodium or potassium metal or with sodium or potassium azide. The cyanide formed can be identified by the Prussian blue test, the demasking of alkali palladium dimethylgloxime (114) or the formation of isonitrile (276). Upon ignition with magnesium powder, carbon is converted into magnesium carbide (196,197). Allylene is formed when the ignition product (magnesium carbide) is treated with acid and is identified by the Illsovay reagent (197) (ammoniacal calcium sulfate containing hydroxylamine hydrochloride and gelatin).

The "Penfield test" (261), a somewhat older, less sensitive test, is based on the formation of barium carbonate when barium hydroxide is exposed to the gaseous combustion products of a sample. The "Penfield test" has the advantage of being somewhat faster and easier to perform than the fusion methods.

1.   Sodium Fusion Test (277)

Place 2–5 mg of sodium metal in a dry test tube, and add an equal amount of dry ammonium sulfate and a few milligrams of the material to be tested. Heat gently to the appearance of a white fume. Allow to cool, and add a drop of water. When the reaction subsides (excess sodium removed), continue to add water until about 0.3 ml of solution is obtained. Add about 2 mg of ferrous sulfate, heat to boiling, cool, and transfer to a spot plate. Take up about 10 μl of clear solution, and transfer to a clean depression (any precipitate is removed by centrifuging). Add a tiny crystal (1–2 mg) of ferric sulfate, and acidify with a few drops of 10% hydrochloric acid. A blue color is a positive test for carbon. Sensitivity is at least 1 μg C.

2.   Sodium Azide Fusion Test (114)

Cautiously heat a small amount (less than 1 mg) of the material with a tenfold amount of sodium azide in a micro test tube. Cool, and add one drop of water. Add one drop of alkali palladium dimethylglyoxime solution (3 N potassium hydroxide saturated with palladium dimethylglyoxime), then add one drop nickel–ammonium solution (0.5 N nickel chloride saturated with ammonium chloride). A red precipitate or pink color is a positive test for carbon. Sensitivity is at least 1 μg C.

3.   Penfield Test (261)

Fuse a mixture of the sample with lead chromate in a hard glass tube. Place a drop of barium hydroxide solution near the mouth of the tube. The appearance of a film of barium carbonate indicates the presence of carbon. A more sensitive modification of this test is to fuse the mixture in a closed tube and break the tube open under the surface of barium hydroxide. Then examine the tip of the capillary under a microscope for the presence of small particles of barium carbonate. This technique is sensitive to at least 2 μg of carbon.

B.   IDENTIFICATION OF VARIOUS FORMS OF UNCOMBINED CARBON

The techniques described in Section V.A identify carbon regardless of its form or state in the sample. Consequently, these tests are of little use if the interests are in identifying free carbon or combined carbon in the presence of one another. A skilled petrographer can easily identify free carbon or graphite and carbides as inclusions in metals or other substances. Techniques have been developed by which a petrographer can identify various forms of carbon in carbonaceous raw material and processed carbon material such as carbon brushes (294). The test described in the previous section can be used to detect free carbon if precaution is taken to remove all combined forms first.

In Section IV.B, means of separating amorphous carbon from graphitic carbon are described. These tests can be used to distinguish between the two forms. Severe problems can be encountered when the graphite content is very low and present in very small crystals which are attached to and intimately mixed with amorphous carbon. More recently an X-ray diffraction procedure has been proposed (300). This technique is based on examination of the (002) diffraction peak. Pure graphite has a strong, sharp peak, while amorphous carbon has a weak, broad band. The calculated interlayer spacing gives information as to the degree of graphitization.

Chemical techniques have been developed that are sensitive to various forms of carbon. Table 7 presents a tabular listing of a number of colorimetric tests used to identify some raw carbonaceous materials. Molybdenum blue is formed when carbonaceous materials other than graphite are heated with molybdenum(VI) oxide (110). Ammonium salts and reducing inorganic compounds give a positive test and must be removed prior to test. With the exception of graphite, carbonaceous material yields potassium iodide and carbon dioxide when heated with potassium iodate at ca. 400°C. A positive starch test for iodine after the reaction mixture is treated with dilute sulfuric acid indicates the presence of carbon other than graphite (112). A combination of tests can be used to identify the forms of carbon present.

VI. DETERMINATION OF CARBON

A. OXIDATION METHODS

With the exception of those materials in which the carbon is present as carbonate, the most common method of determination is based on oxidation to carbon dioxide. This oxidation can be accomplished by wet ashing in acid solution or by direct combustion in oxygen, the latter being far more commonly used. The carbon dioxide liberated through oxidation is separated from contaminants and interfering substances and measured quantitatively by one of the many available techniques. Each of these steps are treated separately and in sequence in the following discussion.

1. Wet Combustion

Wet oxidation methods for converting carbon to carbon dioxide were used extensively until 1920, even in the analysis of metals. The most common oxidant was chromic acid in either sulfuric or phosphoric acid

TABLE 7

Color Reactions for the Testing of Graphite and Other Carbonaceous Materials (1)

Material	Benzol extraction	Behavior on heating	Heated with dilute sodium hydroxide solution	Boiled with dilute nitric acid	Boiled with concentrated nitric acid	Heated with a dilute solution of potassium permanganate	Immersed in molten sodium sulfate
Graphite	Colorless	No change	No solution or reaction	No effect	No reaction	No reaction	No reaction
Retort carbon	Colorless	No change	No solution or reaction	No effect	No reaction	No reaction	Violent reduction to sodium sulfide
Coke	Colorless	No change	No solution or reaction	No effect	No reaction	The permanganate solution is decolorized by formation of carbonates	Violent reduction to sodium sulfide
Anthracite, calcined	Colorless	No change	No solution or reaction	No effect	Very little effect, at best a black coloration	The permanganate solution is decolorized by formation of oxalic acid in small amounts	Violent reduction to sodium sulfide
Anthracite, uncalcined	Colorless	Very small amount of volatile matter	No solution or reaction	No effect	Brown-red solution ammonia darkens color; with calcium chloride and lead acetate solution gives a brown precipitate; the residue is black	Like calcined anthracite	Violent reduction to sodium sulfide

Material							
Lampblack	Slightly yellow, colored with fluorescence from tar content	Very small amount of volatile matter	Faint yellow solution	No effect	Like anthracite	Very slight effect with formation of small amounts of oxalic acid	Violent reduction to sodium sulfide
Bituminous coal	Faintly yellow with deeper fluorescence	Distillation products; tar, stable alkaline aqueous liquor, and hydrogen	No solution or reaction	No effect	Like anthracite	Strong decolorization with formation of large amounts of oxalic acid	Violent reduction to sodium sulfide
Lignite	Bright yellow extract without fluorescence	Distillation products; tar, usually acid or nearly neutral aqueous liquor, and hydrogen	Yellow to brown solution from which dilute acids precipitate brown flocks of humic acid	Red-yellow to orange-red solution; hydrocyanic acid can be detected in the distillate	Red-yellow to orange-red solution; hydrocyanic acid can be detected in the distillate	Strong decolorization with formation of large amounts of oxalic acid	Violent reduction to sodium sulfide
Charcoal, high-calcined	Colorless	No change	Bright yellow solution with no flocks of humic acid	No effect	Like anthracite	Rapid decoloring of solution with formation of small amounts of oxalic acid	Violent reduction to sodium sulfide
Charcoal, low-calcined	Colorless	Small amount of volatile matter	Yellow-brown solution with much humic acid	Slight effect, solution nearly colorless; considerable hydrocyanic acid can be detected in the distillate	Like anthracite	Rapid decoloring with formation of small amounts of oxalic acid	Violent reduction to sodium sulfide

(84,208,176). Another oxidant widely used was the Van Slyke–Folch reagent, which consists of a mixture of chromic oxide, potassium iodate, phosphoric acid, and fuming sulfuric acid (376). This reagent requires an anhydrous medium. A persulfate reagent consisting of potassium persulfate and silver nitrate (catalyst) has also been employed. An advantage of the Van Slyke–Folch and the persulfate reagents is that nitrogen and halogens do not interfere in the measurement.

Incomplete conversion of carbon to the dioxide is the major source of error in most wet-oxidation schemes. Several different catalysts, including mercury (212) and silver (396), have been employed to reduce the formation of carbon monoxide. In another approach, the yield of carbon dioxide was increased by using a hot platinum wire in an air or oxygen atmosphere above the oxidizing solution (38). Hot copper oxide has also been used as a secondary oxidant (81).

With the ready availability of compressed oxygen, wet oxidation methods have fallen into disuse owing to the greater convenience of dry combustion techniques. However, there are certain types of materials where wet oxidation is still the best approach. These include liquid samples, highly volatile compounds, explosive materials, biological samples, samples of rock with noncarbonate carbon (150), and those samples that evolve large amounts of acidic gases such as hydrogen chloride or chlorine.

A sulfuric acid–perchloric acid mixture has also been used for the oxidation of carbon in soils (295). This reagent when used properly can provide a very rapid means of wet oxidation for inorganic carbon.

Wet-oxidation methods are still widely used in the analysis of organic compounds for carbon and are discussed in detail in Part II, Volume XI, p. 383. The techniques and apparatus employed are virtually identical regardless of whether the sample is organic or inorganic in nature.

2.   Dry Combustion

With the ready availability of compressed air and oxygen, the dry-combustion technique has become the method of choice on all but a few materials. Although the resistance-heated tube furnace continues to serve as a valuable tool in the laboratory for many materials, the increased demands placed upon the technique by new and more temperature-resistant materials has led to new or modified combustion techniques. Much of the work in this area has been prompted by the development of metal alloys that are stable at higher temperatures and by advances in technology of the more refractory metals such as tungsten and tantalum. Such materials can be oxidized if sufficiently high temperatures are utilized. Because it is not always possible to attain these temperatures, much effort has been made to develop fluxes and accelerators that promote combustion of the refractory material at a lower temperature. Higher temperatures are obtained in the regular resistance furnace arrangement by using platinum windings or glow bar heating elements. Since the advent of the induction furnace with its inherent advantage of higher attainable temperatures and direct sample heating, the combustion of high temperature alloys and metals has been greatly simplified.

In the sections that follow, we consider the oxidizing atmosphere (air or oxygen), the combustion furnace and related components, and the use of fluxes and accelerators.

a. OXYGEN PURIFICATION

Air or oxygen that has been purified to remove the last traces of carbonaceous material, carbon monoxide, carbon dioxide, and moisture serves as the combustion atmosphere as well as the sweep gas for carrying the evolved carbon dioxide to the measuring system. Purification of the air or oxygen is a desirable feature for all carbon determinations, but it is a vital requirement when low carbon concentrations are to be measured. The accuracy and reproducibility of measurements of low carbon concentrations are greatly influenced by the contaminant blank in the air or oxygen used for combustion. Several schemes for purifying oxygen have been proposed (33,60,386,257). Platinized asbestos at 600°C has been used to oxidize carbonaceous impurities to carbon dioxide (68). One purification system that is convenient because of its ease of handling and care consists of passing the oxygen (99.5%) over hot (300–500°C) copper oxide, then through a carbon dioxide absorber such as "Ascarite," followed by anhydrous magnesium perchlorate to remove water. Such a system if properly maintained can be used to prepare oxygen with a very low reproducible blank.

b. COMBUSTION FURNACE

A combustion train (8) using a nichrome-wound electric furnace and silica combustion tube has long been the standard instrumentation available in most industrial laboratories. Because of its economy, simplicity of operation, and broad utility, this instrument has over the years maintained its usefulness. In many laboratories it still finds extensive use for materials that can be combusted within its temperature capabilities. It is particularly useful where large samples are desirable because of the inhomogeneous nature of the material to be analyzed.

The nichrome-wound electric furnace with a practical operating temperature of 1000°C is of limited value in metal and metal-alloy analysis. Copper and lead are examples of metals that can be analyzed with this apparatus. While this type of furnace has been used for determining carbon in iron and certain low-alloy steels, the trend for analyzing these materials is toward higher-temperature furnaces. This shift to high-temperature furnaces was caused by the incomplete combustion, particularly of high-temperature alloys, at the temperature available in the nichrome-wound furnace.

Some analysts have used the platinum-wound furnace, which can achieve tube temperatures up to 1500°C (146). Others have employed the Globar furnace, which can operate continuously at temperatures up to 1500°C (33). More recently with the advent of commercially available equipment analysts have widely utilized the high-frequency induction furnace (128,157,260,351,396). Temperatures well above 2000°C can be achieved with the high-frequency furnace under proper conditions.

In a resistance furnace the samples are heated by transmission of heat through the combustion tube from an external heating element. Thus any sample material placed in the tube will with time assume the furnace temperature. One drawback of this method of heating is that the combustion tube must be of such material that it can withstand the full furnace temperature. Thus the silica tube used in the nichrome-wound furnace must be replaced with a high-temperature ceramic tube when a platinum-wound or Globar furnace is used. Since the resistance-heated furnace requires several hours warm-up and since ceramic combustion tubes are susceptible to cracking on cooling (especially high-temperature ceramic tubes), it is usually necessary to maintain the furnace at about 800°C when not in use.

The high-frequency induction furnace is based on the principle that an electromotive force is induced in an electrical conductor when placed in the field of a coil carrying an alternating current. This action results in eddy currents, which dissipate as heat part of the electrical energy fed into the coil. The power dissipated in the sample is a function of the intensity and frequency of the electromagnetic field and of the size, shape, and composition of the sample. The temperature attainable depends on this dissipated power and on the mechanism of heat transfer from the object to its surroundings (48,365). Since intervening walls of poorly conducting materials (glass, water, quartz) are neither heated appreciably by or affect the transmission of the electromagnetic field, it is possible to heat an object inside a glass vacuum system to a high temperature while maintaining the walls of the chamber at a low temperature by cooling with either air or water. This is a decided advantage over conventional resistance heaters, since it is possible to use a quartz chamber even though the sample itself held in a ceramic crucible may be heated to over 2000°C. Another advantage of induction heating is the very rapid heating and cooling cycles possible. Since the system returns to practically room temperature between samples, the apparatus can be flushed with oxygen before combustion, thus keeping blanks at a minimum. The time for heating from ambient to operating temperature is a matter of only minutes. A minor advantage is a smaller volume for the combustion tube, which takes less time to sweep free of unwanted gases before analysis. In addition, there is no need to move the sample into the hot zone after purging the system as in the tube furnace method. The chief disadvantage of induction heating is the necessity for sufficient conducting material to be present in the sample to "couple" with the high-frequency electromagnetic field. When non-conducting materials are being analyzed, it is necessary to add iron or some other material that will "couple" so the sample can be fused. Several trials may be required to establish the iron-to-sample ratio on new or unknown materials before satisfactory burns are obtained.

c.   COMBUSTION TUBE

The quartz or silica combustion tubes that were popular when electric furnaces first came into use have been almost completely supplanted by higher-temperature refractories. Quartz tubes are relatively expensive and limited to a maximum temperature of 1100°C. Intermittent use of quartz tube above 1000°C may result in its becoming porous because of devitrification.

Included among the refractory materials used for combustion-tube construction are porcelain, sillimanite, clay, and alundum. In addition to high-temperature stability, the refractory must be gas tight at the operating temperature. Already mentioned is the problem of the ceramics' susceptibility to cracking on cooling after being heated to a high temperature. Combustion tube failures come mainly from fusion with metal oxides that have run out of the sample boat. This can be controlled by the proper choice of boats, sample loadings, boat covers or skids, or tube liners. Ceramics have a certain affinity for absorbing carbon dioxide; thus if very low blanks are to be obtained, they must be preconditioned by heating to a temperature somewhat above that to be used and then protected from the atmosphere.

Platinum has been employed as a combustion tube material; but because of its relatively high cost, it is not widely used.

d.   COMBUSTION BOATS

Combustion boats for use in resistance furnaces may be either of the reusable or "one-sample" type. The reusable boats are much larger and heavier and are used with a lining of granulated alundum so the sample, or burning, fuses with some of the bedding rather than to the boat. Fifty to one hundred combustions per boat can be obtained when used in this fashion.

"One-sample" boats are rapidly gaining in popularity since the elimination of packing and their smaller mass gives better blank control. It is the practice to preburn the crucible immediately before use to reduce the blank. The cost per analysis of the two types of boats is approximately the same.

Included among the many types of material used for combustion boats and crucibles are zirconium, silicate, beryllia, alundum, and clay. Boats made from sheet nickel still find some use because of their low cost. Although they are converted to the oxide after one or two combustions, they still have sufficient strength to be used for a large number of determinations. A bedding such as aluminum oxide is necessary to protect the boat from molten slag. Platinum, because of its expense, is seldom used as the primary sample container. It has been used as a secondary container for holding a refractory crucible (396).

The ceramic crucibles used in the vertical tube high-frequency furnaces are generally usable for only a single determination. Zircon and clay are two of the most commonly used materials for these crucibles. Experience has shown that if a low, reproducible blank is to be obtained, it is necessary to heat the crucible to at least 1100°C before use. This preignition is generally carried out in a tube or muffle furnace for a period of several hours. The crucibles are then stored under such conditions that the pickup of organic vapors, carbon, or carbon dioxide will be unlikely. In some laboratories, because of the need for very low blanks, the crucible is ignited just before use with no more than 5–6 min allowed for cooling before loading the sample and transferring to the furnace.

e. BEDDING, ACCELERATORS, OXIDANTS, AND FLUXES

The materials introduced into the sample boat or crucible other than the sample can be divided into four categories: bedding, accelerators, oxidants, and fluxes. Respectively, these serve the purpose of protecting the boat, providing an additional source of heat, providing additional oxidizing power, and maintaining a fluid or porous slag to enable complete oxidation and escape of the carbon dioxide. Most of the materials used serve in two or more of these categories.

(1) Bedding

Bedding material has been a frequent source of trouble in low-carbon work because of large and erratic blanks. Synthetic aluminum oxide, the most widely used bedding material because it does not fuse at the combustion temperature, formerly suffered from the disadvantage of containing small amounts of alkalies that picked up carbon dioxide from the air and released it at furnace temperatures. This problem has been largely overcome in the material available today. For analysis of low-carbon content materials, bedding material should be eliminated whenever possible. The increasing use of one-use disposable crucibles has made bedding unnecessary and greatly reduces this problem.

(2) Fluxes, Accelerators, and Oxidants

Because it is virtually impossible to describe independently these actions for a given material, they are treated simultaneously. Green et al. (146) have discussed the combustion process and have suggested satisfactory combustion conditions for general application. Almost any metal or alloy will ignite in oxygen if heated to a sufficiently high temperature. When this temperature is beyond the capabilities of existing furnaces, it may be possible to add a material that (1) can act as an accelerator (igniter) by burning rapidly and causing a local temperature increase or (2) acts as a flux by dissolving the oxide skin on the sample, thereby permitting combustion to proceed more easily.

Complete ignition of the sample does not of itself guarantee complete release of the carbon as carbon dioxide into the gas stream. Release of the carbon dioxide is dependent on the sample condition both during and after combustion. These conditions are influenced by the relationship between the melting point of the metal (sample) oxide and the temperature attained in the combustion process. It becomes clear that because of the high melting point of many metal oxides, release of carbon dioxide will be made easier if some fluxing material can be added to make the whole mass (sample and flux) fluid or very porous at the combustion temperature. The most satisfactory condition would be to have the complete mass fluid so that convective stirring could aid the release of carbon dioxide.

The choice of flux and the amount required depends on its capability of producing a fluid or porous melt with the oxide of the sample, the carbon content of the sample, on the sensitivity of the measurement technique, the carbon content of the flux, the attainable furnace temperature, the resistivity to attack of the refractory sample boat, and the method of heating. The sample weight used must be such that, for the carbon content of the sample, a reasonable measurement precision is obtained. As the carbon content of the flux contributes to the total carbon being measured, the quantity of flux used, consistent with a good burn, should be as low as possible since it imposes a limit on the precision attainable.

The high-frequency induction furnace places a further limitation on the choice of flux since the flux must be capable of "coupling" with the electric field. This can often be overcome by using a combination of fluxes. The blank due to traces of carbon in the added "coupling" material can be measured directly and a blank correction can be made, or the material can be preburned prior to the analysis to remove the carbon. Alternately, a molten bath of the flux-coupling material can be used. In this instance, after allowing sufficient time for the removal of the carbon in the flux, the sample can, by appropriate means, be dropped into the still molten bath.

It has been the general practice to use metals rather than oxides as fluxes. By using a metal flux it is possible to take advantage of the heat of combustion of the metal to burn samples that would not be heated to the ignition point at the furnace temperatures available. This is particularly true for refractory materials. In addition, the metal fluxes generally have better "coupling" properties in the induction furnace than their corresponding oxides. Tin, lead, copper, and iron have proved to be the most useful metals for this purpose.

Tin acts mainly as an igniter and helps to form a very porous slag (231). It is available in a state of high purity and has the merit of a low carbon content (about 5 ppm). It burns readily with the production of considerable heat, but the oxide contributes little to fluidity or to "coupling" and is, when used alone, mainly confined to the tube furnace. Tin is most useful for alloys containing much nickel, cobalt, or tungsten (357). A tin and iron flux of equal proportions has been employed in a high-frequency induction furnace for the determination of carbon in refractory metals such as niobium, tantalum, molybdenum, and tungsten (240).

Lead appears to act primarily as a flux and is most useful with iron-rich alloys. Lead burns to an oxide melting below 900°C and is a good flux for use in the tube furnace. Lead oxide "couples" very poorly and is, therefore, of limited usefulness in the induction furnace. It is very difficult to obtain a satisfactory blank from analytical-reagent-grade lead foil by any method of cleaning with abrasives, acids, or solvents. If heated to a bright redness immediately before use, the carbon content of lead can be reduced to about 1 ppm (357).

Oxidants such as lead dioxide were necessary for refractory samples when furnace temperatures were limited to 1100°C. They are still employed by a few chemists (68,186,216,282). The chief disadvantages are the difficulty in obtaining the oxidant free from carbon and the fact that it is quite corrosive to combustion tubes when splattered or spilled. The carbon content of red lead can be reduced to between 0.0003 and

0.0005% carbon on a 2.7 gram sample basis by storing in a muffle furnace at 450°C until used (68).

Copper serves as both an igniter and a flux. Because of the lower melting point of its oxide, it is more effective than iron as a flux in the tube furnace. Copper, if heated to a dull red heat just prior to analysis, gives a low blank value. As both the metal and its oxide are capable of "coupling," it is also suitable for use in the induction furnace.

Electrolytic or other types of low-carbon iron are widely used as accelerators. They ignite readily, provide a high initial temperature, serve as a good flux for many refractory materials, and form iron oxide that probably serves as an extra source of oxygen for the sample. Iron is superior to copper in "coupling" and as an igniter, but has the disadvantage of a higher carbon content (146).

3. Purification of Combustion Products

Combustion of the sample in oxygen provides a separation of the carbon dioxide formed from almost all other elements; the notable exceptions are oxides of sulfur, oxides of nitrogen, and halogens. Although somewhat dependent on the measurement method employed, it is almost always necessary to remove the previously mentioned contaminants from the gas stream to avoid an error in the measurement of the carbon dioxide. The volatilization or mechanical carryover of metal oxides from the combustion boat is another form of contamination that must be removed to avoid error in the measurement of carbon dioxide. This source of interference is generally removed by the packing used to remove other contaminants.

a. OXIDES OF SULFUR

Depending on the conditions of combustion and on the sample composition, the proportions of sulfur dioxide and sulfur trioxide evolved on combustion will vary. In low-sulfur, plain carbon steels the sulfur is evolved principally as sulfur trioxide, while with high-sulfur steels, sulfur dioxide and sulfur trioxide are evolved in about equal proportions (49). Since both of these oxides are acidic, they would be absorbed by the same reagents used to absorb carbon dioxide and thereby give a positive error. The removal of sulfur oxides from the gas stream may be accomplished in a number of ways depending on its concentration. The small amounts of sulfur dioxide that are given off from materials low in sulfur may be satisfactorily removed by concentrated sulfuric acid that has been saturated with chromic acid (49). Materials with high sulfur contents require other absorbents such as chromic acid (34) or potassium permanganate followed by suitable desiccants, or heated platinized silica gel, that will convert the sulfur dioxide to sulfur trioxide (348).

The sulfur trioxide that is formed is not removed by any one absorbent, but is condensed and absorbed during its passage through the liquids or columns of solids in the train. One of the best ways of removing sulfur trioxide is by absorption in concentrated sulfuric acid. The oxidation and absorption step can be conveniently combined in a sulfuric acid solution of chromic acid. The bulk of the moisture absorbed by the gaseous stream is subsequently removed with concentrated sulfuric acid, while the acid mists are retained in an asbestos column. The gas stream is passed through "Anhydrone," which serves to finish the drying and to remove the last traces of sulfur trioxide.

The special apparatus required by this method, the difficulty in obtaining blank-free carbon determinations using these corrosive liquids, and the necessity of renewing the drier frequently are marked disadvantages. Oxidized copper cloth, granulated cupric oxide, and platinized silica gel are efficient substitutes, although these require specific elevated temperatures for their activity.

Dry absorption at room temperature is desirable from the point of simple and inexpensive combustion train design. Granulated zinc and granulated hydrated manganese dioxide have been suggested as absorbents for this method.

While granulated zinc is effective in removing sulfur trioxide, its absorption capacity declines rapidly during use and is not effective in removing sulfur dioxide. Thus its use is limited to low-sulfur-containing materials (49).

Freshly prepared granular hydrated manganese dioxide removes sulfur dioxide by vigorous absorption of the gas followed by chemical reaction. The sulfur oxides are removed completely, and the temporary retention of carbon dioxide is small so that this method is satisfactory for high-speed determination of carbon in various materials. Because of these properties, hydrated manganese dioxide has largely replaced the previously described materials for the removal of sulfur dioxide (157,290,392). The fact that the material is now available commercially in an easy-to-use powder form that requires no pretreatment or maintenance has greatly aided the rapid acceptance of this reagent. Recently the use of lead dioxide for the removal of sulfur dioxide in the combustion method for carbon in iron and steel has been reintroduced (290). Earlier problems of efficiency and carbon dioxide holdup have been overcome by use of thin fibers of lead dioxide deposited on chemically inert quartz sand grains.

Techniques for removal of sulfur oxides have been used to determine sulfur and carbon simultaneously by first absorbing the sulfur dioxide and finally the carbon dioxide (182,217,239,343).

b. HALOGENS

Halogens other than fluorine are removed by passing the combustion gases over silver in the form of fine mesh gauze, silver wool, or colloidal metal. In resistance-furnace applications the silver is placed near the end of the tube where a temperature of 400–600°C is maintained. If only trace quantities of halogen are present, the purification train used to remove sulfur oxides will retain the halogen.

If fluorine is present in appreciable quantities, special precautions must be taken to completely remove it. Magnesium oxide (184,253,366) and sodium fluoride (395) have both been used to retain fluorine. A more detailed treatment of the fluorine removal problem is discussed in Part II, Volume II, p. 333 of this treatise, the section that deals with analysis of organic materials for carbon.

c. NITROGEN OXIDES

Techniques for removing nitrogen oxides are discussed in detail in the analysis of organic material for carbon (Part II, Volume II, p. 336). Nitrogen oxides do not generally present a problem in analysis for carbon in metals and alloys.

d. CARBON MONOXIDE

Combustion at temperatures above 1000°C converts the carbon quantitatively to carbon dioxide, with only traces, at most, of the carbon existing as carbon monoxide. Cook and Speight (68) investigated the completeness of the carbon oxidation to carbon dioxide in a resistance furnace and found a complete absence of carbon monoxide. Trace amounts of carbon monoxide even if present would be completely oxidized by the sulfur-removal system.

4. Techniques for Measuring Evolved Carbon Dioxide

The final step in the determination of carbon by the combustion method is the measurement of the carbon dioxide gas. Since carbon dioxide is a chemically active gas, methods for the determination of carbon differ primarily in the means adopted for the measurement of the quantity of this gas. Thus a variety of methods have been developed that include gravimetric, titrimetric, pneumatic, conductometric, infrared, thermal conductivity, photometric, and mass spectrometric measurement of carbon dioxide or of some reaction product of carbon dioxide.

The gravimetric, titrimetric, and volumetric methods were among the earliest developed and have the greatest utility for carbon contents of 0.1% and higher. With the advent of low-carbon steels (<0.05% carbon) where greater accuracy is required, new methods including the low pressure and conductometric measurements were developed. More recently with the need for decreased analysis time, infrared and thermal conductivity techniques have been described.

a. GRAVIMETRIC

One of the most widely used methods for the determination of carbon in various materials is the gravimetric procedure based on the collection of the carbon dioxide in a solid or liquid reagent. The increase in weight represents the carbon dioxide absorbed. Although liquid absorption reagents such as potassium hydroxide have been used in the past (314), solid absorbers because of their convenience and ease of use are now employed almost exclusively. The simplicity of the apparatus and operation together with its high precision make it ideally suited for routine operation. Soda–lime (sodium hydroxide–calcium hydroxide), which previously was widely used, has been largely supplanted by soda-asbestos (sodium hydroxide on asbestos) (309). The latter gives a higher surface area and is less subject to channeling and packing. Because water is a product of the reaction of carbon dioxide and sodium hydroxide, a layer of dessicant, usually anhydrous magnesium perchlorate, is incorporated in the bulb on the gas exit side. This precaution eliminates any error that would be caused by loss of moisture from the absorption bulb. As solid absorbents will fix any acidic gas or moisture, these must be removed beforehand (see Section VI.A.3).

The technique and precautions taken in preparing the absorption tube for weighing have, over the years, been the source of considerable discussion and modification (160). When the carbon content of the sample is above 0.1% and sufficient sample is available, many of the precautions given in the following are less important, except where the ultimate in accuracy is required. However, when the carbon content of the sample is less than 0.1% or if only a small quantity of sample is available, considerably more effort must be expended to achieve an acceptable level of accuracy. An example serves to point out the difficulties encountered in the latter case.

For a 3 gram sample each milligram increase in weight of the absorption tube is equivalent to nearly 0.01% carbon. Even with proper attention to balance techniques such as taring, counterpoises, elimination of static charges, bulb temperature change, and the like, it is difficult to weigh an absorption bulb or approximately 200 grams consistently within ±0.2 mg, and because each bulb must be weighed twice, it is possible to introduce an error of at least 0.003% from weighing alone. Some improvement can be made by using lighter absorption tubes in conjunction with a microbalance. However, owing to the relatively large oxygen flow used in high-temperature combustion, a more massive absorption bulb is necessary for complete absorption. The microbulbs are generally used in lower temperature combustion with much smaller gas flows.

The controlling factor in low-carbon work is the precision of the blank. The blank, which arises from several causes such as carbon dioxide absorption or desorption by the boat, tube, or flux and the weighing errors, assumes a greater significance for very-low-carbon-containing compounds. The advantages and simplicity of the method have given impetus to the development of modifications capable of yielding greater accuracy. Refinements of the standard method have been directed toward reduction of the blank and increasing the sample weight up to six to eight times normal. The system blank can be reduced to below 0.001% by passing oxygen through the system for 2 days before making a series of measurements. Sample size is increased by successively burning five or six charges and absorbing all the carbon dioxide evolved in a single lightweight absorption bulb (125). A more recent method described the use of up to 20 grams of sample in a single combustion (22).

By taking advantage of the precautions mentioned to achieve a low blank and of the use of large samples, it is possible to obtain precision of the order of 0.0005%. However, the gravimetric method is being replaced by other methods for the analysis of low carbon contents because the newer methods have equal or greater precision and have a much shorter analysis time.

The gravimetric method has been combined with a chemical separation to obtain results of low carbon concentrations. The sample is dissolved in an appropriate solvent, and the residue after filtering is used for the combustion analysis of carbon. By using large weights of sample, a concentration of the carbon can be achieved. Solvents such as copper potassium chloride and copper ammonium chloride have been used in the long-established technique (18). While the method gives satisfactory results for high-purity irons and silicon irons of 0.01% carbon or less, low results are obtained for alloy steels. The low results have been attributed to the formation of colloidal carbon that escapes separation during filtration. The time requirement for the method (sample solution may require 2 hr) severely limits the application of this principle, and more practical procedures are not available for low carbon contents.

b. PNEUMATIC

The amount of carbon dioxide formed from the combustion of carbon can be ascertained from its volume at constant pressure or from its pressure at constant volume. Both principles have been adapted to the determination of carbon.

The volumetric measurement of the carbon dioxide involved in combustion of the sample is one of the oldest and simplest methods for the determination of carbon. The exhaust gas from the combustion tube, after being scrubbed free from sulfur and nitrogen oxides and other possible interfering substances, is collected in a gas burette over a dilute acid solution, and its volume is measured. After passing the gas through a solid or liquid carbon dioxide absorber the volume of oxygen is again measured. The difference between the two measured volumes corrected for prevailing atmospheric pressure and temperature conditions is used to calculate the carbon content of the original sample.

When this technique is used with a resistance-heated combustion furnace a large excess of oxygen is present owing to the relatively high flow rate of oxygen required for combustion. As a result, a special gas burette is needed for accurate work. The lower oxygen flow of high-frequency furnaces makes them ideally suited for this technique (288,-345). The smaller quantity of oxygen permits a more accurate measure of the carbon dioxide. A completely automatic carbon analyzer employing these principles is commercially available.

The oxygen can be eliminated and the precision of the measurement improved by freezing out the carbon dioxide in a capillary trap cooled by liquid nitrogen (120) or liquid oxygen (129). After conversion back to a gas, the carbon dioxide can be measured and absorbed in smaller, more precise burettes.

Another common application of the volumetric technique is the absorption of carbon dioxide in a sodium or potassium hydroxide solution such as in an Orsat-type gas analyzer.

The factors that can affect the precision and accuracy of the volumetric measurement have been discussed (345). In order to obtain precision of the order of $\pm 0.001\%$ carbon in a 1 gram sample it is necessary to take into consideration the following variables: temperature, barometric pressure, water vapor equilibration, burette drainage, and gas solubility. By taking these variables into consideration and making some modifications in the technique, Nesbitt and Henderson were able to obtain results accurate to 0.0003% carbon on a 2 gram sample. The carbon dioxide was absorbed in sodium hydroxide, then treated with sulfuric acid. The resulting carbon dioxide was measured under reduced pressure (273).

Measurement of the pressure of carbon dioxide at a known volume is the basis of the "low-pressure" method that has been employed for the determination of low-carbon-containing steels. The method consists of combusting the sample in purified oxygen at reduced pressure in a closed system and freezing out the carbon dioxide with a liquid air trap. After removal of excess oxygen, the carbon dioxide is vaporized into a known evacuated volume, and the pressure is measured. This method was first described in 1920 by Yensen (398) and later modified and improved by Wooten and Guldner (396), Murray and Ashley (268), Naughton and Uhlig (272), and others (151,269,356).

Pepkowitz and Moak (289) described a precise method for the determination of low concentrations of carbon in metals in which they used a high-frequency Lindberg combustion unit. An examination of the variables present in the high-frequency combustion-volumetric method for the determination of carbon in metals and methods of controlling or correcting for them have been made by Simons et al. (345).

Recent work has been directed towards simplifying the procedures involved. Wells (388) has described a technique whereby the sample is combusted in a stream of oxygen at atmospheric pressure in a tube furnace, followed by removal of interfering gases, and finally condensing the carbon dioxide in a liquid nitrogen trap. Cook and Speight (68) have extended this work by carrying out modifications aimed at even further simplification while retaining the high precision afforded by the basic method of Wells (388).

The low-pressure method is fairly rapid as a result of the recent improvements, but still requires apparatus that is involved, complicated, and relatively expensive to assemble. The services of a competent glass blower are required for construction and maintenance. As a result, the method has not been widely used for general laboratory routine analysis.

Compared with gravimetric and volumetric methods, the low-pressure method has the inherent advantage of extremely high sensitivity. A precision of ±0.0001% carbon on a 2.729 gram sample has been achieved by Cook (68). By close attention to operational details, that is, purification of oxygen, blanks, and so forth, the low-pressure method will give a reproducibility of ±0.0005% carbon or better on 2.729 grams of most low-carbon materials (257). The greatest limitation on sensitivity is set by the magnitude of the instrument blank. The low-pressure method is best suited for nonroutine investigations where the utmost in accuracy and precision are desired.

c. Precipitation and Titrimetric

Carbon dioxide can be determined by passing the combustion gas through a barium hydroxide solution, filtering, and weighing the precipitated barium carbonate. The difficulties encountered in performing these operations without picking up atmospheric carbon dioxide makes this approach of little more than historical interest in the determination of large amounts of carbon. However, the technique finds some use in the determination of large amounts of carbon, especially small amounts of hydrochloric acid provided the amounts present are not sufficient to neutralize the barium hydroxide solution.

The precipitation gravimetric finish has by and large been replaced by a titrimetric procedure. In the conventional titrimetric procedure the carbon dioxide is absorbed in a suitable liquid, which is then titrated either indirectly by an acidimetric-back titration or directly by an alkalimetric procedure. In the direct method the carbon dioxide is collected in a standard solution of an alkali; the excess absorbent is determined by back titration. Either sodium, potassium, or barium hydroxide may be used, the first two yielding a soluble carbonate; the last, yielding an insoluble carbonate, is often preferred because the excess alkali may be back titrated in one operation (94,162,198,400).

Sodium and potassium hydroxide have been used as the absorption reagent in place of barium hydroxide; however, when this is done the back titration cannot be performed in one operation. These reagents yield a soluble carbonate that will react to give bicarbonate before the indicators normally used show a change. An additional titration using an indicator of lower pH range such as methyl orange must be carried out so that the amount of carbon dioxide involved in the determination may be calculated (31).

The need for two titrations may be avoided if the carbonate is precipitated by addition of barium chloride to the absorbing solution (72,161). The excess alkali is then titrated with hydrochloric acid using phenolphthalein as the indicator.

A very useful type of absorber that is easy to construct is described by Pieters (293), who employs barium hydroxide as absorbent for determining about 15 mg of carbon dioxide (semimicro scale). Barium hydroxide, although not as efficient an absorbent for carbon dioxide as alkali hydroxides, may be used with confidence for micro or semimicro quantities of carbon dioxide if an excess of barium chloride is present to depress the solubility of the carbonate by the common ion effect. A combination of sodium hydroxide and barium chloride is the preferred absorber when working with micro quantities of carbon dioxide.

Archer (16) has described a modification of his semimicro wet combustion method for the determination of carbon in which the evolved carbon dioxide is absorbed in a barium hydroxide solution. The excess alkali is then neutralized to the thymolphthalein end point, and the precipitated barium carbonate is titrated directly with standard acid using bromphenol blue as indicator.

A potentiometric titration of the carbon dioxide evolved during combustion (116,117,397), a spectrophotometric titration of parts per million carbon dioxide (233), and a complexometric titration of the carbon dioxide produced by combustion of a low-carbon steel (201) have been described. For the complexometric titration, the carbon dioxide was absorbed in a barium hydroxide solution in 0.1 N sodium hydroxide. The excess barium was titrated with EDTA (ethylene diamine tetracetic acid), without removing the precipitated carbonate, using Eriochrome Black T as an indicator.

The titrimetric method has certain disadvantages that require special attention. Barium hydroxide solutions are relatively poor absorbers of carbon dioxide, particularly at high gas-flow rates and at high carbon dioxide concentrations. Slow combustion in air or oxygen gives better results because dilution of the carbon dioxide favors complete absorption of the carbon dioxide (230). The solution of barium hydroxide must be prepared, stored, and used under conditions that preclude contamination with atmospheric carbon dioxide. In addition, the nature of a barium hydroxide solution makes accurate measurement of the volume somewhat difficult.

The advantages of the method include simplicity of operation and no requirement for precise temperature control or pressure corrections.

It is well known that various organic liquids are more suitable for collection of carbon dioxide than are aqueous solutions of the hydroxides. Absorbents that have been recommended in the literature include dipiperidyl (285), ethanolamine (332), ethylenediamine (363), pyridine (39), and an aniline, ethanol–barium hydroxide mixture (375). Dipiperidyl absorbs carbon dioxide ten times more rapidly than the aqueous absorbents. Ethylenediamine can be made to quantitatively release its absorbed carbon dioxide by acidification with sulfuric acid followed by heating. When an aniline, ethanol–barium hydroxide mixture, is used to absorb carbon dioxide, the excess barium hydroxide is titrated using phenol red as indicator.

Blom and Edelhausen (35) examined numerous organic solvents in a search for one with high capacity for dissolving carbon dioxide and that permitted a good titration end point to be observed. A pyridine and acetone mixture was selected as the absorber. The carbon dioxide in the mixture was titrated with sodium methoxide using thymol blue as indicator. This procedure has been modified by several workers. Grant et al. (145) replaced the pyridine with dimethylformamide and used potassium methoxide in a benzene–methanol mixture as a titrant. Bohn and Krans have described a 6 min determination of carbon and hydrogen using pyridine as the absorber (39). Jones et al. (193,194) and Wilkinson et al. (393) used a mixture of formdimethylamide and monoethanolamine as absorber and tetra-n-butyl ammonium hydroxide as titrant with thymolphthalein indicator. They state that a 2–3 min analysis time is possible for iron and steel samples containing 0.01% carbon.

Braid et al. (47) have examined some of the factors affecting the determination of carbon dioxide by nonaqueous titrimetry. Their investigation resulted in a method that involves absorption of carbon dioxide in a 5% solution of ethanolamine in formdimethylamide, followed by titration with standard tetrabutyl ammonium hydroxide in a benzene–methanol solution to a visual end point with thymolphthalein indicator. Sen Gupta (339) had difficulty adopting the titration technique of Blom (35) for an induction furnace. Acetone either alone (for 0–10 mg C) or in a 1:1 mixture with methanol (for 10–30 mg C) containing 0.6% monoethenolamine and an excess of standard sodium methylate was found to be a better absorbent for carbon dioxide than pyridine. After reaction, the excess sodium methylate was back-titrated by a standard methanolic solution of benzoic acid with phenolphthalein as indicator. This approach was applied to rocks, stony meteorites, and metallurgical samples (339).

Coulometric titration of carbon dioxide on a microscale has been reported (390). The carbon dioxide from combustion of the sample is absorbed in a solution of dry acetone containing 0.5% methanol and saturated with potassium iodide. Thymol blue is used as a visual indicator of the titration end point. A coulometric automatic titrator for determining carbon in metals has also been described (53).

A rapid direct coulometric titration method for the determination of carbon in steel was described by Metters et al. (128). A solution of 3% ethanolamine in isopropanol was used as a carbon dioxide absorbent. The carbon dioxide reacts with ethanolamine to form the ethanolamine salt of 2-hydroxyethylcarbamic acid (342). The isopropanal is reduced with 100% coulometric efficiency to form basic isopropoxide ions, which are then available for the titration of the previously formed 2-hydroxyethylcarbamic acid. The main advantage of the coulometric method is that standardization is by direct reference to Faraday's law, and no standardization of solutions is necessary.

Hobson and Leigh (177) described a precise coulometric titration method for the determination of carbon in steel. The carbon dioxide is absorbed in the cathode compartment of the coulometric cell containing barium perchlorate, thus precipitating barium carbonate and liberating the equivalent amount of hydrogen ions. The original pH is restored by electrolysis. Factors such as temperature control, absorption of carbon dioxide, the pH electrode, the coulometer cell, and combustion conditions were thoroughly investigated to improve the precision of the coulometric procedure. An excellent precision of 1 part per thousand at the 1% carbon level was obtained.

d. CONDUCTOMETRIC

The conductometric method is based on the determination of the change in conductance of an alkali or alkaline earth hydroxide solution when used for the absorption of carbon dioxide produced during combustion of a sample. The change in conductance is caused by the formation of an insoluble or unionizable carbonate. This technique has gained wide acceptance in recent years, particularly in the determination of carbon in steel and other metals. The popularity of this method is due to high precision, rapid, simple operation, and application to samples of very-low-carbon content.

The method was originally proposed by Cain and Maxwell (55), but found little application until modern instrumental techniques were applied (33,134). In recent years it has been applied to the determination of low-carbon contents because of its potentially greater accuracy than that possible with the standard gravimetric or titrimetric procedures (357). A conductometric method for the accurate determination of carbon in low-carbon steels that uses samples containing approximately 100 μg of carbon has been described (76).

The conventional combustion conductometric technique is limited in sensitivity due to the magnitude and variation of the blank, a major part of which arises from the iron and tin flux added to the crucible to aid combustion. A method has been described whereby the sample (tungsten in this case) could be completely combusted by direct induction heating if the sample is in the form of granules of 16–80 mesh. Less than 10 ppm of carbon in tungsten has been determined in this way (142).

Interest in determining parts per million carbon in ultra-high-purity metals has required modification of the procedure to increase the sensitivity of the conductance measurement. By saturating the barium hydroxide conductance solution with barium carbonate prior to the measurement, a twofold increase in sensitivity has been reported (308).

The conductometric method in addition to high sensitivity offers the advantage of being able to monitor the rate of carbon dioxide evolution from the sample during combustion (140). The carbon dioxide evolution is not in a steady stream as expected; the bulk is liberated in the early stage of combustion as the sample burns violently. As the reaction subsides, the carbon dioxide is liberated in small bursts (373).

An automated instrument has been described for the induction furnace combustion-conductometric determination of carbon in metals. The sweep gas purge is regulated by time, and the conductance change is read out using a digital device (167).

The extreme sensitivity of the conductance system to temperature and electrical parameters is overcome by balancing the measuring cell against an identical blank cell in a Wheatstone bridge circuit (316,331).

The use of barium hydroxide has been criticized from time to time because of its poor absorption property for carbon dioxide. This is particularly true when the content of carbon dioxide is high. Sodium hydroxide is useful in extending the method to high quantities of carbon dioxide (340). However, the majority of workers prefer the barium hydroxide conductance solution, particularly for low carbon contents, because of the greater change in conductance for a given amount of carbon dioxide.

As the conductance is determined by a bridge method, it is the proportionate change in conductance in the solution that is of importance rather than the absolute change. The weaker the concentration of the initial barium hydroxide solution, the greater the proportionate change when small amounts of carbon dioxide are absorbed. Thus for low-carbon containing samples, a dilute barium hydroxide conductance solution is essential. This factor must be measured against the difficulty of obtaining complete absorption, particularly from a rapid gas stream. Specially designed absorbers may be necessary to compensate for this factor.

The conductance method has found application not only for analysis of steel and various steel alloys, but also for tungsten (142), uranium (83), boron (223), and many other metals.

Since the amount of carbon dioxide evolved from the sample is small and the gas-flow restriction limits the weight of sample possible, careful attention is necessary to avoid contamination with atmospheric carbon dioxide both during combustion and in preparation and handling of the barium hydroxide solution (378). By observing these precautions and with attention to other details including temperature control of the absorber to 0.1°C maintenance of a low and reproducible furnace blank and appropriate sample preparation and cleaning, it is possible to obtain an analysis reproducibility of ±0.0001% carbon (378).

As in any technique for measuring very low carbon concentrations with high precision, general cleanliness cannot be over emphasized.

e. GAS CHROMATOGRAPHIC

Although gas chromatographic techniques hold great potential for carbon dioxide measurements, they received little emphasis until the early 1960s. However, with the need for faster and more sensitive control analyzers for carbon to keep pace with new developments in the metals industry (particularly the steel-making operation), gas chromatographic detection of carbon dioxide has been exploited.

Workers have used the gas chromatographic techniques to determine carbon dioxide in gas mixtures (352,364), but it was not until 1962 and 1963 that the first reports utilizing this technique for determination of carbon in metals were presented (200,262). Others have utilized this technique for determination of carbon and hydrogen in organics. Combustion of the sample was carried out in a stream of oxygen in one method (91) and in a helium atmosphere using copper oxide as an internal oxidizing agent in another method (361). In both cases, carbon dioxide was condensed in a liquid-nitrogen-cooled trap to remove excess oxygen and to concentrate the carbon dioxide for injection into the gas chromatograph. Both copper oxide and oxygen were used for combustion in another case (96). The unconsumed oxygen is adsorbed by copper, then the gas is mixed with helium and dried. The carbon dioxide is adsorbed on a molecular sieve 5A column, desorbed by heating, and is measured by a thermal conductivity detector. Other workers have used helium momentarily enriched with pure oxygen to combust the organic sample (287). The flash combustion was followed by catalytic oxidation over compressed Cr_2O_3. The combustion gases were passed over silver-treated copper at 640°C to remove excess oxygen and injected into a Poropak Q chromatographic column.

As the thermal conductivities of oxygen and carbon dioxide are so close together, it is the general practice following combustion to remove excess oxygen by various techniques previously outlined (91,96,287,361) and use a carrier gas other than oxygen for the gas chromatographic analysis. However, a few methods using oxygen as carrier gas have been reported (228,379). A disadvantage of using oxygen as a carrier gas is that the thermal conductivity detector must have filaments resistant to oxidation.

Most authors prefer to have the combustion of the sample carried out in an excess of oxygen to ensure complete oxidation (220).

The first published reports utilizing the gas chromatographic technique for determining carbon in metals appeared in 1963 (380). A ferrous metal sample was burned in a stream of oxygen with a high-frequency induction furnace. The gaseous products were passed through a molecular-sieve column, the carbon dioxide being retained on the column. After removing the excess oxygen by purging with argon, the temperature of the molecular sieve column was raised to 275°C by temperature programming, and the carbon dioxide was swept into the thermal conductivity cell with argon as the carrier gas. While the method permits detection of 0.0005% carbon, it is equally useful for determining carbon in samples containing up to 20% carbon.

A method employing the same general principles except that oxygen is used as the carrier gas has been described (228). By using oxygen as both combustion and carrier gas, the time for analysis is reduced. Since this method was developed for rapid control work it was necessary to sacrifice a bit in sensitivity and precision.

In another instance (133), the steel sample was combusted in a classical tube furnace setup, the evolved carbon dioxide was trapped in a U-tube cooled to liquid oxygen temperature to separate it from the carrier gas, and finally after warming the U-tube the carbon dioxide was measured in a standard Pye Argon Chromatograph. The determination of carbon in meteorites (107), metallic beryllium (106), and rocks and minerals (243) by gas chromatography after formation of CO_2 have been reported. Usually molecular sieve, Poropak Q, or silica gel columns are employed.

Commercial instruments are now available that are based on the thermal conductivity measurement of a mixture of carbon dioxide and oxygen (54,226). One of these instruments (226) has recently been evaluated with regard to precision and accuracy (133).

As the thermal conductivity measurement for carbon dioxide is not an absolute measurement, it must be calibrated by use of standards. It has the advantage of being substantially unaffected by environmental factors, such as vibration, humidity, or heat. It is generally necessary to remove sulfur oxides and halogens from the gas stream before analysis of the carbon dioxide. Because of the speed of analysis, relative insensitivity to environment, ease of use, and high detection sensitivity this technique has become especially valuable in the steel industry, where it can be used in very close proximity to the furnace area by relatively untrained operators.

Reaction gas chromatography has been used to determine the carbon dioxide combustion product (181). The carbon dioxide was catalytically reduced to methane. A hydrogen flame ionization detector was used to measure methane. The advantage of the flame ionization detector when compared to the thermal conductivity detector is greater sensitivity.

f. Miscellaneous

The mass spectrometer has been used to replace the analytical train associated with the conventional low-pressure combustion procedure (172). This method not only provides a measure of the quantity of gas involved but at the same time it permits positive identification of the molecular mass. The mass spectrometric technique has been applied to the determination of carbon in small amounts of carbide (199,370) in steel. The high cost of equipment and complexity of operation have virtually precluded the use of this type of instrumentation for routine measuring of evolved carbon dioxide.

Infrared absorption spectrophotometry has been proposed for the determination of carbon dioxide resulting from the combustion of carbon in steel (210,368,374). A completely automatic instrument was described that can be applied to the determination of low concentrations of carbon (368). Less than 0.01% carbon in a sample can be determined with an error of ±0.0005% in 4 min. The gas analyzer is specific for carbon dioxide, and there is, therefore, no need to remove sulfur dioxide or water vapor from the gas stream. By using two analyzers in series with the second adjusted to measure sulfur dioxide it is possible to provide a simultaneous determination of carbon and sulfur. A similar method has been proposed for determination of carbon in organic materials as a replacement for the classical Pregl method (220). The infrared method has also been applied to the determination of carbon in uranium tetrafluoride (344). The presence of fluorides, which cause an interference in standard methods of measuring carbon dioxide, do not influence the infrared measurement. Thus the infrared method is an accurate and specific technique that is unaffected by the usually interfering species.

An absorptimetric determination of small amounts of carbon in iron and steel with Alizarin Yellow R as indicator and with induction-heated combustion has been described (265,266). The evolved carbon dioxide is absorbed in 0.01 N sodium hydroxide containing the indicator. The absorbance of the resulting solution is then measured, and the concentration of carbon is read from a calibration curve.

Other techniques that have been reported but have not been widely used in recent years include turbidimetric (2,310) and nephelometric (370) measurements.

B. EVOLUTION METHODS

The principal use of evolution methods is for the determination of carbon present as carbonate. The method consists of liberating the carbon dioxide through attack by acid and measuring the evolved carbon dioxide using one of the several techniques discussed in Section VI.A.4. Many methods have been proposed by various investigators for the determination of carbonate carbon dioxide; they are too numerous to be reviewed in detail. The methods proposed in recent years are essentially based upon chemical reactions and methods reported in analytical chemistry textbooks and other publications. The improvements are directed toward replacing earlier pieces of apparatus with equipment that is less fragile, more easily maintained and operated, and has increased accuracy achieved by excluding parts and procedures possessing potential sources of error.

Although the basic concept for the evolutionary determination of carbonate carbon dioxide is simple, there are certain precautions that must be taken. Care must be exercised that (a) no carbon dioxide except that in the sample is introduced into the apparatus by the air, distilled water, or other materials used; (b) no carbon dioxide is fixed in the apparatus except in the measuring system; and (c) all other compounds that would interfere with the determination of carbon dioxide are removed (218).

The technique for decomposition of carbonates present in the sample is common for different methods and generally is done by action of dilute acid (sulfuric, phosphoric, or hydrochloric acid). Hydrochloric acid has long been used in spite of its volatility because of the solubility of chlorides. However, phosphoric acid is now widely used particularly in the decomposition of silicate rocks (150,243).

Although certain materials such as calcite give off their carbon dioxide on treatment with cold acid, other materials such as dolomite and siderite do not. Thus it is necessary to add heat to the reaction flask to ensure complete reaction of the sample with the acid. It has been shown (41) that certain minerals such as scapolites containing carbon dioxide are not decomposed by reaction with hydrochloric or phosphoric acid. In these cases addition of hydrofluoric acid is necessary for decomposition. Use of hydrofluoric acid requires a modified absorption train such as that described by Hillebrand (173).

The gravimetric technique for measuring evolved carbon dioxide is by far the most widely used and is treated in the greatest detail. When employing the gravimetric method, the liberated carbon dioxide must be freed from other gases that will interfere in the measurement system. In general, the purification train consists of the following elements: (1) a tube containing sulfuric acid to bear the brunt of the dehydration; (2) a tube containing anhydrous copper sulfate (silver arsenite may also be used) to remove hydrochloric acid (when used as the acid for evolution), chlorine, and hydrogen sulfide; (3) a tube containing the final desiccant, magnesium perchlorate ("Anhydrone" may be used); (4) the carbon dioxide absorption tube containing soda asbestos ("Ascarite" and "Lithasorb" are frequently used) and the desiccant used in tube 3; and (5) a second weighed carbon dioxide absorption tube to act as guard and to indicate when tube 4 is beginning to fail (174). Other absorption trains have been described (150) that differ only slightly.

In addition to liberating and purifying the evolved carbon dioxide it must be quantitatively driven into the absorbent by a current of air that has been freed from carbon dioxide. In the traditional procedure compressed air is used to carry the evolved carbon dioxide from the reaction flask to the absorption tube. The air is freed beforehand from carbon dioxide by passage through soda asbestos or one of the other carbon dioxide absorbing reagents. More recently, techniques have been described that use a recirculating system for the air, thus avoiding the problems associated with the purification of large quantities of air (10–15 liters) used in the conventional method (190).

An additional precaution to be considered in the traditional method is that the unabsorbed gas passing through the absorption tubes must not undergo a permanent change in moisture content. This is usually avoided by drying the gas before it enters the absorption tubes and by including a desiccant in the absorption tube following the soda–asbestos to pick up moisture evolved as a result of the carbon dioxide–sodium hydroxide reaction. This precaution avoids errors caused by gain or loss of moisture in the absorption tube.

Detailed procedures for determination of carbonate carbon in a variety of materials are available in several standard reference texts (150,174,-182). Pieters has summarized results obtained with macro and semimicro titration methods and discussed some of the precautions that must be employed to obtain accurate results (293).

A rather detailed discussion by Krumin and Svanko (218) on the various standard methods of measuring the evolved carbon dioxide is available in the literature. A comparison of gravimetric, titrimetric, gas-volumetric, and pressometric methods by these authors indicates that the gravimetric method is convenient for analyzing samples with a wide range of carbonate carbon dioxide content and exhibits high accuracy and good reproducibility. The titrimetric method gives satisfactory results if the carbon dioxide content is not above 6 mg. Generally, results obtained by the gas-volumetric method were found to be too low. This technique is more useful for samples containing large amounts of carbon dioxide. The pressometric method was found suitable for analyzing samples with both high and low carbonate carbon dioxide. Ease of operation and good accuracy contributed to the utility of this technique.

However, a somewhat time-consuming calibration is required for each new apparatus. These techniques for measuring carbon dioxide are discussed in more detail in Section VI.A.4.

Improvements in the conventional apparatus and in the methods of measurement that have appeared in the literature include the reduction of blank values through the use of a closed-circulation system (190,297,-334), and the use of gas chromatography (58,189) and infrared spectrophotometry (296) for the measurement of evolved carbon dioxide. Because of these improvements, the analysis of very small amounts of carbonate carbon dioxide has been greatly improved. A discussion of the gas chromatographic and infrared method may be found in Section VI.A.4.

Carbon in the form of cyanide has also been analyzed by the evolution technique. Hydrocyanic acid can be removed from an aqueous solution by bubbling air through the solution. The resulting gas is collected and analyzed by gas chromatography for hydrocyanic acid (333). Cyanides in plating wastes have been measured by steam distilling the free hydrocyanic acid from a buffered solution. After addition of sulfuric acid a second distillation releases the complexed cyanide. The cyanide is collected in a sodium hydroxide solution, and the excess sodium hydroxide is titrated with a standard acid solution.

C. OPTICAL METHODS

1. Colorimetric

Colorimetric methods for determining combined carbon in iron and steel were widely used for the first quarter of the twentieth century. However, as electric-furnace-combustion methods were developed, this method fell into disuse. Improvements in spectrophotometers have renewed interest in colorimetric methods.

Most colorimetric methods are based on the formation of a brown-yellow color of unknown and probably indefinite composition that develops when intermetallic compounds of carbon are dissolved in nitric acid. If certain conditions are fulfilled, the depth of color is proportional to the combined carbon content. Theoretically, it should also increase in direct proportion with the amount of sample used, but this is not so in practice. Further, the tint and depth of color vary with a great number of conditions such as the temperature and duration of heating prior to the measurement, the concentration of acid employed, the presence of other elements normally found in the material, for example, phosphorus, and the manner in which the sample has been made and its subsequent physical treatment. Hence it is evident that for reasonably accurate results it is essential to employ standards of closely similar composition and history to the samples under test, and to treat the samples and standards identically during the determination. This color method is less reliable for cast iron than it is for steel, owing to the more complex nature of the former. It is as yet inapplicable in the presence of any element that forms a colored species absorbing in the same region of the spectrum such as nickel or chromium (156).

In this method the sample is heated with nitric acid (32,156,274) or with a mixture of sulfuric, phosphoric, and nitric acids (187). The solution is diluted quantitatively, and the optical density is measured at 375 nm or through a Wratten No. 2 or Ifford No. 601 violet filter. A modification of this technique has been applied to the analysis of titanium (67).

A novel colorimetric method has been suggested (89). The sample, sulfur, and quartz wool are placed in a quartz tube, evacuated, and sealed. The tube is heated to 1000°C for 1 hr. During the heating, the carbon is converted to carbon disulfide. After cooling, the carbon disulfide is extracted with benzene and converted to diethyldithiocarbamate by:

$$CS_2 + HN(C_2H_5)_2 \xrightarrow{\text{NaOH}} S = C \begin{smallmatrix} SH \\ | \end{smallmatrix} \underset{\diagdown}{\overset{\diagup}{N}} \begin{smallmatrix} C_2H_5 \\ \\ C_2H_5 \end{smallmatrix}$$

The colored copper complex is formed by addition of copper ion, and the absorbance is read at 435 nm. The lower limit of detection is about 1–5 ppm.

2. Infrared

Carbon dioxide has absorption bands at 2.7 (pair), 4.4, 13.9, and 15 μm. Carbon monoxide has a pair of absorption bands at 4.6 μm (292). Carbon is converted to the dioxide before analysis by infrared methods generally by combustion (see Section VI.A). Since carbon monoxide is a possible combustion product, some interest in analyzing for carbon monoxide by infrared techniques has been shown (221). Both dispersive and nondispersive infrared analyzers have been used (292,368). The nondispersive analyzer is by far the more commonly used technique for the determination of carbon dioxide. General infrared gas analysis is covered in Part I, Volume VI, Chapter 66 of this treatise.

3. Emission

Owing to carbon's high excitation potential, its principal lines (most sensitive) are in the ultraviolet and far-ultraviolet region of the spectrum (Table 8) (255). Three of the more sensitive lines are below 2000 Å in the less accessible region of the ultraviolet spectrum. With modern vacuum direct-reading spectrometers it is possible to make use of the lines below 2000 Å, but the high cost of vacuum spectrometers and the inconvenience of working in this region has restricted the use of these lines. Those carbon lines located in the parts of the spectrum normally used in emission analysis are of comparatively low sensitivity, require large amounts of energy for their excitation, and are subject to severe interference from other elements such as iron. In an electrical discharge, the effective energy potential is largely dominated by the elements with low excitation potentials, and elements with high potentials are suppressed. Thus when more easily excitable elements (metals) are present, the potentials needed to excite carbon are not reached; consequently, special techniques have been developed to provide the necessary energy to excite the spectral lines of carbon.

a. DIRECT-CURRENT ARC

The determination of carbon, using direct-current arc excitations, has probably not been widespread because of the difficulty of exciting usable carbon lines and the impracticality of making electrodes of sufficiently high ionization potential. Both of these difficulties may be overcome by using copper electrodes and utilizing a cyanogen bandhead as the analysis line.

The cyanogen-violet system with principal bandheads at 3590, 3883, and 4216 Å is the most sensitive of the various cyanogen systems (80).

TABLE 8
Analytically Important Carbon Lines

Line	Lowest useful concentration (%)
4267 C(II)	0.1
3883 CN	0.001
2478 C(I)	—
2296 C(III)	0.02
1930 C	0.003
1657 C	0.05
1548 C(IV)	0.05

CN 3883 Å is usually chosen as it is the most sensitive of the three. Since the excitation potential for the cyanogen-violet system is about 3.2 eV and that of copper, 3.77 eV, the use of copper electrodes ensures the excitation of the cyanogen in a copper arc. The maximum current is about 6 A, since current higher than this would melt the copper electrodes (80,255). Although the relative sensitivity for carbon varies considerably with the matrix, techniques have been developed to minimize the effect (80). A modification of this technique using 30–70 A for 2–5 sec has been made for the analysis of steel (353).

b. Spark Excitation of Carbon Lines Longer than 2000 Å

The carbon line at 2296.9 Å has been used for the analysis of carbon in ferro alloys, particularly steels. The counter-electrode is either copper (251), silver (66,135,194), or magnesium (79,222). Iron counter-electrodes have been found to give erratic results. Copper is the usual choice for the counter-electrode unless, of course, copper is one of the sought elements, in which case silver is employed. For nickel steels the use of magnesium counter-electrodes has been suggested (79,222) to reduce the interference from nickel by suppressing the nickel spectrum.

The excitation of the C(III) 2296.9 Å line will be obtained only by high-energy conditions. The presence of a large excess of ionized iron may cause its partial or complete suppression. This means that a very high effective temperature (in practice, a high current density) must be coupled with adequate quenching of the discharge. It is clear that, in order to obtain the maximum current density with any particular circuit, the spark must be allowed to pass only when the voltage in the secondary circuit is at or near its peak and that the duration of each discharge must be minimized (135). Experimentally the conditions for which the spark best satisfies these requirements are lowest inductance possible (i.e., only residual inductance), 0.005–0.02 μf capacitance, and either an auxiliary spark gap (2–6 mm) or a rotary spark interrupter in series with the analysis spark. The Fe(III) 2295.9 Å line is usually used as the internal standard. A sensitivity of 0.02% can be achieved by these techniques.

Harvey and Mellichamp (158) have described a technique for determination of halogens, carbon, sulfur, phosphorus, and selenium in a single sample. They employ the C(II) 4267.3 Å doublet as the analytical line for carbon. The method involves high-voltage spark excitation at reduced pressures. The sample is pressed into a pellet with silver powder, and a silver rod is used for the counter-electrode. A sensitivity of 0.1% is obtained for carbon.

c. VACUUM ULTRAVIOLET REGION

The lines of carbon shorter than 2000 Å that are useful for analysis are relatively free from interference. The high absorption of the atmosphere in this region makes it essential to work with spectroscopic equipment that has been evacuated or that has been filled with a transparent gas such as hydrogen or helium. The carbon lines at 1930.9 (75), 1657 (132), and 1548.2 Å (311) are used with spark and "sparklike" conditions, such as described in Section VI.C.3.b. Although the most sensitive arc lines of carbon are located in this region, they are not normally used for detection of low quantities of carbon because conditions for excitation in a vacuum or an inert gas favor the emission of spark lines. The availability of modern vacuum spectroscopic equipment makes it practical to use lines in the far ultraviolet for analysis of certain materials such as steel (224,347). In most cases, the sample is sparked using a silver counter-electrode in an excitation chamber filled with argon. Detection limits are thereby extended, and reproducibility of results is improved. The detection limit is about 30 ppm.

4. X-RAY DIFFRACTION

Graphite and diamond can be identified by standard X-ray diffraction methods. Of more interest, however, is distinguishing between graphite and amorphous carbon or quantitative estimation of graphite in materials.

Walker and co-workers (381,382) have suggested a method for determining the relative amounts of graphite and amorphous carbon by examining the shape of the (002) diffraction peak. The (002) region consists of the superposition of the broad and weak amorphous carbon peak and the sharp and strong peak of graphite. The interlayer spacing as calculated from the sharp peak is used to estimate the graphite content of partially graphitized carbon. The broad peak is used to calculate the amount of amorphous carbon.

Frad and Harold (126) have developed a method for quantitative graphite analysis based on the comparison of the height above background of the graphite line (3.37 Å) to the height of an internal standard (fluorite) line (3.16 Å). The sample and fluorite are ground together, and the heights of the two diffraction peaks are determined. The graphite content is determined by comparing the ratio of the height of the graphite peak and the fluorite peak to a calibration curve.

5. X-RAY EMISSION

The only X-ray emission line available for carbon is the K_α line at 44.7 Å. The use of this very-long-wavelength line presents many difficult experimental problems. It has only been since 1963 that any practical use has been made of the line. Henke (166) has described a system for excitation and detection of the carbon K_α fluorescent emission. Soft copper L (13.5 Å) radiation generated in a special X-ray tube with a very thin (1 μm aluminum sheet) window is used for excitation. The analyzing crystal is a lead stearate decanoate film. A flow-proportional counter with an extremely thin window (two 0.1 μm Parlodion films) is used. The sample and analyzing crystals are maintained in a vacuum. A sensitivity of 0.006% for carbon in steel was obtained.

Since the middle 1960s the electron microprobe has become a well-established analytical technique based on the emission and analysis of characteristic X-rays produced by a focused electron beam. Quantitative analysis by the electron microprobe is based on the use of pure homogeneous standards and an X-ray emission process model proposed by Castaing (70). Baird (24) and Ong (281) have discussed the difficulties associated with the quantitative analysis of carbon and other light elements with atomic number less than ten. Henke (164) indicated that the accurate X-ray absorption correction factors are of extreme importance for quantitative analysis of light elements. For carbon, the mass absorption coefficient is high and is not known with sufficient precision (165) for an accurate analysis of carbon less than approximately 5 atom%. If the energy of the primary electron beam is more than 1 keV greater than the ionization energy of the excitation level for carbon, the available absorption correction formulas are not accurate (42). Because of the high absorption coefficient of carbon, only carbon X-rays emitted from the outermost surface layers can emerge from the sample; hence surface conditions have a great effect on the carbon X-ray intensity observed. Therefore, if low electron beam energies (less than 6 keV) are used, if sample surfaces are cleaned prior to and during analysis, if thin detector windows are utilized, and if appropriate diffracting crystals are employed, the quantitative determination of carbon at concentrations higher than 5 atom% is possible.

The lead stearate decanoate analyzing crystal is usually employed for the electron microprobe determination of carbon (119,124,256,263). The counting of the carbon K_α can be achieved by use of pulse height analyzers for energy discrimination (87,263,300). A fine stream of air (90,119) or helium (362) directed at a point of the electron beam impact eliminated the build-up of contamination on the sample surface during the analysis. Hot pressed chromium carbide was found to be an excellent primary carbon standard (119). The electron microprobe has been used for the determination of carbon in ferrous alloys (90,119) and tantalum (263).

A method for determining carbon by measuring the ratio of coherent to incoherent scattering of X-rays has been reported (92). The technique, which is useful for determining hydrocarbons in simple matrixes, employs standard equipment and has the precision and accuracy of conventional microcombustion methods for many types of samples.

D. PETROGRAPHIC METHODS

The most common petrographic method for carbon is the analysis of carbon in steel. The low solubility of carbon in alpha iron (<0.01%) gives rise to a distinctive second phase (pearlite) with increasing carbon content up to 0.80%. An experienced petrographer can estimate the carbon content of steels to better than ±0.5%. Petrographic methods for analysis of carbon have been applied to tantalum (103), tungsten (103), molybdenum (103), and uranium carbide (203,247). Classification of carbonaceous materials have been done petrographically (315).

E. MASS SPECTROMETRIC METHODS

The development and use of mass spectrometers for the analysis of materials is a recent application of the instrument. Historically the mass spectrometer was used for the determination of the relative abundance of isotopes. The slow development of the mass spectrometer as an analytical tool can be traced to its high cost. In 1953, the first ASTM method employing the mass spectrometer was established. The method, D1137-53, included the determination of carbon dioxide in the "Analysis of Natural Gases and Related Types of Gaseous Mixtures" (13). A mass-fraction pattern table is given. Most mass spectrometric methods for carbon are based on the conversion of the carbon into carbon dioxide and analysis of the carbon dioxide. Vacuum fusion (172), dry oxidation (40,299), and wet oxidation (101) techniques have been employed. Oxidation techniques have been covered in Section VI.A.

Under favorable conditions, most commercial mass spectrometers can detect 10–50 vpm (ppm by volume) of a component in a 1 ml sample of a gas mixture (312). The sensitivity for carbon dioxide is usually not as good owing to a high background. This background is caused by residual oxygen in mass spectrometers that produce a beam of CO_2^+ ions, probably formed by the reaction of oxygen with carbide in the filament. Consequently, most methods concentrate the carbon dioxide by freezing with liquid oxygen (101,172) or precipitation as barium carbonate (299). The quantity of gas evolved is calculated, and a portion of the gas is transferred to the mass spectrograph for analysis. Isotopic dilution (40,101) has been used to eliminate the need to determine the volume of gas and has the advantage of not being affected by loss of generated carbon dioxide or by contamination from any source except carbon.

A method for analysis of carbon dioxide in the presence of oxygen has been developed (312). The principle of this method is to estimate the effect of oxygen in the sample by admitting pure oxygen at the same pressure as the partial pressure of oxygen in the sample. The method is useful for samples containing more than 100 vpm carbon dioxide.

Several direct analysis techniques for carbon in solid samples have been developed. In one case, a 10 keV beam of argon ions or protons has been used to bombard a solid sample (60). The bombardment generates a secondary ion emission that is analyzed by mass spectrometry. In a second example, a spark source has been employed to produce charged ions (155). Both of these methods have a sensitivity of less than 0.5 ppm.

F. RADIOCHEMICAL METHODS

1. Assay of Carbon-14

Carbon-14 is a β^- emitter with a half-life of about 5600 yr. It is found as a naturally occurring radioactive isotope of carbon and is synthesized in some quantity.

In nature, carbon-14 is formed by N^{14} (n,p) reaction with neutrons produced in the upper atmosphere by cosmic-ray interactions. The rate of production of carbon-14 by natural means is believed to have been unchanged for at least 15,000 years prior to 1954, when nuclear weapons testing began to perturb the natural levels to a noticeable extent (99).

Carbon-14 exists in an equilibrium concentration in the carbon of living substances. After death, the carbon-14 equilibrium is no longer maintained, making it possible to use the carbon-14 content of organic materials for the purpose of measuring age (99). It is beyond the scope of this discussion to consider the carbon-14 method of dating.

The only practical reaction (61) for production of carbon-14 is N^{14} (n,p) C^{14}. To produce significant quantities of carbon-14, long irradiation with high neutron flux is required. The most common target materials are beryllium and aluminum nitride. Beryllium nitride is the most desirable but also the most expensive. Owing to the difficulties of obtaining carbon-12-free targets, the best product contains only about 80% of its carbon as carbon-14.

Much of the carbon-14 produced is used to synthesize various organic compounds that are utilized in biological research as tagged compounds to follow life processes. Nonbiological applications of carbon-14 tagging have been limited but hold much promise for the future in such applications as the study of very high purity of metals (3), evaluation of the efficiency of processes such as home-laundering procedures (97), and the study of the intake and combustion of fuel in gasoline engines (69).

Counting procedures for gaseous, solid, and liquid samples have been developed. Solid and gaseous samples are much more common, and more general techniques for their preparation have been developed. Virtually all methods of counting have been employed. Liquid scintillation techniques generally are employed when no prior oxidation of the sample is required. Those methods that employ prior oxidation of the sample result in the sample being counted either as a gas (carbon dioxide) or a solid (carbonate). Ion chamber or internal gas-counting methods are used for carbon dioxide. Solid-counting techniques employing windowless flow-proportional counter or Geiger counters are used for solid carbonates.

a. COUNTING WITHOUT PRIOR OXIDATION

There are two general methods employed. Either the sample is dissolved in a suitable solvent, or it is coated on a holder (direct mount). Procedures for direct mounting of fatty acids (104) and biological materials (179) have been reported. Sources of error in the direct mount technique are back-scattering, self-scattering and self-absorption, and control of sample thickness. These problems have been considered (62), and a method of correcting for self-absorption has been suggested (141).

The advantage of liquid counting techniques are convenience of preparation and constant geometry. Disadvantages include the increased self-absorption and the reduced sensitivity caused by the solvent. Since there is no universally applicable solvent, many have been employed including formamide (335,336), toluene (20), dimethyl formamide (336), ethylene glycol (336), and 90% phosphoric acid (336).

b. COUNTING AFTER OXIDATION

Oxidation of the sample to form carbon dioxide is by far the more common technique used for determination of carbon-14. Usually the carbon dioxide is either counted directly in an ion chamber or internal gas counting tube, or it is converted to a solid carbonate and counted by solid-counting techniques. The greatest precision and sensitivity can be achieved with gas-counting methods, particularly with the ion-chamber method; however, the carbonate method offers a much more rapid technique, is less subject to memory effects, requires less equipment, is more easily adapted to automation, and is more easily mastered. A comparison of the sensitivity of different methods for the determination of carbon-14 has been made (52,307).

(1) Oxidation Techniques

Generally, the methods of oxidizing carbon-14 compounds are the same as those already discussed in Section VI.A. Both wet- and dry-combustion techniques have been used, but with the advent of commercially available high-frequency induction furnaces dry combustion is the more common technique. Care must be taken in collection of the carbon dioxide to ensure that a representative sample is obtained. Armstrong et al. (17) have shown that different parts of the molecule can be converted to carbon dioxide at different rates, thus causing a difference in the rate of evolution of carbon-14 with time. This difference in evolution rate results in an inhomogeneous precipitate. As a result, some counting techniques can give incorrect counts. Recently, a sealed-tube method of oxidizing samples has been proposed (336). In this method the sample is placed in a vycor tube; the tube is filled with oxygen, sealed, and heated to 850°C for 5 min. The advantage of this method is the removal of any memory effect.

(2) Collection and Counting of the Carbon Dioxide as a Carbonate

After combustion, the carbon dioxide is purified (see Section VI.A.3) and led into a gas-dispersing tube. The adsorbing liquid is usually barium hydroxide; however, calcium hydroxide, sodium hydroxide, potassium hydroxide (175), phenethylamine (279), and ethanolamine (21,43) have been used. After combustion is complete, the gas-dispersion tube assembly is removed, and the precipitate is collected by filtrating or by centrifuging. When the adsorbing liquid is potassium or sodium hydroxide, barium carbonate is precipitated by addition of barium chloride (175,385). The advantage claimed for this procedure is the ease with which the collection assembly can be washed to ensure a quantitative transfer. Whenever possible an "infinitely thick" (greater than 20 mg/cm^2) sample thickness needs to be made since the lowest layer no longer contributes significantly to the superficial activity. Under these conditions, the superficial activity becomes proportional to the specific activity. Very thin samples (0.1 mg/cm^2 or less) have very little self-adsorption, and the counting rate is proportional to total activity; however, the reproducibility is not as good as the "infinitely thick" samples. Sample thickness between "infinitely thick" and very thin are the least desirable because it is difficult to obtain uniform thickness, the thickness must be measured, and experimentally determined correction factors must be applied. The thickness of sample depends on the available amount of solid and the geometry of the counting system. Whenever practicable, sufficient sample should be taken for combustion so that the amount of precipitate will be adequate to prepare an "infinitely thick" sample for counting.

(3) Counting of Carbon Dioxide as a Gas

When carbon dioxide is to be analyzed in the gas phase, collection is accomplished either by condensation at $-195°C$ or precipitation as barium carbonate. The barium carbonate is filtered and washed; then the carbon dioxide is liberated by addition of sulfuric (188) or perchloric acid (391). In either case, the water is removed from the carbon dioxide by condensation at $-80°C$. When an electrometer (153,188) or scintillation (153) counter is to be used for counting, the purified carbon dioxide is transferred directly to the counting chamber and counted. From pressure-volume measurements, the specific activity of the compound can be determined. More commonly the activity of the carbon dioxide is measured inside a Geiger–Müller counter. The carbon dioxide is mixed with a suitable gas such as carbon disulfide (98,391), methane (346), argon–methane mixture (346), or ethanol (78). Part of the mixture is transferred to the inside of a Geiger–Müller tube and counted. A sensitivity of about 10^{-9} Ci/gram of carbon has been obtained for the electrometer (188, 206) and of about 5×10^{-12} Ci for a Geiger–Müller counter (78).

2. Activation Analysis

Activation analysis can profitably be employed for the determination of carbon when there is need for a nondestructive method, or for low level detection when there are a number of interfering substances present. Activation methods for carbon have found their greatest use in the determination of carbon in metals and high-purity semiconductor materials. Activation of carbon in the sample can be caused by excitation with neutrons, charged particles, or photons. All three types of activation have been employed for the determination of carbon; however, deuteron and proton activation are more commonly used. Photon activation is increasingly used as equipment becomes available because of its greater sensitivity.

a. Neutron Activation

Since the capture of a neutron by C^{12} results in production of stable C^{13}, no analytical methods can be developed from this reaction (264). It is possible with higher-energy neutrons to have $C^{12}(n,2n)C^{11}$. This reaction has a low probability and has found no analytical applications. The reaction $C^{13}(n,\gamma)C^{14}$ with a sensitivity of 4×10^4 μg has been used. Since C^{13} has a natural abundance of about 1%, only low sensitivity can be expected. Neutron activation analysis has been applied to steels (225,354).

b. PHOTON ACTIVATION

The only important photon activation reaction is $C^{12}(\gamma, n)C^{11}$. This reaction has a threshold of 18.74 MeV, and the resulting C^{11} has a half-life of 20.5 min. By using coincidence counting and pulse-height discrimination, much of the background can be eliminated. Typical irradiation times vary from 1 min (30) using the bremsstrahlung spectrum from 40 MeV electrons to 30 min (108) using the 24 MeV Allis–Chalmers betatron. The interference from N^{13}, Cu^{63}, and Fe^{55}, all with half-lives of about 10 min, can be eliminated by following the rate of decay. Sensitivities of 1 ppm with 5% precision (46) for pure materials and of 0.3 ppm with separation by combustion (4) have been claimed. Photon-activation analysis of carbon has been applied to the analysis of lithium (108), lithium hydride (108), beryllium (4,5,30,46), aluminum (4,5), iron (5,235), zirconium (4,5), silver (235), gold (235), molybdenum (25,143,-235), zinc (235), cadmium (235), sodium (236), and nickel (143).

c. CHARGED-PARTICLE ACTIVATION

Activation by proton, deuteron, alpha, and helium-3 particle radiation has been suggested. The reactions are respectively: $C^{12}(p,\gamma)N^{13}$, $C^{12}(d,n)N^{13}$, $C^{12}(\alpha,\alpha n)C^{11}$, and $C^{12}(^3He,\alpha)C^{11}$. Proton activation has been applied to carbon analysis (93,341). Carbon impurities in the range of 2–25 ppm have been determined in metal foils (202) by reaction induced by 14 MeV protons. Carbon impurities in beryllium at the 0.07% level were determined using 8.45 MeV protons (399). Use has been made of 0.6 MeV protons to determine carbon in steel with a sensitivity of 20 ppm without separation (45). Alpha particle activation has been employed for the analysis of high-purity silicon (102,318). Deuteron activation techniques have found much application. General procedures for solid materials have been developed (102,211,291,306,360,377,394) for carbon at the level of 1 ppm. Helium-3 activation of carbon has also been used in recent years (102,148,377).

G. ELECTROCHEMICAL METHODS

Coulometric, potentiometric, and conductometric titrations have been used to determine carbon in the form of carbon dioxide. Since these procedures are usually applied after combustion of the sample, they are discussed in Section VI.A.4.

H.   MISCELLANEOUS METHODS DEVELOPED PRIMARILY FOR DETERMINATION OF CARBON IN IRON OR STEEL

A number of methods based on the determination of some physical property of iron or steel have been developed for the determination of carbon in iron or steel. These methods find their greatest use in steel mills where very rapid analysis with moderate accuracy are needed. These tests are most useful when a large number of nearly identical samples are to be analyzed and in general are very restrictive as to sample variations. As a result, virtually all variables except the carbon content are eliminated. Instruments designed specifically for these analyses are commonly called carbometers. Most carbometers determine the carbon content by measuring the magnetic permeability of the sample. The chief advantage of the technique is the speed with which a nontechnical worker can make analysis. Some disadvantages are the need to know very much about the nature and form of the sample, the limited range of applicability, and the relatively low accuracy.

1.   Magnetic Permeability

When a specimen of hardened steel is placed in a magnetic field, the magnetic flux in the specimen varies with the carbon content (254). A cylindrical ingot of steel to be analyzed is cast with a radius of about 12–13 mm. The ingot is placed in a magnetic field of strength H_1, and the induction B_1 is measured. The procedure is repeated with a lower strength field H_2, and the induction B_2 is measured. The difference, $B_1 - B_2$, in inductance is compared to a calibration curve for the particular magnetic fields, H_1 and H_2, and the carbon content is calculated (241). If the manganese content is greater than 0.3% or if the silicon content is greater than 0.1%, special calibration curves must be used (63). Oxygen, chromium, tungsten, molybdenum, nickel, and aluminum can influence the magnetic permeability measurement (241,284).

2.   Internal Friction

The peaks that carbon and nitrogen cause to occur in the internal friction versus temperature curve are proportional to their respective concentration. At a frequency of 1 Hz these peaks appear at $+36°C$ and $+22°C$, respectively. The sample, in the form of a thread about 1 mm in diameter and about 30 cm long, is made part of a torsion pendulum. The damping is determined by measuring the logarithmic decrement of the free torsional vibration using a very small vibrational amplitude (32). If internal friction is expressed as Q^{-1}, which is equal to $1/\pi$ times the logarithmic decrement, this figure will, by coincidence, give the weight percent of carbon or nitrogen in solid solution (195). Q^{-1}, which is equal to the weight percent of carbon, is calculated at $+36°C$.

3. Thermoelectrical

The thermal electromotive force of steel varies linearly with the change in the carbon content. The sample is cast into a rod (214) or truncated cone (215), quenched, and the thermoelectric electromotive force is measured between a cold and a hot carbon electrode. The carbon content is read off a calibration graph. The procedure takes less than 2 min.

4. Cooling Curve

The solidification interval of cast iron varies with the content of carbon, silicon, and phosphorus (303,367). The variation is linear and follows the relationship: percent C + percent Si/4 + percent P/2 equals the solidification interval. The iron is poured into a mold, and the cooling curve is measured by means of a thermocouple in the center of the mold and a high-speed recording potentiometer. Factors affecting the accuracy of this determination have been outlined (204). The determination can be made in less than 2 min, and an accuracy comparable to conventional chemical analysis is claimed (367).

5. Quenched-steel Hardness

The sample is cast in a special die, removed from the die at $1100°C$, and quenched in ice water (temperature less than $5°C$) in 3–5 sec, and the Rockwell hardness of the sample is determined (298). The carbon content is determined from calibration graphs for various steels.

6. Carbonization

The resistance of an iron-nickel alloy wire varies with the carbon concentration. This property has been used to monitor the carbon content of gases from heat-treating furnaces (372). The resistance of a wire is changed as it dissolves carbon from the gas or as it gives up carbon to the gas depending on the carbon content of the gas. A second similar wire, kept at the same temperature and immersed in a gas of a known low carbon content, compensates for temperature changes.

VII.   ANALYTICAL CHEMISTRY OF PURIFIED CARBON

Purified carbon or graphite, as the name implies, has been given a special treatment so as to yield a material virtually free of those impurities that would otherwise render it unsuitable for applications in the fields of electronics (anodes for ignition rectifiers, radar, and other electronic tubes and for melting crucibles, boats, jigs, and fixtures in the transistor industry), nuclear (reactor moderator), and spectroscopy. Impurities, defined as any element other than carbon that is undesirable in the product may be metallic, nonmetallic, or gases. The impurities may arise from two sources: they are either present in the raw material or they are introduced in the processing operations. Elements such as boron, vanadium, and sulfur are usually present in the raw materials and are not generally introduced in processing. Iron and silicon in addition to being present in raw materials, may be introduced in processing operations. Control of impurities in graphite products is effected in one or a combination of the following ways: selection of raw material, which for nuclear graphite involves choosing materials low in boron and vanadium; thermal purification, which volatilizes most undesirable impurities; and chemical purification, which can be achieved by direct use of a halide-containing gas or by use of a halide salt mixed with the packing around the piece being purified (a high-temperature furnace is required in both cases). Typical impurity levels for graphites purified by the different techniques are compared to a normally graphitized material in Table 9.

TABLE 9
Representative Impurity Levels of Three Classes of Graphite (278)

Impurity	Impurity level (ppm)		
	Unpurified	Thermally purified	Chemically purified
Total ash	1600	540	5
Silicon	94	46	<1
Iron	310	10	<2
Vanadium	30	25	<0.2
Titanium	34	11	1
Aluminum	40	2.5	0.1
Calcium	320	147	1.4
Boron	0.5	0.4	0.06
Sulfur	175	19	10
Gas content (cm³/gram to 1000°C)	0.80	0.50	0.30

It would be a most difficult task indeed if it were necessary to determine the assay of a purified carbon or graphite by direct analysis for carbon. However, the situation for carbon is similar to other highly purified materials where, instead of specifying a certain value of the matrix element, the specifications are written for certain values of elemental impurities. Thus the analytical chemistry of purified carbon becomes the complex problem of determining the concentration of a variety of elements present in trace concentrations.

Among the various analytical techniques that have been used to measure the impurity level in purified graphite are emission or mass spectroscopy, spectrophotometric procedures, and neutron activation and subsequent radiochemical analysis. One of the most useful and informative analyses in the production control of purified carbon is the determination of the total weight of nonvolatile constituents by dry ashing. MacPherson (237) has discussed the ashing of graphite in detail.

A. ASH DETERMINATION

Of great benefit to the analyst is the special property exhibited by carbonaceous materials in that they burn in air or oxygen and that their oxides are gaseous and disappears in the burning process, leaving behind the metallic impurities convenient for analysis. The ashing of carbonaceous materials is a common and simple procedure, but it must be done with care to obtain accurate results. Temperature, impurities, porosity, particle size, and atmosphere all have an effect on the time required for ashing (36). Under identical conditions of atmosphere, temperature, and sample weight, highly purified graphite requires a longer time for complete oxidation than a less pure graphite (Table 10).

Perhaps more important from the analyst's viewpoint is that the temperature used for dry ashing can influence the quantity and quality of

TABLE 10
Oxidation Rates of Graphite in Air (57)

Graphite type	Oxidation temperature (°C)	Time for 1% wt loss (hr)	Time for 6% wt loss (hr)
Purified	500	200	—
	600	7	25
	700	0.6	2
Unpurified	500	6	—
	600	0.8	2.5
	700	0.2	0.7

the resulting ash. Too high a temperature leads to loss of more volatile oxides and in certain cases oxide fusion that can coat carbon particles, preventing complete oxidation. Use of too low a temperature not only greatly increases ashing time, but also may result in incomplete conversion of some impurities to the oxide.

The temperature used for dry ashing purified graphite varies somewhat, depending on the experimenter and the end use of the data. Temperatures from 500°C to 950°C have been utilized. Although oxidation time can be reduced by ashing at high temperatures, one must then contend with the problems mentioned previously.

Fusion of the ash occurs most frequently with graphites that have high vanadium content. If vanadium is present the temperature should not exceed 720°C, since above this temperature any vanadium pentoxide formed will fuse around the graphite particles and thus prolong the ashing time. It has been the practice in the authors' laboratory to ash at 680°C to minimize losses of impurities. However, it is not uncommon to find many literature references that suggest temperatures of 800 and 900°C. The ASTM Method, C 561-69, for ash in graphite recommends a temperature of 950°C (12). Regardless of the temperature chosen, it is important that a given temperature be used when results are to be compared. This is especially true for production-control purposes.

In addition to the quantity of ash, its physical appearance can be meaningful to an experienced analyst. A light-colored ash will generally be highest in calcium, magnesium, aluminum, silicon, or titanium. An ash of red or brown coloration will be highest in iron, vanadium, or manganese. A spectrographic method is often used for qualitative and quantitative analysis of the ash constituents (349).

For chemically purified graphite where total ash impurities are less than 10 ppm, fairly large samples are required (~100 grams). In the case of such high purity graphite, it is not possible to crush or grind the sample to effect homogenization because of the danger of contamination. General practice is to break the sample into suitably sized pieces to fit into the ashing crucible. It is recommended that washed white cotton gloves be worn during this and the weighing operations. Sampling of high-purity nuclear-grade graphite has been discussed by Clark (65).

A typical procedure for ashing purified graphite is as follows:

Place approximately 50–100 grams of the sample in a clean tarred 100 ml platinum dish. Choose the sample to be as nearly representative of the original stock as possible. Ash the sample to constant weight in an electric muffle furnace at 680°C in an oxidizing atmosphere. Pass clean dry air (or oxygen) into the furnace (50–200 ml/min) to speed up the oxidation reaction. A quartz lining in the furnace is often used to reduce danger of contamination. The time required for complete oxidation varies from 1 to 5 days depending on the variables previously mentioned.

To obtain the greatest accuracy for samples of very high purity, it is necessary for the analyst to take the following precautions: (1) calibrate the balance at the time of weighing platinum dishes, sample, and ash; (2) evaluate the weight change of platinum dishes during the ashing procedure by running blanks of empty platinum dishes (from an identical lot of platinum fabrication) along with the samples (65); and (3) properly clean and condition the platinum dishes prior to use.

B. METAL IMPURITIES IN ASH

Only conventional colorimetric methods are considered in this section. Spectrographic procedures are discussed in Section VII.E. Conventional colorimetric methods of analysis for nickel, aluminum, manganese, iron, titanium, chromium, vanadium, and molybdenum are the most widely utilized techniques for determination of these impurities in the ash of carbon and graphite because of the accuracy of these methods. Only the dissolution techniques used on the ash residues of carbon and graphite are presented in detail. Details of the colorimetric procedure can be obtained from the references given.

Tentative methods were first published in 1965 by ASTM for Chemical Analysis of certain impurities in carbon and graphite (9). The current designated method, C 560-69, includes silicon, iron, calcium, aluminum, titanium, vanadium, and boron (12).

The quantity of ash obtained from 50–100 grams of purified carbon or graphite may vary from less than 1 mg to more than 50 mg depending on the purification treatment that was used. Thus it is necessary to exercise some judgment on the quantity of sample ashed and the volume to which the resulting solution is finally diluted.

Procedure

In order to minimize attack on platinum in the fusion procedure, grind and intimately mix the ash from 50 grams of sample with 3 grams of anhydrous sodium carbonate in a smooth agate mortar. Transfer the mixture into a platinum crucible. "Rinse" out the platinum dish used for ashing and the mortar with an additional 2 grams of sodium carbonate. Add the rinsings to the crucible, cover, and heat cautiously until the contents are molten. Fuse for 20 min, with occasional swirling to insure mixing and allow to cool. Add two or three small crystals of KNO_3 (the size of a grain of wheat), and fuse again for 5 min. Allow the melt to cool. Place the crucible in a 250 ml beaker containing sufficient water to cover the crucible and lid. Heat the beaker on a hot plate or steam bath to dissolve the melt. Remove the crucible and lid from the beaker, police, rinse thoroughly, add rinsings to the contents of the beaker, and set platinum ware aside. Add six drops of ethyl alcohol to the solution, heat nearly to boiling, and then digest on a steam bath for 15 min. Filter the hot solution through double Whatman No. 42 filter papers, wash the residue five times with hot 1% sodium carbonate and finally three times with hot water. Retain the filter paper containing the precipitate for subsequent analysis.

The filtrate from the carbonate fusion is neutralized with 6 N sulfuric acid, the volume is reduced to approximately 80 ml, cooled, transferred to a 100 ml volumetric flask, and diluted to volume. Aliquots of this solution are used for determination of chromium, molybdenum, and vanadium, as mentioned in the following (65).

Chromium is determined in a 25 ml aliquot as the red-violet complex with diphenylcarbazide (320). After color development, a period of about 20 min should elapse prior to measuring the optical density. This waiting period permits the vanadium complex if present to fade and minimizes interference effects.

Molybdenum (in a 25 ml aliquot) in the presence of stannous chloride as a reducing agent reacts with thiocyanate to give a red complex. This compound is soluble in a variety of organic liquids and can be determined after extraction into the organic phase (323).

Vanadium is determined in a 25 ml (or less) aliquot according to the phosphotungstate procedure (326).

Aluminum is determined in a 25 ml aliquot as the yellow chloroform-soluble complex with 8-hydroxyquinoline (319). Although this reagent is not specific for aluminum, good sensitivity can be attained by the addition of a masking agent and adequate pH control. For graphite of very low ash content (<50 ppm) it is recommended that a separate sample of 25–50 grams be ashed and the total filtrate from the sodium carbonate fusion be used for the determination of aluminum (65).

Place the filter paper containing the precipitate in the crucible that was used for the carbonate fusion. Char and ash the paper, cool, add 2 ml of water, four drops of sulfuric acid, and about 5 ml of 48% hydrofluoric acid. Evaporate to dryness on a steam bath or hot plate, and heat over a burner at red heat for 5 min. Cool the crucible, and add sufficient potassium pyrosulfate to obtain a good fusion. Fuse, cool, dissolve the melt in 50 ml of water and 2 ml of sulfuric acid. Transfer quantitatively to a 100 ml volumetric flask, and dilute to volume. Aliquots are subsequently removed for nickel, manganese, titanium, and iron determination as indicated in the following.

Nickel is determined in an aliquot containing 5–30 μg of nickel by the dimethylglyoxime colorimetric method (324).

Manganese is determined colorimetrically in an aliquot of not more than 25 ml as permanganate by oxidation with potassium periodate (322).

Titanium is assayed colorimetrically according to the standard hydrogen peroxide procedure (325). Usually a 25 ml aliquot is very satisfactory.

Iron is determined as the orthophenanthroline complex (321).

The size of the aliquot necessary can best be determined from the weight and/or color of the original ash residue. If the weight was large and the ash was red-brown in color, a small aliquot should be taken.

C. NONMETAL CONTAMINANTS

Dry-ashing procedures ignore all contaminants of purified graphite that are volatilized during the oxidation. Spectroscopic products are not generally affected by the small amounts of volatile constituents (such as water) or gases (nitrogen) that are present. However, boron, which is lost during ashing, can affect spectroscopic analysis for boron. Boron is even more harmful in certain nuclear applications. Boron analysis is considered in Section VII.D.

Most volatile impurities have a minimal effect on nuclear graphite applications. However, chlorine can be detrimental if present in sufficient quantity. Hydrogen and carbon monoxide released from graphite can interact with fuel-cladding material, which results in hardening and embrittling of the metal. Carbon monoxide can disproportionate and deposit carbon on cooler-metal heat-exchanger surfaces, which results in an adverse effect on heat-transfer in the system (301).

Volatile impurities exhibit the most damaging effect when graphites are used in electronic applications. These impurities limit the operating life of the electronic tubes in which graphites are used. Generally, chemical or thermal purification substantially reduce the gas content of graphite (Table 9).

The volume and composition of gases evolved when graphite is heated have been measured by a number of workers (19,95,283,301). The equipment necessary to make this measurement consists of a high-frequency induction furnace, a vacuum system with a gauge for measuring pressure rise in the system, and a means of analyzing the gas composition such as mass spectrometry, gas chromatography, or some gas-absorption system. A general procedure for measuring gas content is as follows: place the sample in a closed system, and evacuate the sample at very low pressure for at least 1 hr before testing; heat the sample by induction to a specified high temperature; measure the pressure in the system, and calculate the volume of gas evolved; analyze the evolved gas to determine its composition. The collected gases may be analyzed by mass spectrometry or by gas chromatography.

Major gaseous components evolved from nuclear-grade graphite are hydrogen, carbon dioxide, water, nitrogen, and carbon monoxide. Minor constituents identified have included hydrogen sulfide, sulfur dioxide, methane, and other hydrocarbons.

D.   DETERMINATION OF BORON

Owing to its high neutron cross-section, the determination of boron in nuclear-grade graphite has become a most important specification test. With the advance of production technology to the point where tons of graphite containing only 20 parts per billion boron can be produced, a method of high sensitivity is required. As a result of the high sensitivity required, major problems are encountered due to contamination from air-borne dust, soaps, reagents, and so forth. An additional difficulty is the problem of quantitatively converting the boron in the carbon and graphite to a soluble form. These problems have been discussed in detail (65,354).

A great deal of work has been done to determine the most accurate precise and economical method for determining traces of boron (44,317). Visual colorimetric procedures such as the one used by Smith (350) lack sufficient sensitivity. Naftel (271) developed a technique for trace boron based on the observations of Cassal and Gerrans (59) that oxalic acid promotes formation of a red substance with curcumin. A modification of this work (59) was used by Rodden and Scherer (330) in the early phases of the Manhattan Project. In order to avoid interference when certain impurities, for example, iron, molybdenum, titanium, vanadium, and oxidizing agents, are present in sufficient quantity (111,329) (greater than 1000 ppm total impurity), it is necessary to separate the boron as the methyl borate by means of distillation before developing the colored complex. However, graphite production technology has advanced to the point where impurities in purified graphite are of such low level that the distillation separation can be omitted (73). If it becomes necessary to accurately determine boron in graphites containing inorganic impurities in excess of 1000 ppm, the distillation procedure of Richmond and Telford (305) or Metz (259) may be used.

A thorough study of the determination of microgram and submicrogram quantities of boron, especially with regard to the photometric system (354) and the distillation separation (355), has been reported by Spicer and Strickland. Included in this report is an examination of interference, contamination problems, techniques for boron recovery by distillation, equipment, and detection limit. These authors describe three procedures for successively increasing detection sensitivity (2–15, 0.5–4, and 0.01–0.2 μg B). The methods, all of which involve a methylborate distillation separation, have been thoroughly tested under routine conditions for the analyses of atomic-energy materials.

Hayes and Metcalfe (159) have described a procedure that simplifies the technique by eliminating the evaporation procedure common to all previous curcumin methods. It has also been shown that the distillation separation can be avoided in many instances. Detailed procedures are given for most of the more common nuclear reactor materials (zirconium, boron, magnesium, uranium dioxide, aluminum, steel, and graphite). A colorimetric method for the determination of boron in graphite, coke and pitch, as it is used for production control, is presented in detail (65,56).

Special Apparatus

Spectrophotometer, Beckman DU.
Furnace, quartz-lined Hoskins electric muffle, No. FH-204.
Quartz ware, distillation flask, condenser, crucibles (30 ml, tall form), and containers for boron-free reagents.
Water bath, 100°C, Electric-Magni-Whirl, model 2642, Ace glass.
Steam bath, electric, Fisher No. 15-496.
Mill, Labconco Laboratory Construction Company.

Special Reagents

Methyl and ethyl alcohol, boron-free. Redistill the alcohols in the presence of sodium hydroxide.
Oxalic acid–hydrochloric acid mixture. Dissolve 8.0 grams oxalic acid in 6.0 ml concentrated hydrochloric acid and 40 ml distilled water. Prepare daily as needed.
Curcumin solution. Dissolve 0.01 gram curcumin in 50 ml 95% redistilled ethyl alcohol. Prepare daily as needed.
Calcium hydroxide suspension. Dissolve 2.8 grams boron-free calcium oxide in distilled water to make 1 liter of solution.

Standard boron solution, 1 μg/ml. Make a stock solution by dissolving 0.1428 grams boric acid in distilled water to make 1 liter of solution. Make the standard solution by diluting 10 ml of the stock solution to 250 ml. This solution will contain 1 μg boron/ml.

Procedure

Weigh a 1 gram sample of finely ground graphite, coke, or pitch into a quartz crucible. Cover the sample with 5 ml methyl alcohol and 10 ml calcium hydroxide suspension. Methyl alcohol is used to wet the sample and attain a more intimate mixing of the sample with the calcium hydroxide. Transfer the crucible containing the sample mixture to the steam bath, and evaporate to dryness. The crucible is placed in a cool quartz-lined muffle furnace, the oxygen flow rate in the furnace is adjusted to 100–125 ml/min, and the temperature controller is set to 700°C. The sample is ashed for 12–16 hr. Remove the crucible from the furnace and allow to cool. Add 2 ml each of warm oxalic–hydrochloric acid mixture and curcumin solution. Evaporate the solution at 55°C in the water bath until just dry (this is ascertained by the discontinuation of evolved hydrochloric acid fumes). The evaporation time is critical: evaporation should not exceed 2 hr and 20 min. Leach the residue in the crucible with 95% ethyl alcohol. Filter through a double boron-free 11 cm filter paper into a 25 ml glass-stoppered volumetric flask. Dilute to volume with ethyl alcohol. Blanks are run in triplicate along with the samples owing to the possibility of contamination. They are carefully spaced among the samples throughout all operations. Read the optical density of the samples and blanks against distilled water at 540 nm, using matched 1 cm silica cells in a spectrophotometer.

Aliquots of the standard boric acid solution (1 ml = 1 μg of boron) are taken for preparation of the standard working curve. A minimum of five points covering the range of 0.1–1.5 μg of boron is necessary to establish the curve. The desired quantity of standard borate solution is pipeted into a clean quartz crucible and treated in a manner identical to the samples. Blanks are run along with the standards. Extreme care must be exercised to prevent contamination from boron in reagents, glassware, water, and atmospheric dust.

Note. In the event the sample is pitch, the more volatile fraction is burned off at 200°C prior to adjusting the furnace controller to 700°C. Care must be exercised to assure complete ashing of the sample. This is indicated by a nearly white ash. The flow of oxygen over the sample is essential to attain a thorough ashing. Samples high in iron or vanadium will impart characteristic oxide colorings.

E. SPECTROSCOPIC METHODS

Direct analysis for metallic impurities in purified graphite without ashing can be accomplished by spectroscopic means. The spectroscopic method has been employed for many years as the best method for qualitative analyses of spectroscopic electrode graphite, which is ordinarily produced in small lots. Weinard (384) has described a spectroscopic method for more quantitative analyses of spectroscopic electrodes. However, in the analysis of reactor graphite and other purified graphites produced in large quantity, the problem of inhomogeneity can be severe (238). It is impractical to machine suitable electrodes from a large block of reactor graphite and expect to have a representative sample or expect the sample to be free of contamination from the machining operation. Thus to reduce the inhomogeneity problem, samples should be taken from various positions within the block of reactor graphite and pulverized and milled to a fine powder. Every effort must be made to avoid contamination during each step of the sample preparation.

The determination of boron in graphite powders has been reported by Feldman and Ellenburg (115). They were able to determine boron down to a level of 0.25 ppm with good accuracy above this level. One of their techniques involved the use of a "sifter" electrode through which the powdered sample sifted into the high-voltage spark-discharge area and was excited. Garton (136) has described a dc arc method for determination of boron in graphite, coke, and pitch in the boron range of 0.2–1.5 ppm, where the powdered sample is pelletized with phenol–formaldehyde resin and arced in graphite holders. Rossi and Soldani (313) reported a dc-arc method for boron in the range of 0.1–5.0 ppm using silicon dioxide as an internal standard.

Webb (383) has used the cathode-layer technique to determine impurities in small amounts of radioactive graphite. The ground sample, mixed with an internal standard mix and placed in a specially designed electrode, is burned in a dc arc. The effective concentration range for 17 of the more commonly sought elements is from 10 to 500 ppm.

Other investigators have reported the use of a carrier-distillation technique for the determination of boron in graphite (209,389). Ko (209) used two parts sample to one part copper fluoride carrier to extend the procedure to include, in addition to boron, aluminum, calcium, iron, silicon, titanium, and vanadium. Golab et al. (139) have described a carrier distillation technique that utilizes sodium fluoride to achieve increased sensitivity for those elements with sufficient volatility to be excited by the dc arc. More refractory elements, in particular the rare earths, were determined by the copper-spark method after ashing the sample and dissolving the residue. However, the need for involved sampling deprives the analyst of the advantage of speed inherent in spectrochemical analysis. Methods involving the use of an ashing step before spectrochemical analysis have been described by several investigators (71,136,234). Smith (349) utilized 20 mg of pulverized ash mixed with 80 mg of buffer (20 parts lithium carbonate to 80 parts graphite) to determine 30 trace elements in fuel ash.

A method for the determination of boron and vanadium (0.1 and 4 ppm, respectively, for limits of detection) was reported by Paterson et al. (286). Boron and vanadium were concentrated in calcium oxide by dry mixing 1 gram of calcium oxide with 1 gram of powdered pitch, coke, or graphite and ashing in a boron-free atmosphere at 800°C in the presence of oxygen. In the case of pitch samples, the volatile matter was first allowed to escape through the open door of the muffle furnace at 350°C. After complete oxidation the residue was ground in a mullite mortar to obtain a homogeneous material. Duplicate charges were weighed into pedestal-type spectroscopic electrodes, tamped, center-vented, and placed in a stainless steel block for heating at 800°C for 15 min. Each electrode was immediately arced at 20 A dc for 45 sec. When compared with standards run under the same conditions, analyses adequate for production control can be achieved. D'Silva (88) employing a similar ashing technique and arcing in an argon atmosphere achieved a detection limit of 0.05 ppm boron by pelletizing the oxide after oxidation.

Neutron activation analysis followed by gamma-ray spectroscopy also has been employed for the analysis of impurities in graphite (250,267,-280,403). Advantages of this technique are that smaller levels of impurities can be detected and that the technique is nondestructive. Disadvantages are that an irradiation time as long as 1 week may be required for certain elements and that the equipment is expensive. In one application (280), the sample of graphite and a standard were irradiated in a nuclear reactor. The gamma activities of the trace elements induced in the sample were compared with the gamma activity of the standard by the gamma spectrometer. Thirteen elements in graphite in the concentration range of 10–2000 ppm were determined.

F. NUCLEAR PURITY

Although some measure of the poisoning effect of impurities can be determined by careful chemical analysis, it has become standard practice to measure the net effect of all impurities by a single test that measures the reactivity effect directly. One of two methods is normally employed: the danger-coefficient method (DIH) or the pile-oscillator method (278). A discussion and comparison of these methods has been given by Nichols (275). These methods are used to measure the absorption cross-section of the sample so that this effect in the nuclear reactor can be assessed.

While the success of these methods depends on several factors (65), they have the advantage over other methods of precisely evaluating the total impurities in a sample. This offers the ultimate in production control methods. The chief disadvantage of such methods is that the impurities are not identified.

In the absence of a neutron source, it is not possible to make a direct measurement of the absorption cross section. However, it is possible to estimate the cross-section (purity) of a graphite from a knowledge of the concentration and cross-section of the specific impurities. Strocchi et al. (358) have suggested a method for describing the neutron-absorption properties of impurity carbon. This method has been discussed by Nightingale (278).

VIII.   DETERMINATION AND ANALYSIS OF CARBON COMPOUNDS

A.   CARBON DIOXIDE AND CARBON MONOXIDE

Methods for determining carbon dioxide in conjunction with oxidation methods are discussed in Section VI.A.4. Two of the more common methods used for carbon dioxide are volumetric measurement of the gas by means of an Orsat (168) or similar apparatus or gravimetric measurement (29) of the carbon dioxide absorbed in absorbers containing soda-lime or sodium hydroxide on shredded asbestos ("Ascarite"). A number of techniques have been described (130,169).

Carbon monoxide can be determined by absorbing the gas in suitable absorbents or by oxidizing to carbon dioxide and subsequent determination of the dioxide. For absorption several quantitative methods are employed. In the case of concentrations above 0.2% by volume, the gas that has been previously treated to remove acetylene, ethylene, and other unsaturates is bubbled through an aqueous solution of acidified cuprous chloride, an ammoniacal cuprous salt, or a commercial reagent known as CO-sorbent, in an apparatus of the Orsat-type. The concentration of carbon monoxide in the gas is determined by the reduction in volume by absorption (149,170).

Carbon monoxide is oxidized to the dioxide by air or metal oxides such as cuprous oxide. The reaction with air without a catalyst is slow below 650°C. "Hopcalite" (a mixture of manganese and copper oxides) will catalyze the reaction at room temperature (147).

Small quantities of carbon monoxide may also be determined by passing the gas, previously purified to remove carbon dioxide, unsaturates, and water vapor, through iodine pentoxide at 145–160°C. The evolved carbon dioxide may be determined, or the iodine that is liberated may be absorbed in aqueous potassium iodide and titrated with thiosulfate. The method is claimed to be accurate to ±0.002% (149). "Hoolamite" (a mixture of fuming sulfuric acid and iodine pentoxide) has been employed for the determination of large amounts of carbon monoxide (170).

Carbon monoxide in the presence of carbon dioxide can be determined quantitatively by passing the mixture over "Ascarite" to remove the carbon dioxide, and then converting the carbon monoxide to carbon dioxide by passing over cupric oxide at 700°C. Finally the gas is passed through "Ascarite" to absorb the carbon dioxide formed. Oxygen is used as a flushing gas. Analyses with an average deviation from the mean of ±0.05% are possible (369).

Gas chromatography employing a thermal-conductivity detector with helium as a carrier gas has been used to determine carbon monoxide. Some typical chromatographic columns that have been utilized for separating this gas from other gases are Porapak Q (77,86), molecular sieve (86,242,252), Porapak N and R (242), carbon molecular sieve (402), and coconut charcoal (149). A better sensitivity for determining carbon monoxide by gas chromatography was obtained using a helium-plasma detector (86) and a modified flame-ionization detector (328).

Carbon monoxide is also quantitatively determined in gas mixtures by infrared techniques.

B. CARBON-CONTAINING ACIDS AND SALTS

Carbonates and bicarbonates or acids and salts that are easily converted to carbon dioxide can be determined by those methods described in Section VI.B. Other salts and acids are determined by regular combustion techniques discussed in Section VI.A.

C. CYANIDES

The detection and determination of small quantities of cyanide are important owing to the high toxicity of cyanide material (see Section I.C). If cyanide is the only carbon-containing material in the sample, the carbon is converted to carbon dioxide by combustion, which is the fastest and quickest method of analysis. Many of the methods for cyanide involve the conversion of the cyanide material to hydrogen cyanide. This serves to separate the cyanide from interferents, break up strong cyanide complexes, and concentrate the cyanide. The liberated hydrogen cyanide is analyzed quantitatively. Colorimetric, titrimetric, and instrumental methods have been proposed, for the determination of cyanide. Of these, colorimetric methods are the best for small amounts in the range of 0.1–10.0 ppm.

1. Conversion of Cyanide Containing Materials to Hydrogen Cyanide

Cyanide may be present as hydrogen cyanide, ionic cyanides or as cyanide complexes. If the cyanide is present only as hydrogen cyanide or ionic cyanides such as alkali cyanides or metal cyanides like nickel cyanide, the cyanide can be easily removed as hydrogen cyanide by distillation from acidic solution or by extraction into isopropyl ether. When more stable cyanide complexes such as ferricyanide are present, the cyanide complex must be destroyed. Kruse and Mellon (219) recommend distillation at reduced pressure from a mixture of phosphoric acid, citric acid and Versene.

2. Colorimetric Methods

a. DETECTION AND ESTIMATION

Many reagents offer high sensitivity for the cyanide ion but are unsuitable for quantitative analysis since they either produce unstable color reactions or are nonspecific for cyanide. Some of these reagents are satisfactory for detection or estimation of the cyanide content. Of these, the formation of blood red ferric ferrithiocyanate (by reaction with polysulfide followed by addition of ferric chloride) and Prussian blue are specific, relatively free of interference, and sensitive (<1 μg CN^-). The ferric ferrithiocyanate test is applicable in the presence of sulfide or sulfite (26). A convenient paper test based on the Prussian blue reaction has been developed (85). Filter paper is impregnated with ferrous sulfate and sodium hydroxide. The impregnated test paper if dried in a vacuum and protected from air is stable for at least 10 months. To use, the test paper is exposed to air containing HCN, then developed in 30% sulfuric acid. Less than 400 ppm hydrogen chloride or ammonia, and less than 50 ppm sulfur dioxide or hydrogen sulfide, have been shown not to appreciably affect the color of the stain. Chlorine inhibits the color development. Several indicator tube methods are available commercially (154,373).

b. Quantitative Methods Based on König Reaction

The König reaction, which is the reaction of a cyanogen halide with an aromatic amine and pyridine, has been the basis for a large number of sensitive (0.01–2 ppm) quantitative methods for cyanide. Aldridge (6,7) developed a method in which the cyanide was converted to cyanogen bromide by treatment with bromine. The excess bromide is removed by arsenious acid. An intense orange color is developed when the cyanogen bromide is caused to react with pyridine and benzidine.

Epstein (105) converted the cyanide to cyanogen chloride by oxidation with chloramine-T: the cyanogen chloride is allowed to react with pyridine containing 3-methyl-1-phenyl-5-pyrazolone, and a small amount of bis(3-methyl-1-phenyl-5-pyrazolone). A blue color is formed that was found to be unstable without the bis(3-methyl-1-phenyl-5-pyrazolone).

Numerous methods have been based on the two techniques (Epstein and Aldridge). Those based on Aldridge have the distinct disadvantage of employing the active carcinogen benzidine or related carcinogenic amines. The chief drawbacks to Epstein's procedure is the formation of the volatile cyanogen chloride and the use of the unstable pyrazolones. In spite of these drawbacks, good results can be obtained over the range of 0.01–2.0 ppm. Chief interferents are cyanogen halides and thiocyanate. Bark and Higson (27) have investigated the use of seven heterocyclic amines other than pyridine and the color formation and stability of the dyestuff formed with pyridine and 46 aromatic amines. They concluded that only pyridine was suitable as the heterocyclic amine and recommended p-phenylenediamine for use in place of benzidine or pyrazolone. Murty and Viswanathan (270) used cyanogen bromide with barbituric acid in pyridine.

c. Other Quantitative Methods

Colorimetric methods not based on the König reaction usually are not as sensitive; however, a few appear attractive for cyanide concentrations greater than 1 ppm. Fisher and Brown (118) have reviewed the alkali picrate procedures and developed method sensitive to 1 ppm and accurate to within 2%. The method is based on the reduction of sodium picrate by cyanide to form a colored product. Any substance that reduces the picrate interferes.

Dagnall et al. (75) have described an indirect spectrophotometric method for the determination of cyanide down to 0.2 ppm. It is based on the fact that cyanide prevents the formation of the strongly absorbing ternary complex between silver(I), 1,10-phenanthroline, and bromopyrogallol red in nearly neutral aqueous solution. Mercury(II), iodide, bromide, and thiocyanate interfered. Iodide, bromide, and thiocyanate interference could be overcome by addition of lead nitrate, ammonium sulfate, and barium nitrate.

3. Titrimetric Methods

When the sample contains more than about 20 ppm and is free of interfering substances, titrimetric methods generally have better precision than colorimetric techniques. A very stable silver cyanide complex is formed when silver ion is added to an alkali cyanide solution:

$$Ag^+ + 2CN^- \rightarrow Ag(CN)_2^- \qquad (213)$$

When all the cyanide is consumed a second reaction

$$Ag(CN)_2^- + Ag^+ \rightarrow Ag[Ag(CN)_2]$$

takes place yielding insoluble silver cyanide. Thus the end point is the appearance of turbidity. This method (Liebig's) can be used when other silver complexers such as ammonia are absent.

Volhard's method (131) for cyanide analysis depends upon the fact that silver cyanide is insoluble. The sample solution is treated with an excess of silver nitrate solution, the insoluble silver cyanide is removed by filtration, and the excess silver is determined by back titration with standard thiocyanate solution.

Halogens interfere in both of the preceding methods. For samples containing halogens, a procedure based on the reduction of the cupric ammonium complex to the cuprous ammonium complex (131) is recommended. The solution of cyanide is made basic with concentrated ammonium hydroxide and titrated with a cupric sulfate–ammonium hydroxide reagent until a faint blue color is evident.

Hoffman (178) describes a method sensitive to 0.2 μg of cyanide. The sample is buffered, and an excess of mercury nitrate is added. A drop of diphenylcarbazone is added. A second solution is prepared by adding one drop of diphenylcarbazone to the buffer alone. Mercury nitrate solution is added from a microburette until the color of the two solutions are the same. The difference in the amounts of mercury nitrate is equal to the amount of mercury cyanide formed. The cyanide must be removed by distillation when there are interfering substances.

When a solution of a nickel salt in excess is added to a cyanide solution, the very stable tetracyanonickelate ion is formed (387). The cyanide can be determined by titrating the excess nickel with EDTA. Any cation that forms a stronger complex with cyanide than nickel or any anion which forms a stronger complex with nickel than cyanide will interfere.

4. Other Methods

A simple gas-chromatographic method in which hydrogen cyanide is trapped on silica gel and subsequently desorbed by electrical heating has been reported (154). Hydrogen cyanide gives only about one tenth the normal response on a flame ionization detector. Isbell (185) has reported the use of a triacetin (25% on Chromosorb P) column for the separation of hydrogen cyanide from a number of common inorganic gases including chlorine and cyanogen chloride. Poly(ethanediol adipate) (152) and Poropak Q columns (64) have also been used for separation of hydrogen cyanide.

A polargraphic method has been described (171). The limit of detection is about 0.05 μg/ml, but free chlorine, bromine, iodine, and sulfide ion must be absent. The supporting electrolyte is 0.1 M sodium hydroxide.

A selective ion electrode for cyanide is available commercially. The electrode has a solid state membrane and can be used over a concentration range of 10^{-2} to 10^{-6} M. Sulfide, iodide, bromide, and chloride are anions that interfere. The electrode can be analyzed for direct determination or as an indicating electrode for titrations with silver nitrate. A number of applications have been reported (191,127,137,249).

D. FREE CARBON IN CARBIDES

In the evaluation of carbides, the free carbon content is of great importance (205). The two techniques commonly employed are dissolution of the sample matrix by acid digestion or preferential oxidation of the free or combined carbon.

Hydrochloric acid (37,183) and nitric acid (50) or a nitric acid–hydrofluoric acid mixture (50,109) have been employed to dissolve the metal carbide. The use of hydrofluoric acid in the acid reagent has been shown to be satisfactory for sample dissolution since it does not give low results as previously suggested (50). Some carbides such as uranium dicarbide give a waxy substance, which must be removed by treatment with diethylether (183). In the determination of free carbon in hafnium carbide some undissolved matter that was not uncombined carbon remained after treatment with nitric–hydrofluoric acid reagent. It was found to be soluble in 10% sodium hydroxide; consequently, the sample dissolution technique was modified to include treating the residue with 10% sodium hydroxide followed by wash with hot water, 10% hydrochloric acid, and again with hot water (109).

Studies (28,341,342) have shown that free carbon oxidizes at different temperatures than the carbide and, therefore, can be preferentially oxidized. For silicon carbide, the free carbon begins to oxidize at 515°C, but silicon carbide does not begin to oxidize until 830°C. The optimum temperature for analysis is 725 ± 25°C. When iron carbide is heated to 650–700°C in a hydrogen–steam mixture, the iron carbide is removed as hydrocarbons leaving the free carbon (245).

E. CARBON IN COAL, GRAPHITE, AND COKE

Combustion techniques are used to determine the carbon content of coal, coke, and graphite. The ASTM procedure, D-271-70 (15), is recommended. For more details see Section IX.C.

F. CARBON IN ORGANIC COMPOUNDS

The determination of carbon in organic compounds is covered in Part II, Volume 11, p. 297 of this treatise.

IX. RECOMMENDED LABORATORY PROCEDURES

An attempt has been made to select those procedures of a general nature that will have the widest application to inorganic materials. Additional details of the procedures may be found in the original references.

A. CHEMICALLY COMBINED CARBON (CARBONATES) (130)

This method for the determination of total carbon dioxide is applicable for limestone, dolomite, magnesite, strontianite, witherite, spathic iron ore, carbonates of sodium and potassium, bicarbonates in baking powder, carbon in materials readily oxidized to carbon dioxide by chromic-sulfuric acid mixture, and dissolved carbon dioxide. The method depends upon the evolution of carbon dioxide by a less volatile acid, or the oxidation of carbon. The carbon dioxide is absorbed in caustic and weighed. Figure 1 shows one form of an apparatus suitable for this determination.

Procedure

Place a sample weighing 0.5–2 grams, according to the carbon dioxide content, in the dry decomposition flask (C). Close the flask by inserting the funnel tube (B), and connect it to the condenser as shown in Figure 1. Sweep out the apparatus with a current of dry, purified air before attaching the weighed absorption bottle. This is accomplished by applying gentle suction at the end of the purifying train. Attach the absorption

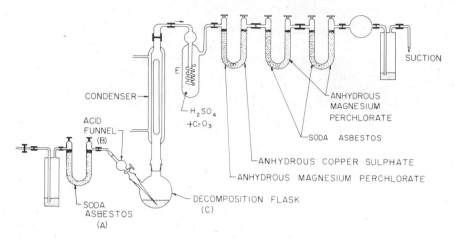

Fig. 1.   Apparatus for determining carbon dioxide.

apparatus. Fill the funnel tube (B) nearly full with dilute sulfuric acid (1:3), the stopcock being closed. Insert the soda–asbestos tube into place as shown in Figure 1. Allow the acid in B to run slowly down on the sample at a rate that evolves gas not too rapidly to be absorbed; 1–2 ml of acid is retained in B to act as a seal.

When the violent action has ceased, heat the solution in C to boiling, and boil for about 3 min. If the sample is baking powder or contains organic matter, protect the decomposition flask from excessive heat by placing a casserole of hot water under it. This prevents charring of the starch or organic matter, which could occur if a direct flame is used. Apply gentle suction to the absorption end of the apparatus and open the stopcock on the funnel tube. This allows the remainder of the acid to flow into the flask (C) and admits a current of air, purified by passing through the soda-asbestos tube in A. The suction should be gentle at first, and then the speed of the flow increased to the full capacity of the absorption bottle. A fairly rapid current is preferred to the older procedure of bubbling the gas through the apparatus at a snail-like pace, but discretion should be used in avoiding too rapid a flow.

In the analysis of baking powders, where foaming is likely to occur, the decomposition flask should be of sufficient capacity to prevent foaming over. A small flask is generally to be preferred for obvious reasons. By gently heating to boiling during the passage of the air, steam assists in expelling any residual carbon dioxide in the flask. When the passage of air is rapid, this boiling should be discontinued.

The increase in weight of the absorption bottle is due to the carbon dioxide of the sample. This procedure gives total carbon dioxide.

B. WET CHEMICAL COMBUSTION (130)

The method depends upon the oxidation of carbon to carbon dioxide when the powdered material is digested with a mixture of concentrated sulfuric acid and either chromic acid, potassium dichromate, or potassium permanganate. The procedure is applicable to oxidation of free carbon, carbon combined in organic substances, liquid samples, highly volatile compounds, explosive materials, biological samples, samples that evolve large amounts of acidic gases such as hydrogen chloride or chlorine, and in certain instances to carbon combined with metals where the substance may be decomposed by reaction with acids. It is of value in determinating carbonates in the presence of sulfides, sulfites, thiosulfates, and nitrites, which would vitiate results were they not oxidized to more stable forms before passing into the soda–asbestos bulb with the carbon dioxide. The technique is not applicable for determining carbon in ferrosilicon, ferrochrom, or tungsten.

The apparatus is identical with that used for determining carbon dioxide in carbonates (Figure 1), with the exception that two additional bulbs are placed after the acid bulb nearest the decomposition flask. The first of these contains glass beads moistened with chromic acid solution; the second is a drying bulb containing concentrated sulfuric acid. The absorption apparatus then follows as shown in the illustration.

Procedure

Place 0.2–1 gram of the powdered material, fine drillings, free carbon, or organic substance in the decomposition flask. If the material is likely to pack, it is advisable to mix with it pure ignited sea sand. Add 5–10 grams of granular potassium dichromate, and sweep the apparatus free of carbon dioxide by passing purified air through it before attaching the absorption apparatus. Weigh the absorption bulb, and attach it to the train. Place concentrated sulfuric acid in the acid funnel, and attach it to the decomposition flask. Allow the acid to flow down on the sample until the funnel is almost empty; then close the stopcock. Place a flame under the flask, when the vigorous action has ceased, and heat the material gently until the reaction is complete and the organic matter or carbon completely oxidized.

Sweep the apparatus free of residual carbon dioxide, by applying suction, the gas being completely absorbed by the soda–asbestos reagent. The increase in weight of the absorption bulb is due to carbon dioxide.

The following additional purifiers are frequently used as indicated: (a) an absorption bulb containing silver sulfate to absorb chlorine and vapors from sulfur compounds; (b) a capillary tube of silica or platinum heated to a dull redness to oxidize any hydrocarbons, carbon monoxide, and so forth, that may be evolved and imperfectly oxidized by the chromic acid.

C. CARBON IN METALS AND ORES

The following two methods for the determination of total carbon depends upon the oxidation of carbon to carbon dioxide when the sample material is combusted in air or oxygen atmosphere. The resulting carbon dioxide, after removal of constituents in the combustion gases that would cause interference, is measured gravimetrically by absorption in a suitable solid absorbent or conductometrically by absorption in a dilute alkali or alkaline earth hydroxide solution whose change in conductance is measured.

1.   Tube Furnace Combustion with Gravimetric Finish (14)

This technique is applicable for the determination of total carbon in iron, low-alloy steels, and other metals or alloys that will combust in an oxygen atmosphere at a temperature below 1100°C. For materials requiring a higher combustion temperature, method 2 is recommended. The gravimetric finish is particularly applicable for higher concentrations of carbon. Two typical arrangements of the apparatus are shown in Figs. 2 and 3.

Procedure

After having properly assembled and tested the apparatus, spread 1–5 grams of the sample (fine chips are preferred, but drillings or millings of 14 and 60 mesh are satisfactory) on the bed material (granular alundum, alkali-free, especially prepared for carbon determinations such as R. R. Alundum is satisfactory) in the alundum boat so that the particles are in intimate contact. An accelerator such as tin should be added to the boat for samples that do not burn readily at the furnace temperature (1000°C) that is used (i.e., alloy steels). Cover the sample with an alundum cover, and introduce the boat into the hot combustion tube. Close the tube and allow the sample to heat for 1–2 min. Then admit oxygen at a rate of

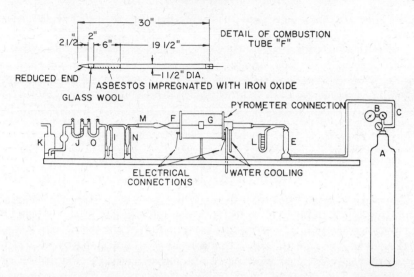

Fig. 2. Typical arrangement for determination of carbon by the direct-combustion method. A, a 100 cubic foot cylinder of oxygen; B, reducing valve; C, rubber tubing; E, tower containing soda–asbestos for removing any CO_2 in the oxygen (a layer of about $\frac{1}{2}$ in. of anhydrous $Mg(ClO_4)_2$ is placed on top for removing traces of moisture); F, combustion tube (the asbestos impregnated with iron oxide is prepared by treating some asbestos with a saturated solution of $Fe(NO_3)_3$, drying, and heating to 1000°C; the treated asbestos is placed lightly in the combustion tube and is not packed); G, electric furnace; L, manometer gage; M, glass tube lightly packed with absorbent cotton to remove solid particles; N, bottle containing 25 ml of H_2SO_4 $(1:4)$ saturated with chromic acid to remove sulfur gases from the gas stream; I, bottle contains 50 ml of H_2SO_4 (sp. gr. 1.84) for removing the bulk of the moisture that passes over from bottle N (where a large number of carbon determinations are being made the acid in this bottle should be changed daily); O, U-tube containing anhydrous $Mg(ClO_4)_2$ (this tube is filled lightly and evenly, so as not to cause packing); J, absorbing bulb containing a 20- to 30-mesh inert base impregnated with NaOH, for absorbing the CO_2. (a layer of glass wool is placed in the bottom and top of the bulb, and the soda-impregnated CO_2 absorbent is covered with a layer of anhydrous $Mg(ClO_4)_2$ approximately $\frac{1}{2}$ in. thickness); K, bottle containing H_2SO_4 (sp. gr. 1.84), may be omitted if the stopcock in J is so manipulated during combustion that no air is drawn through the exit tube.

800–1000 ml/min while the combustion is going on. When combustion is complete (1.5–2 min), reduce the flow of oxygen to 400–500 ml/min, and continue for 6–8 min in order to sweep out the carbon dioxide.

Withdraw the absorption tube filled with oxygen, place it in the balance case for 10 min, open momentarily, and weigh against a similar tube used for counterpoise. The increase in weight represents carbon dioxide.

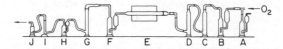

Fig. 3. Alternate arrangement for determination of carbon by the direct-combustion method. *A*, mercury valve; *B*, bottle containing concentrated H_2SO_4; *C*, tower with gooseneck top, containing $CaCl_2$; *D*, tower containing soda–asbestos; *E*, electric furnace with combustion tube and adapters; *F*, bottle containing $KMnO_4$ (50 grams/liter), inserted only for steels with over 0.05% sulfur; *G*, tower containing granulated zinc; *H*, bottle containing concentrated H_2SO_4; *I*, tower containing P_2O_5; *J*, weighed tower containing a 20- to 30-mesh inert base impregnated with NaOH, for absorbing the CO_2.

Remove the boat from the tube, and examine the fusion for evidence of incomplete combustion. If the drillings are not thoroughly fused in a solid pig, the determination should be rejected.

Make a blank determination, following the same procedure, and using the same amounts of all materials except the sample.

2. Induction-Furnace Combustion with Conductometric Finish (14)

This technique is applicable for the determination of total carbon in high-temperature metals and alloys such as zirconium, niobium, tantalum, tungsten, their alloys and alloy steels, as well as those materials mentioned in method 1. This method has found wide application because of the speed and sensitivity of the analysis. The analysis time is 5–8 min. The method is recommended for measuring up to 0.035% carbon, based on a 1 gram sample, and several times that amount if smaller samples are used. Through special modifications, the method may be applied to measuring a few micrograms of carbon.

Figure 4 shows a schematic diagram of an apparatus that is commercially available (manufactured by Laboratory Equipment Company, St. Joseph, Michigan). Oxygen is admitted to the apparatus by means of a diaphragm two-stage regulator valve, through a purifying train consisting of copper oxide at 300°C, A; magnesium perchlorate and Ascarite in trap B; sulfuric acid, magnesium perchlorate, and Ascarite, followed by a flow meter, E. A pressure gauge, C, is inserted, as well as an oxygen reservoir, D.

The furnace consists of a quartz or Vycor tube within which a ceramic crucible is supported on a ceramic pedestal. The pedestal is supported from the furnace base, which can be lowered by simply rotating the weight, G, until the handle is released. The seal of the metal base to the quartz tube is made with a rubber "O" ring.

The analyzer consists of a sensitive Wheatstone bridge having two conductivity dip cells, one as a reference cell, Q_1, and the other a measuring cell, Q_2. Both cells dip into a dilute barium hydroxide solution and are balanced in the bridge circuit. The oxygen containing the carbon dioxide bubbles through a mercury trap, K, and a diffusion frit, L, and into the barium hydroxide contained in the measuring cell. The resistance required to rebalance the bridge is used as a measure of the carbon dioxide absorbed. Since the conductivity of a solution changes with temperature, the cells and analyzer are water jacketed with thermostatic control at a temperature above ambient (40°C).

The barium hydroxide solution is prepared as follows: (1) pass nitrogen gas for 45 min through 17 liters of distilled water, contained in

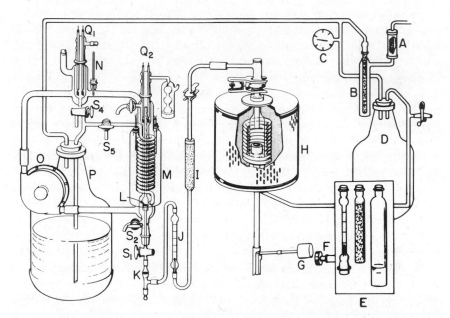

Fig. 4. Carbon apparatus for induction furnace combustion with conductometric finish. Carbon apparatus: *A*, copper oxide, 325°C; *B*, magnesium perchlorate; *C*, pressure gauge; *D*, carboy, oxygen reservoir; *E*, purification train, sulfuric acid, magnesium perchlorate, ascarite, and flow meter; *F*, pinch clamp; *G*, weight, compression of "O" ring; *H*, combustion furnace, induction heater, ceramic crucible supported on a ceramic pedestal; *I*, sulfur trap, manganese dioxide; *J*, flow meter; *K*, diffusion frit with mercury trap; *L*, diffusion frit; *M*, tempering coil for barium hydroxide, water-jacketed and thermoregulated; *N*, thermometer; *O*, water pump; *P*, barium hydroxide solution; Q_1, conductivity cell, reference; Q_2, conductivity cell, measuring: S_1, S_2, S_3, 2-way stopcocks; S_4, S_5, three-way stopcocks.

an 18 liter flask. For best results, use distilled water that has been passed through a cation and anion exchange resin (Amberlite MBI, or equivalent); (2) boil about 1 liter of distilled water for a few minutes to render it free from carbon dioxide. Add 15 grams of barium hydroxide to a portion of the water, and vacuum filter through a fine-glass frit containing some paper pulp. Add this solution to the 17 liters of water; (3) adjust the volume to 18 liters, stopper, shake, and allow to settle for about 1 week before using. When small amounts of carbon are being determined, the barium hydroxide solution may be diluted to 5 grams/18 liters.

Procedure

The procedure is as follows: (1) turn on the induction heater and the conductometric analyzer, and allow sufficient time for the water bath to reach a constant temperature of 40°C; (2) close all stopcocks and pinch clamps. Add dilute sodium hydroxide solution to exhaust trap at top of M, adjust the oxygen with the regulator to a pressure of 2–3 psi; (3) open S_5 to permit barium hydroxide to fill the glass tempering coil and the measuring cell, Q. Close S_5 and S_1, and open S_3 to drain the barium hydroxide down to constant volume. Close S_3. Open S_1 partly to adjust oxygen flow. Open S_2 to drain; (4) open S_4 to reference cell Q_1, and fill to slightly below the water-bath level. Initially both the measuring and reference cells should be flushed out several times. Leave the cell full; (5) close the pinch clamp, F, lower the furnace base, and place the crucible, with sample weighed to the nearest milligram, on the pedestal. The sample, in the form of millings, drillings, or nibbled chips should be dust free and degreased in a suitable solvent. Close the furnace, open the pinch clamp, and then open S_1 slowly until the oxygen flow indicated by the flow meter, J, reads about 300 ml/min; (6) fill the measuring cell with barium hydroxide, and adjust the volume as described in step (3). Adjust oxygen flow and flush for about 4 min; (7) balance the measuring and standard cell on the Wheatstone bridge as indicated by the coincidence of the lines on the oscilloscope; (8) turn on the induction furnace timer so that the sample is fired for 3 min and followed by a 4 min flush. Rebalance the bridge with the decade resistor, and record the resistance value; (9) open S_2 to drain, close F, lower the furnace base, and remove the crucible and burned sample.

A calibration graph is established by burning several standard samples of known carbon content and recording the resistance change of the barium hydroxide solution for each. These resistance changes are plotted against the carbon percentages on coordinate paper and should provide nearly a straight line over the range of 0–0.035% carbon. For example, with standards containing 0.35, 0.07, and 0.035% carbon, a 0.1, 0.5, and 1.0 gram sample, respectively, is used. One scoop of granulated tin (less than 2 grams) is used with each combustion of the standard and the sample.

Burn at least two different NBS samples, weighed to the nearest milligram, covering the range of the unknowns, and plot the resistance change against the carbon content.

Some metals and alloys require an accelerator in order to effect a complete combustion. In this case, the same amount of iron accelerator and tin is burned with the standards as with the unknowns (these accelerators have a low uniform carbon content, and are available from Laboratory Equipment Company, St. Joseph, Mich.). All corrections are built into the standard curve to eliminate the correction for the carbon content of the accelerator. In general, the amount of accelerator used is about 0.5 grams.

A new calibration graph is plotted daily to compensate for day-to-day variations that can shift the curve.

An alternative method for standardization is the use of potassium acid phthalate. Since the NBS steel standards have an appreciable variation in carbon content within a specified sample and are relatively high in carbon content, this especially helpful in the low-carbon levels. By making a standard solution of potassium acid phthalate and dispensing measured amounts into tin capsules (available from Laboratory Equipment Company) by means of a microsyringe, it is possible to prepare about 100 standards in 30 min. Tin capsules are placed in holes of an aluminum drying block, filled with 0.2 cc of solution of different concentrations by means of a microsyringe (0.24 cc volume, Scientific Glass Company, or equivalent), and slowly evaporated to dryness in an oven at 100°C or less. After the solution has been evaporated, the capsules are flattened with a spatula and rolled up tight into a compact form. In this way, standards can be prepared rapidly and stored for future use. It is always necessary to burn an accelerator with this standard in order to provide a sufficient mass of sample to effect adequate "coupling" to the induction heater. When such low carbon standards are run, the barium hydroxide solution used is more dilute (5–7.5 grams/18 liters).

D. CARBON IN GRAPHITE, CARBON, COAL, AND COKE (15)

This method is applicable for the determination of total carbon in graphite, carbon, coal, coke, and related materials. It is also applicable for determining total carbon in soil, rock, minerals, and similar materials. The determination of carbon (and hydrogen) is made by burning a weighed quantity of sample in a closed system and fixing the products of combustion in an absorption train after complete oxidation and purification from interfering substances. This method gives the total percentage of carbon (and hydrogen) in the material as analyzed and includes the carbon in carbonates (and the hydrogen in the moisture and in the water of hydration of silicates). It is often the normal practice to determine hydrogen in the sample at the same time by measuring the water generated during the combustion operation by making a slight change in the absorption train.

Apparatus

The apparatus is similar to that described in Section IX.C.1 with only slight modification. The oxygen purification train consists of a magnesium perchlorate trap for removing moisture followed by a soda-asbestos trap for removing carbon dioxide followed by a second magnesium perchlorate trap. The combustion unit consists of three electrically heated furnace sections, individually controlled, that are mounted on rails for easy movement. The upper part of each furnace is hinged so that it can be opened for inspection of the combustion tube. The three furnace sections are as follows:

Furnace Section 1, nearest the oxygen inlet end of the combustion tube, approximately 13 cm long is used to heat the inlet end of the combustion tube and the sample. It must be capable of rapidly attaining an operating temperature of 850–900°C.

Furnace Section 2, approximately 33 cm in length is used to heat that portion of the tube filled with cupric oxide. The operating temperature is $850 \pm 20°C$.

Furnace Section 3, approximately 23 cm long, is used to heat that portion of the tube filled with lead chromate or silver. The operating temperature is $500 \pm 50°C$.

The combustion tube made of fused quartz or high-silica glass has a nominal inside diameter of from 19 to 22 mm and a minimum total length of 97 cm. The exit end is tapered down to provide a tabulated section for connection to the absorption train. Instrumentation is commercially available that satisfies these requirements.

Arrangement of combustion tube filling is shown in Fig. 5. Since removal of oxides of sulfur and nitrogen and halides is taken care of in the latter portion of the combustion tube, no separate combustion gas-purification train is required.

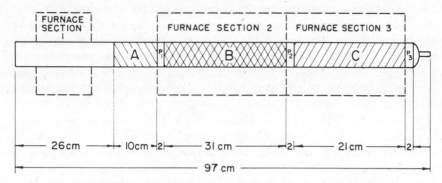

Fig. 5. Arrangement of tube fillings for combustion tube. *A*, clear fused-quartz section (optional) when a translucent quartz tube is used; *B*, cupric oxide filling; *C*, lead chromate or silver filling; P_1, P_2, or P_3, oxidizer copper gauze plugs.

The absorption train consists of a tube containing a solid desiccant for absorbing water followed by a tube containing a solid reagent for retaining carbon dioxide. A guard tube packed with equal volumes of the water absorbent and a solid carbon dioxide absorbent is placed at the end of the absorption train to prevent contamination of the absorption tubes from the atmosphere.

Reagents

   Oxygen, 99.5% purity or better.
   Cupric oxide, (CuO) wire form, dust free.
   Lead chromate, ($PbCrO_4$) approximately 8–20 mesh size.
   Silver gauze, 99.9% silver minimum purity, 20 mesh, made from approximately No. 27 B and S gage wire.
   Copper gauze, 99.0% copper minimum purity, 20 mesh, made from approximately No. 26 B and S gage wire.
   Water absorbent, anhydrous magnesium perchlorate [$Mg(ClO_4)_2$] of approximately 8–45 mesh size, (Anhydrone and Dehydrite).
   Carbon dioxide absorbent, reagents such as "Ascarite," "Caroxite" and "Mikohlite", sodium or potassium hydroxide on an inert base are satisfactory.

Procedure

   After the apparatus has been set up and conditioned, bring the absorption tubes to room temperature near the balance for 15–20 min, vent momentarily to the atmosphere, wipe with a chamois of lint-free cloth in the areas where handled, and weigh to the nearest 0.1 mg. Weigh approximately 0.2 grams (weighed to the nearest 0.1 mg) of air-dry sample ground to pass a No. 60 (250 μm) sieve into a combustion boat. With furnace sections 2 and 3 at specified temperatures and positioned as shown in Fig. 5, place furnace section 1 so that its left-hand edge is about 10 cm from the oxygen inlet end of the combustion tube.

Attach the weighed absorption train to the tube, and push the sample boat into the tube to a point within approximately 2 cm from plug P_1. Close the tube, and adjust the oxygen flow to a rate of 50–100 ml/min (standard temperature and pressure) being the same as used in blanking. Apply full heat to heating section 1 to bring it to an operating temperature of 850–900°C as rapidly as possible.

Move the heater slowly toward the boat so that it completely covers the boat and is brought into contact with heating section 2 in a period of 10–20 min. Allow it to remain in this position for an additional 5–10 min, and then shut off the heat and return the sample heater to its original position. Continue the flow of oxygen through the tube for 10 min, close the absorbers under a positive pressure of oxygen, and detach them from the train. Remove the absorbers to the vicinity of the balance, allow them to cool to room temperature for 15–20 min, vent momentarily to atmosphere, wipe them with a chamois or lint-free cloth in the areas handled, and finally weigh them to the nearest 0.1 mg. While the absorbers are cooling, it is recommended that the ash remaining in the combustion boat be examined for traces of unburned carbon, which, if present, will nullify the determination. The increase in weight in the carbon dioxide absorption tube is due to the carbon of the sample.

REFERENCES

1. Abbott, H. W., "Carbon (Baked and Graphitized)," in R. E. Kirk, and D. F. Othmer, *Encyclopedia of Chemical Technology,* Vol. III, Wiley-Interscience, New York, 1949, pp. 23–25.

2. Agassant, P., and J. L. Andrieux, *Bull. Soc. Chim. Fr.* **1950,** 253 (1950); through *Chem. Abstr.,* **44,** 7712 (1950).

3. Albert, P., *Chim. Ind. (Paris)* **75,** 275 (1956); through *Chem. Abstr.,* **50,** 9258 (1956).

4. Albert, P., C. Engelmann, S. May, and J. Petit, *C. R. H. Acad. Sci.* **254,** 119 (1962); through *Chem. Abstr.,* **56,** 10896 (1962).

5. Albert, P., Proceedings, International Conference on Modern Trends in Activation Analysis, College Station, Tex., 1961, p. 78; through *Chem. Abstr.,* **57,** 9198 (1963).

6. Aldridge, W. N., *Analyst (London),* **70,** 474 (1945).

7. Aldridge, W. N., *Analyst (London),* **69,** 262 (1944).

8. *Amer. Soc. Testing Mater. ASTM Methods for Chemical Analysis of Metals*, 1974, p. 577.
9. *Amer. Soc. Testing Mater. Proc.*, **65**, 311 (1965).
10. *Amer. Soc. Testing Mater. Proc.*, **53**, 387 (1953).
11. *Amer. Soc. Testing Mater. ASTM Stand.*, 1974.
12. *Amer. Soc. Testing Mater. ASTM Stand.*, Part 12, 1974.
13. *Amer. Soc. Testing Mater. ASTM Stand.*, Part 17, 1966.
14. *Amer. Soc. Testing Mater. ASTM Stand.*, Part 12, 1974.
15. *Amer. Soc. Testing Mater. ASTM Stand.*, Part 19, 1974.
16. Archer, E. E., *Analyst*, **79**, 30 (1954).
17. Armstrong, W. D., L. Singer, S. H. Zbarsky, and B. Dunshee, *Science*, **112**, 531 (1950).
18. Arnold, J. O., and F. Ibbatson, *Steel Works Analysis*, 4th edit., Pitman, London, 1919, p. 419.
19. Asher, R. C., *U. S. Atom. Energy Comm. Report TID-7597 (Book 2)*, 1961, pp. 504–522; *Nucl. Sci. Abstr.*, **15**, 19223 (1961).
20. Audric, B. N., and J. V. P. Long, *Nature*, **173**, 992 (1954).
21. Baba, S., T. Konishi, and H. Ido, *Yakugaku Zasshi*, **93**, 532 (1973); through *Chem. Abstr.*, **79**, 29328 (1973).
22. Bagshawe, B., and P. H. Pinder, *Analyst (London)*, **81**, 153 (1956).
23. Bahensky, V., and Z. Zika, *Galvanotechnik*, **53**, 122–123 (1962); through *Anal. Abstr.*, **10**, 2050 (1963).
24. Baird, A. K., *Advan. X-Ray Anal.*, **13**, 26 (1969).
25. Baker, C. A., and D. R. Williams, *Talanta*, **15**, 1143 (1968).
26. Bark, L. S., and H. G. Higson, *Analyst (London)*, **88**, 751 (1963).
27. Bark, L. S., and H. G. Higson, *Talanta*, **11**, 471 (1964).
28. Baron, J., *Chim. Anal.*, **33**, (1951).
29. Beamish, F. E., and W. A. E., McBryde, "Reagents," in C. L. Wilson and D. W. Wilson, *Comprehensive Analytical Chemistry*, Elsevier, New York, 1959, p. 502.
30. Beard, D. B., R. G. Johnson, and W. G. Bradshaw, *Nucleonics*, **17**, 90 (1959).
31. Belcher, R., and A. J. Nutten, *Quantitative Inorganic Analysis*, Butterworths, London, 1955, p. 163.
32. Beneker, J. C., *Chem. Anal.* **22**, 3 (1917).
33. Bennett, E. L., J. H. Harley, and R. M. Fowler, *Anal. Chem.*, **22**, 445 (1950).
34. Blair, A. A., *The Chemical Analysis of Iron*, 8th edit., Lippincott, Philadelphia, 1918, p. 318.
35. Blom, L., and L. Edelhausen, *Anal. Chim. Acta*, **13**, 120 (1955).
36. Blyholder, G. D., and H. Eyring, *J. Phys. Chem.*, **61**, 682 (1957).
37. Bogdanchenko, A. G., *Zavod. Lab.*, **29**, 163 (1963).
38. Boivin, A., *Bull. Soc. Chim. Biol.*, **11**, 1269 (1929); through *Chem. Abstr.*, **24**, 3808 (1930).
39. Bolm, L., and M. H. Kraus, *Z. Anal. Chem.*, **205**, 50 (1964); through *Chem. Abstr.*, **61**, 15336 (1964).
40. Boos, R. M., S. L. Jones, and N. R. Trenner, *Anal. Chem.*, **28**, 390 (1956).

41. Brogström, L. H., *Z. Anal. Chem.*, **53**, 685 (1914).

42. Borovskii, I. B., Proceedings of the 7th National Conference on Electron Probe Analysis, San Francisco, California, July 17–21, 1972.

43. Bossart, R. E., and R. K. Young, *Anal. Chem.*, **44**, 1117 (1972).

44. Bost, W. E., *Chemistry of Boron and Boron Compounds, A Literature Search, U. S. At. Energy Comm., Report TID-3523*, January, 1959; through *Nucl. Sci. Abstr.*, **13**, 4448 (1959).

45. Bowen, H. J. M., and D. Gibbons, *Radioactivation Analysis*, Oxford U. P., London, 1963, p. 38.

46. Bradshaw, W. R., R. Johnson, and D. Beard, *U. S. Dept. Comm., Office Tech. Serv.*, PB 159, 872 (1960); through *Chem. Abstr.*, **60**, 8631 (1964).

47. Braid, R., J. A. Hunter, W. H. S. Massie, J. D. Nicholson, and B. E. Pearce, *Analyst (London)*, **91**, 439 (1966).

48. Brewer, L., *UCRL-653 April 3, 1950 (Revision of CT-1010, Sept. 14)* (1943); through *Nucl. Sci. Abstr.*, **5**, 158 (1951).

49. Bright, H. A., and G. E. F. Lundell, U. S. National Bureau of Standards Journal of Research, **5**, 943 (1930).

50. Methods of Analysis Subcommittee *Brit. Cast Iron Res. Ass. J. Res. Develop.*, **4**, 520 (1953).

51. R. L. Brown *Brit. Coal Util. Res. Ass. Monthly Bull.*, **9**, 97 (1945).

52. Broda, E., and B. Kalab, *Mikrochim. Acta*, **1962**, 128 (1962).

53. Buechel, E., *Chem. Ind. Jahrb., 1962/63*, 21 (1963); through *Chem. Abstr.*, **60**, 15109 (1964).

54. Burtec Instrument Corp., *The Thermocarb Carbon Analyzer for Precise Carbon Analysis on the Furnace Floor*, Pittsburgh, Penn., 1962.

55. Cain, J. R., and L. C. Maxwell, *Ind. Eng. Chem.*, **11**, 852 (1919).

56. Carbon Products Division, Union Carbide Corporation, unpublished data.

57. Carbon Products Division, Union Carbide Corporation, *The Industrial Graphite Engineering Handbook*, New York, 1965, p. 5D.03.02.

58. Carpenter, F. G., *Anal. Chem.*, **34**, 66 (1962).

59. Cassal, C. E., and H. Gerrans, *Chem. News*, **87**, 27 (1903).

60. Castaing, R., and G. Slodzian, *J. Microscopie*, **1**, 395 (1962); through *Chem. Abstr.*, **59**, 12152 (1963).

61. Catch, J. R., *Carbon-14 Compounds*, Butterworths, Belfast, 1961.

62. Catch, J. R., *Carbon-14 Compounds*, Butterworths, Belfast, 1961, p. 91.

63. Chub, G. F., J. N. Treiger, A. P. Savraskaya, and G. A. Balrich, *Zavodsk. Lab.*, **22**, 391 (1956); through *Chem. Abstr.*, **51**, 2469 (1957).

64. Claeys, R. R., and H. Freund, *Environ. Sci. Tech.*, **2**, 458 (1968).

65. Clark, F. E., "Graphite," in C. J. Rodden, *Analysis of Essential Nuclear Reactor Materials, U. S. Atom. Energy Comm.*, Washington, 1964, pp. 541–576.

66. Codell, M., G. Norwitz, and O. W. Simmons, *Anal. Chim. Acta*, **9**, 555 (1953).

67. Convey, J., and J. H. Oldfield, *J. Iron Steel Inst.*, *18*, 580 (1945).

68. Cook, R. M., and G. E. Speight, *Analyst (London)*, **81**, 144 (1956).

69. Cooper, D. E., *Nucleonics*, **15**, 136 (1957).

70. Castaing, R., these de doctorat, Universite de Paris, 1951; Onera Publication Number 55.

71. Cotterill, J. C., United Kingdom A.E.R.E.-AM46, June, 1959; through *Nucl. Sci. Abstr.*, **14**, 80 (1960).

72. Cotton, J. B., *Analyst (London)*, **70**, 446 (1945).

73. Coursier, J., J. Hure, and R. Platzer, *Proc. Internat. Conf. Peaceful Uses Atomic Energy*, **8**, 487 (1956).

74. Croall, G., J. R. Brown, and W. Ramsden, Eighth Colloq. Spectrosc. Internat., Lucerne, Switz., 1959, p. 238 (1960).

75. Dagnall, R. M., M. T. El-Ghamry, and T. S. West, *Talanta*, **15**, 107 (1968).

76. Dailly, D. F., and T. A. Elliott, *J. Chem. Soc.*, **1956**, 3398 (1956).

77. Davidson, E., *Chromatographia*, **3**, 43 (1970).

78. Delibrias, G., *J. Inorg. Nuclear Chem.*, **1**, 238 (1955).

79. Dem'Yanchuk, A. S., *Inzh.-Fiz. Zh.*, **1**, 116 (1958).

80. Dennen, W. H., *Spectrochim. Acta*, **9**, 89 (1957).

81. Dieterle, H., *Arch. Pharm.*, **262**, 35 (1924); through *Chem. Abstr.*, **18**, 2858 (1924).

82. Dijkstra, L. J., *Trans. AIME*, **185**, 252 (1949).

83. Divekar, K. G., K. A. Khasgiwale, T. P. Ramachandram, and M. Sundaresan, *J. Sci. Ind. Res.*, **21B**, 418 (1962); through *Chem. Abstr.*, **58**, 12 (1963).

84. Dixon, B. E., *Analyst (London)*, **59**, 739–743 (1934).

85. Dixon, B. E., G. C. Hands, and A. F. F., Bartlett, *Analyst (London)*, **83**, 199 (1958).

86. Dognall, R. M., D. J. Johnson, and T. S. West, *Spectrosc. Lett.*, **6**, 87 (1973).

87. Dolby, R. M., *J. Sci. Instr.*, **40**, 345 (1963).

88. D'Silva, A. P., T. R. Saranathan, and L. C. Chandola, A.E.E.T./Spec./4; A.E.E.T./ ANA418, Atomic Energy Establishment Trombay, Bombay, India, 1964.

89. Ducret, L., and C. Cornet, *Anal. Chim. Acta*, **25**, 542 (1961).

90. Duerr, J. S., and R. E. Ogilvie, *Anal. Chem.*, **44**, 2361 (1972).

91. Duswalt, A. A., and W. W. Brandt, *Anal. Chem.*, **32**, 272 (1960).

92. Dwiggins, Jr., C. W., *Anal. Chem.*, **33**, 67 (1961).

93. Dzhemard'yan, Y. A., G. I. Mikhailov, and L. P. Starchik, *Zav. Lab.*, **37**, 555 (1971); through *Anal. Abstr.*, **22**, 1426 (1972).

94. Earnshaw, M. R., Ind. Chem. MFR., *33*, 253 (1957).

95. Eggleston, R. R., R. L. Carter, W. J. Greening, and R. E. Durand, U. S. Atomic Energy Comm., Report NAA-SR-Memo-1240, 1955.

96. Ehrenberger, F., *Z. Anal. Chem.*, **267**, 17 (1973).

97. Ehrenkranz, F., and E. H. Jebe, *Nucleonics*, **14**, 96 (1956).

98. Eidinoff, M. L., *Anal. Chem.*, **22**, 529 (1950).

99. Eisenbud, M., *Environmental Radioactivity*, McGraw-Hill, New York, 1963, p. 162.

100. Emeleus, H. J., and J. S. Anderson, *Modern Aspects of Inorganic Chemistry*, 2nd edit., Van Nostrand, Princeton, N.J., 1952, pp. 408–440.

101. Eng, K. Y., R. A. Meyer, and C. D. Bingham, *Anal. Chem.*, **36**, 1832 (1964).

102. Engelmann, C., and A. Marschal, *Radiochem. and Radioanal. Lett.*, **6**, 189 (1972).

103. Engle, E. W., *Trans. Amer. Inst. Min. Met. Eng.*, *71*, 691 (1925); through *Chem. Abstr.*, **20**, 1211 (1926).

104. Entenman, C., S. R. Lerner, I. L. Chaikoff, and W. G. Dauben, *Proc. Soc. Exp. Biol. Med.*, **70**, 364 (1949); through *Chem. Abstr.*, **43**, 4719 (1949).

105. Epstein, J., *Anal. Chem.*, **19**, 272 (1947).

106. Evseeva, T. I., E. P. Cherstvenkova, V. A. Nikol'skii, and V. I. Denisova, *Zavod. Lab.*, **39**, 397 (1973).

107. Fanter, D. L., and C. J. Wolf, *Anal. Chem.*, **45**, 565 (1973).

108. Faraday, B. J., and C. R. Bingham, Rep. NRL (Naval Res. Lab. U.S.) Prog., PB181071, p. 5, 1962.

109. Feick, G., *Anal. Chem.*, **36**, 2198 (1964).

110. Feigel, F., Translated by Oesper, R. E., *Spot Tests in Inorganic Analysis*, Elsevier, New York, 1958, p. 380.

111. Feigel, F., translated by Oesper, R. E., *Spot Tests in Inorganic Analysis*, Elsevier, New York, 1958, p. 342.

112. Feigel, F., translated by Oesper, R. E., *Spot Tests in Inorganic Analysis*, Elsevier, New York, 1958, p. 381.

113. Feigel, F., translated by Oesper, R. E., *Spot Tests in Inorganic Analysis*, Elsevier, New York, 1958, p. 422.

114. Feigel, F., translated by Oesper, R. E., *Spot Tests in Inorganic Analysis*, Elsevier, New York, 1958, p. 423.

115. Feldman, C., and J. Y. Ellenburg, *Anal. Chem.*, **27**, 1714 (1955).

116. Fischer, J., and W. Schmidt, *Z. Erzbergbau Metallhuttenw.*, **9**, 25 (1956); through *Chem. Abstr.*, **50**, 6252 (1956).

117. Fischer, W., and H. Bastius, *Metall*, **14**, 429 (1960); through *Chem. Abstr.*, **54**, 16262 (1960).

118. Fisher, F. B., and J. S. Brown, *Anal. Chem.*, **24**, 1440 (1952).

119. Fisher, G. L., and G. D. Farningham, *Anal. Chem.*, **45**, 2295 (1973).

120. Flament, P., and J. Marot, *Rev. Met.*, **51**, 702 (1954).

121. Florentin, D., *Bull. Soc. Chem.*, **35**, 228 (1924); through *Chem. Abstr.*, **18**, 1469 (1924).

122. Florentin, D., *Chim. Ind. (Paris)*, **11**, 875 (1924); through *Chem. Abstr.*, **18**, 2573 (1924).

123. Foa, M., *Ind. Chem. Min. Met.*, **5**, 2 (1918); through *Chem. Abstr.*, **12**, 889 (1918).

124. Fornwalt, D. E., and A. V. Manzione, *Norelco Reporter*, **13**, 39 (1966).

125. Fowler, R. M., W. G. Guldner, T. C. Bryson, J. L. Hague, and H. J. Schmitt, *Anal. Chem.*, **22**, 486 (1950).

126. Frad, W. A., and P. G. Harold, *Nature*, **180**, 1273 (1957).

127. Frant, M. S., *Plating*, **58**, 686 (1971).

128. Friedrich, K., and H. Engelhardt, *Z. Anal. Chem.*, **249**, 244 (1970).

129. Friedrich, K., and H. Engelhardt, *Z. Anal. Chem.*, **249**, 244 (1970).

130. Furman, N. H., *Standard Methods of Chemical Analysis*, Vol. I, 6th edit., Van Nostrand, Princeton, N. J., 1962, pp. 298–307.

131. Furman, N. H., *Standard Methods of Chemical Analysis*, Vol. I, 6th edit, Van Nostrand, Princeton, N. J., 1962 pp. 760–762.

132. Gale, P., Colloq. Spectroc. Internat., 8th, Lucerne, Switz., 1959, Verlag H. R. Sauereänder, Frankfurt-Am-Main p. 233 (1960).

133. Galwey, A. K., *Talanta*, **10**, 310 (1963); through *Chem. Abstr.*, **58**, 10714 (1963).

134. Gardner, K., W. J. Rowland, and H. Thomas, *Analyst (London)*, **75**, 173 (1950).

135. Garton, F. W. J., *Spectrochim. Acta*, **3**, 68 (1947).

136. Garton, F. W. J., *Spectrochim. Acta*, **9**, 297 (1957).

137. Gillingham, J. T., M. M. Shirer, and N. R. Page, *Agron. J.*, **61**, 717 (1969).

138. Gladshtein, B. M., *Zh. Anal. Kim.*, **11**, 114 (1956); through *Chem. Abstr.*, **50**, 9940 (1956).

139. Goleb, J. A., J. P. Faris, and B. H. Meng, *Appl. Spectrosc.*, **16**, 9 (1962).

140. Goodwin, R. D., *Anal. Chem.*, **25**, 263 (1953).

141. Gora, E. K., and F. C. Hickey, *Anal. Chem.*, **26**, 1158 (1954); through *Chem. Abstr.*, **49**, 1439 (1955).

142. Gordon, W. A., J. W. Graab, and Z. T. Tumney, *Anal. Chem.*, **36**, 1396 (1964).

143. Gorenko, A. F., N. A. Skakun, A. S. Zadovornyi, N. I. Bugaeva, G. M. Shevchenko, and A. P. Klyucharev, *Ah. Anal. Khim.*, **28**, 1227 (1973); through *Anal. Abstr.*, **26**, 2655 (1974).

144. Goto, H., K. Hirokawa, and K. Takada, *Jap. Anal.*, **19**, 1021 (1970); through *Anal. Abstr.*, **22**, 789 (1972).

145. Grant, J. A., J. A. Hunter, and W. M. S. Massie, *Analyst (London)*, **88**, 134 (1963).

146. Green, I. R., J. E. Still, and R. C. Chirnside, *Analyst (London)*, **87**, 530 (1962).

147. Green, R. V., "Carbon Monoxide," in Kirk-Othmer, *Encyclopedia of Chemical Technology*, Wiley-Interscience, Vol. IV, New York, 1964, p. 430.

148. Gross, C., G. Gaetano, T. Tucker, and V. Baker, *J. Electrochem. Soc.*, **119**, 926 (1972).

149. Green, R. V., "Carbon Monoxide," in Kirk-Othmer, *Encyclopedia of Chemical Technology*, Wiley-Interscience, Vol. IV, New York, 1964, pp. 441–442.

150. Groves, A. W., *Silicate Analysis*, 2nd edit., George Allen and Unwin, London, 1951, pp. 109–114.

151. Gurry, R. W., and H. Trigg, *Ind. Eng. Chem. Anal. Ed.*, **16**, 248 (1944).

152. Guseinov, A. G., N. M. Nikolaeva, and S. I. Mekhtiev, *Azerb. Neft. Khoz.*, **2**, 36 (1971); through *Anal. Abstr.*, **23**, 1525 (1972).

153. Hanle, W., K. Hengst, and H. Schneider, *Z. Naturforsch., B*, **7**, 633 (1952); through *Chem. Abstr.*, **47**, 5265 (1952).

154. Hanson N. W., D. A. Reilly, and H. E. Stagg, *The Determination of Toxic Substances in Air*, W. Heffer and Sons, Cambridge, 1966, pp. 139–146.

155. Harrington, W. L., R. K. Skogerboe, and G. H. Morrison, *Anal. Chem.*, **38**, 821 (1966).

156. Harrison, T. S., *J. Soc. Chem. Ind.*, **68**, 49 (1949).

157. Harrison, T. S., and S. A. Marshall, *J. Iron. Steel Inst.*, **207**, 323 (1969).

158. Harvey, C. E., and J. W. Mellichamp, *Anal. Chem.*, **33**, 1242 (1961).

159. Hayes, M. R., and J. Metcalfe, *Analyst (London)*, **87**, 956 (1962).

160. Hayman, D. F., *Ind. Eng. Chem. Anal. Ed.*, **8**, 342 (1936).

161. Heck, A. F., *Soil Science*, **28**, 225 (1929); through *Chem. Abstr.*, **24**, 676 (1930).

162. Heczko, T., *Arch. Eisenhuttenw.*, **25**, 413 (1954); through *Chem. Abstr.*, **49**, 778 (1955).

163.   Heiskanen, S., *Jernkontorets Ann.*, **139,** 78 (1955); through *Chem. Abstr.*, **49,** 6776 (1955).

164.   Henke, B. L., *Advan. X-Ray Anal.*, **8,** 269 (1969).

165.   Henke, B. L., and R. R. L. Elgin, *Advan. X-Ray Anal.*, **13,** 634 (1969).

166.   Henke, B. L., "X-Ray Fluorescence Analysis of Sodium, Fluorine, Oxygen, Nitrogen, Carbon and Boron," in W. M. Mueller, G. Mallett, and M. Fay, *Advances in X-Ray Analysis*, Vol. VII, Pelnum, New York, 1964, p. 460.

167.   Henry, J. J., U. S. Atom. Energy Comm., Y-1356, p. 14 (1961); through *Chem. Abstr.*, **56,** 9419 (1962).

168.   Heron, A. E., and H. N. Wilson, "Gas Analysis," in C. L. Wilson and D. W. Wilson, *Comprehensive Analytical Chemistry*, Elsevier, New York, 1959, p. 258.

169.   Heron, A. E., and H. N. Wilson, "Gas Analysis," in C. L. Wilson and D. W. Wilson, *Comprehensive Analytical Chemistry*, Elsevier, New York, 1959, pp. 240, 268, 270, 294, 298, and 306.

170.   Heron, A. E., and H. N. Wilson, "Gas Analysis," in C. L. Wilson and D. W. Wilson, *Comprehensive Analytical Chemistry*, Elsevier, New York, 1959, pp. 243 and 244.

171.   Hetman, J., *J. Appl. Chem.*, **10,** 16 (1960).

172.   Hickam, W. M., *Anal. Chem.*, **24,** 362 (1952).

173.   Hillebrand, W. F., *U. S. Geol. Surv. Bull.*, **700,** 219 (1919).

174.   Hillebrand, W. F., G. E. F. Lundell, H. A. Bright, and J. I. Hoffman, *Applied Inorganic Analysis*, 2nd edit., Wiley, New York, 1953, pp. 768, 769, 770, 934, and 935.

175.   Hilton, H. W., N. S. Nomura, and S. S. Kameda, *Anal. Biochem.*, **49,** 285 (1972).

176.   Hoagland, C. L., *J. Biol. Chem.*, **136,** 543 (1940).

177.   Hobson, J. D., and H. Leigh, *Analyst (London)*, **99,** 93 (1974).

178.   Hoffman, E., *Z. Anal. Chem.*, **169,** 258 (1959).

179.   Hogness, J. R., L. J. Roth, E. Leifer, and W. H. Langham, *J. Am. Chem. Soc.*, **70,** 3840 (1948); through *Chem. Abstr.*, **43,** 1454 (1949).

180.   Holler, A. C., R. Klinkenberg, C. Friedman, and W. K. Aites, *Anal. Chem.*, **26,** 1658 (1954).

181.   Horton, A. D., W. D. Shults, and A. S. Meyer, *Anal. Lett.*, **4,** 613 (1971).

182.   Horwitz, W., *Methods of Analysis*, 10th edit., Association of Official Agricultural Chemists, Washington, D.C., 1965, p. 25.

183.   Huber, F. E., and D. L. Chase, *Chem. Anal.*, **53,** 14 (1964).

184.   Ingram, G., *Analyst (London)*, **86,** 539 (1961).

185.   Isbell, R. E., *Anal. Chem.*, **35,** 255 (1963).

186.   Ital. Pat. 460,119 (October 19, 1950), R. Cordans, and N. Fantini,; through *Chem. Abstr.*, **46,** 4958 (1952).

187.   Jamieson, A., *Foundry*, **83,** 132 (1955).

188.   Janney, C. D., and B. J. Moyer, *Rev. Sci. Instrum.*, **19,** 667 (1948).

189.   Jeffrey, P. G., and P. J. Kipping, *Analyst (London)*, **87,** 379 (1962).

190.   Jeffrey, P. G., and A. D. Wilson, *Analyst (London)*, **85,** 749 (1960).

191.   Jerry, L., and G. Fischer, *Metalloberflaeche-Angew. Electrochem.*, **26,** 391 (1972).

192. Johansson, A., *Anal. Chem.*, **26**, 1, 183 (1954).

193. Jones, R. F., P. Gale, P. Hopkins, and L. N. Powell, *Analyst (London)*, **91**, 623 (1965).

194. Jones, R. F., P. Gale, P. Hopkins, and L. N. Powell, *Analyst (London)*, **91**, 399 (1966).

195. Josefsson, A., and E. Kula, AIME Metals Trans., **194**, 161 (1952).

196. Jurecek, M., *Mikrochim. Acta,* **1955**, 1088 (1955).

197. Jurecek, M., and M. Nepras, *Mikrochim. Acta,* **1956**, 1762 (1956).

198. Kalina, M. H., and T. L. Joseph, *Blast Furn. Steel Plant,* **27**, 347 (1939).

199. Kamada, H., S. Toda, and T. Nishiya, *Jap. Anal.,* **6**, 146 (1957).

200. Karpathy, O. C., Pittsburgh Conference on Analytical Chemistry and Applied Spectroscopy, March, 1963.

201. Kawahata, M., H. Mochizuki, R. Kajiyama, M. Watanabe, M. Ishii, and K. Kasaki, *Bunseki Kagaku,* **11**, 192 (1962); through *Chem. Abstr.,* **57**, 24 (1962).

202. Kennedy, A. J., J. C. Pacer, A. Sprinzak, J. Wiley, and N. T. Porile, *Nucl. Instrum. Method.,* **101**, 471 (1972).

203. Kerr, W. R., U. S. At. Energy Comm., NAA-SR-Memo-8943 (1963).

204. Khainadzhiev, A., S. Sherilov, and G. Ivanov, *Tr. Nauch. Issled. Inst. Cherna Met.,* **4**, 173 (1970); through *Chem. Abstr.,* **79**, 14279 (1973).

205. Kieffer, R., "Carbides (Industrial Heavy-Metal)," in R. E. Kirk and D. L. Othmer, *Encyclopedia of Chemical Technology,* Vol. IV, Wiley-Interscience, New York, 1964, p. 90.

206. Kigoshi, K., *Jap. Anal.,* **3**, 296 (1954); through *Chem. Abstr.,* **49**, 14496 (1955).

207. Kincaid, J. F., E. L. Stanley, C. H. Beckworth, and F. W. Sunderman, *Am. J. Clin. Pathol.,* **26**, 107 (1956).

208. King, N. J., *Chem. Eng. Mining Rev.,* **24**, 429 (1932); through *Chem. Abstr.,* **27**, 40 (1946).

209. Ko, R., U. S. Atom. Energy Comm., Report HW-66219, July, 1960; through *Nucl. Sci. Abstr.,* **15**, 12836 (1961).

210. Koch, W., and H. Lemm, *Arch. Eisenhuettenw.,* **41**, 427 (1970); through *Anal. Abstr.,* **20**, 3046 (1971).

211. Kohn, A., and J. Doumerc, *J. Phys. Radium,* **16**, 649 (1955).

212. Koide, T., T. Kubota, N. Komuro, and T. Kuroi, *J. Soc. Rubber Ind. (Japan),* **22**, 272 (1949); through *Chem. Abstr.,* **46**, 1792 (1952).

213. Kolthoff, I. M., and E. B. Sandell, *Textbook of Quantitative Inorganic Analysis,* 3rd edit., MacMillan, 1952, p. 458.

214. Korzh, P. D., *Zavod. Lab.,* **27**, 996 (1961).

215. Korzh, P. D., and A. P. Ershova, *Zavod. Lab.,* **24**, 41 (1958).

216. Kovtun, M. S., *Zavod. Lab.,* **14**, 487 (1948); through *Chem. Abstr.,* **43**, 1285 (1949).

217. Kraus, R., *Metall. Giessereitech.,* **3**, 333 (1953); through *Chem. Abstr.,* **48**, 11975 (1954).

218. Krumin, P. O., and D. Svanko, *ASTM Bull.,* **227**, 51 (1958).

219. Kruse, J. M., and M. G. Mellon, *Sewage Ind. Wastes,* **23**, 1402 (1951).

220. Kuck, J. A., J. W. Berry, A. J. Andreatch, and P. A. Lentz, *Anal. Chem.,* **34**, 403 (1962).

221. Kuck, J. A., J. W. Berry, A. J. Andreatch, and P. A. Lentz, *Microchem. J., Symp. Ser.*, **2,** 417 (1962).

222. Kudelya, E. S., *Zavod. Lab.*, **24,** 458 (1958).

223. Kuo, C., G. T. Bender, and J. M. Walker, *Anal. Chem.*, **35,** 1505–9 (1963).

224. Kuz'kin, G. M., and Y. A. Mel'nikov, *Zavod. Lab.*, **39,** 168 (1973); through *Anal. Abstr.*, **29,** 1575 (1973).

225. La Barre, de, F., *Genie Civ.*, **141,** 117 (1964); through *Chem. Abstr.*, **61,** 4934 (1964).

226. LECO Corp., St. Joseph, Michigan, *Carbon Analysis in 50 Seconds,* 1963.

227. Larsen, C. M., and D. N. Glass, NAA-SR-Memo 11495 1965.

228. Lewis, L. L., and M. J. Nardozzi, *Anal. Chem.*, **36,** 1329 (1964).

229. Liggett, L. M., "Carbon (Baked and Graphitized, Manufacture)," in R. E. Kirk and D. F. Othmer, *Encyclopedia of Chemical Technology,* Vol. IV, 2nd editor, Wiley-Interscience, New York, 1964, pp. 189–193.

230. Linder, J., *Mikrochem.*, **20,** 219 (1936); through *Chem. Abstr.*, **30,** 7489 (1936).

231. Long, C., *Chim Ind.*, **45,** 190 (1941); through *Chem. Abstr.*, **38,** 32149 (1944).

232. Louth, G. D., *Anal. Chem.*, **20,** 717 (1948).

233. Loveland, J. W., R. W. Adams, H. H. King, Jr., F. A. Nowak, and L. J. Cali, *Anal. Chem.*, **31,** 1008 (1959).

234. Lundegardh, H., *Metallwirtschaft,* **17,** 1222 (1938); through *Chem. Abstr.*, **33,** 937 (1939).

235. Lutz, G. J., and L. W. Masters, *Anal. Chem.*, **42,** 948 (1970).

236. Lutz, G. J., and D. A. DeSoete, *Anal. Chem.*, **40,** 902 (1968).

237. MacPherson, H. G., First Conference—Analytical Chemistry in Nuclear Reactor Technology, November 4–6, 1957, Gatlinburg, Tennessee, (TID-7555), p. 184; through *Nucl. Sci. Abstr.*, **12,** 16236 (1958).

238. Maillard, P., and C. Ades, *Can. Spectrosc.*, **14,** 17 (1969); through *Anal. Abstr.*, **18,** 3047 (1970).

239. Malissa, H., M. Grasserbauer, and E. Waldmann, *Microchim. Acta,* **3,** 455 (1922).

240. Mallett, M. W., *Talanta,* **9,** 133 (1962).

241. Malmberg, C. J. G., *Jernkontorets Ann.*, **114,** 508 (1930); through *Chem. Abstr.*, **25,** 671 (1931).

242. Marchio, J. L., *J. Chromatogr. Sci.*, **9,** 432 (1971).

243. Marinenko, J., and I. May, Prof. Pap. U. S. Geol. Surv., No. 700-D, D103-D105, 1970; through *Anal. Abstr.*, **22,** 95 (1972).

244. Marion, F., and R. Faivre, *C. R. Acad. Sci.*, **247,** 206 (1958); through *Chem. Abstr.*, **53,** 2926 (1959).

245. Marion, F., and R. Faivre, *Chim. Anal.*, **42,** 355 (1960); through *Chem. Abstr.*, **55,** 1277 (1960).

246. Marion, F., *Bull. Soc. Chem. France,* **1181,** (1958); through *Chem. Abstr.*, **55,** 14162 (1961).

247. Marnoch, K., U. S. Atom. Energy Comm., NAA-SR-Memo-6819 (1961).

248. Martin, J., and E. Haas, *Z. Anal. Chem.*, **259,** 97 (1972).

249. Mascini, M., *Anal. Chem.*, **45,** 614 (1973).

250. May, S., and G. Pinte, *J. Radioanal. Chem.*, **3,** 329 (1969).

251. Mazumder, K. C., and M. K. Ghosh, *Indian J. Phys.*, **23,** 477 (1949).

252. McAllister, W. A., and N. V. Southerland, *Anal. Chem.*, **43**, 1936 (1971).

253. McCoy, R. N., and E. L. Bastin, *Anal. Chem.*, **28**, 1776 (1956).

254. Mee, A. J., *J. Sci. Instrum.*, **20**, 137 (1943).

255. Mellichamp, J. W., "Nonmetallic Elements," in G. L. Clark, *Encyclopedia of Spectroscopy*, Reinhold, New York, 1960, p. 230.

256. Merritt, J., C. E. Muller, W. M. Sawyer, Jr., and A. Telfer, *Anal. Chem.*, **35**, 2209 (1963).

257. Methods of Analysis Committee, *J. Iron Steel Instrum.*, **183**, 287 (1956).

258. Metters, B., B. G. Cooksey, and J. M. Ottaway, *Talanta*, **19**, 1605 (1972).

259. Metz, C. F., U. S. Atomic Energy Comm., Report LA-303, Los Alamos Scientific Laboratory, June, 1945; through *Nucl. Sci. Abstr.*, **10**, 5115 (1956).

260. Millen, E. R., and R. M. Vredenburg, *Iron Age*, **166**, 68 (1950).

261. Mixter, W. G., and F. L. Haigh, *J. Amer. Chem. Soc.*, **39**, 374 (1917).

262. Mooney, J. B., and L. J. Carbini, Pittsburgh Conference on Analytical Chemistry and Applied Spectroscopy, March, 1962.

263. Morabito, J. M., *Anal. Chem.*, **46**, 189 (1974).

264. Morrison, G. H., *Appl. Spectrosc.*, **10**, 71 (1956).

265. Makaewaki, K., *Bunseki Kagaku*, **9**, 427 (1960); through *Chem. Abstr.*, **56**, 10895 (1962).

266. Makaewaki, K., *Bunseki Kagaku*, **9**, 1056 (1960); through *Chem. Abstr.*, **57**, 4017 (1963).

267. Mukhamedshine, N. M., and A. V. Yankovskii, *Zavod. Lab.*, **38**, 1099 (1972).

268. Murray, Jr., W. M., and S. E. Q. Ashley, *Ind. Eng. Chem. Anal. Ed.*, **16**, 242 (1944).

269. Murray, Jr., W. M., and L. W. Niedrach, *Ind. Eng. Chem. Anal. Ed.*, **16**, 634 (1944).

270. Murty, G. V. L. N., and T. S. Viswanathan, *Anal. Chim. Acta*, **25**, 293 (1961).

271. Naftel, J. A., *Ind. Eng. Chem. Anal. Ed.*, **11**, 407 (1939).

272. Naughton, J. J., and H. H. Uhlig, *Anal. Chem.*, **20**, 477 (1948).

273. Nesbitt, C. E., and J. Henderson, *Anal. Chem.*, **19**, 401 (1947).

274. Newberg, H., *Chem. Anal.*, **43**, 93 (1954).

275. Nichols, P. F., *Nucl. Sci. Eng.*, **7**, 395 (1960).

276. Niederl, J. B., and J. A. Sozzi, *Mikrochim. Acta*, 496 (1957).

277. Niederl, J. B., and J. A. Sozzi, *Mikrochim. Acta*, 1512 (1956).

278. Nightingale, R. E., *Nuclear Graphite*, Academic, New York, 1962, pp. 77–80.

279. Nigrovic, V., and A. N. Jarvis, *Anal. Biochem.*, **30**, 403 (1969).

280. Okada, M., and N. Tamura, Rep. Japan Atom. Energy Res. Inst., SAERI-M-4900, 1972; through *Anal. Abstr.*, **25**, 3681 (1973).

281. Ong, P. S., *Advan. X-ray Anal.*, **8**, 341 (1969).

282. Optz'kil, V., V. Nazarenko, and A. Tyulpina, *Zavodsk. Lab.*, **4**, 408 (1935); through *Chem. Abstr.*, **29**, 7862 (1935).

283. Overholser, L. G., and J. P. Blakely, Proceedings of the U. S./U. K. Meeting on the Compatibility Problems of Gas-Cooled Reactors, U. S. Atom. Energy Comm. Report TID-5797 (Bk. 2), 1961, pp. 560–585; through *Nucl. Sci. Abstr.*, **15**, 19978 (1961).

284. Papushev, P. I., and S. A. Pchelkin, *Zavodsk. Lab.*, **30**, 504 (1964).

285. Parkes, D. W., and R. B. Evans, *J. Soc. Chem. Ind.*, **57**, 302 (1938).

286. Patterson, J. E., H. D. Whitehead, and R. K. Bennett, U. S. Atom. Energy Comm., Report Y-810, May, 1950; through *Nucl. Sci. Abstr.*, **11**, 8298 (1937).

287. Pella, E., and B. Columbo, *Mikrochim. Acta, 5*, 677 (1973).

288. Pepkowitz, L. P., and P. Chebiniak, *Anal. Chem.*, **24**, 889 (1952).

289. Pepkowitz, L. P., and W. D. Moak, *Anal. Chem.*, **26**, 1022 (1954).

290. Peterson, W. M., *Anal. Chem.*, **34**, 575 (1962).

291. Pierre, T. B., J. W. McMillan, P. F. Peck, and I. G. Jones, *Radiochem. Radioanal. Lett.*, **14**, 375 (1973).

292. Pierson, R. H., A. N. Fletcher, and E. St. Clair, Gantz, *Anal. Chem.*, **28**, 1218 (1956).

293. Pieters, H. A. J., *Anal. Chim. Acta, 2*, 263 (1948).

294. Pincus, I., and N. J. Gendron, Proc. 4th Conf. Carbon, Buffalo, 1959, p. 687, 1960. Pergamon Press, New York, 1960.

295. Plice, M. J., and J. Lunin, *J. Amer. Soc. Agron.*, **33**, 851 (1941).

296. Pobiner, H., *Anal. Chem.*, **34**, 878 (1962).

297. Pringle, W. J. S., *Fuel*, **42**, 63 (1963).

298. Pyatigorskii, M. G., *Zavod. Lab.*, **22**, 778 (1956); through *Chem. Abstr.*, **51**, 2498 (1957).

299. Rankama, K., and K. J. Neuvonen, *Anal. Chem.*, **20**, 589 (1948).

300. Ranzetta, G. V. T., and V. D. Scott, *Brit. J. Appl. Phys.*, **15**, 263 (1964).

301. Redmond, J. P., and P. L. Walker, Jr., *Nature*, **186**, 72 (1960).

302. Redmond, C. K., A. Ciocco, J. W. Lloyd, and A. W. Rush, *J. Occup. Med.*, **14**, 621 (1972).

303. Redshaw, H. A., and C. A. Pyane, Internat. Foundry Congr., Papers, 29th, Detroit, 1962, p. 89; through *Chem. Abstr.*, **58**, 6489 (1963).

304. Reid, D. D., and C. Buck, *Br. J. Ind. Med.*, **13**, 265 (1956).

305. Richmond, M. S., and R. E. Telford, U. S. Atom. Energy Comm., Report, TID-5241, National Bureau of Standards (1954); through *Nucl. Sci. Abstr.*, **10**, 6112 (1956).

306. Riezler, W., *Z. Naturforsch, A*, **4**, 545 (1949); through *Chem. Abstr.*, **44**, 4798 (1950).

307. Reinharz, M., G. Rohringer, and E. Broda, *Acta Phys. Austr.*, **8**, 285 (1954).

308. Reiser, W., and H. Schneider, *Arch. Eisenhuettenw.*, **32**, 31 (1961); through *Chem. Abstr.*, **60**, 19 (1964).

309. Rodgers, L. J., *Can. Chem. J.*, **3**, 122 (1919).

310. Roller, P. S., G. Erving, Jr., *Ind. Eng. Chem. Anal. Ed.*, **11**, 150 (1939).

311. Romand, J., and R. Berneron, Colloq. Spectrosc. Internat., 9th, Lyons, 1961, **2**, 325 Vol. II, Muray-Print, Paris, France, (1962).

312. Ross, P. J., *Anal. Chem.*, **38**, 1436 (1966).

313. Rossi, G., and G. Soldani, *Analyst (London)*, **97**, 124 (1972).

314. Rottmann, C. J., *Trans. Amer. Electrochem. Soc.*, **37**, 245 (1920).

315. Roush, G. A., *Ind. Eng. Chem.*, **3**, 368 (1911).

316. Roux, A., and A. Viallard, *Bull. Soc. Chim. Fr., 7,* 2777 (1971).

317. Russell, J. J., National Research Council of Canada, Atomic Energy Program, Division of Research, MC 47, 1944; through *Chem. Abstr., 43,* 4596 (1949).

318. Saito, K., Nozaki, T., Tanaka, S., Furukawa, M., and Cheng, H. S., *Internat. J. Appl. Radiation Isotopes, 14,* 357 (1963); through *Chem. Abstr., 51,* 13340 (1963).

319. Sandell, E. B., *Colorimetric Determination of Traces of Metals,* 2nd edit., Wiley-Interscience, New York, 1950, pp. 152–154.

320. Sandell, E. B., *Colorimetric Determination of Traces of Metals,* 2nd edit., Wiley-Interscience, New York, 1950, pp. 260–264.

321. Sandell, E. B., *Colorimetric Determination of Traces of Metals,* 2nd edit., Wiley-Interscience, New York, 1950, pp. 375–378.

322. Sandell, E. B., *Colorimetric Determination of Traces of Metals,* 2nd edit., Wiley-Interscience, New York, 1950, pp. 429–433.

323. Sandell, E. B., *Colorimetric Determination of Traces of Metals,* 2nd edit., Wiley-Interscience, New York, 1950, pp. 455–459.

324. Sandell, E. B., *Colorimetric Determination of Traces of Metals,* 2nd edit., Wiley-Interscience, New York, 1950, pp. 469–474.

325. Sandell, E. B., *Colorimetric Determination of Traces of Metals,* 2nd edit., Wiley-Interscience, New York, 1950, pp. 572–576.

326. Sandell, E. B., *Colorimetric Determination of Traces of Metals,* 2nd edit., Wiley-Interscience, New York, 1950, pp. 607–609.

327. Sayers, R. R., and W. P. Yant, U. S. Bur. Mines Rep. Invest. No. 2476 (1923).

328. Schaefer, B. A., and D. M. Douglas, *J. Chromatogr. Sci., 9,* 612 (1971).

329. Schäfer, H., *Z. Anal. Chem., 110,* 11 (1937).

330. Scherrer, J. A., and C. J. Rodden, U. S. At. Energy Comm., New Brunswick Laboratory, Collected Papers C-37, Report A-133, March, 1941.

331. Schmidts, W., and W. Bartscher, *Z. Anal. Chem., 181,* 54 (1961); through *Chem. Abstr., 55,* 23162 (1961).

332. Schneerson, A. L., and A. G. Liebush, *Zh. Prikl., Khim., 19,* 553 (1949); through *Chem. Abstr., 45,* 2291 (1951).

333. Schneider, C. R., and H. Fruend, *Anal. Chem., 34,* 69 (1962).

334. Schollenberger, C. J., and C. W. Whittaker, *Soil Sci., 85,* 10 (1958).

335. Schwebel, A., H. S. Isbell, and J. V. Karabinos, *Science, 113,* 465 (1951); through *Chem. Abstr., 45,* 6970 (1951).

336. Schwebel, A., H. S. Isbell, and J. D. Moyer, *J. Res. Nat. Bur. Stand., 53,* 221 (1954).

337. Scott, B. F., and J. R. Kennally, *Anal. Chem., 38,* 1404 (1966).

338. Seeley, S. B., "Carbon (Natural Graphite)," in R. E. Kirk and D. F. Othmer, *Encyclopedia of Chemical Technology,* Vol. IV, 2nd edit., Wiley-Interscience, New York, 1964, p. 320.

339. SenGupta, J. G., *Anal. Chim. Acta, 51,* 437 (1970).

340. Serrini, G., and W. Leyendecker, *Metallurgia Ital., 64,* 124 (1972); through *Anal. Abstr., 23,* 3798 (1972).

341. Shabason, L., and B. L. Cohen, *Anal. Chem., 45,* 284 (1973).

342. Sharma, H. D., and M. S. Subramanian, *Talanta, 11,* 655 (1964).

343. Shmelev, B. A., and Y. V. Glushko, *Zavod. Lab.*, **36**, 1188 (1970).

344. Simmons, R. E., and M. H. Randolph, *Anal. Chem.*, **34**, 1119 (1962).

345. Simons, E. L., J. E. Fagel, Jr., E. W. Balis, and L. P. Pepkowitz, *Anal. Chem.*, **27**, 1119 (1955).

346. Sinex, F. M., J. Plazin, D. Clareus, W. Bernstein, D. D. VanSlyke, and R. Chase, *J. Biol. Chem.*, **213**, 673 (1955).

347. Slickers, S. K., and J. Vorpe, *Arch. Eisenhuettenu.*, **43**, 819 (1972); through *Anal. Abstr.*, **25**, 198 (1973).

348. Slocum, H. E., Laboratory, Vol. III, No. 2, Fisher Scientific Co., Pittsburgh, Pa., 1930, p. 24.

349. Smith, A. C., *J. Appl. Chem.*, **8**, 636 (1958).

350. Smith, G. S., *Analyst (London)*, **60**, 735 (1935).

351. Smith, G. F., and G. L. Hockenyos, *Ind. Eng. Chem. Anal. Ed.*, **2**, 36 (1930).

352. Smith, R. N., J. Swinehart, and D. G. Lesnini, *Anal. Chem.*, **30**, 1217 (1958).

353. Sorokina, N. N., and P. A. Kondrat'ev, *Ind. Lab.*, **31**, 1680 (1965).

354. Spicer, G. S., and J. D. H. Strickland, *Anal. Chem. Acta*, **18**, 231 (1958).

355. Spicer, G. S., and J. D. H. Strickland, *Anal. Chem. Acta*, **18**, 523–533 (1958).

356. Stanley, J. K., and T. D. Yensen, *Ind. Eng. Chem. Anal. Ed.*, **17**, 699 (1945).

357. Still, J. E., L. A. Dauncey, and R. C. Chirnside, *Analyst (London)*, **79**, 4 (1954).

358. Strocchi, P. M., S. Noe, and P. Rebora, *Energ. Nucl. (Milan)*, **5**, 815 (1958); through *Chem. Abstr.*, **53**, 13811 (1959).

359. Stuckey, W. K., and J. M. Walker, *Anal. Chem.*, **35**, 2015 (1963).

360. Sue, P., *C. R. H. Acad. Sci.*, **242**, 770 (1956).

361. Sundberg, O. E., and C. Maresh, *Anal. Chem.*, **32**, 274 (1960).

362. Swaroop, B., *Rev. Sci. Instrum.*, **44**, 1387 (1973).

363. Swick, R. W., D. L. Buchanan, and A. Makao, *Anal. Chem.*, **24**, 2000 (1952).

364. Szulczewski, D. H., and T. Higuchi, *Anal. Chem.*, **29**, 1541 (1957).

365. Templeton, D. H., and J. I. Walters, Nat. Nuclear Energy Serv., Div. VIII, I, Anal. Chem., Manhattan Project, p. 644, 1950.

366. Throckmorton, W. H., and G. H. Hutton, *Anal. Chem.*, **24**, 2003 (1952).

367. Thyberg, R., *Jernkontorets Ann.*, **147**, 943 (1963).

368. Tipler, G. A., *Analyst (London)* **88**, 272 (1963).

369. Toensing, C. H., and D. S. McKinney, *Anal. Chem.*, **22**, 1524 (1950).

370. Tratapel, M., Centre Document Siderurg., Circ. Inform. Tech., No. 10, p. 2085, 1957; through *Chem. Abstr.*, **55**, 16267 (1961).

371. Tsuchiya, M., *Jap. Anal.*, **7**, 12 (1958).

372. U. S. Pat. 2,698,222 (Dec. 28, 1954), R. L. Davis, II (to Leeds and Northrup Co.); through *Chem. Abstr.*, **49**, 4491 (1955).

373. U. S. Pat. 2,728,639 (Dec. 27, 1955), P. W. McConnaughey (to Mine Safety Appliances Co.).

374. Updegrove, W. S., and J. M. Baldwin, *Anal. Chem.*, **45**, 2115 (1973).

375. Van Nieuwenburg, C. J., and L. A. Hegge, *Anal. Chem. Acta*, **5**, 68 (1951); through *Chem. Abstr.*, **45**, 3690 (1951).

376. Van Slyke, D. D., J. Folch, and J. Plazin, *J. Biol. Chem.*, **136**, 509 (1940).

377. Vandecasteele, C., F. Adams, and J. Hoste, *Anal. Chim. Acta*, **72**, 299 (1974).

378. Violante, E. J., *Anal. Chem.*, **36**, 856 (1964).

379. Vogel, A. M., and J. J. Quanttrone, Jr., *Anal. Chem.*, **32**, 1754 (1960).

380. Walker, J. M., and C. W. Kuo, *Anal. Chem.*, **35**, 2017 (1963).

381. Walker, Jr., P. L., H. A. McKinstry, and J. V. Pustinger, *Ind. Eng. Chem.*, **46**, 1651 (1954).

382. Walker, Jr., P. L., J. F. Rakszawski, and A. F. Armington, *ASTM Bull.*, No. 208, 52 (1955).

383. Webb, M. S. W., J. C. Cotterill, and T. W. Jones, *Anal. Chim. Acta*, **26**, 548 (1962).

384. Weinard, J., "Purity of 'National' Spectroscopic Electrodes," in W. D. Ashby, *Developments in Applied Spectroscopy*, Vol. I, Plenum, New York, 1962, pp. 137–142.

385. Weisburger, J. H., E. K. Weisburger, and H. P. Morris, *J. Amer. Chem. Soc.*, **74**, 2399 (1952).

386. Welcher, F. J., *Standard Methods of Chemical Analysis*, Vol. I, 6th edit., Van Nostrand, Princeton, N. J., (1962).

387. Welcher, W. F., *The Analytical Uses of Ethylenediamine Tetra-acetic Acid*, Van Nostrand, Princeton, N. J., 1957, p. 255.

388. Wells, J. E., *J. Iron Steel Inst.*, **166**, 113 (1950).

389. Wenzel, A. W., and H. R. Mullin, U. S. Atom. Energy Comm., Report NBL-159, New Brunswick Laboratory, May, 1960, pp. 18–27; through *Nucl. Sci. Abstr.*, **14**, 17811 (1960).

390. White, D. C., *Talanta*, **13**, 1303 (1966).

391. White, Jr., L., Isotopes Division Circular, ID A-2; through *Chem. Abstr.*, **44**, 2410 (1950).

392. Whymark, D. W., and J. M. Ottaway, *Talanta*, **19**, 209 (1972).

393. Wilkinson, T., W. W. Foster, and T. S. Harrison, *Metallurgia Metal Form*, **40**, 61 (1973).

394. Winchester, J. W., and M. L. Bottino, *Anal. Chem.*, **33**, 472 (1961).

395. Wood, P. R., *Analyst (London)*, **85**, 764 (1960).

396. Wooten, L. A., and W. G. Guldner, *Ind. Eng. Chem. Anal. Ed.*, **14**, 835 (1942).

397. Yakovlev, P., and A. I. Orzhekhovskaya, *Sb. Tr. Tsentr. NachnIssled. Inst. Chernoi Met.*, 1963 (31), Pt. 1, p. 144; through *Chem. Abstr.*, **59**, 13336 (1963).

398. Yensen, T. D., *Transact. Amer. Electrochem. Soc.*, **37**, 227 (1920).

399. Zadvornyi, A. S., A. F. Gorenko, A. P. Klyucharev, P. V. Serykh, and N. A. Skakun, *Radiokhimiya*, **13**, 781 (1971); through *Anal. Abstr.*, **23**, 3742 (1972).

400. Zelenetskil, B. P., *Zavodsk. Lab.*, **5**, 1391 (1936); through *Chem. Abstr.*, **32**, 2491 (1938).

401. Ziegler, N. A., *Transact. Amer. Electrochem. Soc.*, **56**, 231 (1929).

402. Zlatkis, A., H. R. Kaufman, and D. E. Durbin, *J. Chromatogr. Sci.*, **8**, 416 (1970).

403. Zmijewska, W., and H. Sorantin, *J. Radioanal. Chem.*, **8**, 83 (1971).

ARSENIC

By Richard F. Skonieczny, *Herman Kiefer Health Complex, Department of Health, Detroit, Michigan and* Richard B. Hahn, *Department of Chemistry, Wayne State University, Detroit, Michigan*

Contents

I. INTRODUCTION

Arsenic, a metal unknown to the ancients, is considered to be one of the elements of the alchemists. The sulfides of arsenic, orpiment and realgar or sandarac, are mentioned by the Greeks and Romans even prior to the Christian era, and Pliny records the presence of these sulfides in gold and silver mines. The Greek name, *arsenikon* (bold, valiant, masculine), which was given at first to the sulfides because of their reactivity with other metals, was transferred to the element itself when it was isolated.

The honor of isolation of the element is accredited to Albert the Great in 1250, Paracelsus in the sixteenth century, and Shroeder in 1649, so the fact remains that no one definitely knows who isolated arsenic.

The early writers describe the medicinal properties of the arsenic compounds, but fail to mention its poisonous character. The Chinese, however, were thoroughly acquainted with its poisonous properties and had devised means of detecting its use in suspected poisonings. They also used arsenic in fields and rice plantations to kill mice and insects. Orpiment was popular with artists as a pigment. The histories of the Middle Ages record the extensive use of arsenic compounds by the professional poisoners (216).

A. OCCURRENCE

Arsenic, the forty-seventh most abundant element, composes about 2×10^{-4} wt% of the earth's crust (134a), including the hydrosphere and lithosphere, and is present to about 1.5×10^{-6} % in sea water (149). Native arsenic is found in many localities, but the quantities are so small that the deposits are of no economic importance. However, the element, as well as the ores, is usually found in association with ores of antimony, cobalt, nickel, lead, and silver, and metallic arsenic is obtained as a by-product in the treatment of the ores of the other elements. The principal arsenic minerals are arsenopyrite, $FeAsS$; realgar, As_2S_2 or As_4S_4; and orpiment, As_2S_3. The latter two are fairly abundant in Europe and Asia where they are mined for their commercial importance. A comprehensive list of arsenic-bearing minerals is available in the literature (210).

B. PRODUCTION OF ARSENIC

Since the greatest portion of the arsenic supply is obtained in the form of arsenic trioxide (white arsenic), a by-product from the smelting of arsenic-bearing ores of other metals, the chief metallurgical interest lies in the method of recovery and treatment of the crude flue dusts that are as much as 30% arsenic trioxide by weight (156). The crude flue dust is mixed with small amounts of iron pyrite or galena to reduce the formation of arsenites, then roasted in a reverberatory furnace, and the resultant gases are passed through a series of brick cooling chambers where the oxide is condensed. The temperature varies from 220°C in the first chamber to about 100°C in the last, and the condensed product varies from an amorphous black mass containing 95% arsenic trioxide at the hot end, to a white, crystalline, higher-grade product in the 180–120°C zone, and finally to a white, low-grade powder at the cold end. If higher purity is desired, the crude oxide is re-refined, and a product that is 99% arsenic trioxide or better is obtained.

This method of recovery and treatment of the flue dusts is applied in the smelting of the copper ores of Montana, the arsenopyrites, the cobalt ores, and the Swedish arsenical gold ores. Since the Swedish ore contains 10% or more of arsenic, the arsenic trioxide from this source alone could almost supply the world demand, but the price of the crude oxide is not high enough to cover the freight charges to the world markets. At present, the Swedish arsenic trioxide is being stored in large concrete storehouses until either a profitable market or a cheap method of disposal is found.

Metallic arsenic is produced by reduction of the arsenic trioxide by heating with charcoal in cast iron or steel retorts, or by direct sublimation from arsenopyrite or leucopyrite in the absence of air. The former method is preferred since a purer product is obtained.

Details of various furnaces and procedures used in the production of arsenic trioxide, metallic arsenic, and other arsenic compounds are available (156,177).

C.   TOXICOLOGY AND INDUSTRIAL HYGIENE

The toxic qualities of white arsenic were recognized as early as the eleventh century, and the mere mention of poisoning in the present age brings arsenic to the mind of most laymen. The hazardous compounds of arsenic are mainly arsine, arsenic trioxide (white arsenic), arsenic trichloride, arsenites, and arsenates. Arsenic trisulfide and arsenic metal are considered to be less hazardous because of their insolubility. The threshold limits of arsine (202), arsenic (202), and arsenic trioxide (42) have been established at 0.2, 0.5, and 0.25 mg/m^3 of air, respectively. The presence of arsenic in excess of 0.01 mg/liter in drinking water constitutes grounds for the rejection of the supply (209).

The possibilities of arsenic poisoning in industry are numerous. Toxic fumes or dusts are encountered in the manufacture of organic chemicals where arsenic-contaminated acids are used: the manufacture of zinc chloride and sulfate, enamelware, dyestuffs and dyestuff intermediates, paints, insecticides, fungicides, and drugs. Smelting of arsenical and arsenic-bearing ores, electroplating and galvanizing, production of hydrochloric and sulfuric acids, felt hat and leather industries all involve processes where some form of toxic arsenic product is produced. In general, any process that involves the evolution of hydrogen gas is a potential source of arsine poisoning.

Exposure to harmful concentrations of the more hazardous compounds may result in skin ulceration, injury to mucous membranes, and perforation of the nasal septum (80). In the case of arsine, which is considered to be principally a nerve and blood poison, there is often a delay of a day or more before the symptoms appear. The victim has a feeling of malaise, difficulty in breathing, severe headache, giddiness, fainting fits, nausea, vomiting, and gastric disturbances.

Thorough discussions of the hazardous industries, symptoms of poisoning, methods of detection and estimation, corrective measures, handling and storage precautions are available (42,57,73,83,136).

II. PROPERTIES OF ARSENIC

A. PHYSICAL PROPERTIES

Several allotropic forms of arsenic are known, but at 20°C only the gray form is stable. The metastable yellow form rapidly changes to the gray, semimetallic form at low temperatures, and instantaneously in sunlight at room temperature. Brown and black amorphous forms are also known. The metallic form is a steel-gray, crystalline solid with a brilliant luster. It is a good conductor of heat, a poor conductor of electricity, fractures coarsely crystalline, and crystallizes as hexagonal.

The specific gravities of the crystalline, black amorphous, and the yellow cubic forms are $5.727^{14°}$, $4.71^{20°}$, and $2.026^{18°}$, respectively. The mean specific heats are: crystalline Sm, 0.083, amorphous Sm, 0.0758. Arsenic melts in the absence of oxygen at 814°C (36 atm) and sublimes at 615°C (88,106,177).

Naturally occurring arsenic consists of one isotope, As^{75}. Other isotopes with mass numbers 70–74, 76–79, and 81 are known (76,132), but only one of these radioactive isotopes, As^{76}, half life of 26.8 hr, is used extensively for tracer work (67,86,87,170,174,184,219).

B. ELECTROCHEMICAL PROPERTIES

Standard electrode potentials of analytical significance are summarized in Table 1. The successive equilibrium constants for H_3AsO_4 in aqueous solution are $K_1 = 2.5 \times 10^{-4}$, $K_2 = 5.6 \times 10^{-8}$, and $K_3 = 3 \times 10^{-13}$, while K_1 for H_3AsO_3 is 6×10^{-10} (105). Polarographic data is presented in Table 6 of Section VI.C.

C. OPTICAL PROPERTIES

Numerous emission spectrochemical methods have been developed for the determination of arsenic in metals, alloys, oxides, and biological materials. Spark methods, as well as ac and dc arc techniques, have

TABLE 1 (105)

Standard Electrode Potential Data for Arsenic

Reaction	$E°$ (V)
In acid solutions:	
$As^+ + e^- = As^0$	-2.923
$As^0 + 3H^+ + 3e^- = AsH_3$	-0.60
$HAsO_2$ (aq) $+ 3H^+ + 3e^- = As^0 + 2H_2O$	$+0.247$
$H_3AsO_4 + 2H^+ + 2e^- = HAsO_2 + 2H_2O$	$+0.559$
In basic solutions:	
$As^0 + 3H_2O + 3e^- = AsH_3 + 3OH^-$	-1.43
$AsO_2^- + 2H_2O + 3e^- = As^0 + 4OH^-$	-0.68
$AsO_4^{3-} + 2H_2O + 2e^- = AsO_2^- + 4OH^-$	-0.67

been found to be suitable (see Section VI.D). The sensitive lines of arsenic used for its detection and determination are given in Table 2.

D.   CHEMICAL PROPERTIES

The chemical properties and reactions of arsenic are discussed thoroughly in several inorganic chemistry reference books (148,177,210). Only properties that are of analytical importance are considered.

Arsenic is in the fifth main group of the Periodic System along with nitrogen, phosphorus, antimony, and bismuth. Its principal oxidation states are 3+ and 5+, and the minor states are 3−, in arsine and the arsenides and 2+ in realgar, As_4S_4.

On exposure to dry air, arsenic undergoes no reaction, but in the presence of moisture, arsenic trioxide is slowly produced. When heated, it burns with a bluish-white flame, producing arsenic trioxide and emitting a garlic-like odor under these conditions. Concentrated nitric acid and aqua regia oxidize arsenic to arsenic acid, while dilute nitric and concentrated sulfuric acids and boiling caustic alkalis oxidize it to arsenious acid. Dilute sulfuric acid has no effect, and hydrochloric acid in presence of air has only a feeble effect, yielding the trichloride. The trichloride is produced more readily by heating arsenic in the presence of chlorine, resulting in a vigorous, inflaming reaction. Many other elements combine similarly with arsenic when heated. Arsenic is strongly amphoteric, forming arsenites and arsenates, as well as more complex acid derivatives.

In the most stable arsenic compounds the oxidation state is 3+, except for the arsenates, where it is 5+. The As^{3+} cation exists to some extent only in strongly acid solutions; under less acid conditions the tendency is toward hydrolysis, so that the anionic form predominates.

Some common reactions of arsenite are: In neutral or basic solution no precipitate forms with hydrogen sulfide, but in hydrochloric acid solution a yellow precipitate of arsenious sulfide forms. The precipitation is incomplete in the presence of citric acid and other organic compounds. The arsenious sulfide dissolves in ammonium, lithium, sodium, and potassium hydroxides, and in alkaline monochloracetate solution, yielding a colorless solution that consists of a mixture of arsenites and thioarsenites. Acidification of this solution with acetic acid causes the reprecipitation of arsenious sulfide. Nitric acid as well as a mixture of ammonium hydroxide and hydrogen peroxide dissolve the sulfide and oxidize the arsenic to the arsenate form. Treatment of the sulfide with ammonium sulfide yields the thioarsenite; treatment with yellow ammonium sulfide (polysulfide) yields the thioarsenate. Arsenite is precipitated by silver nitrate, calcium nitrate, and lead acetate to give yellow Ag_3AsO_3, soluble in nitric acid and ammonium hydroxide, white $Ca_3(AsO_3)_2$, soluble in strong acids, and white $Pb_3(AsO_3)_2$, soluble in nitric acid and sodium hydroxide. In neutral solution arsenites react with cupric sulfate to form $Cu_3(AsO_3)_2 \cdot xH_2O$, soluble in ammonium hydroxide and dilute acids, but in basic solution arsenite is oxidized to arsenate, and reddish-orange Cu_2O is formed (Fehling's test). Arsenites are oxidized to arsenates by chloric, bromic, iodic, and nitric acids. Oxidation also takes place with potassium ferricyanide in alkaline solution, iodine in neutral or alkaline solution, and chromates in sodium bicarbonate solution. In the latter case, chromic arsenate is formed. Any oxidizing agent stronger than nitric acid will bring about the oxidation.

TABLE 2 (72)
Sensitive Lines of Arsenic

| Wavelength | Excitation potential | Intensities | | Sensitivity[b] |
		Arc[a]	Spark	
2898.71	6.7	25r	40	—
2860.452	6.6	50r	50	—
2780.197	6.7	75R	75	U5
2456.53	6.5	100r	8	U4
2370.77	6.7	50r	3	—
2369.67	6.7	40r	—	—
2349.84	6.6	250R	18	U3
2288.12	6.7	250R	5	U3

[a] r indicates narrow self-reversal; R, wide self-reversal.

[b] U1, U2, and so forth indicate the order of decreasing sensitivity for the neutral atom. The absence of a U1 line signifies that the most sensitive lines lie outside the spectral range of 10,000–2000 Å.

Arsenites are reduced to brown, flocculent arsenic by a fresh stannous chloride solution that is at least 25% hydrochloric acid. Hypophosphites in concentrated hydrochloric acid also will reduce the arsenite ion to free arsenic. Arsenites are reduced to arsine by zinc in dilute sulfuric acid and by aluminum in sodium hydroxide solution. Arsine reacts with concentrated aqueous silver nitrate to produce yellow $Ag_3As\cdot3AgNO_3$, and with dilute silver nitrate solution to form free silver. Arsenites are not precipitated by ammonium molybdate in nitric acid solution, but when treated with magnesia mixture, do form $Mg_3(AsO_3)_2$, soluble in ammonium hydroxide and ammonium chloride ($Mg_3(AsO_4)_2$ is not soluble in these reagents). Arsenious oxide, As_2O_3, reacts with alkali hydroxides and carbonates to form soluble arsenites.

Some common reactions of arsenate are: Hydrogen sulfide precipitates the yellow arsenic sulfide slowly from a cold hydrochloric acid solution, more rapidly from a hot solution. The arsenic sulfide, As_2S_5, is insoluble in concentrated hydrochloric acid, but dissolves readily in lithium or ammonium hydroxides, ammonium sulfide, and ammonium carbonate to form mixtures of sulfoxyarsenates and thioarsenates. Acidification of these latter products results in the reprecipitation of the arsenic sulfide. The sulfide reacts with ammonium hydroxide–hydrogen peroxide reagent with the formation of sulfate, and with nitric acid with the formation of free sulfur. Arsenates react with silver nitrate to form chocolate brown Ag_3AsO_4, soluble in nitric acid and ammonium hydroxide, with barium chloride to form white $BaHAsO_4$, soluble in hydrochloric or nitric acid, and with calcium nitrate to form white $Ca_3(AsO_4)_2$, soluble in dilute acids.

With magnesia mixture, a white crystalline precipitate of $MgNH_4AsO_4$ is formed, soluble in hydrochloric and even in acetic acid. In presence of an excess of ammonium molybdate in a hot nitric acid solution, insoluble yellow $(NH_4)_3AsO_4\cdot12MoO_3$ forms, soluble in ammonium or sodium hydroxide. Arsenate reacts with copper sulfate in neutral solution, precipitating as $Cu_3(AsO_4)_2$, soluble in ammonium hydroxide and dilute acids. Arsenates are reduced readily to arsenites by many reducing agents in hydrochloric acid solution. Oxalic acid, however, will not reduce arsenate, but concentrated hydrobromic, hydroiodic acid, or potassium iodide in acid solution will reduce it to arsenite. Arsenates are reduced to free arsenic by concentrated hydrochloric acid solutions of hypophosphites or fresh stannous chloride. Arsine is the reduction product if arsenate is treated with zinc in dilute sulfuric acid. The reactions of arsine with silver nitrate have been discussed previously. Arsenic oxide, As_2O_5, forms soluble arsenates in solutions of alkali hydroxides and carbonates.

The addition of ammonium molybdate to a very dilute arsenate solution results in the formation of a yellow heteropoly molybdodiarsenate solution that can be reduced with a suitable reducing agent to the strongly colored "molybdenum blue." Both forms are used in the photometric determination of arsenic.

Arsenic forms complexes with a number of organic reagents including alizarin, rhodanine, diphenylcarbazone, zinc complex of toluene-3,4-diol, quercetin, sodium ethylxanthate, diethylammonium diethyldithiocarbamate, silver diethyldithiocarbamate, and sodium diethyldithiocarbamate. The complexes find use in the separation, detection, and/or determination of arsenic (see Sections IV.C, V, and VI.D).

III. SAMPLING OF ARSENIC-CONTAINING MATERIALS

Since arsenic is normally present as a minor constituent or as a contaminant in the material being analyzed and the types of material subjected to analysis for arsenic are numerous, a discussion of sampling methods applicable to all the types of samples would require several chapters. For metals, alloys, and ores of metals, the standard sampling techniques are adequately discussed in the chapters devoted to the major constituent elements and in *A.S.T.M. Methods for Chemical Analysis of Metals* (5).

Standard sampling procedures for fresh and dried fruits, drugs, plants, and soils have been developed by the Association of Official Agricultural Chemists. Sampling precautions, sample containers, sampling instruments, and procedures for foods and food products are adequately described (84,89).

IV. SEPARATION AND ISOLATION OF ARSENIC

In preparing a sample for the determination of its arsenic content, the volatility of arsenic trichloride dictates that certain precautions be taken while getting the material into solution. It should be kept in mind that arsenious solutions should not be boiled, since a loss is apt to occur unless provisions are made to prevent this. Arsenic trioxide in the presence of sulfuric acid alone is not volatilized appreciably at temperatures under 200°C, but the volatility rises rapidly above this temperature. If the arsenic is in the quinquevalent state, there is less danger of loss.

The dissolution of arsenic-bearing materials has been quite thoroughly studied, and many procedures for the various types of samples have been compiled (5,8,59,158,178).

A. DECOMPOSITION, DISSOLUTION, AND OTHER PRELIMINARY TREATMENT OF INORGANIC MATERIALS

A number of dissolution methods are in use for the treatment of arsenic-bearing ores, alloys, soils, silicate rocks, and arsenic compounds such as paint pigments, germicides, disinfectants, and insecticides. Fuming, concentrated, and dilute nitric acids, aqua regia, sulfuric, hydrofluoric, and hydrobromic acids have been used in various procedures. Mixed acids such as nitric and hydrochloric acids in the ratio of 1:1, nitric and sulfuric acids, nitric and/or sulfuric acid with potassium chlorate, and nitric and perchloric acid have been employed. Other techniques involve the use of nitric acid and 30% hydrogen peroxide, sulfuric acid in the presence of potassium hydrogen sulfate or potassium pyrosulfate, and hydrochloric acid with cuprous chloride.

For certain dissolution, basic treatment with alkali hydroxides or carbonates, or hydrogen peroxide oxidation in sodium hydroxide medium have been recommended.

Dissolution of some insoluble inorganic samples is obtained through several dry methods. Fusion mixtures that are commonly used are magnesium metal, a 1:1 mixture of potassium carbonate and nitrate, sodium peroxide, sodium carbonate with a 2% sodium nitrate content, and sodium sulfite heptahydrate that has been recommended for fusion of sulfur (51).

The dissolution of lead and lead alloys (5,113), carbon steel, cast iron, ferromanganese, and ferrous alloys (5) may be carried out with dilute nitric acid. The addition of an equal volume of hydrochloric acid to ferrous alloys and steels with a high chromium content is beneficial. The solution obtained is evaporated to dryness at not over 130°C, cooled, and taken up in concentrated hydrochloric acid. A nitric acid–hydrogen peroxide treatment is suggested for technical iron (212).

Concentrated nitric acid is recommended for brass, bronze, and bearing metals (5), and zinc metal and zinc smelting residues (33), while fuming nitric acid or bromine water and nitric acid is preferred for pyrite ores and arsenopyrites (158). When complete decomposition of the sulfide ores, the copper, and the lead and zinc base alloys is obtained, sulfuric acid is added, and the heating continued until fumes of sulfur trioxide are evolved. The samples may then be treated according to the requirements of the procedure selected for analysis.

Copper (26), arsenical copper (185), and copper-base alloys (26,185) are oxidized with a mixture of sulfuric and nitric acids, the solutions evaporated to near dryness, and the residue dissolved in water. Subsequent treatment depends on the method of analysis selected.

The ASTM procedures (5) for the determination of arsenic in lead, tin, and antimony, as well as alloys of these elements, in white-metal bearing alloys, and in lead- and tin-base solders involve the use of a mixture of sulfuric acid and potassium hydrogen sulfate or potassium pyrosulfate. Preliminary decomposition with nitric acid prior to treatment with sulfuric and potassium hydrogen sulfate is recommended for special brasses and bronzes.

Steel, iron, and isolated steel residues may be dissolved in a mixture of nitric acid and perchloric acids (19). Silicon may be treated with a mixture of sodium hydroxide and hydrogen peroxide, and the resulting solution acidified with hydrochloric acid and evaporated under oxidizing conditions (86). Silicon and germanium samples may also be attacked with oxalic acid and hydrogen peroxide (115). Cobalt minerals are decomposed by treating with concentrated hydrochloric acid and cuprous chloride in a distillation apparatus. The arsenic trichloride distills and is absorbed in water (195). A mixture containing an iron salt in addition to the reagents used for the attack of the cobalt minerals has been applied to the dissolution of metals and alloys (18).

A hydrofluoric acid treatment is suggested for granite and serpentine. The sample is attacked with hydrofluoric acid at 150°C for 1 hr in a small steel bomb lined with teflon (112).

Various arsenic compounds that are subjected to analysis require special attention. Hot 10% sodium hydroxide solution is recommended for lead arsenate (59) and alkali hydroxides or carbonates for arsenic trioxide. Nitric acid can also be used since it converts the arsenic trioxide to the pentoxide, which is soluble in water.

The determination of acid-soluble arsenic in soils involves a Kjeldahl digestion with nitric and sulfuric acids in the presence of potassium chlorate (8) or a distillation from 48% hydrobromic acid containing a small amount of bromine (218).

Several fusion techniques have been employed for the decomposition of inorganic compounds containing arsenic. Heating of arsenic compounds with magnesium results in the formation of magnesium arsenide, As_2Mg_3, from which arsine may be liberated by acid (90).

Scott (59) suggests that silicate minerals be fused with a 1:1 mixture of potassium carbonate and nitrate, while Sandell (134b) prefers sodium hydroxide–sodium peroxide or simply a sodium hydroxide fusion. With the latter method no more than 5% of the arsenic is lost during the decomposition. The method is considered acceptable, considering the minute amounts of arsenic usually occurring in silicate rocks (0.000n%).

A rapid, field Gutzeit test for arsenic in soils has been described that requires a potassium hydroxide fusion in a nickel crucible (4). A sodium peroxide fusion in a nickel or iron crucible is recommended for pyrites and arsenides (30).

The rapid increase in the use of platinum catalysts in the petroleum industry and the knowledge that the presence of arsenic impurities inhibits the catalytic efficiency have resulted in the development of methods of analysis that require fusion. Both alkali hydroxides (170) and sodium peroxide (110) have been successfully used for this purpose.

B. DECOMPOSITION, DISSOLUTION, AND OTHER PRELIMINARY TREATMENT OF ORGANIC MATERIALS

The majority of dissolution procedures for organic compounds require some form of Kjeldahl digestion with nitric and sulfuric acids. Variations of the method involve the addition of hydrogen peroxide, carbohydrates, anhydrous sodium sulfate, ammonium persulfate, or perchloric acid to the previously mentioned acids.

Digestions with aqua regia, fuming nitric acid, and with mixtures of sulfuric acid and hydrogen peroxide, or anhydrous sodium sulfate, sulfuric acid, and carbohydrate have been investigated. The method of digestion that should be employed is dependent on the accuracy and speed of analysis desired and, most of all, on the nature of the sample. When a wet method of digestion is used, it is essential to maintain either a low temperature or oxidizing conditions. If the sample is permitted to char, the arsenic may be reduced from the pentavalent to the more volatile trivalent form. This possibility can be eliminated by maintaining an excess of nitric acid or other oxidant at all times when organic matter and halogens are present. If extreme difficulties are encountered during wet ashing, the substance must be submitted to dry ashing. The general opinion is, however, that wet ashing is simpler and shorter.

Dry-ashing procedures for organic compounds involve the use of magnesium, magnesium oxide or nitrate, a mixture of magnesium oxide and calcium hydroxide, or a sodium carbonate–sodium peroxide mixture. For some samples a Parr bomb fusion with sodium peroxide is preferred. Direct ignition of the sample even at temperatures as low as 450°C results in large losses of arsenic (70a).

A strong oxidizing attack with a Kjeldahl nitric–sulfuric acid digestion is recommended for most organic samples (6a,8,63a,146a,158,178,189). The addition of perchloric acid to this acid mixture is suggested for difficultly oxidizable material. Fuming nitric acid digestion prior to the addition of sulfuric acid and the subsequent evaporation to fumes of sulfur trioxide is standard procedure for iron–arsenic tablets (8) and suggested for the decomposition of tobacco (178).

Hydrogen peroxide has been used in conjunction with strong acids for the oxidation of many organic substances (40a). Hydrogen peroxide–sulfuric acid procedures have been devised for the digestion of quanidino-type nitrogen compounds (186) and naphthas (110). Digestion with nitric and sulfuric acids, followed by a treatment with 30% hydrogen peroxide, is used in the oxidation of petroleum fractions (119). A hydrogen peroxide–nitric acid treatment is suggested for animal tissue (178). The tissue is added in small portions to cold 30% hydrogen peroxide, and after the initial frothing subsides, the mixture is heated gently in a water bath. Concentrated nitric acid is then added in small amounts until almost complete decomposition is obtained. The residue is treated with magnesium nitrate, and the sample is evaporated to dryness and ashed at a low temperature. The residue is taken up with acid, and the sample is analyzed according to the procedure selected. As is seen, this is a combination of wet- and dry-ashing techniques.

A recent investigation indicated that wet digestion with a mixture of sulfuric, nitric, and perchloric acid or with hydrogen peroxide and nitric acid to complete oxidation was not necessary for full recovery of arsenic from various animal tissues. A preliminary digestion with a 2:1 water–hydrochloric acid mixture for 15 min was found to be sufficient (94). Refluxing of the mixture resulted in a 50% loss of arsenic, the loss being due to the refluxing of sulfides (hydrogen sulfide or mercapto compounds) and the subsequent formation of insoluble sulfides or arsenic rather than to the distillation of arsenic trichloride. The procedure was successfully applied to urine, tissues, and many food products.

A comparison study (11) was made of eight various methods for the decomposition of organic compounds and the determination of the arsenic content. The oxidation mixtures studied were (a) anhydrous sodium sulfate, sulfuric acid, and carbohydrate (the carbohydrate was in the form of cigarette paper on which the samples were weighed); (b) hydrogen peroxide, sulfuric acid, and hydrazine sulfate; (c) fuming nitric acid and carbohydrate, followed by sodium sulfate, sulfuric acid, and additional carbohydrate; (d) fuming nitric acid and carbohydrate, followed by hydrazine sulfate; (e) potassium sulfate, starch, and sulfuric acid (AOAC method (8)); and (f) potassium permanganate and dilute sulfuric acid (USP XII, arsphenamine procedure (198)). Experimental data showed that the precision available with the techniques was in the same order as given here and that high accuracy was obtained with methods a, b, and c, intermediate with d, and low with e and f. Technique f, the arsphenamine procedure, has been discontinued with the fourteenth revision of the USP (199).

The eventual selection of a procedure for decomposition is dependent on the composition of the material. The applicability of the following digestants is adversely affected by the presence of halogens (11): sulfuric acid; sulfuric acid and sodium or potassium sulfate; sulfuric acid, sodium sulfate, and carbohydrate; sulfuric acid and potassium nitrate; and sulfuric acid and hydrogen peroxide. Mixtures affected by bromine and iodine are sulfuric and nitric acids, potassium permanganate and sulfuric acid, sulfuric and fuming nitric acids.

Decomposition with the aid of ammonium persulfate is not recommended because highly variable results are obtained. In the absence of halogens digestion by sulfuric acid, sodium sulfate, and carbohydrate gives excellent results and is recommended for research purposes. For halogen compounds and for routine analysis treatment with hydrogen peroxide, sulfuric acid, and hydrazine sulfate is adequately accurate and precise, very rapid, and easy to perform. Complete recovery of ring arsenic is obtainable only with the fuming nitric acid–carbohydrate treatment, which is carried to completion by the addition of anhydrous sodium sulfate and sulfuric acid (11).

Magnesium, magnesium oxide, and magnesium nitrate have been found suitable for the dry ashing of organic compounds. Magnesium nitrate has been suggested for the mineralization of eggs, fresh and dried liver, muscle tissue, skin (48), and vegetable matter (178). Saturated magnesium nitrate solution is added to the sample, the mixture is evaporated to dryness in an oven at 105°C, and ignited overnight in a furnace at 600°C. A study showed that the recovery of arsenic was dependent on the amount of magnesium nitrate added. If no or only a small amount of magnesium nitrate was added, a low recovery of arsenic was obtained, but as the amount was increased, the recovery was almost 100% (48). Magnesium nitrate is also suitable for the decomposition of sodium methyl arsenate in pharmaceuticals (190). After heating the treated sample for 1 hr at 100–110°C, the residue is ignited to redness for 15 min to destroy the organic material and to convert the arsenic to magnesium pyroarsenate. The arsenic is then determined volumetrically.

The arsenic in organic pharmaceuticals and biological material may be transformed to the arsenide by heating with magnesium (90). Arsine can be liberated from the magnesium arsenide with acid and collected in a suitable absorption solution.

Mineralization of coal has been obtained by heating with a mixture of magnesium oxide and calcium hydroxide (41). The prepared sample is ignited at 675°C for 70–90 min and the arsenic determined colorimetrically as the molybdenum blue complex.

A sodium carbonate–sodium peroxide fusion is suggested for organic powders (59). The material is fused in a nickel crucible and cooled, and the arsenic is converted to a chloride with hydrochloric acid in the presence of an oxidizing agent.

Carius oxidation (178) and Parr bomb decomposition (14) are highly recommended for organic samples. In the Carius procedure, a sample of blood serum, spinal fluid, or dry tissue is sealed in a tube in a bomb furnace and heated at 260°C for 2 hr. The contents of the tube are evaporated with sulfuric acid, and the residue is taken up in water.

Bomb decomposition is the most rapid and efficient method for organic samples containing arsenic. The charge is made up usually of the sample, sucrose, and sodium peroxide and is well mixed by shaking in the closed bomb. The mixture is collected in the bottom of the cup by tapping on a table, and the charge is ignited by heating with the tip of a small, hot flame for approximately 40 sec. The bomb is air cooled for a short time and immersed in cold distilled water. The fusion is extracted with hot water, and the solution is concentrated by evaporation and acidified with hydrochloric acid.

C. SELECTIVE SEPARATION METHODS

Arsenic may be separated from other elements by precipitation, distillation as a halide, volatilization as arsine, liquid–liquid extraction, or by adsorption. Both chromatographic and ionic exchange procedures are included under adsorption methods.

1. Precipitation Methods

Several precipitation processes may be suitably applied to the separation of arsenic from other elements. Elemental arsenic may be precipitated by such reducing agents as stannous chloride, hypophosphite, and copper in acid solution. Of these reagents, the hypophosphite is considered to be a more sensitive and satisfactory reagent and has been used extensively (1, 7, 21, 26, 28, 44, 46, 47, 58, 82, 98, 104, 151, 173, 185, 200, 201). Only selenium, tellurium, mercury, and the noble metals (26, 104) are precipitated along with the arsenic, and their presence in moderate amounts does not interfere in most methods for the determination of arsenic. Silica and other insoluble substances must be filtered off prior to reduction. The recovery of elemental arsenic is not quantitative, but averages about 95% for samples containing 0.1–0.7 mg of arsenic (26). Since the loss is small and the method of analysis is standardized empirically, the incomplete recovery becomes unimportant.

Reduction with hypophosphite has been successfully applied to separation of arsenic from copper and copper-base alloys (26,185), organic compounds (173), and germanium (82). Tin, nickel, lead, iron, manganese, silicon, bismuth, aluminum, phosphate, and chloride do not interfere. Nitrates must be removed prior to reduction (185), and if copper is not present, it should be added to catalyze the reaction (158).

Stannous chloride in hydrochloric acid is sometimes used for the precipitation of elemental arsenic. It has been employed in the determination of arsenic in iron and steel (222), and in ores and concentrates (166).

Reduction with chromous chloride is also possible (161), but not recommended because the reagent is inconvenient to handle and relatively expensive.

Small amounts of arsenic may be effectively separated from solution by coprecipitation with ferric hydroxide, manganese dioxide, or magnesium ammonium phosphate. The ferric hydroxide method has been applied to the separation of arsenic from natural waters, silicates, and biological materials (191). After filtration, the precipitate is digested with nitric acid, the arsenic is converted to the xanthate, and a further separation is made by extraction with carbon tetrachloride. Coprecipitation with ferric hydroxide has been applied in a polarographic procedure for the determination of arsenic in zinc metal and zinc-smelting residuals containing more than 0.1% cadmium (33).

Manganese dioxide is used as a collector in the precipitation of arsenic during its determination in lead and lead alloys (5,114). After dissolution of the sample in 1:3 nitric acid, the arsenic, antimony, and tin are oxidized to their higher oxidation states with potassium permanganate and then coprecipitated with manganese dioxide that forms upon the addition of manganous nitrate to the solution. If desired, a further separation can be made by distillation.

In the determination of arsenic in rectifier-grade selenium, a separation is made by quantitative precipitation of elemental selenium with sulfur dioxide. The coprecipitation of arsenic is negligible, and if larger amounts of arsenic are present, it can be precipitated and separated as the magnesium ammonium arsenate. The only interference likely to be encountered in rectifier-grade selenium would be phosphorus (147).

The precipitation of the sulfides of arsenic from dilute (0.3 N) hydrochloric acid solution along with the other members of the hydrogen sulfide group as well as molybdenum, selenium, tellurium, gold, platinum, and palladium is well known. The sulfides of arsenic, antimony, tin, and molybdenum are separated from the copper subgroup by treatment with an alkaline solution or an excess of ammonium or alkali sulfide, in which the tin subgroup will redissolve and enter solution as complex polysulfide ions. A small amount of mercuric sulfide may also dissolve if the treatment is prolonged. The arsenic, antimony, and tin are reprecipitated upon acidification of the polysulfide solution with acetic acid, and the arsenic is separated by treatment of the sulfides with 6–8 N hydrochloric acid in which only the arsenious sulfide is insoluble (183).

To insure complete precipitation of arsenic, the usual practice is to make the solution initially 0.6 N in hydrochloric acid, heat to boiling, and saturate with hydrogen sulfide. At this higher acid concentration, the arsenate ion is converted to arsenious sulfide through a series of complex reactions. The solution is cooled, diluted to reduce the acidity *to* 0.25 N, and again treated with hydrogen sulfide to complete the precipitation of the more soluble sulfides of group II.

A 1% lithium hydroxide solution with 5% potassium nitrate separates cleanly all mixtures of sulfides of the metals of the hydrogen sulfide group when heated just to boiling (77). Since the polysulfides have a tendency to dissolve cupric sulfide and small amounts of mercuric sulfide and to render some cadmium sulfide colloidal, and since the lithium hydroxide solution is so efficient, pleasant to use, relatively less expensive, and more specific, the latter reagent is highly recommended for general laboratory use in place of the usual ammonium sulfide.

Arsenic can be precipitated as the sulfide and separated from tin(IV) by complexing the tin with hydrofluoric acid and from tungsten(VI) by complexing with tartrate.

Arsenic(III) and arsenic(V) may be precipitated quantitatively as the sulfides by thioacetamide from hot acidic solution or from an ammonia-cal solution that is acidified (55,56). For 15–160 mg of arsenic sulfide an average deviation of ±0.2 mg (mostly positive) was observed. A study of the reactions of As(III) and As(V) with thioacetamide and a comparison with the hydrogen sulfide–pressure method showed some interesting results (24). Although at least 30 min were required for the quantitative precipitation of 500 mg of As(V) by the hydrogen sulfide method, only 6 min were required for complete precipitation using thioacetamide. It was also found that 99.9% of reduction to As(III) was obtained at 90°C in approximately 1 min. The reaction rate was shown to be dependent on acid concentration, being highest at the highest concentration. The reduction rate was found also to be dependent on the thioacetamide concentration, being fastest at the higher concentrations.

Thioformamide has been suggested as a quantitative reagent for precipitating of arsenic(III) and arsenic(V) sulfides from 1 and 6 N hydrochloric acid solutions, respectively. The sulfides precipitate rap-idly, are easily filtered, can be washed with water, dried at 105–115°C, and weighed (61). This procedure can find use in separations from metals whose sulfides are soluble in the acid concentrations used.

Arsenic present in solutions of copper can be separated and determined by precipitation of the copper with ammonium thiocyanate in presence of sodium hypophosphite. The filtrate containing the arsenic is treated with concentrated hydrochloric acid and heated to produce carbonyl sulfide, which precipitates the arsenic as arsenious sulfide (145).

2. Volatilization Methods

Arsenic is readily volatilized as trichloride, tri- or pentabromide, and arsine. Distillation as the trichloride from hydrochloric acid solution in the presence of a reductant such as hydrazine sulfate is the most common micro and macro separation used and is an integral part of many ASTM procedures (5). Ferrous salts and/or cuprous chloride are suitable alternative reducing agents (8,18,165,195). The presence of minor amounts of bromide is beneficial to reduction (79,117,153), and the presence of sulfuric acid is not detrimental. Nitrates and other strong oxidizing agents must be absent.

Arsenic trichloride boils at 130°C, but it is quite volatile at 110°C. Germanium tetrachloride, bp 86°C, comes over with arsenic, but stannic chloride, bp 115°C, can be held back by complexing with phosphoric acid. Any tungsten (VI) present is also complexed by phosphoric acid. If large amounts of phosphoric acid are used, a small amount may be carried over mechanically, and a second distillation is recommended. This precaution is suggested if the arsenic is to be determined as the molybdenum blue complex. Antimony trichloride, which boils at 220°C, is appreciably volatile at 130°C, and, therefore, if antimony is known or suspected to be present, the distillation temperature must be controlled and kept below 107°C. Germanium and antimony do not interfere with the molybdenum blue determination.

Several procedures suggest the passage of a slow stream of nitrogen or carbon dioxide to facilitate the volatilization of arsenic (30,158), but Sekino (165) states that such treatment is not beneficial.

In most procedures the distillate is collected in cold water, but in a study conducted to ascertain the causes of arsenic losses, the findings showed that the losses occurred during the distillation and could be eliminated by collecting the distillate in an oxidizing medium, preferably one containing hydrogen peroxide (53).

Sometimes the arsenic is distilled as the bromide from an acidic bromide solution, but there is no general agreement as to the composition of the volatile compound. Some researchers (12,117,152,170) are of the opinion that arsenic pentabromide is the volatile substance, but others (30,86) believe that it is arsenic tribromide. Distillation is performed from concentrated hydrobromic acid containing bromine (152) or from concentrated sulfuric acid containing bromide (12,117,170). Selenium and germanium accompany arsenic.

Distillation as the bromide has been utilized in neutron-activation procedures for the determination of traces of arsenic in hydrocarbon-reforming catalysts (170) and silicon semiconductors (86). In both cases carrier arsenic is added prior to the distillation, and the arsenic is precipitated from the distillate as the element by treating with a hypophosphite.

A bromide-separation procedure for arsenic in canned fruit utilizes a cupferron–chloroform extraction of interfering tin prior to distillation (12).

A study (75) made of the volatility of the halides of arsenic from various acid mixtures showed that 100% recoveries were possible from hydrobromic–sulfuric, hydrobromic–phosphoric, hydrochloric–sulfuric, and hydrobromic–perchloric acids if the arsenic was present in the 3+ oxidation state. Complete volatilization was obtained for the 5+ state with the same acid mixtures with the exception of hydrochloric–sulfuric acid from which only a 5% recovery was obtained.

Separation of arsenic as arsine is more rapid and more convenient than by halide distillation. This procedure, essentially the Gutzeit method, is often subjected to criticism, but the criticism relates to the nonuniformity of the stains produced during the actual test and not to the efficiency of evolution of arsine. All of the arsenic is volatilized in the absence of lead and mercury, and only a slight interference by antimony is encountered under limited conditions (85). Large amounts of heavy metals cause serious interference with evolution, but reasonably moderate amounts of iron, chromium, and molybdenum can be tolerated because they increase the evolution of hydrogen. In the presence of small amounts of silver and lead, heating of the solution is recommended. Copper in greater than small amounts prevents complete evolution of arsine, while nickel and cobalt cause low results. Selenium does not accompany the arsenic in the volatilization. Because of heavy metal interference, the arsine-evolution method is recommended mostly for only decomposed organic matter, silicates, and inorganic materials in which heavy metals are not present in appreciable amounts.

Several methods are used for the evolution of arsine. Usually it is evolved by the action of zinc in hydrochloric or sulfuric acid solution. Pure aluminum and sodium hydroxide have been suggested to give arsine since no stibine is evolved with these reagents (54). Zinc–sodium hydroxide and tin–hydrochloric acid systems have been studied and shown to produce no arsine (27).

Sodium borohydride, $NaBH_4$, has been used to evolve arsine. This reagent usually is added in the form of solid pellets or as an aqueous solution to the acid solution containing arsenic. Arsine is rapidly evolved, but is contaminated with the volatile hydrides of antimony, bismuth, germanium, selenium, tellurium, and tin. (162b,174a,201b). Arsenic then is usually determined by atomic absorption.

The arsine is absorbed in sodium hypobromite, mercuric chloride followed by bromine oxidation, mercuric chloride containing an excess of potassium permanganate (159), iodine (48), and silver diethyldithio-carbamate in pyridine (90,211,212). Arsine can also be decomposed in a hot quartz tube and the metallic arsenic dissolved in concentrated nitric acid. Great care must be exercised in the selection of the absorption medium and apparatus because the chief source of error is incomplete absorption.

3.   Liquid–Liquid Extraction Methods

Trivalent arsenic forms a xanthate that can be extracted from hydro-chloric acid solution with carbon tetrachloride (95). Other metal xan-thates, including antimony, that are extracted along with arsenic are washed out of the carbon tetrachloride by concentrated hydrochloric acid containing some stannous chloride, leaving the arsenious xanthate in the organic phase. Aluminum, manganese, zinc, lead, mercury, cadmium, and bismuth do not interfere with the extraction. The carbon tetrachloride solution is evaporated to dryness, the residue is dissolved in bromine water, and the resulting quinquevalent arsenic is determined by the molybdenum blue method.

Sodium ethylxanthate has been found to be effective in the separation of arsenite from arsenate ion, but has proved to be unsuitable for the separation of unsubstituted phenyl derivatives of these acids (35). However, ethane-1:2-dithiol in carbon tetrachloride has been found to be very satisfactory for the separation of phenylarsenious from phenylar-sonic acid and of arsenious from arsenic acid in aqueous solutions and urine dialysates.

A more complete extraction is said to be obtained with diethylammon-ium diethyldithiocarbamate (188). As in the xanthate method, the arsenic must be trivalent and in a solution 5–6 N in hydrochloric acid. Any pentavalent arsenic present is reduced with 20% potassium iodide, and the resulting solution is treated with some 5% sodium bisulfite, diluted, and shaken vigorously with several portions of a 1% diethylammonium diethyldithiocarbamate in chloroform solution. The chloroform layer which now contains all the arsenic is washed free of any entrained droplets that might contain phosphate by shaking with 1 N sulfuric acid. An aqueous arsenic solution can be obtained by evaporating the chloroform and oxidizing the residue with bromine water or perchloric acid. Complete extraction is claimed over the range, 1–10 N sulfuric acid (less than 0.2 μg not extracted). Lead is not extracted from a solution more than 2 N in hydrochloric acid, and zinc, cadmium, nickel, and iron are not extracted. However, copper, mercury, and bismuth carry over with the arsenic in this procedure.

Wyatt (221) devised a scheme using diethylammonium diethyldithiocarbamate that permits the removal of copper and bismuth and eliminates any possible interference from these ions in subsequent analysis. Since the higher oxidation states of arsenic, antimony, and tin do not form complexes with the diethylammonium diethyldithiocarbamate, they can be oxidized and retained in aqueous solution while copper and bismuth are extracted. After extraction, the tin subgroup is reduced, extracted, and determined using normal methods.

The diethylammonium diethyldithiocarbamate extraction procedure has been successfully applied to the separation of arsenic from germanium and silicon (115). The arsenic is separated from most of the germanium and completely from phosphorus, silicon, zirconium, tungsten, niobium, tantalum, selenium, and gold. Traces of germanium that carry over are removed by distillation with hydrochloric acid.

The diethyldithiocarbamate extraction has been used in the separation of arsenic in samples of granite and serpentine (112). After attacking the mineral with hydrofluoric acid at 150°C in a Teflon-lined steel bomb, the arsenic is extracted with the reagent from the strong hydrofluoric acid solution. Polyethylene apparatus is used for all the manipulations.

Arsenic(III) can be extracted from a strong hydrochloric acid solution (10.5–11 N) by shaking with chloroform (13). For analysis the arsenic is returned to the aqueous phase by shaking the chloroform extract with water. Germanium and extremely large amounts of selenium interfere with the procedure.

A chloride method is suggested for the extraction of traces of arsenic in germanium (66). A strong hydrochloric acid solution of the sample is treated with hydrogen peroxide to oxidize the arsenic to pentavalent state and the germanium is extracted by shaking with benzene. After discarding the organic phase, concentrated hydrobromic acid is added to the aqueous solution containing the arsenic, and the mixture is extracted with several portions of benzene. The arsenic can be returned to the aqueous phase by shaking with water containing a small amount of hydrazine hydrochloride that reacts with the bromine produced during the reduction of arsenic. This procedure is applied to the separation of radioactive arsenic produced by neutron activation of a germanium target.

Arsenic also may be extracted as the iodide. The solution is made 2–8 M in sulfuric acid and 0.05–1.0 M in KI. The iodide is extracted into toluene, but germanium, mercury, tin, and lead are also extracted along with the arsenic (24a).

Up to 10 mg each of the molybdate complexes of arsenate, phosphate, and silicate can be separated by successive extraction (3). A 1:3 butanol–chloroform mixture is used for phosphate, a 1:1 butanol–ethyl acetate mixture followed by chloroform for arsenate, and butanol for silicate after acidification of the aqueous solution with nitric acid. Successive extractions with isobutyl acetate and n-butanol have been used to separate molybdophosphates and molybdoarsenates (135a).

Another scheme for the separation of the heteropoly(molybdo) acids of arsenic, silicon, and phosphorus has been devised but not tried (40). The phosphate complex is extracted quantitatively with isoamyl acetate, and the arsenate complex is separated from the silicate by adding sufficient ethanol to give a 17% alcoholic solution and extracting with isoamyl acetate. The silicate complex remains in the alcoholic phase.

Several solvent extraction procedures for arsenic have been compiled by Morrison and Freiser (130).

4. Adsorption Methods

A number of schemes for the separation of arsenic from other metallic ions by paper chromatography have been devised. Some of these separations are summarized in Table 3.

Several other materials have been investigated by Sen as adsorbents for the chromatographic separation of arsenic from other ions in the hydrogen sulfide group. A summary of the separations and materials studied is given in Table 4.

An interesting adsorption method is used for the separation of arsine and stibine from other gases during gas sampling (69). The gas sample is passed through an adsorption tube filled with silica gel impregnated with a 1% solution of silver nitrate on which the arsine and stibine are collected. The silica gel is transferred to a beaker and extracted with several portions of a solution that is 1 M in hydrochloric and tartaric acids. The arsenic and antimony are then determined polarographically.

TABLE 3

Separation of Arsenic and Other Ions by Paper Chromatography

Ions separated	Solvent	Reference
$Fe(CN)_6^{3-}$, $Fe(CN)_6^{4-}$, S^{2-}, AsO_4^{3-}, PO_4^{3-}, I^-	Butanol, 95% ethanol, water (2:2:1)	—[a]
As^{3+}, Sb^{3+}, Sn^{2+}	Ethyl ether	—[b]
H_2S group	Butanol satd. with 3.5 N HCl	—[c]
Simple mixtures containing arsenic	Butanol–water plus benzoyl acetone; collidine–water; dioxane plus pyridine	—[d]
H_2S tin group	Dry ethyl acetate satd. with 2% HCl	—[e]
AsO_3^{3-}, AsO_4^{3-} from other anions	Electrochromatographic separation in neutral ammonium acetate solution	—[f]
H_2S tin group	Triethanolamine plus ammonium tartrate	—[g]

[a] DeLoach, W. S., and C. Drinkard, *J. Chem. Educ.*, **28**, 461 (1951).
[b] Anderson, J. R. A., and A. Whitley, *Anal. Chim. Acta*, **6**, 517 (1952).
[c] Pfeil, E., G. Ploss, and H. Saran, *Z. Anal. Chem.*, **146**, 241 (1955).
[d] Pollard, F. H., J. F. W. McOmie, and I. I. M. Elbeih, *J. Chem. Soc.*, **1951**, 466 (1951).
[e] Burstall, F. H., G. R. Davies, R. P. Linstead, and R. A. Wells, *Nature*, **163**, 64 (1949).
[f] Cetini, G., *Ann. Chim. (Rome)*, **45**, 216 (1955); through *Chem. Abstr.*, **49**, 13818 (1955).
[g] Maki, M., *Jap. Anal.*, **4**, 21 (1955); through *Chem. Abstr.*, **50**, 4716 (1956).

Ion-exchange methods have been applied in several procedures requiring the separation of arsenic from other ions. In the preparation of radioactive isotopes of arsenic by the deuteron bombardment of germanium targets, carrier-free arsenic activities are separated from the germanium and gallium by anion exchange from 0.5–2.5 M hydrofluoric acid (162). The germanium and gallium are absorbed, while the arsenic passes through the column in a few column volumes. The arsenic must be in the 3+ oxidation state because the 5+ state is retained in the column.

An anion exchange procedure has been developed for the separation of tin(II), arsenic(III), and antimony(III) (96). The ions are precipitated as sulfides, dissolved in sodium polysulfide to form thio anions, and the solution is passed through an anion-exchange resin in the hydroxide form. Tin is eluted with 0.5 N, arsenic with 1.2–2.5 N, and antimony with 3.5 N potassium hydroxide.

Arsenic in insecticides is determined in the presence of iron and copper after separation with a cation exchange resin (133). After oxidation to arsenate, the acidity is adjusted to 0.3 N in hydrochloric acid, and the solution is passed through the column. Within 5 min, all

TABLE 4

Chromatographic Separation of Arsenic on Diverse Materials

Ions separated	Column material	Solvent	Reference
Bi–As; As–Sb; As–Sn; Hg–As; As–Sn–Sb	CaSO$_4$ pencils	Dil. acid	—[a]
Hg–As; Cu–As; Bi–As	CaSO$_4$ sticks	Dil. acid	—[b]
As–Sb–Sn	CaSO$_4$ rods	Dil. HCl	—[c]
As–Sb–Sn	Shredded jute plant stalks	Aq. or dil. acid soln.	—[d]
As–Sb–Sn	Asbestos millboard	Aq. soln.	—[e]

[a] Sen, B. N., *Z. Anorg. Allg. Chem.*, **273**, 183 (1953).
[b] Sen, B. N., *Anal. Chim. Acta,* **12**, 154 (1955).
[c] Sen, B. N., *Z. Anorg. Allg. Chem.*, **268**, 99 (1952).
[d] Sen, B. N., *Z. Anorg. Allg. Chem.*, **279**, 328 (1955).
[e] Sen, B. N., *Aust. J. Sci.,* **13**, 49 (1950).

the arsenate filters through and is titrated with thiosulfate solution. The accuracy and precision are satisfactory compared to results obtained with the AOAC method.

V.   DETECTION AND IDENTIFICATION OF ARSENIC

Because of its amphoteric properties, arsenic is found by identification of both the cations and anions. In the usual scheme of analysis it is precipitated by hydrogen sulfide from a dilute acid solution as the yellow sulfide, As$_2$S$_3$. Strong acidification and heating are required to obtain precipitation if the arsenic is present originally in the pentavalent state. In this way, arsenic is separated from the elements of the ammonium sulfide group, the alkaline earths, and the alkali metals. Arsenic trisulfide, as well as the sulfides of antimony and tin, are readily dissolved in ammonium polysulfide or the alkali hydroxides and are thus separated from the copper division in the analytical scheme. The classical ammonium polysulfide separation is being rapidly replaced by potassium, sodium, or lithium hydroxide procedures.

Arsenic sulfide is separated from the other members of the arsenic division by dissolving the other sulfides in 8 *M* hydrochloric acid, or by dissolving the arsenic sulfide in ammonium carbonate, leaving the other sulfides in the residue. After dissolving the remaining arsenious sulfide in nitric acid or acidifying of the carbonate solution, the arsenic is oxidized to the pentavalent state with hydrogen peroxide. The presence of arsenic is confirmed by precipitation of the white, granular magnesium ammonium arsenate or the chocolate brown silver arsenate.

Arsenic is reduced by nascent hydrogen in acid solution to arsine, a colorless gas with a garlic-like odor. This gas decomposes upon heating to form metallic arsenic and burns in air with a bluish flame to form arsenious oxide. These reactions serve as the basis for the Marsh test.

In the Gutzeit test, a modification of the Marsh test, the arsine is detected by means of silver nitrate or a mercury(II) salt. With concentrated silver nitrate, a yellow color due to $Ag_3As \cdot 3AgNO_3$ is obtained, while with a dilute solution, a black deposit of metallic silver is observed.

In the Fleitmann test, aluminum and sodium hydroxide are substituted for the zinc and sulfuric acid of the Marsh and Gutzeit tests, thereby eliminating interference by antimony. Since arsenates do not give the test, a preliminary reduction to arsenite is required.

The Reinsch test involves the deposition of a gray film of arsenic on bright copper foil from a boiling hydrochloric acid solution of trivalent arsenic. The deposit is thought to be an arsenide. Antimony, bismuth, mercury, selenium, and tellurium also are deposited on the copper, but through the use of suitable solvents and testing reagents identification and estimation is possible.

The Bettendorf test and the hypophosphite test depend on the formation of metallic arsenic. In the former, stannous chloride is used to reduce the arsenic in hot, concentrated hydrochloric acid solution, while in the latter, a hypophosphite, usually hypophosphorous acid (Bougault's reagent), is used for the same purpose.

Arsenic can be detected spectrographically in the order of 5 ppm. The sensitive lines for arsenic are summarized in Table 2.

A number of inorganic and organic reagents have been used for the spot test detection of micro amounts of arsenic. A summary of the important data for some of these spot test reagents is given in Table 5.

VI. DETERMINATION OF ARSENIC

Of the numerous methods developed for the determination of arsenic only a few have been investigated thoroughly and found suitable for the estimation of the element. Titration with a standard bromate solution using methyl orange as the indicator has been adopted widely as a standard method for the determination of macro amounts of arsenic. Another titration procedure that has received equal acceptance is the iodine method of oxidation of arsenite in a basic medium using starch as the end point indicator.

TABLE 5

Tests for Detection of Arsenic

Reagent	Medium	Color of complex	Identification limit (ng As)	Remarks	Reference
Alizarin	Filter paper, ammoniacal	Red violet	6.4 (As(V)) 0.64 (As(III))	Hg, Bi, Cr, Ca, Mg give identical color	—[e]
Auric chloride	H_2SO_4 + Zn to give arsine	Blue to Blue-red	0.5/drop	As(III) and As(V)	—[d]
Chromate	Acid	Green Cr(III)		Applies to As(III); Sn(II), Fe(II), SO_3^{2-}, NO_2^-, $Fe(CN)_6^{4-}$, $Fe(CN)_6^{3-}$, interfere	—[a]
Diphenylcarbazone	Filter paper, ammoniacal	Pink	0.64	Applies to As(III); Sb, Co, Zn interfere	—[e]
Mercurous chloride	Dil. HCl	Brown-black	3/10 ml	As(III) and As(V)	—[f]
Mercurous chloride	Arsine from acid soln.	Yellow	0.5	As(III) and As(V); P interferes	—[b]
Palladium chloride	Dil. HCl	Brown-black	0.3/10 ml	As(III) and As(V)	—[f]

Reagent	Conditions	Color	Sensitivity	Interferences	Ref.
Rhodanine	Filter paper, ammoniacal	Green	6.4	Fluorescence in uv, As(III) and As(V)	[e]
Silver diethyldithiocarbamate pyridine	Arsine from acid soln.	Pink	0.5	As(III) and As(V); P, H_2S interfere	[g]
Silver nitrate	H_2SO_4 + Zn to give arsine	Gray	1/drop	As(III) and As(V); Sb, H_2S, PH_3, Hg interfere	[d]
Silver nitrate	Dil. acetic acid soln.	Red-brown	6/drop (arsenious acid)	Chromates, ferricyanide interfere	[d]
Sodium thiosulfate–cupric acetate	Filter paper, ammoniacal	Brown	6.4 As(V) 0.64 As(III)	Sb interferes	[e]
Stannous chloride	Concd. HCl	Brown-black	1/drop	As(III) and As(V); noble metals, Hg interfere	[d]
Zinc dithiol	3 N HCl	White	0.5/ml	As(V); Ge, Re, Os interfere	[c]

[a] Batalin, A. Kh., *Zh. Anal. Khim.*, **5**, 123 (1950); through *Chem. Abstr.*, **44**, 4820 (1950).
[b] Bennett, E. L., C. W. Gould, Jr., E. H. Swift, and C. Niemann, *Anal. Chem.*, **19**, 1035 (1947).
[c] Clark, R. E. D., *Analyst (London)*, **82**, 760 (1957).
[d] Feigl, F., *Spot Tests in Inorganic Analysis*, 5th edit., Elsevier, New York, 1958, pp. 99–103.
[e] Heisig, G. B., and F. H. Pollard, *Anal. Chim. Acta*, **16**, 234 (1957).
[f] Sakuraba, S., *J. Chem. Soc. Japan, Pure Chem. Sect.*, **73**, 501 (1952); through *Chem. Abstr.*, **47**, 2629 (1953).
[g] Vasak, V., and V. Sedivec, *Chem. Listy*, **46**, 341 (1952); *Chem. Abstr.*, **47**, 67 (1953).

Through the years, the Gutzeit method for the estimation of micro and semimicro amounts of arsenic has been the subject of many investigations. Scores of arsine generators have been devised, various reagents have been used to react with the arsine, and different materials have been suggested as the indicator media. A thorough discussion of all the modifications proposed would require a chapter in itself. However, because of the ease of separation of arsenic from other elements and the great sensitivity and precision obtainable with the molybdenum blue procedure, very small amounts of arsenic are preferentially determined colorimetrically by the latter method. If the present trend continues, the molybdenum blue method will probably displace the Gutzeit procedure. Another colorimetric method, although not as well known as the molybdenum blue, is worth more consideration than it is getting at present. This method, which has been devised in Europe, depends on the development of a pink to red color as a result of a reaction between arsine and silver diethyldithiocarbamate in pyridine. It is a method that may be used for precise estimation of very small amounts of arsine with relatively few interferences.

Spectrochemical and atomic absorption methods may be advantageously applied, especially in routine metal, alloy, metallic oxide, biological material, and water analysis.

A. PRECIPITATION AND GRAVIMETRIC DETERMINATION

Although arsenic may be separated readily from most metals except tin by precipitation as the sulfide from a concentrated hydrochloric acid solution, the method is seldom used for the determination of arsenic since the precipitate is rarely pure, especially if hydrogen sulfide is the precipitant. Thioacetamide (24,56) has been investigated as a substitute precipitant, and quantitative recoveries of arsenic are reported for 15–160 mg of arsenic sulfide with an average deviation of ±0.2 mg. Similar results are obtainable with thioformamide (61).

Arsenic may be determined accurately by weighing as the magnesium pyroarsenate, but the method is rarely employed because the procedure is tedious and subject to many interferences from cations as well as anions. Arsenate may be precipitated in the presence of arsenite, selenite, and selenate with magnesia mixture, and after standing 12 hr, the precipitate of magnesium ammonium arsenate is filtered off on a Gooch crucible and ignited to the pyroarsenate at 850–900°C.

B. TITRIMETRIC METHODS

1. Precipitation Reactions

Only a few titrimetric precipitation methods have been developed, but they have not been widely applied. Arsenate may be precipitated with an excess of silver nitrate and the excess determined potentiometrically with standard potassium chloride solution after the removal of the precipitate (108,109). The stability of the excess silver nitrate is determined by the potential of a silver electrode. The limit of absolute error is reported to be ±0.2% for arsenic. The method has a limited application because of interference from halides and other anions precipitated by silver ion.

Arsenate can be precipitated as the quinolinearsenimolybdate from 0.9–1.6 N hydrochloric acid (126). After filtering and washing, the precipitate is dissolved in excess standard base, 26 equivalents of sodium hydroxide being consumed on solution and the excess base being titrated with standard acid to the phenolphthalein end point. The results are reported to be good to 0.01 mg arsenic.

Direct determination of 0.05–0.10 gram arsenate in the presence of oxidants such as iodates, bromates, and chromates is possible by titration of an alcoholic, ammoniacal, ammonium chloride solution with magnesium sulfate (10). The end point is the formation of a violet-colored magnesium complex with Eriochrome T indicator.

2.   Redox Reactions

Arsenic may be determined by oxidation of As(III) or reduction of As(V). Although many redox reagents have been suggested and used, arsenic is usually determined by oxidation of As(III) with bromate in acid solution or with iodine in weak alkali.

a.   BROMATE OXIDIMETRY

Although bromate oxidation is normally slow, it is catalyzed by the high chloride concentration from the hydrochloric acid used in the arsenic determination. The optimum acidity at the end point is 1.2–3.5 N with respect to hydrochloric acid, and a titration temperature of 80°C is recommended (93). All ions that can be oxidized by bromate in acid solution will interfere. To eliminate most of the interference the solution may be oxidized with bromine to convert the arsenic and antimony to pentavalent state, the excess bromine driven off on a water bath, and the arsenic and antimony reduced with sulfur dioxide to the trivalent state. After this treatment the only interference is from Fe(II), V(IV), Sb(III), and iodide. If a halide distillation is used for separation, only a slight interference may be encountered from antimony. In the presence of a small amount of iodine chloride, arsenic may be titrated preferentially in the presence of antimony with potassium bromate (142).

The indicator commonly used is methyl orange, the slightest excess of the titrant slowly oxidizing the dye to a colorless compound that unfortunately cannot be reduced back to the colored form. If methyl orange is used and the acidity is not within the recommended limits, a sluggish response is obtained. In some procedures no indicator is added, and the end point is simply the first appearance of a yellow color due to free bromine. Quinoline yellow (15), fuchsin (143), 1-naphtoflavone (206,221), p-sulfonamidochrysodine hydrochloride (206), and p-ethoxy-chrysoidine (97) have been suggested as reversible indicators. Trypa-flavin (65), rivanol, and harmine (64) have been recommended as fluorescent indicators.

Arsenious acid may be titrated amperometrically with potassium bromate using rotating or vibrating platinum electrodes (71,101,187). With 0.001 N arsenic solutions an accuracy of 0.1% can be obtained (71), and the absolute error appears to depend much less on the concentrations of the solutions than on the total volume of solution, or, more likely, the total surface area (101).

The end point of an arsenite titration with bromate may be determined photometrically by measuring the amount of excess bromine. The measurements are made in the range, 360–270 nm, where the absorbance is linearly dependent on the concentration of the bromine (193). Another photometric end point that makes use of the absorbance of the chloro complex of antimony is applied in the simultaneous determination of arsenic and antimony (193). In this determination, the initial absorbance of the system remains constant, begins to drop as all the arsenic is oxidized and the antimony begins to be oxidized, and finally rises again as excess bromine forms after the complete oxidation of antimony.

b. IODIMETRY

Another frequently used method for the volumetric determination of arsenic is the oxidation of arsenite with a standard iodine solution (5,8). Prior to titration, the arsenic solution is adjusted to pH 8 with sodium bicarbonate, and the solution is titrated to the faint blue of iodo-starch. Trypaflavin (65), rivanol, and harmine (64) have been suggested as fluorescent indicators for the iodimetric titration of arsenic. The main interfering ions are those that are stable in acid solution, but become reductants in alkaline, that is, Fe(II), V(IV), and so forth. Tungsten(VI) also interferes.

Organic arsenic compounds such as triphenyl arsine also may be titrated with standard iodine in aqueous acetic acid solution. The end point is detected potentiometrically (159a).

Arsenic may be determined in the presence of copper by complexing the latter with an alkali citrate solution and titration of the arsenic in a neutral media with standard iodine solution (213). After acidification, the copper complex breaks down, and the iodine equivalent of copper is liberated and titrated. Alternatively, the copper may be determined first by precipitation with ammonium thiocyanate in the presence of sodium dihydrogen phosphite (145). The filtrate is then treated with concentrated hydrochloric acid and heated to produce carbonyl sulfide, which precipitates the arsenic as arsenious sulfide. The arsenic may then be determined iodimetrically.

Coulometrically generated iodine has been used by a number of investigators (37,50,144) for the determination of arsenic. The end point in the procedures is determined either amperometrically (137,144,187) or spectrophotometrically (50). With the former method, an accuracy of approximately ±0.2% is claimed for the range of 64–1200 μg of arsenic and an average error without regard to sign of 0.6 μg. In the spectrophotometric end point method, an average error of 0.5% is reported for the 50–100 μg range, and 2.1% for the 10–50 μg range. Below 10μg, the error increases rapidly, but this may possibly be overcome by using smaller quantities of solutions and the uv region, where iodine has a greater molar absorbance. Coulometrically generated bromine also has been used for the microtitration of arsenic(III). Bromine is generated with an efficiency of 99.9% using a vitreous carbon electrode. An intergrating digital milliameter is used to measure the number of coulombs. Three to fifteen micrograms of arsenic were determined with an accuracy of ±2% (87a).

c. CERATE OXIDIMETRY

Cerate solutions have been used for the determination of micro amounts of arsenic (23,60,133a,175), using a trace of osmium tetroxide or iodide as catalyst. Smith (175) titrated arsenic with 0.001 N cerate solutions in 2 N perchloric acid with nitroferroin as indicator. With this procedure 150–270 μg of arsenious oxide is determinable with an average error of 0.57%. Diphenylamine-type indicators also may be used along with osmium tetroxide or iodine as catalyst (61a). Potentiometric procedures are reported by Pribil (140,141) that are suitable for the simultaneous determination of Cu, Sb, and As in castings and alloys; Pb, Sn, Cd, Zn, and other constituents of the alloys do not interfere with the determination.

Accurate and rapid photometric end-point procedures are also available. These procedures make use of the strong absorbance of the ceric ion at 320 nm, where the cerous, arsenite, and arsenate ions do not absorb. Because of the strong absorption of the ceric ion, a very dilute solution can be used for the microtitration of arsenic (23). From 66.7 μg to 33 mg can be determined with a 10^{-4} N ceric sulfate solution in an average error range of 0.18–0.11%, equivalent to an accuracy of one to two parts per thousand.

Another photometric end-point procedure utilizes coulometrically generated ceric ion as the titrant (60). This procedure is especially good for running a series of determinations in the same vessel without discarding the solution from a previous sample. The results obtained compare favorably with coulometric halogen procedures. From 5 to 900 μg of arsenic are easily determined with an absolute error of the general order of 1 μg.

d. MISCELLANEOUS REDOX METHODS

A number of other oxidizing agents are used for the determination of arsenic. Good results are possible with a potentiometric potassium permanganate titration if proper alkalinity is chosen and telluric acid is present (81). Under proper conditions, as little as 2 μg of arsenic may be determined with a maximum error of 2%.

A standard Mn(III) solution stabilized with pyrophosphate may be used for titrimetric oxidation of arsenic (16,157). The use of potassium iodate as a promoter (157) and aqueous barium diphenylamine sulfonate indicator is recommended (16) for this titration.

Potassium metaperiodate has been investigated as a titrant for arsenic. The use of a mixed indicator of diphenylamine and methyl orange, which produces a blue color at the end point, is advised (194). A potentiometric end point using bright platinum foil immersed in the solution being titrated coupled with a saturated calomel electrode through an agar–agar potassium chloride bridge is also feasible (171).

Sodium hypobromite is a suitable titrant for alkaline arsenite solutions. Diphenylamine sulfonic acid and 2-aminodiphenylamine sulfonic acid-4 are recommended as suitable visual indicators (217), and a 0.01% luminol solution is suggested as a luminescent indicator (45). Standard hydrogen peroxide and standard hypochlorous acid solution may also be used as titrants for arsenious oxide with luminol as the indicator.

Chloramine (138), chloramine T (36), and chloramine B (172) have been suggested as economical substitutes for iodine in the determination of arsenic. Iodine monochloride is added as a catalyst, and p-ethoxychrysoidin hydrochloride or Brilliant Carmoisin are used as indicators. From 50 to 250 mg of arsenious oxide can be determined quantitatively using these titrants.

Procedures have been studied and described for the titration of As(III) in quantities from 30 to 1000 μg with an average error without regard to sign of less than 0.5 μg using coulometrically generated bromine (131,187) or chlorine (52,187) and an amperometric end point.

TABLE 6
Polarographic Characteristics of Arsenic

Supporting Electrolyte	$E_{1/2}$ vs SCE[a]	I	Reference
As(III)			
0.1 M HCl, 0.001% methylene blue	—	12.6	—[c]
1 M HCl, 0.0001% methylene blue	−0.43	6.04	—[d−f]
	−0.67	12.00	—[g,h]
12 M HCl	>0	3.94	—[h,l]
1 M HNO$_3$, 0.01% gelatin	−0.7	—	—[g]
7.3 M H$_3$PO$_4$	−0.46	—	—[j]
	−0.71	—	
0.05 M H$_2$SO$_4$, 0.001% methylene blue		12.8	—[c]
0.5 M H$_2$SO$_4$, 0.01% gelatin	−0.7	—	—[g,k]
	−1.0	8.4	
1 M H$_2$ tartrate, 1 M HCl	−0.40	4.32	—[e,i]
	−0.67	—	
2 M HOAc, 2 M NH$_4$OAc, 0.01% gelatin	−0.92	—	—[m,n]
0.5 M Na$_2$ tartrate, pH 4.5, 0.01% gelatin	−1.0	—	—[g]
0.1 M NH$_3$, 0.1 M NH$_4$Cl	−1.71	—	—[n]
1 M Na$_3$ citrate, 0.1 M NaOH	(−0.31)	—	—[j]
0.5 M KOH, 0.025% gelatin or 0.002% methylene blue	(−0.26)	−3.82	—[g,o−r]
0.5 M Na$_2$ tartrate, pH 8.8, 0.01% gelatin	NR[b]	—	—[g]
1 M Na$_2$ tartrate, 0.8 M NaOH	(−0.31)	−2.87	—[g,p]
0.1 M EDTA, pH 6–8	−1.6	—	—[s]
As(V)			
11.5 M HCl	>0	—	—[h]
Dil. HCl, H$_2$SO$_4$, NaOH or acetate buffers	NR	—	—[d,h,m,q]

[a] Half-wave potential in parentheses denotes anodic wave.

[b] NR = not reducible.

[c] Everest, D. A., and G. W. Finch, *J. Chem. Soc.*, **1955**, 704 (1955).

[d] Bambach, K., *Ind. Eng. Chem. Anal. Ed.*, **14**, 265 (1942).

[e] Haight, G. P., Jr., *Anal. Chem.*, **26**, 593 (1954).

[f] Kacirkova, K., *Collect. Czech. Chem. Commun.*, **1**, 477 (1929); through *Chem. Abstr.*, **24**, 1012 (1930).

[g] Lingane, J. J., *Ind. Eng. Chem. Anal. Ed.*, **15**, 583 (1943).

[h] Meites, L., *J. Amer. Chem. Soc.*, **76**, 5927 (1954).

[i] Haight, G. P., Jr., *J. Amer. Chem. Soc.*, **75**, 3848 (1953).

[j] Meites, L., *Polarographic Techniques*, Wiley-Interscience, New York, 1955, p. 252.

[k] Coulson, R. E., *Analyst (London)*, **82**, 161 (1957).

[l] Kolthoff, I. M., and J. J. Lingane, *Polarography*, Wiley-Interscience, New York, 1941.

[m] De Sesa, M. A., D. N. Hume, A. C. Glamm, Jr., and D. D. De Ford, *Anal. Chem.*, **25**, 983 (1953).

[n] Pribil, R., Z. Roubal, and E. Svatek, *Collect. Czech. Chem. Commun.*, **18**, 43 (1953).

[o] Bayerle, V., *Rec. Trav. Chim.*, **44**, 514 (1925).

[p] Cozzi, D., and S. Vivarelli, *Anal. Chim. Acta*, **5**, 215 (1951).

[q] Kolthoff, I. M., and R. L. Probst, *Anal. Chem.*, **21**, 753 (1949).

[r] Vivarelli, S., *Anal. Chim. Acta*, **6**, 379 (1952).

[s] Valenta, P., and P. Zuman, *Chem. Listy*, **46**, 478 (1952); *Chem. Abstr.*, **46**, 10953 (1952).

C. POLAROGRAPHY

Both anodic and cathodic waves characterize the polarographic behavior of arsenic, the shape and type of wave being dependent on the
nature of the supporting electrolyte. Of these two waves the anodic is
considered to be eminently more suitable for analytical purposes since it
is not affected by other common metals. Polarographic half-wave
potentials and diffusion current constants for arsenic in various base
solutions appear in Table 6.

A two-step reduction of arsenious acid at a dropping mercury electrode from a 1 M hydrochloric acid solution was first reported by
Kacirkova (92). The first wave is not quantitative and is thought to result
from a reduction to elemental arsenic, but the height of the second wave
is proportional to the concentration of the arsenic and is believed to be
due to further reduction to arsine.

In later experiments by Lingane (111), the presence of the two steps in
hydrochloric acid solution has been confirmed, and in the presence of
0.01% gelatin as maximum suppressor a fairly well-defined but slightly
distorted first wave is obtained, the height of which is proportional to
the arsenious acid concentration. A more recent investigation (124) of
the polarographic behaviour in 1 M hydrochloric acid media in the
absence of a maximum suppressor gave a fairly well-defined wave that
was not as distorted as Lingane's gelatin suppressed wave. This wave is
quantitative for arsenic concentrations between 0.018 and 0.9 mM, but
above the upper limit the height of the wave becomes constant and
independent of arsenic concentration, hydrochloric acid content, and
temperature changes. This phenomenon is thought to be the result of the
formation of a film of adsorbed elementary arsenic on the surface of the
mercury drop. The second wave is also well defined up to a concentration of 0.05 mM; above this concentration an odd double maximum is
obtained. In the presence of 1.5×10^{-4}% methylene blue this maximum
is suppressed, and a diffusion current constant of 12.0 for both steps is
obtained. This is almost twice the value for the first step alone and is
proposed as proof that the reduction of trivalent arsenic proceeds in two
successive three-electron steps.

Well-defined steps are obtained from both 0.5 M sulfuric and 1 M
nitric acid solutions plus 0.01% gelatin (111). The diffusion-current
constants for the first wave are 8.4 and 8.8, respectively, for the two
base solutions and are almost identical with the value obtained from the
hydrochloric acid medium. A 0.001% methylene blue solution is a better
suppressing agent than gelatin (49) and produces steps that are suitable
for analytical purposes.

Arsenite is not reduced from strong alkaline, neutral, or alkaline tartrate solutions, while only an ill-defined wave is produced from an acidic tartrate base solution. These waves are definitely unsuitable for analytical application. However, there are possibilities of using alkali and alkaline tartrate solutions for the determination of such ions as bismuth, lead, and cadmium in the presence of high concentrations of arsenic.

An anodic wave due to the oxidation of trivalent arsenic to its pentavalent state is produced from highly alkaline solutions (100). Similar anodic waves are obtained from alkaline solutions containing a high concentration of citrate or tartrate salts. The pronounced maximums that occur are only slightly suppressed by gelatin but are completely eliminated by thymolphthalein (34). The regular step that results is suitable for analytical purposes, since it is not affected by other common metals.

Quinquevalent arsenic is not reduced from dilute hydrochloric and sulfuric acids, sodium hydroxide, or acetate buffer media (100). A double-step reduction of pentavalent arsenic has been reported from 11.5 M hydrochloric acid solution (124), the first step of which is not well defined due to its merging with the anodic step from the dissolution of the electrode mercury, while the second is irreversible. The first wave is equivalent to a five-electron step and the second to a three-electron reduction from elemental arsenic to arsine. A review of various polarographic methods for arsenic is given in reference 7b. Methods for the determination of arsenic in ferrotungsten (41b), selenium (41a), and steel (192a) are discussed in the noted references. Certain organic compounds of arsenic give characteristic polarograms and are discussed in reference 215a. Parts per billion of arsenic may be determined by differential pulse polarography (126a), which is one of the most sensitive polarographic methods for arsenic.

Anodic stripping voltammetry and differential pulse anodic stripping voltammetry were used to determine nanogram quantities of arsenic. A 1 M hydrochloric acid or a 1 M perchloric acid solution is used as supporting electrolyte. Gold electrodes are superior to platinum. Both methods have a detection limit of 0.02 ng As/ml. Copper and mercury interfere since they have overlapping peaks (57b).

D. PHOTOMETRIC METHODS

Of the colorimetric methods that have been devised for the determination of arsenic only the Gutzeit, molybdenum blue, and the silver diethyldithiocarbamate can be considered as being of major importance.

However, the first mentioned is being rapidly displaced by the latter two in the more recent procedures for the estimation of arsenic in a variety of materials.

1.   Gutzeit Methods

The well-known Gutzeit test depends on a preliminary reduction of the arsenic to arsenite and the eventual evolution of arsine from a hydrogen generator over sensitized paper strips impregnated with mercuric chloride or bromide. The resultant stain is then compared to stains produced by standard amounts of arsenic submitted to the same treatment (4a,8). Better results are obtainable through a modification that involves the use of cotton thread instead of the paper strip, especially if the thread is impregnated with mercuric bromide, which is superior to mercuric chloride as an absorbing reagent (25,78). The estimation of the arsenic level requires an indirect comparison involving the length of stain produced versus the amount of arsenic present. A sensitivity of 1 $\times$ 10^{-5} mg of arsenic is obtainable and 1 $\times$ 10^{-4} mg of arsenic is determinable. Standard bromide stains will keep for about 6 months, but chloride stains fade after 1 week.

Impregnated paper discs are superior to strips and threads as far as reproducibility of results is concerned (4,119,160). From 4×10^{-5} to 1×10^{-3} mg of arsenic can be readily determined if the hydrogen is evolved at room temperature in a system in which the pressure is reduced to 0.25–0.50 atm and the gases drawn through a sensitized disc having an exposed circular area with a diameter between 0.125 and 0.25 in. The stain may be intensified with ammonia and compared to a standardized photo step scale. The stains can also be treated with cadmium iodide (189), whereby the stains turn brown and are easier to compare. Standard artificial spots made from chrome yellow and deep chrome yellow have been developed for field use (4). The reflectance of the mercuric bromide test spot at 440 nm against a clean unsensitized disc is also used for the estimation of arsenic. This method requires a standard curve of the reflectance readings, expressed as absorbance, versus the total micrograms of arsenic (119).

Because of the high sensitivity of the Gutzeit method, it is considered to be the method of choice for very minute amounts of arsenic. However, if other than minute amounts are to be determined, the molybdenum blue is more precise and more accurate. In addition, the Gutzeit method has several disadvantages that tend to turn the analyst to other methods, especially if only occasional samples are run. Most of the procedures are troublesome to carry out and require extreme care to maintain constant conditions. For example, the hydrogen-evolution rate, the temperature of both the hydrogen generator and the absorption apparatus, and the construction of the apparatus itself have an effect on the results. A constant temperature is necessary to control the moisture content of the absorbents. Lower temperatures in the generator produce shorter stains and more intense colors, while lower temperatures in the absorption apparatus result in faint, long stains that are less definite. A difference in the apparatus results in a difference in the size and shape of the stain (150). Methods of preparation of the sensitized media and the positioning in the absorbing apparatus are other important factors. Some procedures call for special electrolytic generators for hydrogen (135) or gold chloride as the absorbing reagent (74,127).

If arsenic is to be determined in organic material that is wet ashed, the sample must be submitted to an arsenic trichloride distillation prior to the determination. A similar distillation is recommended if mercury, bismuth, copper, iron, or selenium are present since these elements reduce the sensitivity if present in appreciable amounts. Pyridine derivatives adversely affect the adherence of tin to the zinc, thereby slowing down the evolution of hydrogen and preventing a complete conversion of arsenic to arsine.

Although granular zinc is usually recommended in the standard procedures, the same accuracy is obtainable with the other forms provided the conditions are standardized. A method has been described for the preparation of zinc pellets with a constant exposed surface area (63). Stannous chloride is normally added to sensitize the zinc and to absorb any iodine liberated. Copper is suggested as an alternative catalyst (197). Iron is often used if arsenic is to be determined in the presence of antimony and phosphorus. Granular aluminum has also been used. With hydrochloric acid alone, an uneven evolution of hydrogen results but is stabilized upon the addition of stannous chloride. Pure aluminum and sodium hydroxide (54) have been used to detect and determine arsenic in the presence of antimony, phosphorous acid, sulfide, selenium, and tellurium.

2. Colorimetric Methods Involving the Molybdate Ion

Arsenate, phosphate, and silicate react with ammonium molybdate to form the corresponding heteropoly(molybdic) acids, which, when treated with an appropriate reducing agent, are reduced to the highly colored molybdenum blue complex in which the molybdenum is present in a lower valence state. In the case of arsenate, the molybdenum blue compound is thought to be $H_7[As(Mo_2O_7)_5OMo_2O_5]$ (196). Since phosphate is similar in reaction to arsenate, its separation from arsenic is required. The removal of silicate is not essential, but its absence is desirable, even though the silicate reacts less easily than the arsenate. Because of the high sensitivity of the molybdenum blue, the method has been extensively investigated, and since procedures for small amounts of arsenic have been developed that are more accurate than the Gutzeit methods, the latter are being replaced rapidly by the former. Through special submicro techniques 0.1 μg of arsenic may be estimated (32).

The choice of reducing agent should be such that only the molybdenum in the complex and not in the excess ammonium molybdate is reduced. Stannous chloride (39,41,103,154,168,188,205,220), a reduced molybdate reagent (95,220,223), and hydrazine sulfate (12,20,29,48,53,-79,94,110,112,115–117,120,128,147,159,191,192) are suitable reagents, the latter being preferred. Hydrazine sulfate produces a stable blue color with an absorption maximum at 840 nm (192) that shows no appreciable loss after 24 hr. Although heating is required to develop the color with this reducing agent, the time of heating does not affect the intensity of the color as long as the length of heating has been sufficient to develop the maximum color. Hydroquinone, although used in phosphate determinations, produces a color that does not conform to Beer's law (214). With stannous chloride a yellowish hue develops, and the resulting color varies with the concentration of the stannous chloride and the time of heating. To keep the formation of the yellowish hue at a minimum, only the minimum amounts of both the molybdate and reducing reagents and the maximum sulfuric acid that permits the full development of color should be used. If the acidity is too low, any silicate present will react and form the molybdenum blue, and, at times, even the molybdate from the reagent will produce a blue. If the acidity is too high, then the maximum color from arsenic will not develop. The optimum final concentrations for the ammonium molybdate and sulfuric acid are approximately 0.05% and 0.25 N, respectively (29).

Various methods have been utilized in the preliminary isolation of arsenic in the molybdenum blue procedures. Most procedures depend on distillation as arsenic trichloride or volatilization as arsine. Although antimony, germanium, and selenium distill along with arsenic, antimony and selenium do not react with the molybdate reagent, while germanium does to a limited extent. Other methods require an extraction as the xanthate with carbon tetrachloride (95,191) or an extraction with diethylammonium diethyldithiocarbamate in chloroform (112,115,188). The carbamate extracts more completely than the xanthate.

Photoelectric measurements have been made at various wave lengths such as 610, 620, 625, and 725 nm with an accuracy of ±5% down to 1 μg, where it is ±10%. The sensitivity is 0.2 μg of arsenic, and the method is definitely preferable to an iodimetric titration for samples containing less than 4 mg of arsenic (22). At 840 nm, the wavelength of maximum absorption, the system has a molar absorptivity of 25,400 liters/m/cm, and Beer's law applies over the range, 0–3 ppm (20).

Spectrophotometric procedures for the determination of arsenic as the 12-molybdoarsenic (215) or molydovanadoarsenic (9,68) acid have been described. The 12-molybdoarsenic acid method is convenient for the determination of arsenic in quantities smaller than 1 mg and requires no heating and no waiting for maximum color development. Larger amounts of soluble silica and concentrations of phosphate slightly greater than those of arsenic being determined are tolerable. There is conformance to Beer's law for the range, 0–600 μg, with a slight negative deviation above 600, but still applicable up to 1000 μg. The molar absorptivity is 5100 liters/m/cm at 370 nm, about one-fifth as sensitive as the heteropoly blue.

Another 12-molybdo acid method, used for the simultaneous determination of arsenic, phosphorus, and germanium, is based on the chromatographic separation of the complex anion spots using butyl alcohol with 10% (vol/vol) nitric acid as the developing solvent (37). The total yellow color is spectrophotometrically measured directly on the filter paper via a continuous scanning technique. A nearly linear relationship exists between the mean peak area and the weight for 0–10 μg arsenic.

The molybdovanadoarsenic acid complex provides the means for a rapid, convenient, but less sensitive method for the determination of arsenic. Beer's law applies for 1–30 ppm at 400 nm, where the molar absorptivity is 2540 liters/m/cm (68). The yellow hue is stable for 24 hr. Through acid control a simultaneous determination of arsenic and phosphorus is possible (9). The phosphorus is determined at a nitric acid acidity of 1.6 N, and the total arsenic and phosphorus at an acidity of 0.2 N. By means of three calibration curves, the concentrations of both elements are determined. From 0 to 0.10% arsenic has been successfully determined in copper-base alloys with this procedure.

Successive selective extractions of the heteropoly(molybdo) acids of phosphate, arsenate, and silicate with organic solvents and subsequent colorimetric determination have been described. In one scheme (3) the phosphate complex is extracted with butanol–chloroform, the arsenate with butanol–ethyl acetate–chloroform, and silicate with butanol at a higher acidity. The other scheme, suggested but not fully investigated

(40), depends on the extraction of the heteropoly(molybdophosphate) with isoamyl acetate and the arsenate with isoamyl acetate from a 17% ethanol solution, leaving the silicate in the aqueous phase. The isoamyl acetate extracts obey Beer's law when the absorbances are measured at 323 nm.

The molybdenum blue method has been used for the determination of arsenic in steel (57a) and in natural waters (87b).

3.   Colorimetric Methods Involving Silver Diethyldithiocarbamate

Arsine reacts with silver diethyldithiocarbamate in pyridine to give a red coloration that allows for the detection of 0.5 μg of arsenic (91,211). A solution of 1-ephedrine in chloroform may be substituted for pyridine as a solvent for silver diethyldithiocarbamate (101a). The sensitivity and accuracy are as good as with pyridine. Hydrogen sulfide is one of the substances that is likely to interfere, but it can be readily removed by passing the generated gases through a scrubber containing glass wool impregnated with lead acetate. Stibine, which develops a different shade, can be present in a concentration as high as 0.5 μg antimony/ml of sample and not interfere in the amount of arsenic found. The system follows Beer's law to 20 μg of arsenic in 3 ml of absorbing solution if measured at 560 nm (91). Since the method is simpler than molybdenum methods, and is considered more sensitive than the Gutzeit, it has become a recommended analytical method for the determination of arsenic in air (31), in water (4a), and in steel (18a).

In the determination of arsenic in technical-grade iron, an accuracy of ±0.0004% arsenic is reported when weighing 0.1 gram samples with an arsenic content in hundredths of a percent, and ±0.006% if the arsenic content is in tenths of a percent (212). Adaptations of the silver diethyldithiocarbamate method have been devised for the determination of arsenic in naphtha on the parts per billion (ppb) level. Precision and accuracy of about 3% relative or 0.3 ppb, whichever is greater, have been reported for a method depending on a chromatographic separation of arsenic on sulfuric acid impregnated silica gel (139). Another procedure for naphtha samples has a standard deviation of 1 ppb at 8 ppb and 4 ppb at 78 ppb with a relative error of 12–6% at the levels, respectively (2).

4.   Miscellaneous Colorimetric Methods

A number of other colorimetric methods have been proposed, but none have attracted wide interest because of the development of the

more accurate and precise molybdenum blue and silver diethyldithiocarbamate procedures. Some of these early methods depend on a colorimetric comparison or a photometric measurement of colloidal arsenic
(167,207) or colloidal arsenious sulfide stabilized with either gum arabic
(125) or gelatin (62). Under optimum conditions, 1–12 μg arsenic/ml can
be determined as colloidal arsenic with an accuracy up to 3% relative.

Another method (17) requires the addition of potassium iodide and
sodium sulfite to the arsenic trichloride distillate and the comparison of
the resulting turbidity with that of standards containing known amounts
of arsenic. By means of a colorimeter as little as 2 mg of arsenic can be
determined.

A colorimetric adaptation of the Levvy method for arsenic is reported
(121). In this modification, the metallic silver liberated by the absorbed
arsine is determined by oxidizing it with an excess of a standard solution
of ceric sulfate, the excess of which is determined colorimetrically.
Quantities of arsenic ranging from 0 to 100 μg may be determined with
an average standard error of 1.8 μg, while quantities ranging from 100 to
700 μg may be determined with an average standard error of 4 μg. Since
the system deviates from Beer's law, a calibration curve is required.

5.   Nephelometric Methods

When arsine is absorbed in a silver nitrate solution, metallic silver is
the reduction product. After filtration the silver is dissolved with nitric
acid, precipitated as the chloride, and determined nephelometrically.
This procedure has been submitted to an extensive investigation (102)
and shown to be unsuitable. Experimental data has indicated that all the
arsenic is not converted to arsenite and that some silver is reduced by
the stream of hydrogen produced during the generation of arsine.
Occlusion of arsenic by the silver precipitate is also a factor. The
reduction of silver by hydrogen is not as great in an ammoniacal solution
as in neutral medium, but still the results are such that an indirect
nephelometric determination of arsenic is not feasible.

The Bougault method serves as a basis for several nephelometric
procedures for arsenic. Some require a stabilizer such as gum arabic or
borax–rosin (203). The gum-arabic-stabilized standards require daily
preparation, while the borax–rosin series shows no change after 1
month.

Arsenic is determined in copper alloys using Bougault's reagent. The
absorbance of the samples and standards are measured at 612 nm, and a
calibration curve is prepared from the standard data. The colloid is
stable for 8 hr if the arsenic content is 0.5% or less. The results are good

compared to other methods. Sn, Ni, Pb, Fe, Mn, PO_4^{-3}, Si, Bi, Al, and Cl^- do not interfere, but nitrate must be removed (185).

6. Emission Spectroscopy

Although arsenic has a number of sensitive lines that are suitable for its spectrochemical detection and determination (Table 2), the lines at 2349.84 and 2860.45 are most often used since they suffer the least from interference. Suitable methods have been developed for the determination of arsenic in various alloys (Table 7) and are reported in detail by the American Society for Testing Materials.

High-frequency plasma sources have been used for the excitation of samples, and arsenic is determined using the 189.0 nm line (94b). A microwave-induced plasma coupled to a Videcon detector also has been used (58a). Several trace elements may be determined simultaneously using this technique.

7. X-ray Emission and Fluorescence

Each chemical element emits characteristic X-rays when excited by a beam of electrons or by white X-rays. X-ray spectra are less complex

TABLE 7 (6)

Tentative and Suggested Methods for the Spectrochemical Determination of Arsenic in Metals and Alloys

Material	Method	Range (%)	ASTM	Designation
Admiralty metal	Cast pin, dc arc	0.025–0.1	E-2	SM 5-8
Wrought copper alloys	dc arc technique	0.003–0.80	E-2	SM 5-2
Lead-base alloys	Dry powder, dc arc	0.001–0.1	E-2	SM 6-1
Antimonial lead alloys	Point-to-plane, spark technique	0.01–0.6	E-2	SM 6-6
Type metal alloys	Point-to-plane, spark technique	0.02–0.50	E-2	SM 6-8
Lead-base bearing alloys	Cast pin, spark technique	0.09–0.8	E-2	SM 6-7
Pig leads	dc arc technique	0.001–0.05	E-2	SM 6-2
Lead–tin solders	Dry powder, dc arc	0.02–0.2	E-2	SM 6-5
Low-alloy steel	Point-to-plane, ac arc	0.02–0.2	E-2	SM 9-5
Tin	Cast pin, spark technique	0.02–0.2	E-2	SM 6-9
Tin alloys	Cup graphite arc	0.001–0.3	E51–43T	
Tin-base alloys	Dry powder, dc arc	0.005–0.02	E-2	SM 6-4

than those in the visible and ultraviolet: hence are easier to interpret, and less interference occurs. The theory and practice of X-ray analysis is discussed in references 18b and 109a. This technique has proved useful for the analysis of a variety of substances, especially alloys. Arsenic may be determined in the range, 1–500 μg, but frequently a preliminary separation (149a) of arsenic is required owing to self-absorption of the desired X-rays by heavier elements in the sample. This technique has been used in the determination of arsenic in steels (211a), copper-base alloys (23a), and other substances (62a,68a,149a).

8. Atomic-Absorption Spectroscopy

Atomic-absorption spectroscopy is now widely used for the determination of a large number of elements. Methods for more than sixty different elements have been published. Atomic absorption, frequently referred to as AA, is the preferred technique for the analysis of complex mixtures, because the method is very specific and frequently no chemical separations are necessary.

In general, the sample is introduced into a flame in the form of a spray from a solution or as a gas, and the amount of radiation absorbed by a specific spectral line of the vaporized element is measured. Methods involving techniques other than a flame to vaporize the sample are called "flameless atomic absorption." All methods require an emission source of the element. These sources are commonly called "lamps". The amount of radiant energy from a specific wavelength absorbed by the vaporized element is proportional to its concentration in the original sample. A calibration curve prepared from a series of known standards is used to determine the concentration of the desired element in the unknown sample. The theory and practice of atomic absorption spectroscopy is discussed in Part I, Volumes 6 and 11 of this Treatise and in references 30a, 38a, 42a, and 143b.

The determination of arsenic by atomic absorption has been the subject of many investigations. Arsenic has several characteristic emission (and absorption) lines at 189.0, 193.7, 197.2, 200.3, and 228.0 nm (94c, 150a). Unfortunately most of these lie in the extreme ultraviolet region, which gives rise to several problems. One of the main sources of interference is the molecular absorption of flame gases in this region. Molecular hydrogen produced by the decomposition of the hydrocarbon fuel or other organic compounds strongly absorbs radiation of wavelength less than 200 nm (150a). High background noise also is characteristic of this spectral region. Several studies using various gas mixtures indicate the 193.7 nm line is the most desirable (94a,94c,120a,150a,172a).

Although it is slightly less sensitive than other lines, it is less affected by scattered light than the 189.0 nm line (124a).

Various mixtures of gases have been studied for use in the AA determination of arsenic. Unfortunately, the more commonly used mixtures are the least sensitive. Air–acetylene flames give the least sensitivity, with a detection limit of about 0.25 ppm of arsenic (94a,94c) using aqueous solutions. Sensitivity may be increased four- to fivefold by using 80–90% ethyl alcohol solutions (115a). Nitrous oxide–acetylene flames give somewhat better sensitivity with lower backgrounds (67a,94a).

Increased sensitivities and lower backgrounds are attained using nitrogen–hydrogen-entrained air (7a) and argon–hydrogen-entrained air mixtures. These mixtures now are the most commonly used. Increased sensitivity also is attained by using an electrodeless discharge lamp as emission source in place of the usual hollow cathode lamp. As little as 2 ng of arsenic can be detected using this technique (11a,124a,162a).

Another technique that has enhanced the sensitivity considerably is the introduction of arsenic into the flame in the form of arsine gas rather than from a solution in the form of a spray. Arsine is generated either by the conventional Gutzeit method using zinc metal plus acid (30b, 65a, 67a, 134c), or by using sodium borohydride (20a, 52a, 162b, 174a, 201b, 210a). Studies indicate the sodium borohydride method is superior to the zinc–acid method (162b), since it gives higher yields of arsine and less pretreatment of the sample is required. Gaseous hydrides of Sb, Se, Te, Bi, Pb, Sn, and Ge also are generated using this technique (174a,201b), and they may cause some suppression of the arsenic line.

Sensitivity also is increased by trapping the arsine in a liquid-nitrogen cold trap (67a,75a,96a,134c) or in a balloon (30b,117a,125a,162b), then introducing the entire sample into the flame. By combining all these techniques arsenic is easily determined in the nanogram range (134c).

Flameless atomic absorption also has been used for the determination of arsenic. The sample, usually in the form of arsine gas, is introduced into a quartz tube heated to 700–1000°C, and the absorption is measured (30b,65a,213a). Graphite furnaces also have been used (11a,96a, 120b,125a,144a). Flameless methods tend to be more sensitive than flame methods but are hampered by interference from calcium, silica, and phosphates (125a). Atomic-absorption techniques have been applied to the determination of arsenic in water (20a,65a,92a,125a,162b), in biological samples (115a,134c), tobacco (67a), steel (144a), and various copper and zinc alloys (11a).

Indirect atomic-absorption methods also have been devised. Arsenic is converted to arsenate ions, then treated with molybdate to form 12-

molybdoarsenic acid, which is extracted into an organic solvent. This is aspirated into the flame, and molybdenum is read at 313.3 nm. This method overcomes the difficulties arising from the measurement of arsenic lines in the ultraviolet region (37a,143a,221a). The sensitivity of the indirect method is increased by using a flameless technique (156a).

Atomic-fluorescence techniques also have been used for the determination of arsenic. The sample is introduced into the flame as gaseous arsine, the arsenic atoms are excited with an electrodeless lamp (206a,219a), and the fluorescence is measured using the 193.7 nm line. This method is 5–30 times more sensitive than atomic absorption, and as little as 0.0001 μg arsenic/ml can be determined (201a).

Direct flame-emission spectroscopy for arsenic is insensitive and usually is not used (38a). Dean and Fues (38b), however, employed this technique to determine arsenic in organic compounds such as triphenylarsine. The compounds were dissolved in benzene and aspirated into a fuel-rich acetylene–oxygen flame and read at 235.0 nm. The detection limit was 2.2 μg arsenic/ml.

E. NEUTRON ACTIVATION ANALYSIS

Neutron activation is one of the most sensitive means for the estimation of arsenic. Compared with the graphite ac arc, the copper spark, amperometric titrations, and the most sensitive colorimetric reaction, the activation analysis is 100,000, 50,000, 4,000, and 1,000 times more sensitive, respectively (123). However, disadvantages that limit the use of neutron activation are the scarcity of nuclear reactors for neutron activation and need of special equipment to handle the activated samples.

Arsenic is monoisotopic, consisting 100% of the isotope ^{75}As. This is transformed to radioactive ^{76}As by bombardment with thermal neutrons by a n-γ reaction with a cross-section of 36.8 barns. Arsenic-76 decays with a half-life of 26.4 hr, emitting 0.4 and 2.97 MeV beta particles and gammas of 0.56 and 1.21 MeV energy.

Numerous procedures have been developed for the activation analysis of arsenic in a variety of materials. The activation periods in these methods vary from 1 day to 1 week, and the subsequent treatment of the sample depends on the nature of the material. In the case of biological samples, a Gutzeit separation has been used, followed by an estimation with a Geiger tube that accepts liquid samples (176). Methods reported for arsenic in hydrocarbon-reforming catalysts (170), silicon semiconductors (86,129), tungsten (87), and germanium oxide (174) utilize a halide distillation for separation and ammonium hypophosphite reduction for the final isolation of the arsenic.

Neutron activation also serves as a basis for the detection of arsenic poisoning in biological matter without destroying the sample. In suspected human cases, the hair can be submitted to activation and then examined in 2 mm sections for radioactive arsenic (67).

Gamma-ray spectrometry has been used to determine arsenic in a variety of substances. The sample is activated by neutron bombardment, and the 0.56 MeV gamma activity is measured. A separation of the [76]As prior to measurement is frequently required owing to interferences from [82]Br and [64]Cu. Gamma-ray spectrometry has been used for the determination of arsenic in rocks and soils (102a,185a), water (145a), biological materials (102b,157a), environmental samples (134d,195a,218a), crude oils (211b), and copper and brass (67b). As little as 1 ppb of arsenic is easily determined in these materials.

VII.   ANALYSIS OF ARSENIC COMPOUNDS

The number of arsenic compounds that are completely analyzed is understandably small because of their limited uses. Also, the method of manufacture of arsenic and arsenic trioxide by sublimation yields highly pure forms and makes extensive analysis of these products unnecessary. However, standards have been established with respect to minimum requirements for arsenious oxide, arsenic oxide, sodium monohydrogen arsenate, and sodium arsenite (146,155,199).

The most important arsenic compound, arsenious oxide, is analyzed for matter insoluble in ammonium hydroxide, residue on ignition, chloride, sulfide, antimony, lead, and iron. If the sample is not of primary standard quality, then an assay for arsenic is desirable.

For the determination of the ammonium-hydroxide-insoluble portion, 10 grams of the sample are refluxed with 100 ml of 1:2 ammonium hydroxide until dissolved. If a residue remains, the solution is filtered, and the residue is washed with warm dilute ammonium hydroxide, dried at 105°C, and weighed.

For residue on ignition, 5 grams of the sample is slowly ignited to constant weight in a platinum or silica dish under a well-ventilated hood. The residue is weighed and saved for the iron determination. If any darkening occurs during ignition, an indication of organic contamination, the sample is considered unsuitable as a primary standard.

The residue obtained on ignition is warmed with 1:1 hydrochloric acid, taken up with concentrated hydrochloric acid, and diluted. A suitable aliquot is taken, diluted, and treated with 30–50 mg of ammonium persulfate and 3 ml of 30% ammonium thiocyanate. Any red color produced should not be darker than that in a control containing the permissible concentration of iron and undergoing the same treatment. A more precise estimation of iron can be made by comparing with a series of standards similarly treated. If larger amounts of iron are present, a volumetric determination by the stannous chloride–dichromate method is in order.

Minute amounts of chloride impurity are determined turbidimetrically with silver nitrate. If the test is a specification check, a standard containing the maximum allowable concentration of chloride is run simultaneously with the sample. If more precise results are required, then either a series of standards may be run or one of the other methods suitable for micro chloride determination may be used.

To meet the sulfide specification, 1 gram of sample is dissolved in sodium hydroxide and treated with one drop of 10% lead acetate solution. The color produced should be the same as that from an equal volume of sodium hydroxide and lead acetate.

Antimony is determined as the Rhodamine B complex. After extraction with toluene (or benzene), the resulting color is compared with that of one or more standards treated in the same manner as the sample. The number of standards prepared depends on the type of information required, that is, whether it is a specifications check or a percentage determination.

Lead and the other heavy metals are separated as the dithiozonates or as sulfides, and the amount is estimated by comparison with standards having known amounts of lead.

An assay for trivalent arsenic is made by titration with standard iodine solution, using starch as the indicator.

Reagent-grade arsenic oxide must meet specifications as follows: assay, water-insoluble residue, chloride, nitrate, sulfate, alkalies and earths, trivalent arsenic, heavy metals, and iron.

The assay requires an iodometric procedure with standard thiosulfate solution as the titrant and starch as the indicator.

The nitrate level is estimated from the decolorization rate of indigo carmine by the sample in presence of chloride in a sulfuric acid media. A complete discharge of the blue color by a 0.25 gram sample in less than 10 min indicates the presence of more than 0.02% nitrate in the sample.

Sulfate is determined gravimetrically as barium sulfate.

The procedure for the determination of alkalies and earths requires a preliminary evaporation treatment with hydrobromic acid and a subsequent removal of the heavy metals with hydrogen sulfide, followed by evaporation of the filtrate to dryness and ignition.

The chloride, trivalent arsenic, heavy metals, and iron determinations are the same as described for arsenious oxide.

Similar tests are described for reagent-grade sodium monohydrogen arsenate and sodium arsenite. The details of these tests and the maximum tolerance levels for the impurities are readily available from the previously mentioned references.

VIII. DETERMINATION OF ARSENIC IN SPECIFIC MATERIALS

A. BIOLOGICAL SAMPLES

The detection and determination of arsenic in biological materials and food products has been the subject of many investigations, and a complete survey of materials examined is impractical. Numerous procedures (79,90,91,102b,107,115a,117,120,128,134c,157a,158,160,179,191) are reported as being suitable for the determination of arsenic in biological matter, including plant and animal tissue. In the plant category specific procedures are cited for vegetable matter, foliage, and tobacco (67a,178) and plants in general (8,134).

Arsenic is estimated by the molybdenum blue procedure in blood and bone (178,192), animal tissue (48,94,134,178,192), liver (48), skin (48), and urine (42,94,192). It is also determined colorimetrically by the silver diethyldithiocarbamate method in blood, urine, and liver (211).

Spectrochemical procedures have been devised for bone, liver, stomach, and kidney (204). Activation analysis may be used for its determination in hair (67) and tissue (176).

B. FOODS AND BEVERAGES

An extensive amount of literature exists on the estimation of arsenic in various foods (8,59,84,94,178,179). Representative procedures include those for baking powders (43,59,178,179); gelatin, sugar, malt, and grain (178); beer (189); fruits and vegetables (12,48,94,179); meats and eggs (48,94,179); and milk (94). The majority of these analyses are based on the formation of the heteropoly blue complex of arsenic.

C. METALS AND ALLOYS CONTAINING ARSENIC

Arsenic is variously determined in a large number of metals and alloys. Titration methods involving iodimetry, iodometry, ceratimetry, and bromatimetry are still highly recommended, even though the more rapid and accurate photometric heteropoly molybdate and silver diethyldithiocarbamate methods have been developed. The procedures recommended by the American Society for Testing Materials (5) are predominantly titration methods.

Excellent general references on procedures for arsenic bearing metals and alloys are readily available (5,59,158,178,179). Some specific methods found in these and other references are carbon steel and cast iron (18a,19,57a,59,99,144a,158,165,178,179,192a,211a,212); ferromanganese and high chromium steel (158,178); ferrous alloys (5,178); brasses, bronzes, and white-metal bearing alloys (5,11a,59,67b,158,178); copper (26,59,67b,158,178,179,213) and copper-base alloys (9,23a,26,145,185); arsenical copper (179,185); lead (5,114,158,178,179) and lead alloys (5,18,58,114,158,178,179,193); selenium (147); germanium and silicon (115,179); and tin and tin alloys (5,58,179).

Activation analysis has been used for arsenic in tungsten (87), silicon (86), and platinum (170). Polarographic procedures for iron and steel (222) and zinc metal and zinc-smelting residuals (33) are reported.

A listing of a number of spectrochemical procedures for specific alloys appears in Table 7. The *Spectrochemical Abstracts* (180-182,208) and the *Index to the Literature on Spectrochemical Analysis* (122,163,164) should be consulted for further information on specific spectrochemical procedures.

D. ROCKS, MINERALS, ORES, AND REAGENTS

In this group of materials, just as in all the other groups, the chief reason for analyzing for arsenic is the determination of the amount of contamination. In one instance, however, the analysis for arsenic serves as an aid in location of a more valuable ore. Through experience it has been observed that cobalt and arsenic tend to be found together in areas rich in cobalt ores (4,195). Since the arsenic concentrations seem to be greater closer to the surface than those of cobalt because of the higher retention of the arsenic by soils, a high arsenic content is a fairly good indication of a possible cobalt ore deposit located at a greater depth.

Other representative procedures for materials in this grouping are soils (4,8,102a,158,178,179,185a), silicate minerals (8,53,112,158,178,179,191), pyrites and other sulfides (38,59,158,178), sulfur and talc (178), coal and coke (41,70), and phosphates (59).

The acids that are used for digestion of samples for arsenic determination require an analysis for arsenic content. Suitable procedures are available for phosphoric (59,178,224) and nitric, hydrochloric, and sulfuric (59) acids.

E. PHARMACEUTICALS AND ORGANIC COMPOUNDS

A number of pharmaceuticals are arsenicals and as such must be submitted to an assay of the arsenic content. Numerous procedures (8,11,90,91,169) are available for their analysis.

Many methods and procedures are suitable for the determination of arsenic in organic compounds (38a, 38b, 59, 90, 91, 108, 109, 158, 178, 186, 188,206,221). Since a high arsenic level in petroleum fractions and naphthas is detrimental to the activity of the catalysts used in cracking, specific methods have been developed for its determination (2,110,119,-139,178,211b).

F. WATER

According to established drinking-water standards (209), the presence of more than 0.05 mg arsenic/liter constitutes sufficient grounds for rejection of the water supply. Suitable methods for analysis are outlined in a number of references (4, 8, 20a, 65a, 87b, 92a, 94, 125a, 145a, 162b, 178, 191, 204).

G. AIR

Because of the relatively high toxicity of arsenic, arsine, and other arsenic compounds, procedures for the determination of low concentrations in air are necessary (42,69,83,134a,195a,218a). The initial concentration of the sample is either through adsorption on silica gel impregnated with silver nitrate or absorption in a solution of silver nitrate or potassium iodide–sulfuric acid.

IX. LABORATORY PROCEDURES

The laboratory procedures consist essentially of two phases: (1) the separation of the arsenic from the other elements and (2) the determination of the arsenic content. The combinations used depend on the type and amount of material subjected to analysis.

A. PROCEDURES FOR SEPARATION

A prior separation of arsenic is required in most procedures for its determination. In general, the separation is achieved by distillation as either the trichloride or tribromide or by volatilization as arsine.

1. Procedure for Trichloride Distillation

Transfer the sample (inorganic material, dry ash, or portion of a digested solution low in water content) using 50 ml of concd. hydrochloric acid to an all-glass distilling apparatus such as that of Scherrer. For subsequent colorimetric analysis the sample should preferably contain 0.02–5 mg of arsenic and must be free from strong oxidizing agents such as nitric acid. Add 2 ml of concd. hydrobromic acid and 10 ml of a concd. hydrochloric acid solution containing 1.0 grams of hydrazine sulfate. Insert a thermometer into the well of the distilling flask, and dip the end of the condenser into a beaker containing 40 ml of cold water. Bubble carbon dioxide or nitrogen through the solution at two to three bubbles per second, and begin to boil gently until the temperature reaches 111°C. After approximately 15 min only about 10 ml of solution will remain in the distilling flask. Remove the receiver containing the distillate before turning off the heat, and rinse the delivery tube with water.

2. Procedure for Tribromide Distillation

This procedure is useful for separation of arsenic from such diverse materials as soil, digestates of organic samples, material isolated from pyrites, and so forth. Germanium and selenium are isolated along with the arsenic.

Transfer the sample to the distillation flask, and add 10 ml of a 1:10 bromine–hydrobromic acid solution a few milliliters at a time with shaking in order to avoid loss because of frothing from possible presence of carbonate. Addition of this reagent is continued until a definite excess is present, the amount depending on the amount of organic matter in the sample. Add sufficient additional concentrated hydrobromic acid to increase the volume of this reagent to 75–100 ml. Only hydrobromic acid that is completely decolorized by sulfur dioxide is suitable.

Connect the flask to the condenser, place 2–3 ml of saturated bromine water in the receiver, and support the receiver so that the adapter dips below the surface of the reagent. During the distillation the first few milliliters of distillate should carry over a few milliliters of bromine, but if this does not happen, add more bromine water to the receiver. Heat the distilling flask gently at first and eventually with full heat until 30–50 ml of distillate have collected. Add 50 ml more of the bromine–hydrobromic acid solution to the distillation flask, and repeat the distillation.

Treat the distillate with sulfur dioxide gas until the bromine is decolorized. Add 0.5 grams of hydroxylamine hydrochloride, stopper loosely, and heat on a steam bath for 1 hr. If the sample contains selenium, let the solution stand overnight at room temperature to precipitate the selenium as a black or red precipitate. Filter off the precipitate with an inorganic filter, and wash it with concentrated hydrobromic acid containing a small amount of hydroxylamine hydrochloride. Analyze the residue for selenium if desired. Treat the filtrate, which contains the arsenic and germanium, with 10 ml of concentrated nitric acid, and evaporate to approximately 25 ml. Cool, add 5 ml of concentrated sulfuric acid, and evaporate to sulfur trioxide fumes. Dilute the cooled sample with water, and use it for the arsenic determination.

3.   Procedure for Arsine Evolution

The generator, Fig. 1, is a 50 ml Erlenmeyer flask fitted with a one-hole rubber stopper containing tube A, which has one or two plugs of glass wool impregnated with lead acetate. A short piece of rubber tubing is used to connect tube A to the delivery tube B, which is drawn to a tip having an orifice of about 0.5 mm. The absorption vessel C is made from a 10 ml test tube and is drawn so that the tapered portion will contain 1.35 ml of absorbing solution in a depth of 6–7 cm. A collar (D) with an inside diameter about 1 mm greater than the outside diameter of the delivery tube is put in the absorption vessel to break up the gas bubbles and provide more absorption surface.

Transfer 25 ml of the prepared sample containing 15 μg of arsenic or less to the generator flask, and add sufficient hydrochloric acid so that

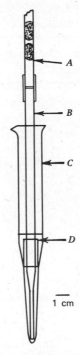

Fig. 1.   Arsine generator.

its total volume is 5 ml, then add 2 ml of potassium iodide solution (15 grams/100 ml of water) and 0.5 ml of stannous chloride solution (40 grams $SnCl_2 \cdot 2H_2O$/100 ml of concentrated hydrochloric acid). Allow the mixture to react at room temperature for 15–30 min to assure the complete reduction from the pentavalent to trivalent form.

Add 1.0 ml of an aqueous iodine–potassium iodide solution (0.25% I_2–0.4% KI) and 0.2 ml of a 4.2% sodium bicarbonate solution into the absorption tube. Stir with the end of the delivery tube; then connect this tube to the tube inserted through the rubber stopper and lower the drawn-out tube into the absorption vessel.

Add 2.0 grams of zinc rapidly to the flask, immediately connect the delivery tube stopper, and bubble the gases through the solution for 30 min without heating the flask. At this time some iodine should remain in the solution. Disconnect the delivery tube, and leave it in the absorption tube. This solution can be used for arsenic determination.

B. PROCEDURES FOR DETERMINATION

1. Titrimetric Determination of Arsenic

a. BROMATE METHOD

The procedure given in the following is suitable for the estimation of 4–80 mg of arsenic. The upper limit can be extended through the use of a more concentrated bromate solution.

Procedure

After the separation of the arsenic as the trichloride, the distillate will contain the arsenic in trivalent form and probably a sufficient amount of hydrochloric acid to permit direct titration. If the distillate has less than 10% by volume of hydrochloric acid, then add enough more to increase the concentration to at least this value.

Heat the distillate to 80–90°C, but do not boil since some of the arsenic trichloride may be otherwise lost. Titrate while hot with 0.01 or 0.05 N potassium bromate, using methyl orange as indicator, which should preferably be added near the end point of the titration. The end point should be approached slowly, since the reaction requires several seconds for completion and the end point is easily overstepped. Run a blank containing a similar amount of hydrochloric acid and indicator.

b. IODIMETRIC METHOD

The iodimetric procedure is especially suitable for the determination of arsenic in Paris Green. In the absence of nitrates it is applicable to lead, calcium, and magnesium arsenates; zinc arsenite; and Bordeaux mixtures with arsenicals.

Procedure

After separation by volatilization as arsenic trichloride, transfer an aliquot to a 500 ml conical flask, neutralize with 10 M sodium hydroxide, using phenolphthalein indicator, and cool the solution. Add 10 grams of finely powdered sodium hydrogen carbonate and 2 ml of starch indicator. Titrate the mixture with standard 0.1 N iodine solution until the blue iodo-starch color persists for one minute.

2. Photometric Determination of Arsenic

During the past twenty years the molybdenum blue reaction has been widely applied to the determination of arsenic, but in recent years the use of silver diethyldithiocarbamate for arsenic has increased so rapidly that some consider it to be the reagent of choice for arsenic in parts per billion.

a. MOLYBDENUM BLUE METHOD

Procedure

Add 10 ml of concentrated nitric acid to the arsenic trichloride or tribromide distillate, and evaporate it to dryness. Heat the residue for 1 hr at 130°C in an oven to remove the last traces of nitric acid. Add 10 ml of hydrazine sulfate–ammonium molybdate solution for each 30 μg of arsenic, and heat the solution on a steam bath or in a water bath for 15 min. Prepare the reagent by mixing just before use 10 ml portions of solutions A and B and diluting to 100 ml with water. Prepare solution A by dissolving 1.0 grams of ammonium molybdate in 10 ml of water, and add 90 ml of 6 N sulfuric acid. For solution B dissolve 0.15 grams of hydrazine sulfate in 100 ml of water. Solutions A and B can be added separately if desired.

If the approximate amount of arsenic is not known, add 10 ml of the hydrazine sulfate–molybdate solution, heat to develop the color, and determine the transmittance of the solution. If the transmittance indicates an arsenic content greater than 3 μg/ml, add 10 or 15 ml of reagent, develop again, and obtain the transmittance. This treatment is repeated until a concentration of less than 3 μg/ml is obtained. After sufficient reagent has been added, cool, dilute to volume with hydrazine sulfate–

molybdate solution; obtain the transmittance of the solution with a photometer, using a filter with a maximum transmittance at 700 nm or above (840 nm is the optimum). Construct a standard curve by treating known amounts of arsenic the same way as the unknown. Results should be corrected by running a blank through all the steps of the procedure.

If the arsenic was separated by the arsine method, take the solution collected in the absorption vessel, add 5 ml of the ammonium molybdate–hydrazine sulfate reagent, a drop of fresh 5% sodium metabisulfite solution, and mix thoroughly. This treatment reduces the iodine in the solution. Heat in a water or steam bath for 15 min, cool, transfer to a 10 or 25 ml volumetric flask, and dilute with water.

Obtain the transmittance as described previously, using as a reference a solution prepared by mixing the iodine–iodide–bicarbonate reagents and treating with molybdate–hydrazine sulfate–sulfite as in the procedure for the samples.

Prepare a standard curve by treating known amounts of arsenic with iodine–iodide–bicarbonate and developing the color as above. Run a blank through the entire procedure.

b. Silver Diethyldithiocarbamate Method (118, 211)

The most commonly used apparatus for the generation and absorption of the arsine for this method is that based on the recommendations of the American Conference of Government Industrial Hygienists (211), Fig. 2. A standard Pyrex Gutzeit generator connected to a glass delivery tip that extends into the absorbing solution in a Kahn tube has also been used with success. Another type consists of a Kipp-type funnel tube with one or two bulbs in the loop and a thistle top. All glass apparatus should be initially cleaned with hot sulfuric acid, rinsed first with water and then acetone. If the apparatus is subsequently reserved for arsenic analysis, the acid wash need not be repeated.

During the reduction, 3 ml of a 0.5% silver diethyldithiocarbamate in pyridine solution are placed into the absorbing tube, and the scrubber containing the lead acetate impregnated glass wool is attached to it. After the samples have stood 15 min, the ground glass joints are lubricated with "Nonaq" stopcock grease, 3 grams of granulated zinc are added to the flasks, and the receiving tube is connected immediately. Thirty minutes is allowed for the complete evolution of the arsine.

After the expiration of this time the absorbing solution is transferred to a 1.0 cm square cell, and the absorbance is measured at 560 nm with a spectrophotometer.

A standard curve is obtained by treating in a similar manner a series of aliquots containing precisely known amounts of arsenic ranging between

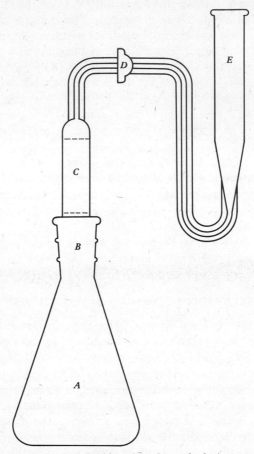

Fig. 2.   Generator for silver diethyldithiocarbamate method. *A*, generator, 125 ml Erlen-meyer; *B*, 19/38; *C*, scrubber, lead acetate on Pyrex wool; *D*, 12/2 ball joint; *E*, absorber, 12 ml heavy-wall centrifuge tube.

0 and 15 μg. A plot of the absorbances versus the micrograms of arsenic is used for the calibration curve.

Moderate amounts of metallic cobalt, mercury, nickel, silver, and palladium and large amounts of copper, chromium, and molybdenum interfere with the evolution of arsine. The salts of these metals also interfere. Antimony interferes with the color development, since stibine reacts with the silver reagent to form a red color having a maximum absorbance at 510 nm.

REFERENCES

1. Agnew, W. J., *Analyst (London)*, **68,** 111 (1943).

2. Albert, D. K., and L. Granatelli, *Anal. Chem.,* **31,** 1593 (1959).

3. Alekseev, R. I., *Zavod. Lab.,* **11,** 122 (1945); through *Chem. Abstr.,* **40,** 1115 (1946).

4. Almond, H., *Anal. Chem.,* **25,** 1766 (1953).

4a. American Public Health Association, *Standard Methods for the Examination of Water and Wastewater,* 13th edit., A. P. H. A., New York, (1971), pp. 62–64.

5. American Society for Testing Materials, *A.S.T.M. Methods For Chemical Analysis of Metals,* A.S.T.M., Philadelphia, Pa., 1956.

6. American Society for Testing Materials, *Methods for Emission Spectrochemical Analysis,* A.S.T.M., Philadelphia, Pa., 1957.

6a. Analytical Methods Committee—The Chemical Society (London), *Analyst (London),* **100,** 54 (1975).

7. Anderson, C. W., *Ind. Eng. Chem. Anal. Ed.,* **9,** 569 (1937).

7a. Ando, A., M. Suzuki, F. Fuwa, and B. L. Vallee, *Anal. Chem.,* **41,** 1974 (1969).

7b. Arnold, J. P., and R. M. Johnson, *Talanta,* **16,** 1191 (1969).

8. Association of Official Agricultural Chemists, *Official and Tentative Methods of Analysis,* 8th edit., A.O.A.C., Washington, 1955.

9. Baghurst, H. C., and V. J. Norman, *Anal. Chem.,* **29,** 778 (1957).

10. Bakacs, E., and L. Szekeres, *Magy. Kem. Folyo.,* **62,** 296 (1957); *Chem. Abstr.,* **52,** 15342 (1958).

11. Banks, C. K., J. A. Sultzaberger, F. A. Maurina, and C. S. Hamilton, *J. Amer. Pharm. Ass. Sci. Ed.,* **37,** 13 (1948).

11a. Barnett, W. B., and E. A. McLaughlin, Jr., *Anal. Chim. Acta.,* **80,** 285 (1975).

12. Bartlett, J. C., M. Wood, and R. A. Chapman, *Anal. Chem.,* **24,** 1821 (1952).

13. Basl, Z., Z. Plasil, and R. Stangl, *Rudy,* **5,** 1 (1957); through *Chem. Abstr.,* **52,** 12666 (1958).

14. Beamish, F. E., and H. L. Collins, *Ind. Eng. Chem. Anal. Ed.,* **6,** 379 (1934).

15. Belcher, R., *Anal. Chim. Acta,* **5,** 30 (1951).

16. Belcher, R., and T. S. West, *Anal. Chim. Acta,* **6,** 322 (1952).

17. Bertiaux, L., *Bull. Soc. Chim. Fr.,* **11,** 547 (1944).

18. Bertiaux, L., *Chim. Anal.,* **32,** 269 (1950).

18a. Bhargava, O. P., J. E. Donovan, and W. G. Hines, *Anal. Chem.,* **44,** 2402 (1972).

18b. Birks, L. S., *X-ray Spectrochemical Analysis,* Wiley-Interscience, New York, 1959.

19. Bohnstedt, U., and R. Budenz, *Z. Anal. Chem.,* **159,** 95 (1957).

20. Boltz, D. F., and M. G. Mellon, *Anal. Chem.,* **19,** 873 (1947).

20a. Braman, R. C., L. L. Justen, and C. C. Forebock, *Anal. Chem.,* **44,** 2195–2197 (1972).

21. Brandt, L., *Chem.-Ztg.,* **37,** 1445, 1471, 1496 (1913).

22. Brekhstedt, A., *Lab. Prakt.,* **1939,** 21 (1939); through *Chem. Abstr.,* **33,** 5767 (1939).

23. Bricker, C. E., and P. B. Sweetser, *Anal. Chem.*, **24**, 409 (1952).

23a. Burke, K. E., and M. M. Yanak, *Anal. Chem.*, **41**, 963 (1969).

24. Butler, E. A., and E. H. Swift, *Anal. Chem.*, **29**, 419 (1957).

24a. Byrne, A. R., and D. Gorenc, *Anal. Chim. Acta,* **59**, 81 (1972).

25. Cahill, E., and L. Walters, *Ind. Eng. Chem. Anal. Ed.*, **14**, 90 (1942).

25a. Caldas, A., *Anal. Chim. Acta,* **45**, 532 (1969).

26. Case, O. P., *Anal. Chem.*, **20**, 902 (1948).

27. Catoggio, J. A., *Rev. Fac. Cienc. Quim.*, **20**, 121 (1945); through *Chem. Abstr.*, **42**, 2887 (1948).

28. Challis, H. J. G., *Analyst (London)*, **66**, 58 (1941).

29. Chaney, A. L., and H. J. Magnuson, *Ind. Eng. Chem. Anal. Ed.*, **12**, 691 (1940).

30. Charlot, G., and D. Bezier, *Quantitative Inorganic Analysis,* Wiley, New York, 1957, p. 342.

30a. Christian, G. D., and F. J. Feldman, *Atomic Absorption Spectroscopy*, Wiley-Interscience, New York, 1970.

30b. Chu, R. C., G. P. Barren, and P. A. W. Baungarner, *Anal. Chem.*, **44**, 1476 (1972).

31. Committee on Recommended Analytical Methods, *Manual of Analytical Methods, Determination of Arsenic in Air,* American Conference of Governmental Industrial Hygienists, Cincinnati, Oh.

32. Cordebard, H., and L. Louis, *Anal. Chim. Anal,* **27**, 204 (1945).

33. Coulson, R. E., *Analyst (London)*, **82**, 161 (1957).

34. Cozzi, D., and S. Vivarelli, *Anal. Chim. Acta,* **5**, 215 (1951).

35. Crawford, T. B. B., and G. A. Levvy, *Biochem. J.*, **40**, 455 (1946).

36. Cushnir, C., *Rev. Fac. Cienc. Quim.*, **23**, 15 (1948); through *Chem. Abstr.*, **47**, 12102 (1953).

37. Damon, J. M. O., and M. G. Mellon, *Anal. Chem.*, **30**, 1849 (1958).

37a. Danchik, R. S., and D. F. Boltz, *Anal. Lett.*, **1**, 901 (1968).

38. Davies, W. C., and C. Key, *Analyst (London)*, **72**, 17 (1947).

38a. Dean, J. A., *Flame Emission and Atomic Absorption Spectrophotometry*, Dekker, New York, 1969.

38b. Dean, J. A., and R. E. Fues, *Anal. Lett.*, **2**, 105 (1969).

39. Deniges, G., *C. R. H. Acad. Sci.*, **171**, 802 (1920).

40. DeSesa, M. A., and L. B. Rogers, *Anal. Chem.*, **26**, 1381 (1954).

40a. Down, J. L., and T. T. Gorsuch, *Analyst (London)*, **92**, 398 (1967).

41. Edgcombe, L. J., and K. H. Gold, *Analyst (London)*, **80**, 155 (1955).

41a. Elenkova, N. G., and R. A. Tsoneva, *Anal. Chim. Acta,* **62**, 435 (1972).

41b. Elenkova, N. G., and R. A. Tsoneva, *Talanta*, **22**, 480 (1975).

42. Elkins, H. B., *The Chemistry of Industrial Toxicology,* 2nd edit., Wiley, New York, 1959.

42a. Elwell, W. T., and J. A. F. Gidley, *Atomic Absorption Spectrophotometry,* Macmillian, New York, 1962.

43. Elzanowski, L., *Rocz. Panstwowego Zakladu Hig.*, **2**, 1 (1951); through *Chem. Abstr.*, **49**, 10116 (1955).

44. Engel, R., and J. Bernard, *C. R. H. Acad. Sci.*, **122**, 390 (1896).

45. Erdey, L., and L. Buzas, *Magy. Tud. Akad. Kem. Tud. Oszt. Kozlem.*, **5**, 279 (1954); through *Chem. Abstr.*, **49**, 9431 (1955).

46. Evans, B. S., *Analyst (London)*, **54**, 523 (1929).

47. Evans, B. S., *Analyst (London)*, **57**, 492 (1932).

48. Evans, R. J., and S. L. Bandemer, *Anal. Chem.*, **26**, 595 (1954).

49. Everest, D. A., and G. W. Finch, *J. Chem. Soc.*, **1955**, 704 (1955).

50. Everett, G. W., and C. N. Reilley, *Anal. Chem.*, **26**, 1750 (1954).

51. Fainberg, S. Yu., and G. A. Taratorin, *Zavod. Lab.*, **9**, 1223 (1940); through *Chem. Abstr.*, **37**, 2296 (1943).

52. Farrington, P. S., and E. H. Swift, *Anal. Chem.*, **22**, 889 (1950).

52a. Fernandez, F. J., *Atom. Absorp. Newslett.*, **12**, 93 (1973).

53. Finkel'shtein, D. N., and G. N. Kryuchkova, *Zh. Anal. Khim.*, **12**, 196 (1957); through *Chem. Abstr.*, **52**, 157 (1958).

54. Fischer, R., and T. Langhammer, *Mikrochem. Mikrochim. Acta*, **34**, 203 (1949).

55. Flaschka, H., *Z. Anal. Chem.*, **137**, 107 (1952).

56. Flaschka, H., and H. Jakobljevich, *Anal. Chim. Acta*, **4**, 486 (1950).

57. Flury, F., and F. Zernik, *Schädliche Gase*, Springer, Berlin, 1931, pp. 173–92.

57a. Fogg, A. G., D. R. Marriatt, and T. Burns, *Analyst (London)*, **97**, 657 (1972).

57b. Forsberg, G., J. W. O'Laughlin, and R. G. Megargie, *Anal. Chem.*, **47**, 1586 (1975).

58. Freeman, J. H., and W. M. McNabb, *Anal. Chem.*, **20**, 979 (1948).

58a. Fricke, F. L., O. Rose, and J. A. Caruso, *Anal. Chem.*, **47**, 2018 (1975).

59. Furman, N. H., Editor, *Scott's Standard Methods of Chemical Analysis*, 5th edit., Vol. I, Van Nostrand, New York, 1939.

60. Furman, N. H., and A. J. Fenton, Jr., *Anal. Chem.*, **28**, 515 (1956).

61. Gagliardi, E., and A. Loidl, *Z. Anal. Chem.*, **132**, 33 (1951).

61a. Gandekota, M., and G. G. Rao, *Anal. Chim. Acta*, **65**, 231 (1973).

62. Gaudy, F. V. M., and M. P. Antola, *An. Asoc. Quim. Argentina*, **24**, 164 (1936); through *Chem. Abstr.*, **31**, 5716 (1937).

62a. Gilmore, J. T., *Anal. Chem.*, **40**, 2230 (1968).

63. Goldstone, N. I., *Ind. Eng. Chem. Anal. Ed.*, **17**, 797 (1946).

63a. Gorsuch, T. T., *Analyst (London)*, **84**, 147 (1959).

64. Goto, H., and Y. Kakita, *J. Chem. Soc. Japan*, **63**, 470 (1942); through *Chem. Abstr.*, **41**, 3010 (1947).

65. Goto, H., and Y. Kakita, *J. Chem. Soc. Japan*, **64**, 515 (1943); through *Chem. Abstr.*, **41**, 3392 (1947).

65a. Goulden, P. D., P. Brooksbank, *Anal. Chem.*, **46**, 1431 (1974).

66. Green, M., and J. A. Kafalas, *J. Chem. Phys.*, **22**, 760 (1954).

67. Griffon, H., and J. Barbaud, *C. R. H. Acad. Sci.*, **232**, 1455 (1951).

67a. Griffin, H. R., M. B. Hocking, and D. G. Lowery, *Anal. Chem.*, **47**, 229 (1975).

67b. Grimanis, A. P., and A. G. Souliotis, *Analyst (London)*, **92**, 549 (1967).

68. Gullstrom, D. K., and M. G. Mellon, *Anal. Chem.*, **25**, 1809 (1953).

68a. Gunn, E. L., *Anal. Chem.*, **29,** 184 (1957).

69. Haight, G. P., Jr., *Anal. Chem.*, **26,** 593 (1954).

70. Hall, R. H., and H. L. Lovell, *Anal. Chem.*, **30,** 1665 (1958).

70a. Hamilton, E. I., M. J. Minski, and J. J. Cleary, *Analyst (London)*, **92,** 257 (1967).

71. Harris, E. D., and A. J. Lindsey, *Analyst (London)*, **76,** 650 (1951).

72. Harrison, G. R., *MIT Wavelength Tables*, Wiley, New York, 1939.

73. Henderson, Y., and H. W. Haggard, *Noxious Gases*, ACS Monograph Series, No. 35, 2nd edit., Reinhold, New York, 1943.

74. Hinsberg, K., and M. Kiese, *Biochem. Z.*, **290,** 39 (1937).

75. Hoffman, J. I., and G. E. F. Lundell, *J. Res. Nat. Bur. Stand.*, **22,** 465 (1939).

75a. Holak, W., *Anal. Chem.*, **41,** 1712 (1969).

76. Hollander, J. M., I. Perlman, and G. T. Seaborg, *Rev. Modern Phys.*, **25,** 469 (1953).

77. Holness, H., and R. F. G. Trewick, *Analyst (London)*, **75,** 276 (1950).

78. How, A. E., *Ind. Eng. Chem. Anal. Ed.*, **10,** 226 (1938).

79. Hubbard, D. M., *Ind. Eng. Chem. Anal. Ed.*, **13,** 915 (1941).

80. Hunter, D., *Industrial Toxicology*, Oxford at the Clarendon, England, 1944.

81. Issa, I. M., and I. M. Elsherif, *Rec. Trav. Chim.*, **75,** 447 (1956); *Chem. Abstr.*, **50,** 13649 (1956).

82. Ivanov-Emin, B. N., *Zavod. Lab.*, **13,** 161 (1947); through *Chem. Abstr.*, **42,** 480 (1948).

83. Jacobs, M. B., *The Analytical Chemistry of Industrial Poisons, Hazards, and Solvents*, 2nd edit., Wiley-Interscience, New York, 1949.

84. Jacobs, M. B., *Chemical Analysis of Foods and Food Products*, 3rd edit., Van Nostrand, New York, 1958, pp. 6–11.

85. Jacobs, M. B., and J. Nagler, *Ind. Eng. Chem. Anal. Ed.*, **14,** 442 (1942).

86. James, J. A., and D. H. Richards, *Nature*, **175,** 769 (1955).

87. James, J. A., and D. H. Richards, *Anal. Chim. Acta*, **15,** 118 (1956).

87a. Jennings, V. J., A. Dodsen, and A. Harrison, *Analyst (London)*, **99,** 145 (1974).

87b. Johnson, D. L., and M. E. Q. Pilson, *Anal. Chim. Acta*, **58,** 289 (1972).

88. Jones, W. N., Jr., *Inorganic Chemistry*, Blakiston, Philadelphia, 1947, pp. 485–86.

89. Joslyn, M. A., *Methods in Food Analysis*, Academic, New York, 1950, pp. 29–47.

90. Jurecek, M., and J. Jenik, *Chem. Listy*, **48,** 1771 (1954); *Chem. Abstr.*, **49,** 4443 (1955).

91. Jurecek, M., and J. Jenik, *Chem. Listy*, **49,** 264 (1955); *Chem. Abstr.*, **49,** 8032 (1955).

92. Kacirkova, K., *Collect. Czechoslov. Chem. Commun.*, **1,** 477 (1929); through *Chem. Abstr.*, **24,** 1012 (1930).

92a. Kan, K., *Anal. Lett.*, **6,** 603 (1973).

93. Kew, D. J., M. D. Amos, and M. C. Greaves, *Analyst (London)*, **77,** 488 (1952).

94. Kingsley, G. R., and R. R. Schaffert, *Anal. Chem.*, **23,** 914 (1951).

94a. Kirkbright, G. F., and L. Ransom, *Anal. Chem.*, **43,** 1238 (1971).

94b. Kirkbright, G. F., A. F. Ward, and T. S. West, *Anal. Chim. Acta*, **64,** 353 (1973).

94c. Kirkbright, G. F., and P. J. Wilson, *Anal. Chem.*, **46**, 1414 (1974).

95. Klein, A. K., and F. A. Vorhes, Jr., *J. Ass. Offic. Agr. Chem.*, **22**, 121 (1939).

96. Klement, R., and A. Kühn, *Z. Anal. Chem.*, **152**, 146 (1956).

96a. Knudson, E. J., and G. D. Christian, *Anal. Lett.*, **6**, 1039 (1973).

97. Kolthoff, I. M., *Anal. Chem.*, **22**, 65 (1950).

98. Kolthoff, I. M., and E. Amdur, *Ind. Eng. Chem. Anal. Ed.*, **12**, 177 (1940).

99. Kolthoff, I. M., and C. W. Carr, *Anal. Chem.*, **20**, 728 (1948).

100. Kolthoff, I. M., and R. L. Probst, *Anal. Chem.*, **21**, 753 (1949).

101. Konopik, N., and K. Szlaczka, *Oesterr. Chem.-Ztg.*, **52**, 205 (1951).

101a. Kopp, J. F., *Anal. Chem.*, **45**, 1786 (1973).

102. Krepelka, J. H., and J. Fanta, *Collect. Czechoslov. Chem. Commun.*, **9**, 47 (1937); *Chem. Abstr.*, **31**, 6995 (1937).

102a. Kronberg, O. J., and E. Steinnes, *Analyst (London)*, **100**, 835 (1975).

102b. Lacroix, J. P., and J. Steines, *Anal. Lett.*, **6**, 565 (1973).

103. Lambie, D. A., *Analyst (London)*, **74**, 260 (1949).

104. Langlois, J., and Ch. Morin, *Bull. Sci. Pharmacol.*, **45**, 482 (1938).

105. Latimer, W. M., *The Oxidation States of the Elements and Their Potentials In Aqueous Solutions*, 2nd edit., Prentice-Hall, Englewood Cliffs, N.J., 1952.

106. Latimer, W. M., and J. H. Hildebrand, *Reference Book of Inorganic Chemistry*, 3rd edit., MacMillan, New York, 1951, pp. 219–21.

107. Levvy, G. A., *Biochem. J.*, **37**, 598 (1943).

108. Levy, R., *Bull. Soc. Chim. Fr.*, **1956**, 517 (1956).

109. Levy, R., *C. R. H. Acad. Sci.*, **238**, 2320 (1954).

109a. Liebhafsky, H. A., H. G. Pfeiffer, E. H. Winslow, and P. D. Zemang, *X-ray Absorption and Emission in Analytical Chemistry*, Wiley, New York, 1960.

110. Liederman, D., J. E. Bowen, and O. I. Milner, *Anal. Chem.*, **30**, 1543 (1958).

111. Lingane, J. J., *Ind. Eng. Chem. Anal. Ed.*, **15**, 583 (1943).

112. Lounamaa, K., *Z. Anal. Chem.*, **146**, 422 (1955).

113. Luke, C. L., *Ind. Eng. Chem. Anal. Ed.*, **12**, 97 (1940).

114. Luke, C. L., *Ind. Eng. Chem. Anal. Ed.*, **15**, 626 (1943).

115. Luke, C. L., and M. E. Campbell, *Anal. Chem.*, **25**, 1588 (1953).

115a. Lunde, G., and P. E. Paus, *Anal. Lett.*, **7**, 363 (1974).

116. Maechling, E. H., and F. B. Flinn, *J. Lab. Clin. Med.*, **15**, 779 (1930).

117. Magnuson, H. J., and E. B. Watson, *Ind. Eng. Chem. Anal. Ed.*, **16**, 339 (1944).

117a. Manning, P. C., and F. Fernandez, *Atom. Absorp. Newslett.*, **10**, 86 (1971).

118. American Conference of Governmental Industrial Hygienists, *Determination of Arsenic in Air*, Cincinnati, Oh., 1956.

119. Maranowski, N. C., R. E. Snyder, and R. O. Clark, *Anal. Chem.*, **29**, 353 (1957).

120. Maren, T. H., *Ind. Eng. Chem. Anal. Ed.*, **18**, 521 (1946).

120a. Maruta, T., and G. Sudoh, *Anal. Chim. Acta*, **77**, 37 (1975).

120b. Massman, H., *Z. Anal. Chem.*, **225**, 203 (1967).

121. McChesney, E. W., *Anal. Chem.*, **21**, 880 (1949).

122. Meggers, W. F., and B. F. Scribner, *Index to the Literature on Spectro-Chemical Analysis,* Part I, 2nd edit., American Society for Testing Materials, Philadelphia, 1941, pp. 1920–1939.

123. Meinke, W. W., *Science,* **121,** 177 (1955).

124. Meites, L., *J. Amer. Chem. Soc.,* **76,** 5927 (1954).

124a. Meins, O., and T. C. Rains, *Anal. Chem.,* **41,** 952 (1969).

125. Mertens, V., *J. Pharm. Belg.,* **23,** 497, 529 (1941).

125a. Mesman, B. B., and T. C. Thomas, *Anal. Lett.,* **8,** 449 (1975).

126. Meyer, S., and O. G. Koch, *Z. Anal. Chem.,* **158,** 434 (1957).

126a. Myers, D. J., and J. Osteryoung, *Anal. Chem.,* **45,** 267 (1973).

127. Mokranjac, M. St., and B. Rasajski, *Acta Pharm. Jugoslav.,* **2,** 9 (1952); through *Chem. Abstr.,* **47,** 440 (1953).

128. Morris, H. J., and H. O. Calvery, *Ind. Eng. Chem. Anal. Ed.,* **9,** 447 (1937).

129. Morrison, G. H., and J. F. Cosgrove, *Anal. Chem.,* **27,** 810 (1955).

130. Morrison, G. H., and H. Freiser, *Solvent Extraction in Analytical Chemistry,* Wiley, New York, 1957, pp. 193–194.

131. Myers, R. J., and E. H. Swift, *J. Amer. Chem. Soc.,* **70,** 1047 (1948).

132. *Nuclear Data,* NBS Circular 499 plus three supplements, United States Department of Commerce, National Bureau of Standards, Washington, D.C. 1950.

133. Odencrantz, J. T., and W. Rieman III, *Anal. Chem.,* **22,** 1066 (1950).

133a. Ohlweiler, O. A., J. O. Meditsch, and C. M. S. Paitinicki, *Anal. Chim. Acta,* **63,** 341 (1973).

134. Oliver, W. T., and H. S. Funnell, *Anal. Chem.,* **31,** 259 (1959).

134a. Onishi, H., and E. B. Sandell, *Geochim. Cosmochim. Acta,* **7,** 1 (1955).

134b. Onishi, H., and E. B. Sandell, *Microchim. Acta,* **1953,** 34 (1953).

134c. Orheim, R. M., and H. H. Bovee, *Anal. Chem.,* **46,** 921 (1974).

134d. Orvini, O., T. E. Gills, and P. D. LaFleur, *Anal. Chem.,* **46,** 1294–1297 (1974).

135. Osterberg, A. E., and W. S. Green, *J. Biol. Chem.,* **155,** 513 (1944).

135a. Palkins, P., *Anal. Chim. Acta,* **47,** 225 (1969).

136. Patty, F. A., *Industrial Hygiene And Toxicology,* Wiley-Interscience, New York, 1949, pp. 565–572.

137. Pitts, G. N., Jr., D. D. DeFord, T. W. Martin, and E. A. Schmall, *Anal. Chem.,* **26,** 628 (1954).

138. Poethke, W., and F. Wolf, *Z. Anorg. Allg. Chem.,* **268,** 244 (1952).

139. Powers, G. W., Jr., R. L. Martin, F. J. Piehl, and J. M. Griffin, *Anal. Chem.,* **31,** 1589 (1959).

140. Pribil, R., *Chem. Listy,* **37,** 205 (1943).

141. Pribil, R., *Proc. XIth Internat. Congr. Pure Appl. Chem.,* **5,** 893 (1947); through *Chem. Abstr.,* **48,** 5017 (1954).

142. Pribil, R., and J. Cihalik, *Chem. Listy,* **44,** 224 (1950); *Chem. Abstr.,* **45,** 5562 (1951).

143. Raikhinshtein, Ts. G., and T. V. Kocherygina, *Zh. Anal. Khim.,* **2,** 173 (1947); through *Chem. Abstr.,* **43,** 6935 (1949).

143a. Ramakrishma, T. V., J. W. Robinson, and P. W. West, *Anal. Chim. Acta*, **45**, 43 (1969).

143b. Ramirez-Munoz, V., *Atomic Absorption Spectroscopy and Analysis*, Elsevier, New York, 1968.

144. Ramsey, W. J., P. S. Farrington, and E. H. Swift, *Anal. Chem.*, **22**, 332 (1950).

144a. Ratcliffe, D. B., C. S. Byford, and P. B. Osman, *Anal. Chim. Acta*, **75**, 457 (1975).

145. Ray, H. N., *J. Proc. Inst. Chem. (India)*, **18**, 108 (1946); through *Chem. Abstr.*, **41**, 3011 (1947).

145a. Ray, B. J., and D. L. Johnson, *Anal. Chim. Acta*, **62**, 196 (1972).

146. *Reagent Chemicals*, American Chemical Society Specifications 1955, American Chemical Society, Washington, D.C., 1956.

146a. Reay, P. F., *Anal. Chim. Acta*, **72**, 145 (1974).

147. Reed, J. F., *Anal. Chem.*, **30**, 1122 (1958).

148. Remy, H., *Treatise on Inorganic Chemistry*, Vol. I, Elsevier, New York, 1956, pp. 650–652.

149. Remy, H., *Treatise on Inorganic Chemistry*, Vol. II, Elsevier, New York, 1956, p. 642.

149a. Reymont, T. M., and R. J. Dubois, *Anal. Chim. Acta*, **56**, 1 (1971).

150. Riou, P., and J. J. Pare, *Ann. ACFAS (Ass. Can. Fr. Avan. Sci.)*, **7**, 76 (1941); through *Chem. Abstr.*, **40**, 1107 (1946).

150a. Robinson, J. W., R. Garcia, G. Hindman, and P. Sleveu, *Anal. Chim. Acta*, **69**, 203 (1975).

151. Robinson, R. G., *Analyst (London)*, **65**, 159 (1940).

152. Robinson, W. O., H. C. Dudley, K. T. Williams, and H. G. Byers, *Ind. Eng. Chem. Anal. Ed.*, **6**, 274 (1934).

153. Rodden, C. J., *J. Res. Nat. Bur. Stand.*, **24**, 7 (1940).

154. Rogers, D., and A. E. Heron, *Analyst (London)*, **71**, 414 (1946).

155. Rosin, J., *Reagent Chemicals and Standards*, 3rd edit., Van Nostrand, New York, 1955.

156. Roush, G. A., "Arsenic," in Kirk and Othmer, *Encyclopedia of Chemical Technology*, Vol. 2, Wiley-Interscience, New York, 1948, pp. 113–118.

156a. Rozenblum, V., *Anal. Lett.*, **8**, 549 (1975).

157. Saito, K., and N. Sato, *J. Chem. Soc. Japan, Ind. Chem. Sect.*, **55**, 59 (1952); through *Chem. Abstr.*, **47**, 9848 (1953).

157a. Samsahl, K., *Anal. Chem.*, **39**, 1480 (1967).

158. Sandell, E. B., *Colorimetric Determination of Traces of Metals*, 2nd edit., Wiley-Interscience, New York, 1950, pp. 175–192.

159. Sandell, E. B., *Ind. Eng. Chem. Anal. Ed.*, **14**, 82 (1942).

159a. Sandhu, S. S., S. S. Pahil, and K. D. Sharma, *Anal. Chim. Acta*, **56**, 154 (1971).

160. Satterlee, H. S., and G. Blodgett, *Ind. Eng. Chem. Anal. Ed.*, **16**, 400 (1944).

161. Schatko, P. P., *Zavod. Lab.*, **10**, 423 (1941); through *Chem. Abstr.*, **38**, 2898 (1944).

162. Schindewolf, U., and J. W. Irvine, Jr., *Anal. Chem.*, **30**, 906 (1958).

162a. Schmidt, F. J., J. L. Royer, and S. M. Muir, *Anal. Lett.*, **8**, 123 (1975).

162b. Schmidt, F. J., and J. L. Royer, *Anal. Lett.*, **6**, 17 (1973).

163. Scribner, B. F., and W. F. Meggers, *Index to the Literature on Spectrochemical Analysis*, Part II, American Society for Testing Materials, Philadelphia, Pa., 1947, pp. 1940–1945.

164. Scribner, B. F., and W. F. Meggers, *Index to the Literature on Spectrochemical Analysis*, Part III, American Society for Testing Materials, Philadelphia, Pa., 1954, pp. 1946–1950.

165. Sekino, M., *Bull. Inst. Phys. Chem. Res. (Tokyo)*, **22**, 71 (1943); through *Chem. Abstr.*, **41**, 5812 (1947).

166. Sevryukov, N. N., and M. A. Vinokurova, *Zavod. Lab.*, **6**, 427 (1937); through *Chem. Abstr.*, **31**, 7789 (1937).

167. Shakhov, A. S., *Zavod. Lab.*, **11**, 270 (1945); through *Chem. Abstr.*, **40**, 1107 (1946).

168. Shapiro, M. Ya., *J. Appl. Chem. (U.S.S.R.)*, **16**, 330 (1943); through *Chem. Abstr.*, **39**, 471 (1945).

169. Shaw, R., B. E. Riedel, and W. E. Harris, *Can. Pharm. J., Sci. Sect.*, **90**, 303 (1957).

170. Shipman, G. F., and O. I. Milner, *Anal. Chem.*, **30**, 210 (1958).

171. Singh, B., A. Singh, and R. Singh, *J. Indian Chem. Soc.*, **30**, 147 (1953).

172. Singh, B., and K. C. Sood, *Anal. Chim. Acta*, **11**, 313 (1954).

172a. Slavin, W., C. Sebens, and S. Sprague, *Atom. Absorp. Newslett.*, **4**, 341 (1965).

173. Sloviter, H. A., W. M. McNabb, and E. C. Wagner, *Ind. Eng. Chem. Anal. Ed.*, **14**, 516 (1942).

174. Smales, A. A., and B. D. Pate, *Anal. Chem.*, **24**, 717 (1952).

174a. Smith, A. E., *Analyst (London)*, **100**, 300 (1975).

175. Smith, G. F., and J. S. Fritz, *Anal. Chem.*, **20**, 874 (1948).

176. Smith, H., *Anal. Chem.*, **31**, 1361 (1959).

177. Smith, W. C., "Arsenic," in D. M. Liddell, *Handbook of Nonferrous Metallurgy*, Vol. II, 2nd edit., McGraw-Hill, New York, 1945, pp. 94–103.

178. Snell, F. D., and C. T. Snell, *Colorimetric Methods of Analysis*, Vol. II, 3rd edit., Van Nostrand, New York, 1949, pp. 172–200.

179. Snell, F. D., C. T. Snell, and C. A. Snell, *Colorimetric Methods of Analysis*, Vol. IIA, Van Nostrand, New York, 1959, pp. 107–120.

180. Someren, E. H. S., *Spectrochemical Abstracts*, Vol. II, Adam Hilger, London, 1941, pp. 1938–1939.

181. Someren, E. H. S., *Spectrochemical Abstracts*, Vol. III, Adam Hilger, London, 1947, pp. 1940–1945.

182. Someren, E. H. S., and F. Lachman, *Spectrochemical Abstracts*, Vol. IV, Hilger and Watts, London, 1955, pp. 1946–1951.

183. Sorum, C. H., and H. A. Wolf, *J. Chem. Educ.*, **27**, 614 (1950).

184. Starke, K., *Naturwissenschaften*, **28**, 631 (1940).

185. Steele, M. C., and L. J. England, *Analyst (London)*, **82**, 595 (1957).

185a. Steinnes, E., *Analyst (London)*, **97**, 241 (1971).

186. Stickler, W. C., *Anal. Chem.*, **24**, 1219 (1952).

187. Stock, J. T., *Amperometric Titrations*, Wiley-Interscience, New York, 1965.

188. Strafford, N., P. F. Wyatt, and F. G. Kershaw, *Analyst (London)*, **70**, 232 (1945).

189. Stringer, W. J., *J. Inst. Brewing*, **60**, 249 (1954).

190. Suarez, L. R., *An. Fac. Quim. Farm.*, **4**, 138 (1955); through *Chem. Abstr.*, **50**, 9211 (1956).

191. Sugawara, K., M. Tanaka, and S. Kanamori, *Bull. Chem. Soc. Japan*, **29**, 670 (1956); through *Chem. Abstr.*, **51**, 2457 (1957).

192. Sultzaberger, J. A., *Ind. Eng. Chem. Anal. Ed.*, **15**, 408 (1943).

192a. Susic, M. V., and M. G. Pjeocic, *Analyst (London)*, **91**, 258 (1966).

193. Sweetser, P. B., and C. E. Bricker, *Anal. Chem.*, **24**, 1107 (1952).

194. Syrokomskii, V. S., and S. I. Melamed, *Zavod. Lab.*, **16**, 131 (1950); *Chem. Abstr.*, **44**, 6759 (1950).

195. Tanaka, N., *J. Chem. Soc. Japan*, **64**, 443 (1943); through *Chem. Abstr.*, **41**, 3395 (1947).

195a. Tanner, J. E., M. H. Friedman, and G. E. Holloway, *Anal. Chim. Acta*, **66**, 456 (1973).

196. Tartakovskii, V. Ya., *Zavod. Lab.*, **4**, 750 (1935); through *Chem. Abstr.*, **30**, 983 (1936).

197. Taylor, G., and J. H. Hamence, *Analyst (London)*, **67**, 12 (1942).

198. *The Pharmacopeia of the United States of America*, Twelfth Revision, USP XII, Mack, Easton, Pa., 1942.

199. *The Pharmacopeia of the United States of America*, Fourteenth Revision, USP XIV, Mack, Easton, Pa., 1950.

200. Thiele, J., *Justus Liebigs Ann. Chem.*, **263**, 361 (1890).

201. Thiele, J., *Justus Liebigs, Ann. Chem.*, **265**, 55 (1891).

201a. Thompson, K. C., *Analyst (London)*, **100**, 307 (1975).

201b. Thompson, K. C., and D. R. Thomerson, *Analyst (London)*, **99**, 595 (1974).

202. Threshold Limit Values for 1960, *A.M.A. Arch. Environ. Health*, **1**, 140 (1960).

203. Thuret, J., *Ann. Fals. Fraudes*, **32**, 328 (1939); *Chem. Abstr.*, **34**, 1585 (1940).

204. Tomiyama, T., *Sci. Crime Detect.*, **8**, 74–76 (1955); through *Chem. Abstr.*, **49**, 11760 (1955).

205. Truog, E., and A. H. Meyer, *Ind. Eng. Chem. Anal. Ed.*, **1**, 136 (1929).

206. Tsuda, K., and S. Sakamoto, *J. Pharm. Soc. Japan*, **61**, 217 (1941); through *Chem. Abstr.*, **44**, 8819 (1950).

206a. Tsujii, T., and K. Kuga, *Anal. Chim. Acta*, **72**, 85 (1974).

207. Tsyvina, B. S., and B. M. Dobkina, *Zavod. Lab.*, **7**, 1116 (1938); through *Chem. Abstr.*, **33**, 2063 (1939).

208. Twyman, F., *Spectrochemical Abstracts*, Vol. I, Adam Hilger, London, 1938, pp. 1933–1937.

209. U.S. Public Health Service Drinking Water Standards, 1961, *J. American Water Works Assoc.*, **53**, 935 (1961).

210. Vallance, R. H., "Arsenic," in J.A.N Friend, *A Text-Book of Inorganic Chemistry*, Vol. VI, Part IV, Griffin, London, 1938.

210a. Van Loon, V. C., and E. J. Brooker, *Anal. Lett.*, **7**, 505 (1974).

211. Vasak, V., and V. Sedivec, *Chem. Listy*, **46**, 341 (1952); *Chem. Abstr.*, **47**, 67 (1953).

211a. Vassilaros, G. L., *Talanta*, **18**, 1057 (1971).

211b. Veal, D. J., *Anal. Chem.*, **38**, 1080 (1966).

212. Vecera, Z., and B. Bieber, *Slevarentsvi*, **12**, 366 (1956); through *Chem. Abstr.*, **51**, 5626 (1957).

213. Verma, M. R., and Y. P. Singh, *J. Sci. Ind. Res.* (*India*), **13B**, 709 (1954); *Chem. Abstr.*, **49**, 6024 (1955).

213a. Vijan, P. J., and G. R. Wood, *Atom. Absorp. Newslett.*, **13**, 33 (1974).

214. Visintin, B., and N. Gandolfo, *Ann. Chim. Appl.*, **33**, 111 (1943); through *Chem. Abstr.*, **38**, 6229 (1944).

215. Wadelin, C., and M. G. Mellon, *Analyst* (*London*), **77**, 708 (1952).

215a. Watson, A., and G. Svehla, *Analyst* (*London*), **100**, 573 (1975).

216. Weeks, M. E., *Discovery of the Elements*, 6th edit., Journal of Chemical Education, Pa., 1956, pp. 92–95.

217. Willard, H. H., and G. D. Manalo, *Anal. Chem.*, **19**, 167 (1947).

218. Williams, K. T., and R. R. Whetstone, *U.S. Dept. Agr. Tech. Bull.*, No. 732, 4 (1940).

218a. Wilshire, F. W., J. P. Lambert, and F. E. Butler, *Anal. Chem.*, **47**, 2399 (1975).

219. Wilson, J., and R. Dickinson, *J. Amer. Chem. Soc.*, **59**, 1358 (1937).

219a. Winefordner, J. D., V. Svoboda, and L. L. Cline, *Accounts Chem. Res.*, **4**, 259 (1971).

220. Woods, J. T., and M. G. Mellon, *Ind. Eng. Chem. Anal. Ed.*, **13**, 760 (1941).

221. Wyatt, P. F., *Analyst* (*London*), **80**, 368 (1955).

221a. Yamamota, Y., T. Kumamura, and Y. Hayash, *Talanta*, **19**, 1633 (1972).

222. Yana, M., H. Mochizuki, R. Kajiyama, and Y. Koyama, *Bunseki Kagaku*, **5**, 160 (1956); through *Chem. Abstr.*, **51**, 9410 (1957).

223. Zinzadze, C., *Ind. Eng. Chem. Anal. Ed.*, **7**, 230 (1935).

224. Zusser, E. E., *Zavod. Lab.*, **12**, 630 (1946); *Chem. Abstr.*, **41**, 1947 (1947).

ANTIMONY

By RICHARD B. HAHN, *Department of Chemistry, Wayne State University, Detroit, Michigan*

Contents

I. INTRODUCTION

A. HISTORY

Antimony was known in ancient times but probably only in the form of its sulfide, stibnite. This colored compound was used as a cosmetic and pigment. Its use is mentioned in the Bible (IV Kings: 9,30), "But Jezabel hearing of his coming in, painted her face with stibnic stone." It was also used by the Egyptians as an eye paint. Stibnite was called $\sigma\tau\mu\mu\iota$ by the Greeks and stibium by the Romans. It later received the name "antimonium," probably derived from the Arabic, and this name was ultimately applied to the free metal. Vases and tablets of antimony metal have been found in Egyptian ruins. The Greeks and Romans also may have been acquainted with it, but confused it with lead (35,153).

Antimony minerals and compounds were widely used by the alchemists in their attempts at transmutation and as medicinals. Its major medicinal use at this time was as an emetic. The preparation of metallic antimony and several of its alloys was discussed by the Benedictine monk, Basil Valentine, who published a book *The Triumphal Chariot of Antimony* during the fifteenth century (112). The preparation of the metal and its use in preparing "bookseller's alloy" (type metal) was discussed by Georgius Agricola during the sixteenth century. In 1707 Nicolas Lemery published his famous *Treatise on Antimony* in which he describes the metal, its minerals, various compounds of antimony, and their uses.

B. OCCURRENCE AND USES

Antimony is widely distributed in the earth's crust, and its minerals are found on all the continents. Richest deposits are found in China, the Republic of South Africa, Bolivia, U.S.S.R., and Yugoslavia. These sources together account for 85% of the world's production (100). Antimony composes about 1×10^{-5} wt% (0.1 ppm) of the earth's crust. It also has been found in trace amounts in meteorites (87).

The principal mineral of antimony is the trisulfide, Sb_2S_3, called stibnite or antimonite. As stibnite weathers and oxidizes, it is transformed into the oxide, Sb_2O_3, which occurs as the mineral, valentitnite or "white antimony." Antimony frequently is found combined with lead, nickel, copper, and silver as antimonides. Examples of these are NiSb, breithauphite, and Ag_2Sb, discrasite. Various mixed sulfides and thio minerals are known. NiSbS, ullmannite; Ag_3SbS_3, pyargyrite; $4Cu_2S \cdot Sb_2S_3$, tetrahedrite; and $2PbS \cdot Sb_2S_3$, jamesonite are a few examples. Antimony is occasionally found as the free element often in isomorphous mixtures with arsenic, this mineral being called allemontite. Other minerals are given in references 11, 145, and 149.

Antimony finds use both in its metallic form and as its oxide, sulfide, and certain other salts. The principal use of the metal is in the production of antimonial lead, a binary alloy of antimony and lead containing 5–12% antimony. In this alloy antimony functions to increase its strength and to inhibit chemical corrosion. Most of the antimonial lead is used to manufacture grids in the widely used lead–sulfuric acid storage battery. It is also used in cable sheath, tank linings, and roofing sheets. Other important alloys of antimony are antifriction bearing metal, also known as babbitt or "white metal," which is composed of lead, tin, antimony, and copper; type metal (lead, antimony, tin); and pewter (tin, antimony).

Compounds of antimony have a wide range of industrial applications. The oxide, Sb_2O_3, is used as a white pigment in paints, enamels, and ceramic glazes. It is used, along with silica, in manufacturing certain glasses. It is added to fabrics as a finishing agent and adds to the life of the material by filtering out the destructive ultraviolet wavelength of sunlight. The oxide along with the trichloride mixed with chlorinated paraffin is used to make fabrics fire resistant. The so-called pentasulfide is used as a vulcanizing agent in rubber, and the trisulfide is used in the manufacture of "tracer" bullets. Potassium antimonyl tartrate (tartar emetic) is used as a mordant in certain cloths. At one time several metallo-organic compounds of antimony were used as medicinals, but these are rarely used now because of their toxicity.

C. PRODUCTION OF ANTIMONY

The major source of antimony is stibnite, Sb_2S_3. Commercial ores range from low grades containing only 1–2% antimony to grades that approach pure Sb_2S_3, containing 71.5% antimony. The lower-grade ores are concentrated by hand sorting, tabling, or by milling, followed by floatation. Intermediate grade ore containing 5–25% antimony frequently are roasted to form volatile Sb_2O_3, which is deposited in condensation chambers.

Antimony sulfide is concentrated from high-grade ores by a liquidation process. The ore is heated in a furnace until a temperature of 550–600°C is reached. At this point Sb_2S_3 melts and runs to the bottom of the furnace, where it is tapped and collected in containers. This material, called "crudium" or "crude antimony," is used directly in many industrial processes.

Antimony metal is produced from the oxide by reduction with coke in a reverberatory furnace using sodium carbonate as flux. It also is produced from rich sulfide ores and from crudium by melting the sulfide, then adding fine, iron metal scrap. Iron metal reacts with the sulfide to form free antimony and ferrous sulfide. Antimony metal, being more dense than ferrous sulfide, sinks to the bottom of the furnace, where it is drawn off. Antimony metal is produced directly from intermediate-grade ores (10–30% antimony) by reduction with coke in a water jacketed blast furnace. A problem encountered in blast furnace production is the loss of antimony by volatilization of Sb_2O_3. This loss is minimized by using low blast pressure, a high smelting column, and relatively large quantities of flux and slag. Detailed discussions of these various processes are given in reference 149.

Unrefined antimony contains arsenic, sulfur, iron, copper, and lead as impurities. These impurities are eliminated by oxidizing and slagging agents. The metal is fused, sodium carbonate and sodium sulfate are added, and the mixture is heated. Under these conditions, arsenic goes into the slag as sodium arsenate, sulfur as sodium sulfide, and iron and copper form sulfides arising from partial reduction of the sodium sulfate. All these impurities are removed in the slag. Another method consists in blowing air through molten antimony metal. These refining operations are usually carried out in a small reverberatory furnace.

Historically the purity of antimony metal has been judged by the appearance of fern-like radiating crystal patterns on the surface of the casting. These patterns are called "stars," and the "starring" of antimony is an important step in the final production of the metal. Actually the appearance of such a structure does not actually indicate the relative purity, but results from slowly cooling the melt under a protective cover composed of sodium carbonate, potassium carbonate, and borax.

D. TOXICOLOGY AND INDUSTRIAL HYGIENE

There was disagreement among earlier workers concerning the toxicity of antimony and its compounds, some considering it very toxic and others considering it quite innocuous. These inconsistent findings probably arose from the fact that many antimony compounds at that time were contaminated with varying amounts of arsenic. In modern technology, arsenic is almost entirely removed from antimony and its compounds, and modern opinion holds that antimony compounds are much less toxic than arsenic compounds. An exhaustive study concerning the toxicity of antimony was made by Fairhall and Hyslop and is compiled in reference 28. Their findings indicate the following:

The most toxic of all antimony compounds is stibine, SbH_3, a gas. This compound is formed when antimony metal or any of its compounds reacts with nascent hydrogen. This may occur when zinc, iron, or aluminum metals containing antimony are treated with a nonoxidizing acid such as dilute sulfuric or dilute hydrochloric acid. During the generation of hydrogen gas by the reaction of the metal with the acid, antimony is converted to stibine. Similar results are obtained if a soluble antimony compound is added to a mixture of any metal plus acid generating hydrogen gas. Stibine also may be formed during the charging of lead–sulfuric acid batteries where antimony is present in the grids. In general, any process involving the evolution of hydrogen gas is a potential source of stibine poisoning.

The physiological action of stibine is similar to that of arsine. It attacks the central nervous system and the blood. The gas has a hemolytic effect on the blood and produces characteristic, morphological changes in the red blood cells. Symptoms of stibine poisoning are headache, nausea, weakness, slow breathing, and weak and irregular pulse. Some of the antimony is rapidly eliminated from the body, but part of it is retained in the red blood cells and other organs of the body. The threshold limit of stibine is given as 0.2 mg/m^3 of air.

Next in order of toxicity are the soluble compounds of antimony, the most common being tartar emetic, potassium antimonyl tartrate. The symptoms of acute antimony poisoning following the ingestion of a soluble antimony compound are nausea accompanied by violent vomiting and diarrhea. This is followed by great muscular weakness and collapse. Respiration is slow and irregular, and the body temperature and blood pressure are depressed. Very few cases of fatal antimony poisoning have been recorded. Because of its emetic effect most of the ingested antimony is expelled. Some of the antimony is absorbed by the digestive tract and enters into the blood stream. It is not concentrated to any great extent by any organ, although higher concentrations may appear in the liver. It is excreted fairly rapidly by the kidneys and unlike lead the effects are not cumulative. Continued ingestion of small doses of antimony may cause symptoms of dryness and soreness of the throat, loss of appetite, nausea, and a tired feeling.

The insoluble compounds of antimony such as the oxides and sulfides are relatively innocuous. Animals have been fed relatively large amounts of these compounds with little ill effects. Workers handling them for many years have suffered no ill effects. Continued exposure to dusts of these compounds, however, may cause skin irritation, irritation of the mucuous membranes, and loss of appetite. Antimony metal is slightly more toxic than the sulfides and oxides, but not as nearly as toxic as soluble compounds. In general the order of toxicity is stibine, soluble compounds, antimony metal, antimony sulfides, and the antimony oxides being least toxic. Trivalent compounds usually are somewhat more toxic than the pentavalent compounds. The toxic effects are not cumulative.

Major precautions in the laboratory consist in providing good ventilation if there is any possibility of generating stibine gas, and to avoid the ingestion of soluble antimony compounds.

II. PROPERTIES OF ANTIMONY

A. PHYSICAL PROPERTIES

The atomic number of antimony is 51, and its atomic weight is 121.75 (C = 12). Two naturally occurring stable isotopes are known with atomic masses 120.90 (57.25%), 122.90 (42.75%). Approximately 25 unstable, radioactive isotopes are known varying in atomic mass from 112 to 125. Most of these have short half-lives except ^{124}Sb (60.9 days) and ^{125}Sb (2.8 yr). Both isotopes decay by beta-gamma emission and are frequently used in tracer work.

Antimony is a silver-grey metal of moderate hardness (3 on Moh's scale) but rather brittle. It can be easily ground and powdered at room temperature. It exists in several allotropic forms. Ordinary grey, alpha, or metallic antimony forms crystals belonging to rhombohedral class of the hexagonal crystallographic system with three neighboring atoms at a distance of 2.87 Å and three others at 3.37 Å. Its density at 25°C is 6.68. A yellow modification may be prepared by passing oxygen gas into liquid antimony hydride at $-90°C$. A black form is obtained by rapidly cooling antimony vapor. These forms are unstable and revert to the ordinary grey form on standing. An extremely active amorphous or vitreous also is known. This form is spontaneously flammable upon exposure to air.

Antimony metal melts at $630.5°C$ and boils at $1640°C$. The latent heat of fusion is 38.84 cal/gram. There is only a slight volume change upon solidification. Earlier investigators concluded that there was a shrinkage, but more recent measurements show an expansion of 0.95%. The average specific heat of solid antimony is 0.05 cal in the range of 0–100°C. The thermal conductivity of metallic antimony is approximately 0.044 gram·cal/cm/sec. The electrical resistance is 8.28×10^{-4} ohm·cm at 15°C. Various other physical constants are given in reference 113.

B. CHEMICAL PROPERTIES

1. Oxidation States

Antimony is classified in Group III-A of the periodic table along with nitrogen, phosphorus, arsenic, and bismuth. Its principle oxidation states are 3− as in SbH_3 and in antimonides, 3+ as in $SbCl_3$, and 5+ as in Sb_2O_5. Some authorities postulate a 4+ state as in the stable oxide Sb_2O_4; however, others consider this as a stoichiometric mixture of the 3+ and 5+ oxides. The most common oxidation state is the 3+ form.

2. Reactions of Metallic Antimony

Free antimony metal is not very active chemically, standing below hydrogen in the electrochemical activity series. It does not react with dilute mineral acids such as hydrochloric or sulfuric acid. It is not oxidized to any extent by air at room temperature, but upon heating, finely divided antimony metal will unite rapidly with oxygen to form Sb_2O_3. It will also react with free chlorine and the other halogens at room temperature to form the corresponding halides. When heated, antimony will unite with sulfur, forming Sb_2S_3. It forms alloys with most of the nonferrous metals and will unite with certain metals to form antimonides.

Nitric acid reacts with antimony to produce a mixture of insoluble oxides, mainly Sb_2O_4 and also Sb_2O_3 and Sb_2O_5, the exact composition depending upon the concentration of the nitric acid and the temperature. In the process nitric acid is reduced to NO and NO_2.

Free antimony is readily dissolved by aqua regia, forming a mixture of soluble tri- and pentachlorides. It is also attacked by hot concentrated sulfuric acid to form the soluble sulfate, $Sb_2(SO_4)_3$. It is slowly dissolved by concentrated hydrochloric acid, but only in the presence of air. The addition of an oxidizing agent such as hydrogen peroxide or bromine to the hydrochloric acid will hasten the dissolution. Antimony metal is readily dissolved by a mixture of nitric and hydrofluoric acid, and also by a mixture of nitric and tartaric acids. It is kept in solution owing to the formation complex anions. These are discussed in Section II.B.4.

3. Simple Compounds

Antimony forms several binary compounds: the oxides Sb_2O_3, Sb_2O_5, and Sb_2O_4; the sulfides Sb_2S_3 and Sb_2S_5; the halides SbX_3 and SbX_5 being the most important. Recent investigations (74a) using X-ray diffraction and Mössbauer spectroscopy indicate that antimony pentasulfide does not contain the Sb^{5+} ion, but rather is a nonstoichiometric compound of Sb_2S_3 united with variable amounts of sulfur. SbH_3, a gas, is formed when hydrogen gas is generated in the presence of antimony metal or its compounds. Antimony also forms antimonides with various metals, for example, cesium antimonide, Cs_3Sb, which is used in photoelectric cells.

Antimony forms a limited number of salts in which antimony is the cation. The trivalent compounds in general are more stable than the pentavalent compounds. In addition to the halides and sulfides mentioned above, antimony forms a soluble sulfate, $Sb_2(SO_4)_3$. The nitrate is unstable and cannot be isolated, but an insoluble basic nitrate $SbONO_3$ is known.

A number of compounds are known containing antimony(III) in the form of the antimonyl, SbO^+ cation. Most of these are formed by hydrolysis. If antimony trichloride, for example, is treated with water, a white insoluble precipitate of antimony oxychloride or antimonyl chloride SbOCl is formed. The other halides behave similarly. Antimony(V) compounds usually hydrolyze completely and precipitate insoluble antimonic acid, H_3SbO_4.

One of the most common compounds of antimony is potassium antimonyl tartrate, commonly called "tartar Emetic," usually written as $[K(SbO)C_4H_4O_6]_2 \cdot H_2O$. This compound is prepared by dissolving antimony(III) oxide in potassium hydrogen tartrate. It is the most important commercial soluble compound of antimony. It is readily soluble in water and does not hydrolyze readily owing to the complexation of antimony by the tartrate ions. Recent investigations indicate the above formula is incorrect, and it should be $K[C_4H_4O_6Sb(OH)_2]$.

Antimony(III) hydroxide is definitely amphoteric, being converted into the antimonite anion, $[SbO_2]^-$ or $[Sb(OH)_4]^-$, by strong bases.

Antimony(V) hydroxide does not exist as such, but is known in the form of antimonic acid, $H[Sb(OH)_6]$ or H_3SbO_4. Polymeric forms such as $[Sb_3O_{10}]^{-5}$ and $[Sb_4O_{13}]^{-6}$ also exist. The potassium salt of antimonic acid, $K[Sb(OH)_6]$, is used as a reagent for sodium, precipitating insoluble $Na[Sb(OH)_6]$ (146). Older literature refers to this compound as potassium pyroantimonate, $K_2H_2Sb_2O_7 \cdot 6H_2O$, but recent investigations favor the former structure (122).

Another important class of compounds are those containing antimony as thiosalts. These compounds are formed by dissolving antimony sulfide in sodium sulfide, or any other soluble sulfide. The thioantimonite, $[SbS_3]^{3-}$, and thioantimmonate, $[SbS_4]^{3-}$, anions are formed. The free acids are unstable and decompose to form antimony sulfide and hydrogen sulfide. Recent investigations have shown these anions are more complex than the formulas given above, $[SbS_2(SH)_2(OH)_2]^{3-}$ being postulated.

4. Complex Ions and Compounds

Both Sb(III) and Sb(V) form complex ions and complex compounds with the halides. If antimony pentachloride or antimony trichloride is treated with an excess of concentrated hydrochloric acid, a variety of chloro complexes are formed. Crystals of $HSbCl_6$ and salts corresponding to the acids, H_2SbCl_7 and H_3SbCl_8, have been isolated. Some examples of these are $CrSbCl_8 \cdot 10H_2O$, $RbSbCl_6$, and $MgSbCl_7 \cdot 9H_2O$. Double salts of antimony(III) chloride are formed by mixing the corresponding chloride solutions in hydrochloric acid. These usually crystallize out as slightly soluble substances and are employed in chemical microscopy for the identification of certain cations. Examples of these are $2KCl \cdot SbCl_3$ and $BaCl_2 \cdot SbCl_3 2H_2O$. Many others are tabulated in reference 145. Although $SbCl_4$ has not been isolated, many double salts containing it are known, such as $2CsCl \cdot SbCl_4$, which may be considered as a compound of Sb(III) plus Sb(V). Complexes are formed also with other halide ions: iodoantimonous acid, for example, is used in the spectrophotometric determination of antimony. Antimony metal readily dissolves in a mixture of nitric and hydrofluoric acids owing to the formation of a stable fluoride complex, $HSbF_6$.

Antimony forms soluble complexes with α-hydroxy organic acids such as citric and tartaric acids, which are useful in preventing the precipitation of hydroxides and basic salts. Antimony also forms a weak complex with EDTA, which is easily decomposed by the addition of tartaric acid or hydrofluoric acid.

Important complexes from the standpoint of the analytical chemist are those formed by treating insoluble antimony sulfides with a strong base or with a solution of a soluble sulfide such as sodium sulfide. Under these conditions antimony is converted to a soluble thio anion of the type, $[SbS_3]^{3-}$, or oxothio anion as $[SbOS_2]^{3-}$. These are stable only in basic solution and are decomposed upon acidification. The separation of certain sulfides in the classical qualitative analysis scheme utilizes these reactions.

5. Organic Compounds of Antimony

Both Sb(III) and Sb(V) form a variety of organic compounds. The best known is tartar emetic, in which antimony is combined with tartaric acid through oxygen linkages. Similar compounds are formed with other α-hydroxy acids. A number of compounds are known, however, in which there is a direct carbon to antimony bond, among these are the trialkyl and triaryl stibines in which the hydrogen atoms of stibine are substituted by organic groups such as the methyl or the phenyl group. Compounds such as R_3SbO or R_3SbCl_2, the trialkyl or aryl stibnic oxide or chloride also are known (18). Stibonium compounds as Me_4SbI, tetramethyl stibonium iodide, have been prepared. Tetraphenyl stibonium chloride has been used as an analytical reagent to precipitate anions such as permanganate and perchlorate (159). At one time many of these compounds were investigated for pharmacological use, especially for treatment of trypanisome infections. These have been largely abandoned, however, in favor of other less toxic drugs.

6. Chemistry in Aqueous Solution

Antimony does not exist as a simple cation in aqueous solution, but as partially hydrolyzed species such as $Sb(OH)^{2+}$ in the case of Sb(III) ions or $Sb(OH)_2^{3+}$ with Sb(V) ions. The number of hydroxyl groups associated with the antimony is dependent on the pH of the solution. Owing to the ease of hydrolysis, solutions containing antimony should be rendered strongly acidic to prevent precipitation of insoluble basic salts. When a solution of antimony trichloride, for example, is diluted with water it rapidly hydrolyzes and forms a series of insoluble compounds varying from SbOCl to $Sb_4O_5Cl_2$. Complete hydrolysis finally results with the formation of the hydrated oxide, $Sb_2O_3 \cdot xH_2O$. These compounds will redissolve in strong hydrochloric acid to form chloroantimonous acid, $HSbCl_4$. Other Sb(III) salts undergo hydrolysis in a similar manner.

Sb(V) compounds undergo even more drastic hydrolysis, finally resulting in the precipitation of insoluble antimonic acid, H_3SbO_4. As with Sb(III), the Sb(V) ion forms soluble chloro complexes; chlorantimonic acid, $HSbCl_6$, for example, is found in 10–12 N hydrochloric acid solution. When the hydrochloric acid concentration is decreased, partially hydrolyzed species such as $[Sb(OH)Cl_5]^-$ are formed. In 6 N acid $[Sb(OH)_2Cl_4]^-$ is present; $[Sb(OH)_3Cl_3]^-$ and $[Sb(OH)_4Cl_2]$ are found in solutions less than 6 M in acid. The rates of hydrolysis is rapid even at room temperature.

Both $SbCl_3$ and $SbCl_5$ are appreciably volatile, and losses may occur when hydrochloric acid solutions are evaporated. Experiments by Buckheit showed about 10% of the antimony is lost when 100 ml of a 1:1 hydrochloric acid solution of $SbCl_3$ was evaporated down to 10–15 ml in a beaker and as much as 50% volatilized if evaporation was continued until the solution became a syrup (80).

When acid solutions containing Sb(III) are made alkaline with ammonium hydroxide or other weak bases, a precipitate of the corresponding hydrous oxide, $Sb_2O_3 \cdot xH_2O$, is proved. Drying in air gives a compound approaching $Sb(OH)_3$, but it is doubtful whether the hydroxides exist as well defined compounds. Treatment of antimony salts with an excess of a strong base results in the formation of soluble anionic species. With Sb(III) and excess sodium hydroxide, sodium antimonite, $NaSbO_2 \cdot 3H_2O$, is formed, and being sparingly soluble it crystallizes out. The corresponding potassium compound, however, is soluble and does not precipitate out.

Antimony(V) compounds form salts of antimonic acid, $HSb(OH)_6$ (also written as H_3SbO_4), in strongly basic solution. Solutions of potassium antimonate, $KSb(OH)_6$, react with sodium ions to form insoluble sodium antimonate, $NaSb(OH)_6$, which is used for the gravimetric determination of sodium.

When acid solutions of antimony are treated with hydrogen sulfide the corresponding insoluble sulfides, Sb_2S_3 and Sb_2S_5, are precipitated. These sulfides dissolve in sodium or ammonium sulfide to form soluble thio anions, $[SbS_2]^-$ or $[SbS_4]^{3-}$. They also are soluble in bases forming a series of oxythio complexes such as $[SbOS_3]^{-3}$. When these solutions are acidified, the thio complexes are decomposed and the corresponding insoluble sulfides reprecipitated. These reactions form the basis of the separation and identification of antimony in the classical qualitative analysis scheme.

Antimony(III) is readily oxidized in acid solution to Sb(V) by strong oxidizing agents such as bromate or permanganate. The bromate oxidation forms the basis of a widely used volumetric method that is discussed in detail in Section VI.B.1. The oxidation potential of the Sb(III)–Sb(V) couple is strongly influenced by the pH of the solution, and in neutral or basic solution Sb(III) is easily oxidized even when a mild oxidant such as free iodine; hence it may be determined by iodometric titration.

With strong reducing agents such as Cr(II) antimony ions are reduced to the free metal. Free metals such as zinc also will reduce antimony compounds to the free element. If hydrogen gas also is formed concurrently from the reaction of the metal with acid, a portion or all of the antimony may be converted to gaseous stibine, SbH_3. All of these reactions have been employed for the detection and/or determination of antimony.

C. ELECTROCHEMICAL PROPERTIES

Antimony metal may be deposited by electrolysis from acid solution, but losses occur owing to partial reduction to stibine gas that is volatilized at the cathode (81). This occurs both at a platinum or mercury cathode. Electrolytic reduction of the 5+ state to the 3+ state is also possible, which forms the basis of certain polarographic methods (see Section VI.C for a discussion of polarographic methods). Standard electrode potentials of analytical importance are summarized in Table 1.

An antimony metal electrode in contact with Sb(III) ion in acid solution does not give reproducible potentials (106). Erratic results are obtained also in alkaline solution. A metallic antimony–antimony trioxide electrode has been used for the determination of pH, the potential varying linearly with pH (124). Since the advent of the glass electrode, however, this is seldomly used.

TABLE 1
Standard Reduction Potentials

Reaction	$E°$ (V)
In acid solution	
$SbH_3 (g) = Sb^0 + 3H^+ + 3e$	0.51
$2Sb^0 + 3H_2O = Sb_2O_3 + 6H^+ + 6e$	-0.152
$Sb^0 + H_2O = SbO^+ + 2H^+ + 3e$	-0.212
$Sb_2O_3 + 2H_2O = Sb_2O_5 + 4H^+ + 4e$	-0.692
$2SbO^+ + 3H_2O = Sb_2O_5 + 6H^+ + 4e$	-0.581
$Sb_2O_4 + H_2O = Sb_2O_5 + 2H^+ + 2e$	-0.48
$SbO^+ + 2H_2O = Sb_2O_4 + 4H^+ + 2e$	-0.68
In basic solution	
$Sb^0 + 4OH^- = SbO_2^- + 2H_2O + 3e$	0.66
$SbO_2 + 2OH = SbO_3^- + H_2O + 2e$	0.589
$Sb^0 + 2S^{2-} = SbS_2 + 3e$	0.85
$SbS_2^- + 2S^{2-} = SbS_4^{-3} + 2e$	0.6

D. OPTICAL PROPERTIES

The various ionic species of antimony are colorless; hence most colorimetric methods for antimony are based upon colored species formed by coupling with a colored organic molecule as, for example, the well-known Rhodamine B method. Iodoantimonous acid absorbs both in the visible and in the ultraviolet, and this forms the basis for certain spectrophotometric methods (Section VI.D). Although some compounds of antimony are highly colored, for example, the sulfides, they are insoluble and are of little use as spectrophotometric methods.

When antimony is excited by an arc or spark, emission occurs and characteristic spectra are obtained. This is the basis of emission spectrographic methods, which are discussed in Section VI.E. Little excitation takes place in the flame, but useful atomic absorption methods have been proposed and are discussed in Section VI.F.

III. SAMPLING OF ANTIMONY-CONTAINING MATERIALS

A variety of materials containing antimony may be submitted for analysis, each posing its own, individual sampling problem. Most commonly encountered are alloys, ores, chemical compounds of antimony, and organic substances containing antimony. Each is discussed separately. A general discussion concerning the theory and techniques of sampling is given in Section I-4 of the Treatise.

A. ALLOYS

The sampling of antimony metal and alloys containing antimony is discussed thoroughly by the ASTM (5). Following is a summary of their recommendations.

Metals in the form of bars, ingots, slabs, rods, plates, and so forth are best sampled by taking drillings equally spaced between the ends of the piece along a diagonal line. The drill should be approximately $\frac{1}{2}$ in. in diameter, and drilling should be through the total thickness of the material. Alternately the material may be milled or sawed through a section from side to center or through an entire cross-section. Clippings may be taken from materials that are too thin to be conveniently drilled or milled.

After collecting the chips, they should be examined for foreign substances such as bits of iron from the tools. A magnet may be used to remove this foreign matter. The chips should be uniform in size, and fine, dustlike material should be avoided.

B. ORES AND MINERALS

Representative pieces may be crushed in a jaw crusher, then ground, and milled to fine powder. No special precautions are necessary since most antimony minerals are nonhydroscopic and are resistant to oxidation by air at room temperature. Techniques employed for crushing, grinding, and so forth are discussed in Section I-4 of the Treatise.

C. CHEMICAL COMPOUNDS

The chemical compounds of antimony vary widely in their properties. The most common ones, for example, the oxides, sulfides, and potassium antimonyl tartrate (tartar emetic), are stable and nonhydroscopic in air, and a "grab" sample may be used if the material is homogeneous. A few compounds, especially the halides, are hydroscopic, and prolonged exposure to moist air should be avoided during the sampling process.

D. ORGANIC SUBSTANCES AND BIOLOGICAL FLUIDS

The usual sampling procedures are followed. The problem with these samples is not so much in the sampling, but in the destruction of the organic material prior to analysis of the sample. These problems are discussed in Section IV.B in detail.

IV. SEPARATION AND ISOLATION OF ANTIMONY

A. DECOMPOSITION, DISSOLUTION, AND OTHER PRELIMINARY TREATMENT OF INORGANIC MATERIALS

1. Antimony Metal

Antimony metal and alloys composed largely of antimony metal are best dissolved in hot, concentrated sulfuric acid or in hydrochloric acid plus some oxidizing agent such as nitric acid, free bromine or iodine, potassium chlorate, and so forth. In the process the metal is converted to soluble Sb(III) or Sb(V) compounds.

2. Alloys of Antimony

Tin-base alloys are dissolved by heating with concentrated sulfuric acid. The soluble sulfates of tin and antimony are formed.

Alloys of antimony, tin, and lead are treated with concentrated sulfuric acid. In this procedure antimony is oxidized mostly to the 3+

state. Mixtures such as hydrochloric acid plus bromine also are effective.

Alloys high in lead content such as antimonyl lead are dissolved in concentrated hydrochloric acid saturated with bromine. If only traces of antimony are present, it may be dissolved in dilute nitric acid.

If the preceding treatments fail, many alloys may be dissolved in a mixture of hydrofluoric and nitric acid, employing, of course, a platinum or plastic container.

3. Sulfide Ores

Sulfide ores may be dissolved in a hydrochloric acid or in a mixture of hydrochloric and nitric acids or by digestion with concentrated sulfuric acid. After cooling, the sulfuric acid solution is diluted with hydrochloric acid or with a sodium tartrate solution to avoid hydrolysis of the antimony.

4. Minerals Containing Antimony Oxides

Minerals containing the oxides of antimony usually may be dissolved by heating with a mixture of concentrated sulfuric acid and potassium hydrogen sulfate. Those insoluble in acids must be fused to render them soluble. Sodium hydroxide and sodium hydrogen sulfate are recommended as fluxes. After fusion, the melt is dissolved in hydrochloric acid.

Geological samples to be analyzed by atomic absorption may be rendered soluble by heating with solid ammonium iodide. The sample (80 mesh or finer), weighing about 0.25 gram, is mixed with 0.5 gram of solid ammonium iodide, then placed in a test tube and heated until all the ammonium iodide has sublimed. The sublimate, which also contains all the antimony in the form of SbI_3, is dissolved in 2 M hydrochloric acid, and the resulting solution is used directly for atomic absorption measurements (101a).

B. DECOMPOSITION, DISSOLUTION, AND OTHER PRELIMINARY TREATMENT OF ORGANIC MATERIALS

Most organic compounds of antimony are dissolved by "wet ashing," which consists of digesting the material with a mixture of concentrated nitric and sulfuric acid. Solid potassium sulfate sometimes is added to elevate the boiling point. Last traces of carbon and other organic material may be destroyed by the dropwise addition of 30% hydrogen peroxide to the boiling mixture. After cooling the mixture is diluted with appropriate reagents and analyzed.

If the material is not decomposed by wet ashing, fusion techniques may be employed. The most common technique is fusion with a mixture of sodium carbonate and sodium peroxide. Other mixtures have been recommended but have no particular advantages.

C. SELECTIVE SEPARATION METHODS

Antimony may be separated from other elements by precipitation, distillation as the chloride, volatilization as stibine, liquid–liquid extraction, or by various chromatographic techniques. These are discussed in the following sections.

1. Precipitation Methods

The most commonly employed precipitation method for antimony consists in precipitating the sulfide from dilute acid solution. This is used in the classical qualitative analysis scheme. A separation is achieved from zinc, iron, cobalt, and so forth, but not from other elements of the acid hydrogen sulfide group such as lead, tin, bismuth, arsenic, and so forth. Both hydrogen sulfide and thioacetamide have been used as precipitants (13,31).

The sulfides of arsenic, antimony, and tin are soluble in ammonium sulfide, alkali sulfides and polysulfides, and in solutions of strong bases such as sodium or potassium hydroxide. These reagents have been used to separate the sulfides of arsenic, antimony, and tin from most other sulfides that are insoluble in these reagents.

Antimony sulfide may be separated from arsenic sulfide by treating the mixed sulfide with 8–10 M hydrochloric acid, which dissolves the antimony sulfide but not the arsenic. Antimony may be separated from tin by complexing the latter with oxalic acid or hydrofluoric acid, then treating the mixture with hydrogen sulfide, which precipitates antimony sulfide leaving tin in solution.

Sulfide precipitations are of interest mainly from a historical standpoint. They suffer from the disadvantages that they are not very selective and they are bulky and tend to carry down many impurities. Sulfide precipitates usually are not of stoichiometric composition, frequently containing variable amounts of free sulfur; on drying some oxidation occurs, and the resulting precipitate is not suitable for weighing in a gravimetric determination. Because of these difficulties the sulfide separation is recommended only if other methods are not suitable. Detailed discussions of sulfide precipitations are given in references 13 and 67.

Antimony(III) forms a precipitate with pyrogallic acid that gives a good separation from arsenic but not from other metals. Oxine likewise precipitates antimony(III) at pH of 6, but many other substances also are precipitated at this pH. Thonalid (β-aminonaphthoylthioglycolic acid) precipitates antimony from basic solution. Using tartrate and cyanide as masking agents, separation is achieved from iron(II), cobalt, chromium(III), cerium(III), and titanium(IV), but not from lead, bismuth, zinc, cadmium, or gold (67).

When alloys containing antimony metal are treated with nitric acid, insoluble oxides of antimony are precipitated. This serves to separate antimony from metals forming soluble nitrates but not from tin, which also forms an insoluble oxide. Arsenic and phosphorus also are carried down with the precipitate. This cannot be employed successfully as a quantitative method owing to the partial solubility of antimony oxides in nitric acid.

Traces of antimony may be separated and concentrated by coprecipitation with manganese dioxide. Hydrochloric acid and other reducing agents must be absent. The technique consists in adding manganese(II) ions to the dilute acid solution containing antimony, heating, and adding permanganate solution. Under these conditions, manganese(II) is oxidized to insoluble, hydrated, manganese dioxide, while permanganate is reduced to the same substance. The reactions may be carried out in dilute nitric, sulfuric, or perchloric acid solution. Coprecipitation is quantitative even in strongly acid solution (about 1.5 N) (67), but lower acidities are usually recommended (76,111). After separation of the precipitate by filtration or centrifugation, it usually is dissolved in 6 N hydrochloric acid containing hydrogen peroxide. Manganese does not interfere in most methods for antimony, and it then may be determined by an appropriate method, either colorimetrically or titrimetrically.

Antimony also can be separated by precipitation as the free metal. Two techniques are employed: antimony ions are reduced to the free metal using a chemical reductant or by electrodeposition. One technique consists in using a more active free metal to displace antimony from its compounds in solution. Cadmium, tin, copper, and iron have been used (67). These methods, however, introduce foreign, metal ions into the solution, and difficulties are encountered in attempting to isolate the free antimony. Other metals below hydrogen in the displacement series, for example, silver, mercury, and so forth, are also displaced.

Certain homogeneous reductants also will reduce antimony ions to the free metal. Among these are sodium dithionite, hypophosphorus acid, and chromium(II) ions (67). As with the heterogeneous reductants, the other less active ions also are precipitated as the free metal. In general these methods serve only to separate antimony from those metals above hydrogen in the displacement series.

2. Electrodeposition Methods

Many studies have been made on the separation and determination of antimony by electrodeposition. Both constant current and controlled potential methods have been investigated. These are reviewed in reference 72. Deposition both from acidic and basic media are possible. Many acids have been used, including sulfuric, hydrochloric, phosphoric, pyrophosphoric, tartaric, and various mixtures of these (36). Deposition from acid solution at constant current has not been too successful because of the difficulty in obtaining adherent deposits and complete depositions. The best method is that proposed by Norwitz (107), which consists in using a mixture of hydrochloric and sulfuric acid with hydroxylamine hydrochloride as depolarizer. A maximum of 0.4 grams of antimony can be determined. Copper, cadmium, tin, arsenic, lead, bismuth, and silver interfere.

Deposition from basic solution also has its difficulties. The most common medium is a sodium sulfide solution in which antimony exists as the soluble thioantimonite anion. Polysulfides interfere in the deposition, but these are removed by adding potassium cyanide, which reacts with polysulfides to form thiocyanates. Solutions must be heated to obtain a good deposit. Results usually are high owing to the inclusion of oxygen and sulfur compounds (147).

Increased specificity can be obtained using controlled potential techniques, which is discussed in detail by Lingane (72). Separation of antimony from tin is accomplished both in acid and in basic solution. Reynolds (123) was able to separate antimony, tin, and bismuth. Hayakawa gives a procedure for the electrolytic separation and determination of antimony, copper, lead, and tin in nonferrous alloys (49).

3. Volatilization Methods

a. STIBINE

Stibine and the halides of antimony are volatile, and this property has been used in separations. Most volatile is stibine, SbH_3, a gas at ordinary temperatures with a boiling point of $-17°C$. As discussed in Sections I.D and II.B, stibine is generated when a solution containing antimony is treated with zinc and dilute hydrochloric or sulfuric acid. The original technique was developed by Marsh and by Gutzeit for the detection of small amounts of arsenic or antimony. Quantitative evolution of stibine is possible only under special conditions, which are not convenient. The zinc metal must be amalgamated, the temperature must be kept below 20°C, and the hydrochloric acid concentration must be not less than 9 M. A period of $1\frac{1}{2}$ hr is required then for quantitative evolution (163). The stibine method, however, is useful in certain radiochemical analyses where quantitative recoveries are not required (44). Webster and Fairhall developed a procedure where 98% recoveries were achieved with 25–50 μg of antimony (152). Low recoveries are usually obtained with organic or biological samples.

Stibine may be generated also by treatment of antimony containing solutions with sodium borohydride. Braman et al. (12a) determined 0.005–0.2 mg of antimony in water samples using this technique. Helium was used as carrier gas, and the measurement was made spectrographically by passing the gases into a dc discharge chamber.

Enhanced sensitivity in the atomic absorption determination of antimony was achieved by the use of stibine (162a). Antimony was converted to stibine by the use of zinc and 12 N sulfuric acid. The stibine was mixed with hydrogen and argon and then burned in an air–acetylene flame. Absorbance was read at 231.2 nm.

b. THE HALOGENS

Table 2 lists the boiling points of the halides of arsenic, antimony, and tin. These data indicate that several compounds are volatile enough to be distilled at ordinary atmospheric pressures and that there is enough difference in the boiling points of certain compounds of arsenic, antimony, and tin to achieve a quantitative separation by fractional distillation. Several studies have been made on the fractional distillation of the halides of arsenic, antimony, and tin (50,96,130), and this technique is recommended by the ASTM (5) for the separation of these elements in alloys. The apparatus (see Fig. 1), and the method is discussed in detail

TABLE 2
Boiling Points of the Halides of Arsenic, Antimony, and Tin

	Fluoride	Chloride	Bromide	Iodide
As(III)	60	130	221	403
As(V)	−53	d[a]	d	d
Sb(III)	319	224	275	400
Sb(V)	150	d	d	d
Sn(II)	603	652	620	717
Sn(IV)	705	114	201	340

[a] d = decomposes.

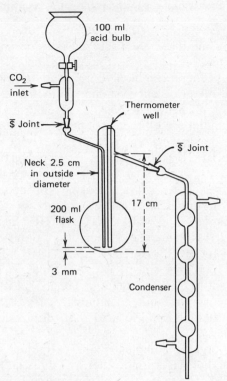

Fig. 1.   Apparatus No. 5 for determination of arsenic and antimony by distillation.

in Section IV.C.2 of "Arsenic" and in Section IX.B.3. The general method was developed by Scherrer and forms the basis of most commonly used procedures (130). The substance to be analyzed is dissolved in sulfuric acid or a mixture of sulfuric and nitric acids, fumed to expel the nitric acid, and hydrochloric acid is added. Hydrazine is added to maintain arsenic and antimony in the 3+ oxidation states, and phosphoric acid is added to prevent the distillation of tin(IV) chloride. A current of carbon dioxide is passed through the apparatus, and arsenic trichloride is distilled off at 110°C. The distillate is absorbed in water or dilute acid. The temperature is then raised to 155–165°C, and antimony trichloride is distilled. A mixture of hydrobromic and hydrochloric acid is finally added, the temperature raised to 140°C, and tin IV bromide is distilled. Arsenic, antimony, and tin then may be determined in the separate distillates by an appropriate method, iodometric methods being most commonly employed.

4. Liquid–Liquid Extraction Methods

Many solvent extraction methods have been proposed for the separation of antimony, and these are discussed in references 84, 98, and 157. Much attention has been given to the extraction of halide, ion-association complexes. Ethers commonly are used as extractants in these systems. In hydrofluoric acid systems, antimony is only partially extracted along with elements such as tin, arsenic, and germanium. Only niobium and tantalum are extracted completely. Better separations are attained in chloride systems where Sb(V) is extracted completely along with As(III), Ga(III), Ge(III), Au(III), Fe(III), Hg(II), Mo(VI), Nb(V), Pt(II), Po(II), Pa(V), Tl(III), and Sc(III). Antimony(III) is only partially extracted. More selectivity is attained using a bromide system. Using a 5 *M* solution of hydrobromic acid and ethyl ether as extractant, over 95% of Sb(V) is extracted along with Au(III), Ga(III), In(III), Tl(III), Sn(II), Sn(IV), and Fe(III) (10). Iodide systems are the most selective, Sb(III), Cd(III), Au(III), In(III), Pb(II), Sn(II) and Tl(III) being extracted completely (19,56). Increased selectivity sometimes is achieved by using benzene as solvent in place of ethers (8,119,155). Little or no extraction of antimony occurs with other commonly used extractants such as thiocyanate (9), alkylphosphoric acid (85), dithizone and trialkylphosphine oxides (158).

A variety of organic complexes are extractable among these are rhodamine B, methyl violet, and other azo dyes (128), antipyrene (98), and cupferron (39). The extraction of rhodamine B is discussed in detail in Section V.D.2.

5. Adsorption Methods

A variety of substances have been used as adsorbants for the separation of antimony, among these are filter paper, alumina, and organic ion-exchange resins. Each of these is discussed separately.

a. PAPER CHROMATOGRAPHY

This technique usually consists in evaporating a drop of solution of the material to be separated near the end of a filter paper strip, then dipping the end into some solvent, usually a mixture of a lower alcohol, water, and acid. As the solvent moves along the paper the various ions are carried along at different rates; hence a separation is achieved. After a certain period, the paper is removed, dried, and sprayed with a color-forming reagent to locate the position of the various ions. The ratio of the distance traveled by the desired substance to the distance traveled by the solvent front is known as the R_f value. Details of this method are discussed in reference 70. Several studies have been made on the separation of As, Sb, and Sn (the tin group in classical qualitative analysis). In most cases excellent separations were obtained (3,16,114). The R_f value of antimony usually lies between that of arsenic and tin. In using isopropyl alcohol with 10% 5 N hydrochloric acid, for example, As(III) is 0.66, Sb(III) is 0.77, and Sn(II) is 0.88 (16).

Sb(III) may be separated from Sb(V) using paper chromatography. The solvent consists of a mixture of acetic acid, water, and ethyl acetate. Sb(V) moves with the solvent front, whereas Sb(III) has an R_f value of 0.65–0.71. Hg(II), Cu(II), Sn(II), Al(III), Cr(III), and Mn(II), however, give R_f values close to Sb(III) (118).

The ring oven technique also may be used to separate antimony from a variety of cations (156). The solution is evaporated at the center of a filter paper that is placed on the ring oven. After treatment with dilute sulfuric acid and potassium iodide, the paper is treated with a benzene–ethyl alcohol mixture that carries along the antimony. When the paper is treated with molybdophosphoric acid a blue ring indicates antimony. Thallium, selenium, tellurium, and tungsten interfere.

b. ALUMINA AND OTHER INORGANIC MATERIALS

Alumina has been widely used as an adsorbant in the separation of organic compounds. It also is useful in certain inorganic separations (138). Alumina columns have been employed in separating Sb(III) from As(III). The separation is effected by elution with dilute hydrochloric acid. Colored bands may be produced by treating the column with water saturated with hydrogen sulfide. Antimony may be separated from arsenic and tin by eluting the column with a solution of sodium sulfide. Arsenic is eluted first, and antimony then may be removed by eluting the column with concentrated ammonium hydroxide. Selenium, tin, tellurium, and molybdenum may be separated from antimony by washing the column with ammonium sulfide solution (70). A more complete discussion including other inorganic ion exchangers is given in references 70, 138, and 141.

Calcium sulfate and asbestos also have been used as adsorbants. A complete separation of arsenic, antimony and tin was achieved by elution with dilute hydrochloric acid (132,133).

c. ORGANIC ION-EXCHANGE RESINS

Organic ion-exchange resins are widely used for the separation of various ions. The theory, techniques, and applications are discussed thoroughly in references 55, 68, and 127. There are two general types of resins: cation exchangers and anion exchangers. Since antimony usually exists as some anionic complex in solution, anion exchangers are commonly employed in the separation of antimony. Krause and Nelson made an extensive study of anion exchange using Dowex-1. They found that Sb(III) in 1 M hydrochloric acid solution is strongly adsorbed by the resin, while Sb(V) is only weakly adsorbed. A separation of Sb(III) and Sb(V), however, is difficult owing to the slow rate of desorption of Sb(V) from the resin. If the eluting acid is too dilute, hydrolysis occurs and complications are encountered. Other elements forming stable chloro complexes also are adsorbed, but separations are possible by eluting with various concentrations of hydrochloric acid. Separations of As(III), Sb(III), and Sn(IV) are possible (64) as well as Sb(V), Te(IV), and Sn(IV) (129).

Fluoride systems also have been studied, and the anion-exchange properties of elements are similar to the chloride systems. Faris found that Sb(III), Sb(V), Sn(IV), and As(V) are strongly adsorbed, while As(III) is not adsorbed (29). Oxalate complexes also have been investigated. Using a system 3 M in hydrochloric acid and 0.1 M in oxalic acid Smith and Reynolds were able to separate Sn(IV), Sb(V), and Te(IV). Tellurium is eluted first using 0.1 M oxalic acid, then antimony with sodium oxalate, and finally tin using dilute sulfuric acid (137). Other organic acids also have been used for elution. It is possible to separate Sb(V) and Sn(IV) from an Amberlite IRA-400 column using ammonium malonate as elutrient. Tin(IV) is retained on the column, while the antimony(V) is eluted (21).

An unusual anion-exchange procedure was developed for the separation of As(III), Sb(III), and Sn(IV). These ions are precipitated as insoluble sulfides, dissolved in sodium polysulfide to form thio anions, and the basic solution passed through a Dowex-2 column in the hydroxide form. Tin is eluted with 0.5 M potassium hydroxide, arsenic with 1.2 M potassium hydroxide, and antimony with 3.5 M potassium hydroxide (60).

Cation exchangers also have been used for the separation of antimony, but are less useful than anion exchangers owing to the ease of hydrolysis of antimony in dilute acid solutions that must be employed for elution. The separation of arsenic, antimony, tin, and bismuth has been achieved using Wolfatit-P, a sulfonic acid exchanger in the hydrogen form. Arsenic passes through, while antimony, tin, and bismuth are retained by the resin. Bismuth is eluted using a 6% solution of ammonium thiocyanate in 0.5 M sulfuric acid. Antimony and tin are eluted with dilute hydrochloric acid (82). Antimony(III) may be separated from tin(II) on Amberlite IR-120 by elution with 2% tartaric acid in 0.1 M hydrochloric acid. Antimony(III) may be separated from antimony(V) by elution with dilute oxalic acid (59).

V. DETECTION AND IDENTIFICATION OF ANTIMONY

Both Sb(III) and Sb(V) are precipitated as sulfides from dilute acid solution using hydrogen sulfide or thioacetamide as precipitant (13,31). Lead, mercury, bismuth, copper, cadmium, arsenic, tin, germanium, molybdenum, selenium, and tellurium also are precipitated. These elements constitute Group II, the acid–hydrogen sulfide group, of the classical scheme of qualitative analysis. The sulfides of antimony, arsenic, tin, germanium, molybdenum, selenium, and tellurium may be separated from the other sulfides of Group II by dissolving them in a variety of reagents. Ammonium sulfide; sodium sulfide; the hydroxides of sodium, potassium, and lithium; and various soluble polysulfides have been recommended. This treatment converts the insoluble sulfides into various soluble thio anions that constitute division B or the Tin Subgroup of Group II. The thioanions are decomposed by adding a slight excess of dilute acid, which causes reprecipitation of the corresponding insoluble sulfides. The sulfides of antimony and tin are separated from arsenic sulfide by treatment with concentrated hydrochloric acid. Arsenic sulfide remains undissolved, while antimony and tin sulfides go into solutions as soluble chloro complexes. Antimony is separated from tin by adding oxalic acid and then treating the mixture with hydrogen sulfide or thioacetamide. Tin remains in solution as a complex oxalate, while antimony is precipitated as a characteristic red-orange sulfide. A detailed discussion and procedure is given in references 89 and 154.

The evolution of stibine forms the basis of several methods for the detection of antimony. A few drops of the sample is added to a mixture of zinc and dilute sulfuric acid. In the original method developed by Marsh, the gases are led through a glass tube that is heated. Both arsine and stibine decompose, and the corresponding free metal is deposited on the walls of the tube. Antimony may be distinguished from arsenic by adding a solution of sodium hypochlorite to the deposit. Arsenic metal dissolves, while antimony remains unchanges (145).

A modification of this method was developed by Gutzeit. The gases are allowed to contact a filter paper moistened with silver nitrate solution. Both arsine and stibine cause an immediate darkening owing to the formation of metallic silver (154). The test, however, does not distinguish between arsenic and antimony.

If the gases are passed over a filter paper moistened with a solution of iodine dissolved in potassium iodide, stibine will form an orange deposit of SbI_3, while arsine forms yellow AsI_3. This test, however, is not as sensitive as the Gutzeit test, and the dark brown color of free iodine may mask the color of antimony triiodide.

Antimony also may be detected by Reinsch's test, which consists of making the solution 2–6 M in hydrochloric acid, then dropping a strip of bright copper metal and heating to boiling. If antimony is present a grey to black film will appear on the copper. However, As, Hg, Bi, Ag, Au, Pt, and Pd also are deposited.

The most specific test for antimony consists in the use of Rhodamine-B as a spot test. This dye forms a characteristic purple color with Sb(V) ions in acid solution with little interference from other ions. The method is discussed in detail under Photometric Methods in Section VI.D.1. The following procedure is recommended:

Place one to two drops of the unknown solution in a spot plate, and make it strongly acid by adding two to four drops of concentrated hydrochloric acid. Add 1–2 mg of solid potassium nitrite or sodium nitrite, and stir thoroughly using the tip of a glass rod to insure oxidation of Sb(III) to Sb(V). Add five to ten drops of a 0.01% solution of Rhodamine-B in water. Stir. If antimony is present the bright red dye changes to violet. If much antimony is present a violet precipitate may form.

This test is sensitive to 0.5 μg of antimony. Gold and thallium give similar colors, likewise molybdates and tungstates. Large amounts of mercury also give a similar coloration. Lead, tin, bismuth, cadmium, zinc, iron, and the other common cations do not interfere.

Emission spectroscopy also may be used for the detection of antimony. The technique recommended by the ASTM consists in mixing the powered sample with graphite powder and lithium carbonate as an internal standard, packing it into a graphite cup electrode, and exciting with a dc arc. The spectra are recorded on a photographic plate, and the lines are measured on a microphotometer. Antimony is identified by the Sb 2877.92 line. The concentration range is 0.10–50.0%. Lines for the identification of many other elements are given. The method may be used for alloys, minerals, and other commercial products (6). Ahrens recommends the use of Sb 2598.06 line for the identification of antimony in minerals. A detection limit of 0.01% is attainable (2). A more detailed discussion of emission spectrographic methods is given in Section VI.E.

VI.   DETERMINATION OF ANTIMONY

A.   PRECIPITATION AND GRAVIMETRIC DETERMINATION

Although many insoluble compounds of antimony are known, most of them are unsuitable for a gravimetric determination. They either are too soluble for accurate results, for example, the oxides and hydroxides of antimony, or their composition is variable and they cannot be converted to a suitable form for weighing. Antimony is usually weighed either as the trisulfide, Sb_2S_3, or as the tetroxide, Sb_2O_4. Difficulties are encountered in using these compounds, and unless certain precautions are taken, erroneous results will be obtained. Various precipitants and forms for weighing are discussed in the following paragraphs. The thermogravimetric properties of several precipitates is discussed by Duval (23).

1. Precipitation and Weighing as the Sulfide

Antimony may be precipitated either as the trisulfide or the pentasulfide. Either hydrogen sulfide gas (13), thioacetamide (31), or ammonium thiocyanate (121) may be used as precipitant. Many studies have been made on the precipitation of antimony sulfide using hydrogen sulfide, and these are summarized in reference 13. Precipitation as the pentasulfide is not recommended. If the solution is less than 2.5 N in acid hydrolysis occurs, and basic salts are precipitated along with the pentasulfide. Partial reduction usually occurs during precipitation with the formation of free sulfur that is occluded in the precipitate. On drying at 110°C some decomposition occurs, and on heating to 280–300°C in a current of carbon dioxide the Sb_2S_5 is converted completely into Sb_2S_3 with the formation of free sulfur. Because of these difficulties it is recommended that Sb(V) be reduced to Sb(III) prior to precipitation.

Antimony trisulfide may be precipitated as two different varieties, a red, amorphous form or a black, crystalline form. The latter is preferred because less coprecipitation of foreign ions occur and it is more constant in composition and less subject to errors arising in drying than the red form. High acidity and high temperature favor formation of the black form. When the solution is less than 1 N in hydrochloric acid the red variety is formed regardless of temperature. If the acid concentration is 2 N or greater and the solution is heated, the black variety is obtained. Completeness of precipitation is dependent both on the temperature and on the acid concentration. In 1 N hydrochloric acid precipitation is complete up to 80°C, 2 N up to 60°C, 3 N up to 40°C, 4 N up to 20°C, and 5 N precipitation is complete only if the solution is maintained at 0°C. When these limits are exceeded losses occur. Most procedures recommend an acid concentration of 2–3 N while the solution is heated almost to boiling and hydrogen sulfide bubbled in at a fairly rapid rate. The solution is then allowed to cool while being saturated with hydrogen sulfide to obtain complete precipitation.

In precipitation with thioacetamide the following technique is recommended (31). The solution is adjusted to 2–3 N hydrochloric acid, an excess of a 2% solution of thioacetamide is added, and the solution is heated to boiling. At first a precipitate of the red form of Sb_2S_3 is obtained, but heating is continued until this is transformed into the black modification. The solution is diluted with water, and a small additional amount of thioacetamide added. A slight red precipitate forms at this point, but it is quickly changed to the black form as the solution is allowed to stand until it is lukewarm.

Antimony trisulfide also may be precipitated using a saturated solution of ammonium thiocyanate (121). The solution containing antimony is made approximately 2 N in hydrochloric acid, heated to boiling, and the ammonium thiocyanate solution is added. The mixture is boiled for approximately 1 min and allowed to settle for 5 min, then diluted with an equal volume of water before filtration. Duval (23) observed that the antimony trisulfide obtained by this method is much purer than that obtained by precipitation with hydrogen sulfide and is stable in the temperature range from 170 to 292°C.

Filtration and washing techniques are the same whether hydrogen sulfide, thioacetamide, or ammonium thiocyanate is used as precipitant. The antimony trisulfide is transferred to a sintered glass filter, then washed with water followed by alcohol. Duval recommends that the precipitate be dried at 176°C in air for 10 min. Longer drying at this temperature results in oxidation of some of the sulfide to the oxide giving low results. Better results are obtained if the precipitate is dried in a current of carbon dioxide at 280–300°C. Using this technique a maximum error of 0.2% was observed with samples ranging from 10 to 500 mg of antimony.

An alternate method consists of treating the antimony trisulfide precipitate with nitric acid, then igniting to Sb_2O_4 for weighing. By this technique the necessity for drying in a current of carbon dioxide is avoided. This is recommended, however, for relatively small amounts of antimony. The precipitate also may be dissolved in acid and Sb(III) titrated with a standard oxidant as discussed in Section VI.B.

2. Organic Precipitants

Antimony(III) ions in dilute hydrochloric acid react with aqueous pyrogallic acid to form a precipitate, $(SbOH)C_6H_3O_3$. The precipitate is washed with water, then dried at 110°C, or may be ignited to Sb_2O_4 at 800–975°C for weighing. The method has little to recommend it, since many other ions also are precipitated with pyrogallol. It does serve, however, to separate antimony from arsenic (30). Gallic acid also may be used as precipitant with results similar to pyrogallic acid (67). Tannin likewise precipitates antimony, but the composition of the precipitate is variable and the method is not exact (67).

Oxine has been recommended for the determination of antimony (116), but conflicting reports appear in the literature (67). If the pH is too low precipitation is incomplete, and at higher pH the precipitate is contaminated with basic salts and is variable in composition. This method is not recommended.

Antimony also may be precipitated as triethylenediamine chromium(III) thioantimonate. In this method antimony is first converted to the thioantimonate anion by addition of sodium sulfide and polysulfide. The precipitant is triethylenediamine chromium(III) chloride which is prepared by the addition of ethylenediamine to a solution of chrom alum and ammonium chloride (139). The precipitate has the composition, $[Cr(en)_3]SbS_4 \cdot 2H_2O$, and for accurate work it should be dried in a vacuum desiccator. If dried in air at temperatures up to 165°C, losses amounting to approximately 0.3% occur (23). The method is useful in the range, 0.5–50 mg Sb.

B. TITRIMETRIC METHODS

Most titrimetric methods for antimony involve the oxidation of Sb(III) to Sb(V). Three methods are widely used. The first involves the oxidation of Sb(III) in strong hydrochloric acid solution using standard potassium bromate, the second employs standard iodine as titrant in neutral or slightly basic solution, and the third involves the use of potassium permanganate as oxidant. McNabb and Wagner (93) made a thorough study of these three methods and found them equally accurate. Each, however, has certain advantages and disadvantages. These methods are discussed in detail in the following paragraphs.

1. Titration with Standard Bromate

This method was devised by Gyory (46) and since then has been the subject of many studies. These are discussed thoroughly in reference 67. The reaction of Sb(III) with bromate normally is slow, but it is catalyzed by chloride ions and the reaction rate also is increased by heating and employing strongly acid solutions. A temperature of 80°C usually is recommended for the titration, although titrations at room temperature are possible if the acid concentration is high enough. If the titration is carried out in hot solution optimum acidity is 1.2–3.5 N in hydrochloric acid. Ions oxidized by bromate interfere. To eliminate most of these interferences the solution is treated with bromine and then boiled to eliminate the excess bromine. It is then reduced with sulfurous acid and the excess removed by boiling. After this treatment the only interfering ions are As(III), Fe(II), and V(IV). High concentrations of ammonium ion also interfere, since it is partially oxidized by bromate in hot acid solution. The method is useful for the determination of macro amounts of antimony and also with slight modification (58) to microanalysis.

A variety of indicators has been recommended for the detection of the end point. Most of these are the irreversible type that are bleached by the excess bromate or bromine formed in the hot solution after the end point. Gyory (46) originally recommended methyl orange as indicator, and this is still widely used. Methyl red, indigo, indigo carmine, benzopurpurin 4B, and fuchsine also have been recommended (42,67). Belcher investigated several reversible indicators including α-naphtho-flavone, p-ethoxychrysaidine, fuchsin, and apomorphine (8a). Of these α-naphthoflavone (131) is the best.

Titrations at room temperature were studied by Zintl and Wattenberg, who used a potentiometric method for detection of the end point (164). Smith and May (136) also studied titrations at room temperature and recommend Brilliant Ponceux 5R, Bordeaux, and Naphthol blue-black as irreversible redox indicators. Successful titrations were obtained in solutions 0.6–4.0 N in hydrochloric acid.

Amperometric titrations with bromate as titrant also have been studied using either a rotating or vibrating platinum electrode (72). The results are comparable to results from the other methods.

2. Iodimetric Methods

Both direct and indirect iodimetric methods have been proposed for antimony. Antimony(III) may be oxidized to Sb(V) by free iodine in neutral or slightly basic solution. In strongly acid solution Sb(V) will liberate free iodine from soluble iodides. Although both reactions have been employed for the titrimetric determination of antimony the former is more widely used. Both are discussed in the following paragraphs.

Titration of Sb(III) with standard iodine was first developed by Mohr (97), who added tartrate to hold the antimony in solution, then made the solution basic with sodium carbonate and titrated it with standard iodine using starch as indicator. Fresenius substituted sodium bicarbonate for the carbonate and thereby increased the accuracy. This method is still in use today. Some investigators believe the accuracy is increased if a slight excess of iodine is added, then back titrated with standard thiosulfate. The major interference in this method is As(III), which also is oxidized by iodine.

In most procedures Sb(V), if present, is reduced by the addition of excess sulfite to an acid solution, and the excess sulfurous acid is destroyed by boiling. Problems encountered here are the possible loss of $SbCl_3$ on boiling or reoxidation of Sb(III) by air in the hot solution. To overcome these difficulties, McCay (90) recommends mercury metal as reductant. The sample is made 4 N in hydrochloric acid, then shaken for 1 hr in a stoppered flask with mercury metal. Under these conditions, As(V) is not reduced. McCay recommended that Sb(III) then be titrated with standard bromate.

Evans recommended the use of sodium hypophosphite as reductant (26). This reagent reduces Sb(V) to Sb(III), while As(III) and As(V) are reduced to the free element which may be filtered off. Antimony(III) is then titrated with standard iodine, which does not react with hypophosphite in the cold.

Indirect iodimetric methods also are possible, but these methods are subject to more error than direct iodimetric methods. The procedure consists in oxidizing Sb(III) to Sb(V) in hydrochloric acid solution using free bromine or potassium chlorate, then boiling to destroy the excess oxidant. The solution is treated with excess potassium iodide, and the liberated iodine is titrated with standard thiosulfate (57). One complication arises from the formation of a deep yellow complex, $SbCl_3 \cdot 3KI$, which obscures the endpoint. Errors of $\pm 1.0\%$ may be expected in this method as compared with $\pm 0.2\%$ or less in the direct iodimetric method. The method is useful, however, for the determination of Sb(V) in mixtures of Sb(III) and Sb(V). The preliminary oxidation step is omitted, and the Sb(V) is determined by this method.

3. Titration with Permanganate

Antimony(III) may be titrated in hydrochloric acid solution using standard potassium permanganate. Various concentrations of hydrochloric acid are recommended ranging from 0.2 to 3.0 N (80). If the acid concentration is too low, the reaction is slow and low results are obtained. If the concentration of hydrochloric acid is too great, high results are obtained owing to the reaction of permanganate ions with chloride ions. The solution must be kept cool to decrease this side reaction. This error also may be decreased by adding phosphoric acid and manganous sulfate to the mixture before titration. In permanganate titrations the endpoint is detected by the color of the excess permanganate or by the oxidation of iodine to iodine monochloride at the endpoint. In spite of the difficulties the method enjoys some use and is recommended by the ASTM for the determination of antimony in certain alloys (5).

4.   Titration with Ce(IV)

The use of Ce(IV) as a standard oxidant was studied in detail by Willard and Young (160,161), who demonstrated that a variety of reducing substances could be determined with it. The reaction with As(III) is very slow; hence using appropriate conditions Sb(III) can be titrated in the presence of arsenic. In strong acid the Sb(III) oxidation is favored, hence by using a solution 4–6 N in hydrochloric acid, the interference of small amounts of As(III) is eliminated, but errors are observed when the amount of arsenic becomes equal to or exceeds that of antimony. The endpoint is observed potentiometrically. After the antimony endpoint, iodine monochloride, which catalyzes the As(III) oxidation, may be added, and the titration may be continued to determine the arsenic (37,47,117).

In solutions free of arsenic the titration is simpler. In this situation the endpoint may be determined visually using ferroin or iodine monochloride as indicator. The iodine monochloride end point also may be used in titrations with iodate and permanganate. The technique consists in adding iodine monochloride to solutions that must be 2.5 N or more in hydrochloric acid. A glass-stoppered flask is used, and a few milliliters chloroform or carbon tetrachloride added. Before the endpoint the iodine monochloride is reduced to free iodine, which imparts a violet color to the chloroform. The flask is shaken vigorously after the addition of each portion of oxidant, and the endpoint is reached when the violet color is bleached owing to the reoxidation of iodine to iodine monochloride.

5.   Other Oxidants

Other oxidants used for the titration of Sb(III) are dichromate (102), iodate (4,135), hypochlorite (41), and ferricyanide (110). These possess no particular advantage over the others discussed previously and are not used extensively.

6.   Miscellaneous Redox Methods

Other miscellaneous redox methods include the reduction of Sb(V) with titanium(III) using a potentiometric end point or neutral red or safranine as visual indicators (52). Air must be excluded, and the temperature must be above 60°C. Antimony(III) and (V) may be titrated with standard solutions of chromium(II) using a potentiometric endpoint (73). Two breaks are observed, one corresponding to the reduction of Sb(V) to Sb(III) and the second from Sb(III) to Sb^0.

7. Coulometric Titrations

Brown and Swift (14) titrated Sb(III) with electrolytically generated bromine using an amperometric end point. The method was used to determine 10–1000 μg of antimony in solutions 2 N in HCl and 0.2 N in KBr.

Dunlap and Shults (22) devised a controlled coulometric titration for the determination of either Sb(V) or Sb(III)´ or both. For Sb(V) the supporting electrolyte was 0.4 M in tartaric acid and 6 M in hydrochloric acid. A potential of 0.21 V (SCE) was used for the Sb(V)–Sb(III) reduction at the mercury cathode and 0.35 V for the Sb(III)–Sb^0 step. For solutions containing only Sb(III) the supporting electrolyte was 0.4 M tartaric acid and 1 M hydrochloric acid, and a potential of -0.28 V was employed. The method was useful for the range 0–10 mg. Most common anions and cations do not interfere except Cu(II).

Wise and Williams (162) employed a coulometric titration for the determination of antimony in glass. The sample is dissolved in hydrofluoric acid, then evaporated to dryness. The residue is dissolved in sodium tartrate, the solution adjusted to pH 7 using borax. Sodium iodide is added as supporting electrolyte, and the titration is carried out at 6.43 mA.

8. Complexometric Titrations

Although antimony forms many complexes, for example, with tartrate, fluoride, chloride, and so forth, these have not been useful for the complexometric titration of antimony. EDTA forms a weak complex with Sb(III), and this was utilized by Takamoto (144) for its complexometric determination. An excess of EDTA is added, and the solution is adjusted to pH 3.5 with an acetate buffer. Acetone is added until the solution contains approximately 60% acetone. Potassium thiocyanate is added as indicator, and the excess EDTA is backtitrated with standard cobalt chloride. This titration was employed by Cheng and Goydish (17) for the analysis of complex tellurides of antimony and bismuth. The EDTA complex of antimony is decomposed on the addition of fluoride, tartrate, or citrate.

C. POLAROGRAPHY

Most polarographic studies involve the reduction of trivalent antimony to the metal. Antimony(III) is readily reducible at the dropping mercury electrode in dilute nitric, hydrochloric, sulfuric, and perchloric acid solution. Polarograms also have been obtained for the reduction of

Sb(III) in acidic, neutral, and basic tartrate solution and in the presence of a variety of other complexing anions (74). Well-defined cathodic waves are obtained even in 1 N sodium or potassium hydroxide solutions, which are useful for the determination of trivalent antimony in the presence of arsenite ions (63). Anodic waves arising from the oxidation of Sb(III) to Sb(V) also may be observed in basic solution (63).

In dilute mineral acids, well-defined steps corresponding to reversible, three-electron reductions are obtained with diffusion current constants of approximately 5.0. It is necessary to add 0.01% gelatin as a maximum suppressor. In sulfuric acid solution, a small secondary wave is observed corresponding to partial reduction of Sb^0 to stibine (95). In sulfuric acid alone, Sb(III) produces only an ill-defined wave. If thiocyanate is added, however, an almost reversible wave is obtained for Sb(III) (56a). In solutions 1 M in sulfuric acid and 0.01 M in thiocyanate the current is proportional to concentration in the range, 2×10^{-5} to $5 \times 10^{-4} M$ for dc polarography and 5×10^{-6} to $1 \times 10^{-4} M$ for ac polarography. No maxima are observed, and the wave is not affected by a thousandfold excess of arsenic or a tenfold excess of iron, bismuth, and lead. In 1 N hydrochloric acid the half wave potential is -0.15 V (vs SCE) with a diffusion current constant of 5.54. In 1 N nitric acid the corresponding values are -0.30 and 5.10; in 1 N sulfuric acid, -0.32 and 4.94. The wave in nitric acid solution has a smaller slope, indicating the reduction of SbO^+ ion present in this medium is not perfectly reversible.

In dilute acid solution Sb(V) is not reduced at the dropping mercury electrode. The reduction of pentavalent antimony was first reported by Kraus and Novak (65), who were studying solutions 8 M in hydrochloric acid. A comprehensive study was made by Lingane and Nishida (74), and they attributed the difficulty in obtaining a reduction wave for Sb(V) to a small rate of reaction and a very large overvoltage or activation energy. If the solution is greater than 4 N in hydrochloric acid or if it is 6 N in perchloric acid and at least 0.2 N hydrochloric acid a stepwise reduction is observed; first of Sb(V) to Sb(III) followed by Sb(III) to Sb^0. In 6 N hydrochloric acid containing 0.005% gelatin the first wave starts at zero applied emf, and the second half-wave is 0.257 versus SCE. Both diffusion currents are proportional to the concentration of Sb(V). The reducible specie in this system is the $[SbCl_6]^-$ ion. In solutions 6 M in perchloric acid there is no Sb(V) reduction wave before the hydrogen wave.

Pentavalent antimony reduces in a single step to the metal in solutions that are 1 N in hydrochloric acid and 4 M in potassium bromide. Under these conditions As(V) does not reduce, hence this method is useful for the determination of Sb(V) in the presence of arsenate ions (71).

Polarography has been used for the determination of antimony in a variety of materials. Page and Robinson (109) investigated a number of organic compounds containing antimony such as sodium antimony thiosalicylate, potassium antimonyl tartrate, sodium antimony(V) gluconate, and several others. The organic compound is dissolved in 1 N hydrochloric acid. If antimony is in the 5+ state, it is reduced with sodium sulfite then boiled to remove the excess sulfite. It is unnecessary to destroy the organic material, since the antimony(III) wave occurs before those produced by organic substances. The solution is adjusted to 1 M in hydrochloric acid, gelatin is added, and the polarographic wave occurring at $E_{1/2} = 0.15$ V is used to determine the amount of antimony. The concentration of the solutions must be adjusted to the range, 0.05–0.001% antimony. Bismuth also may be determined by the same procedure.

Many applications involve the determination of antimony in alloys (33). In aluminum alloys antimony must be separated by distillation as the chloride because copper and iron give waves that interfere (95). The method is satisfactory for antimony in amounts up to 0.5%. A direct determination of antimony is possible in certain magnesium alloys. The alloy is dissolved in hydrochloric acid with the aid of a few drops of nitric acid. The nitric acid is removed by evaporation, the residue is then dissolved in dilute hydrochloric acid, and the antimony wave starting at zero applied potential is used. Copper, tin, and nickel must be absent (95).

The polarographic determination of traces of antimony and other impurities in refined copper was investigated by Eve and Verdier (27). The copper is dissolved in nitric acid, fumed with sulfuric acid, diluted, and the copper removed by electrodeposition. The solution is made 0.1 M in sodium fluoride, and the pH is adjusted to approximately 3. A polarogram is run, and under these conditions waves for antimony and lead are observed. The amounts are determined by comparison with standard curves made from solutions containing known amounts of lead and antimony.

Tamba and Vantini (144a) determined antimony in steels and cast iron. The sample is dissolved in nitric acid, and the antimony is separated by coprecipitation with manganese dioxide. This precipitate is dissolved by fuming with sulfuric acid, and the resulting solution is acidified with dilute dydrochloric acid. Ascorbic acid and phosphoric acid are added, and the solution is polarographed, using gelatin as a maximum supressor. Antimony III is reduced giving a well-defined peak at −170 mV versus SCE.

Several procedures have been devised for the determination of antimony in lead-base alloys. Zotta developed a rapid method for the analysis of trace constituents in refined lead (165). The alloy is dissolved in 7 M nitric acid, sulfuric acid is added to precipitate most of the lead, and the supernatant liquid is treated with ammonium hydroxide, which forms a precipitate. The precipitate is removed and dissolved in dilute sulfuric acid. Sodium citrate is added, and the polarogram is run. The first wave is iron, followed by bismuth, then antimony, and finally by one for the residual lead. The method is good only for antimony in amounts less than 0.02%. If the amount is greater than this, a portion remains insoluble in the nitric acid.

A better procedure was devised by Cozzi (20). The lead alloy is heated with sulfuric acid until completely disintegrated. It is then diluted and the lead sulfate filtered off. An aliquot is taken and evaporated to dryness. The residue is dissolved in sodium tartrate, then made basic with sodium hydroxide. The antimony is determined from the anodic antimony wave starting at -0.1 V versus SCE. Basic fuchsine is used as maximum suppressor.

A detailed study of the determination of antimony in lead was made by Hourigan and Robinson (53). The sample is dissolved in dilute hydrochloric acid with the addition of bromine. The solution is cooled, and potassium chloride is added, which causes precipitation of lead chloride. The solution is adjusted to 5 M in hydrochloric acid, gelatin is added, and the antimony wave starting at zero applied potential is used. The wave height must be compared with a standard made from pure lead and standard potassium antimonyl tartrate solution.

Toren determined antimony in water in the part per billion range using cathode-ray (also called single sweep) polarography (146a). A 250 ml sample of water is acidified with 0.5 ml conc. sulfuric acid and evaporated down to 0.5 ml. This sample is diluted with 10 ml of 6 M hydrochloric acid and transferred to a polarographic cell. At higher concentrations of antimony (10 ppm) direct measurement with a forward sweep gave the expected curve with the antimony wave at -0.15 V versus mercury pool. At lower concentrations however, interference of a prewave caused by reduction of a mercury chloride film on the surface of the mercury drop was observed. By applying a voltage sweep from -0.4 to $+0.1$ V followed by reverse voltage sweep, an anodic current corresponding to the stripping of antimony metal from the drop was obtained. In addition to avoiding the mercury chloride interference, the sensitivity of the determination was increased because the anodic peak corresponds to the oxidation of all antimony in the drop regardless of its initial valence state. If tin is present the antimony current-voltage curve has a steeply sloping base line. Copper gives an anodic peak prior to antimony and interferes if present in excess. Bismuth interferes completely.

Gilbert and Hume (40a) determined antimony in sea water by anodic-stripping voltametry. A mercury-coated graphite electrode was used. The sample was first made 1 M in hydrochloric acid, and bismuth was plated out at -0.4 V. This gives a stripping peak at -0.2 V without interference from antimony. The sample is made 4 M in hydrochloric acid, and bismuth plus antimony is plated out. A stripping peak is determined then at -0.2 V that is proportional to the sum of bismuth plus antimony. Antimony is then determined by difference.

Anodic-stripping voltametry was used also for the determination of traces of antimony in copper (148a). After dissolution of the sample the antimony was separated by distillation at 165°C from HBr solution. This separation was necessary owing to the interference of large amounts of copper.

Anodic-stripping voltametry was also applied to the determination of antimony in zinc metal (69a). The sample is dissolved in nitric acid, then fumed with sulfuric acid, and finally acidified with hydrochloric acid. After deoxygenation with nitrogen gas the sample is electrolyzed at -0.32 V SCE using a hanging-mercury-drop electrode. A reverse scan is then made for the determination of antimony.

Details of these methods and others are discussed in reference 95. The principles and techniques of polarography are discussed in Volume IV of the Treatise and also in references 62 and 94.

D. COLORIMETRIC METHODS

Many reagents have been proposed for the colorimetric determination of antimony. Two of these, iodoantimonous acid and Rhodamine-B, enjoy widespread use, and they are discussed in the following paragraphs.

1. Iodoantimonous Acid Method

This method was devised by Fouchon (32), who reported that antimony(III) reacts with potassium iodide in dilute sulfuric acid solution to form a yellow complex, $KSbI_4$. Precise conditions for applying this method to the determination of antimony in the 10–500 μg range were developed by McChesney (91). He found that the optimal concentration of sulfuric acid at the time of color development is 8% by volume and the optimal concentration of potassium iodide is 8% by weight. When the concentration of potassium iodide is reduced to 0.8%, no antimony complex is formed but bismuth gives a strong color. Using a 5.6% concentration of iodide, the bismuth and antimony colors are exactly additive. By adjusting the concentration to 8% sulfuric acid and 5.6%

potassium iodide at the time of color development, he devised a suitable method for both bismuth and antimony in the same sample. An essential constituent in this method is ascorbic acid or sodium hypophosphite, which serves as an antioxidant and protects the sample from extraneous color arising from the reduction of Sb(V) to Sb(III) by potassium iodide. It also prevents the formation of free iodine by air oxidation and protects the reagent from extraneous oxidizing agents. Although this method does not have quite the sensitivity of rhodamine B procedures, it is essentially free of interferences and is simple in manipulative detail.

The sensitivity of the method was increased sevenfold by Elkind et al. (25), who measured the absorbance at 330 nm instead of 425 nm as recommended by McChesney. The method has been adapted to the determination of antimony in copper base alloys by Holler (51), who used a combination of sodium hypophosphite solution and a starch–iodide sodium thiosulfate additive to prevent the formation of free iodine in the solution at the time of color measurement. Many alkaloids and organic bases form highly colored insoluble iodoantimonates. This was used in a modification of the iodoantimonite method involving the formation of insoluble, yellow pyridine iodoantimonate, which is kept in suspension using gum arabic. The method was used for the determination of 0.001–0.05% antimony in tin and is relatively insensitive to arsenic. The method also has been adapted to the determination of antimony in copper and its alloys. These modifications, however, offer no special advantages over the iodoantimonous acid method.

2.   Rhodamine-B Method

The formation of a colored complex by the reaction of Rhodamine B and Sb(V) ions was first observed by Eigriwe (24), who used it as a qualitative spot test for antimony. Rhodamine B or tetraethylrhodamine has the following structure:

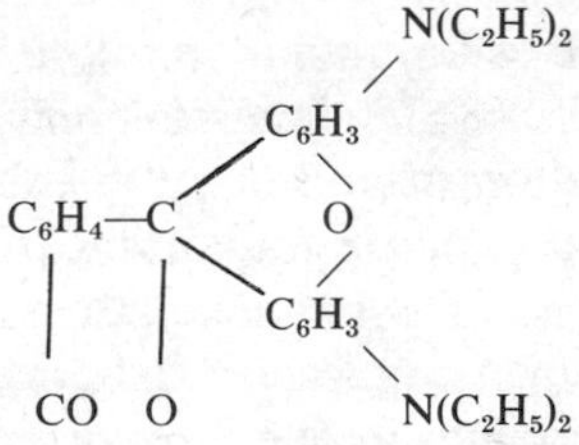

$$N(C_2H_5)_2$$
$$C_6H_3$$
$$C_6H_4-C \quad O$$
$$C_6H_3$$
$$CO \quad O \qquad N(C_2H_5)_2$$

In acid solution the basic amine sites are protonated, and RH^+ and RH_2^{2+} (R = Rhodamine B) species are formed. In strong hydrochloric acid solution these unite with chloro anions of certain metals to form red-violet, ion association complexes of the type, $[RH][MeCl_x]^-$, and $[RH_2]^{2+}[MCl_x]_2^-$. Gallium(III), gold(III), thallium(III), iron(III), and antimony(V) form such complexes, which are soluble in organic solvents such as ether or benzene. Tungsten(VI) also forms a colored complex, but this is not extracted by benzene. The complex formed by antimony(V) in strong hydrochloric acid solution is $RHSbCl_6$. This complex is insoluble in aqueous media, but it is suitable for a colorimetric method because it is in the form of a colloidal suspension. With large amounts of antimony a precipitate is found. The complex is extractable into benzene and organic ethers (86,128) to form a stable solution that obeys Beer's law in the range, $0.1-300$ μg.

Several conditions must be met for a successful Rhodamine B determination. The solution must be strongly acid, usually 6 M or greater to prevent hydrolysis. A high chloride ion concentration also is required in order to form the chloro complex, and an oxidant must be added to insure that all antimony is in the 5+ oxidation state. The excess oxidant must be removed, since it may destroy the colored complex or oxidize the Rhodamine B to a colored specie that interferes, and finally the colored complex must be separated from the excess reagent, which also is highly colored. Many procedures have been proposed, and the most important of these are discussed in the following paragraphs.

The first quantitative method was developed by Frederick (34), who used it for the determination of $0.1-300$ μg of antimony. The determination was carried out in a mixture of sulfuric acid and hydrochloric acid with ceric sulfate as oxidant and free bromine to bleach the excess Rhodamine B. The complex was solubilized by the addition of ethanol, and the absorbance was measured at 530 nm. No extraction was employed.

The method was improved by Webster and Fairhall (152), who added an extraction step. Subsequent studies have shown that Ce(IV) is the best oxidant and that an extraction improves the method by separating the complex from the excess reagent and interfering ions. In the organic phase the complex is in a soluble form, rather than a colloid as in aqueous media, and better results are obtained (45,128).

The nature of the Rhodamine B-antimony(V) complexes and their behavior on extraction was clarified in studies by Ramette and Sandell (120) and Neuman (104,105). In 6 M hydrochloric acid the predominant specie is $RHSbCl_6$, but there is also present about 10% $RH_2[SbCl_6]_2$, which also is extracted. As the acid strength is increased another complex $RH_2ClSbCl_6$ is formed. The amount of $RHSbCl_6$ extracted falls off with increasing acidity. On the other hand, if a high acidity is not maintained partially hydrolyzed species such as $[Sb(OH)Cl_5]$ are formed, and these do not form colored complexes with Rhodamine B (120). To avoid hydrolysis the solutions must be kept cool, 25°C or less (148,150).

Another difficulty is encountered in the incomplete oxidation of antimony to the 5+ state. Maren (86) pointed out that under certain conditions Sb may be present in solution in the Sb(IV) oxidation state, which is not oxidized to Sb(V) on treatment with Ce(IV). If this is suspected the solution should be treated with sulfurous acid prior to the determination to reduce Sb(IV) to Sb(III), which is then easily oxidized by Ce(IV).

Because of these complexities it is evident that a successful Rhodamine B determination is dependent upon strict observation of all details of the procedure. The order of addition of the reagents also is critical. In spite of these difficulties the Rhodamine B method is recommended for the determination of small amounts of antimony in a variety of substances including rocks, soils (151), alloys (77–79,148) and biological materials (86). A detailed review of this method is given in reference 128.

3.   Miscellaneous Colorimetric Methods

a. INORGANIC COMPLEXES

Bromide ions form a colored complex with Sb(V), but this is much less sensitive than the iodide complex and many common cations and anions interfere (103). The measurement of the stain produced on mercuric chloride paper by stibine evolved from zinc and acid, as in the Gutzeit method for arsenic, has been used for the determination of traces of antimony (38). Arsenic interferes, and the accuracy of the method is questionable owing to the incomplete evolution of stibine (162). Heteropoly blue methods also have been used for the colorimetric determination of antimony. Two techniques are employed. Antimony(III) ion reduces molybdotungstophosphoric acid forming a blue color, but this method is not specific (128). The other method requires a separation of antimony by distillation, after which a molybdoantimonate is formed that is then reduced with ascorbic acid to form a heteropoly-blue. This method requires no extraction and is useful for the determination of antimony in alloys (88).

Microgram amounts of antimony may be determined colorimetrically by formation of the colloidal sulfide. Gum arabic is used for suspension (43).

b. ORGANIC COMPLEXES

In addition to Rhodamine-B, several other organic dyestuffs form colored ion-association complexes with antimony. Most widely used are methyl violet (31a,69,134), brilliant green (15,31b,40,140), crystal violet (142), and phenylfluorone (66). The sensitivities and interferences are similar to those encountered in the Rhodamine-B method. These are discussed in reference 128.

A fluometric method for the determination of submicrogram amounts of antimony was developed by Filer (30a). The reagent used was 3,4,7-trihydroxyflavone. The fluorescence of the Sb(III) complex is excited at 422 nm and measured at 475 nm. Measurements are made in perchloric acid solution using phosphate as a masking agent. The detection limit is 0.04 μg of antimony.

E. EMISSION SPECTROSCOPY

The characteristic spectral line of antimony may be obtained by excitation either by an arc or with a spark. Table 3 lists the sensitive lines for antimony. Most sensitive of these is the 2068.38 line, but its wavelength lies just below the usual working range. Most commonly used are 2877.915, 2598.062, and 2528.535 (48). Silicon 2528.516 interferes with the last; hence this line should not be used in the analysis of minerals or of any substance containing even a trace of silicon.

In running a spectrographic analysis the intensity of the line is usually compared with a neighboring line of the major constituent. Alternately a known amount of some internal standard may be added. Arsenic is recommended as the best standard for antimony and vice versa (2). Emission spectroscopy is recommended for the determination of antimony in a variety of alloys (6) and in certain minerals (2). The theory and practice of emission spectroscopy is discussed in reference 2, and procedures for the spectrographic analysis of many common alloys is given by the ASTM in reference 6.

TABLE 3
Sensitive Lines for Antimony (48)

Wavelength (Å)	Excitation potential	Intensities	
		Arc	Spark
3267.50	5.8	150	150
3232.50	6.1	150	250
2877.92	5.3	250	150
2598.06	5.8	200	100
2528.54	6.1	300	200
2311.47	5.3	150	50
2175.89	5.7	300	40
2068.38	6.0	300	3

F.   ATOMIC ABSORPTION

When antimony compounds are volatilized in a flame very little excitation occurs; hence flame photometric methods are insensitive and usually are not employed for the determination of antimony. Atomic-absorption methods, however, are accurate and sensitive and have been used for the determination of antimony in a variety of alloys. A comprehensive study was made by Mostyn and Cunningham (99), and their findings are summarized in the following paragraphs.

A hollow cathode source was employed, and the sample was burned in an air–acetylene mixture. Four possible absorbing wavelengths were investigated. The line, 2127.4 Å, proved too weak. The most sensitive line was 2175.8 Å, but less noise was observed at 2311.5 Å; hence this line is recommended for most analytical use. The signal at 2068.3 Å was much noisier than the others and is not recommended. The sensitivity at 2175.8 Å was found to be 1.4 ppm Sb/1% abs., and at 2311.5 the sensitivity was 2.0 ppm Sb/1% abs. Using a Perkin-Elmer Model 303 the optimum operating conditions were determined. These are summarized in Table 4.

No interference was caused by a variety of common elements such as Cd, Pb, Sn, Fe, Zn, and so forth at the 1000 ppm level. A slight interference was observed with copper using the 2175.8 Å line, but this is eliminated by using the 2311.5 Å line. No significant depression of antimony absorption was produced by any of these metals at the 1000 ppm level. At the 10,000 ppm level, however, lead caused high results using the 2175.8 Å line and low results using the 2311.5 Å line. This interference may be eliminated by adding approximately the same concentration of lead ion to the standard as is in the unknown sample. In alloy analysis it is recommended that the major constituent be added along with antimony to the standard.

TABLE 4
Optimum Operating Conditions

Wavelength	Sb 2311.5 Å (or 2175.8 Å)
Slit opening	1 mm
Lamp current	2 mA
Air supply	25 psi
Acetylene supply[a]	8 psi
Sample uptake	3.5 ml/min
Scale expansion	× 2
Response time	1

[a] The burner should be adjusted for maximum absorption.

It was found that the acid concentration of the sample influenced the antimony absorption. In hydrochloric acid, for example, 100 ppm of antimony gave an absorption of 0.281 in 2 N acid and 0.245 in 5 N acid. Even greater changes were observed in nitric and sulfuric acid solutions. The authors recommend that all unknowns and standards be adjusted to an acid strength of 2 N. Standards were prepared by dissolving a known weight of antimony tetroxide in concentrated hydrochloric acid then diluting to volume making the final solution 5 N in hydrochloric acid, or by dissolving a known weight of tartar emetic in water.

The method was checked by determining the antimony content of a variety of alloys ranging from Duraluminum containing 0.05% antimony to a lead–tin–antimony alloy containing 10.4% antimony. Excellent agreement between values obtained by chemical analysis and values by atomic absorption was obtained in all samples. The method was used also to determine the antimony sulfide content in percussion caps, and again excellent agreement between values by chemical analysis and atomic absorption were observed.

In samples with low antimony content, extraction into methyl isobutyl ketone using ammonium pyrrolidine dithiocarbamate as the complexing agent was investigated. The organic layer is separated and sprayed directly into the flame. A sensitivity of 1 ppm/1% abs. was obtained using this technique.

Yanagisawa et al. (162c) recommend direct extraction into methyl isobutyl ketone without the use of added chelating agents. The sample is made 6 M in hydrochloric acid; the antimony is oxidized by addition of sodium nitrite and then is extracted into methyl isobutyl ketone. The organic phase is aspirated into the flame, and absorbance is read at 2175.9 Å. This line was found to be approximately twice as sensitive as the 2068.3 or the 2311.5 line. A similar technique was used by Meranger and Sommers (94a) for the determination of antimony in titanium dioxide. Burke (14b) extracted the iodo complex of antimony into a 5% solution of TOPO in methyl isobutyl ketone. The absorbance was read at 2176 Å. This method was applied to the determination of microgram amounts of antimony in aluminum-, iron-, and nickel-base alloys. Microgram quantities of antimony can be concentrated by coprecipitation with manganese dioxide, which is dissolved in hydrochloric acid and hydrogen peroxide. The resulting solution is used directly for the atomic-absorption determination at 2176 Å (14a).

Type metal containing 3–17% antimony may be analyzed after dissolution in a mixture of fluoboric and nitric acids. Readings are made at 2175.8 Å (42a). All of these investigators used air–acetylene mixtures in the burner. Nicolas (101a) used atomic absorption for the determination of antimony in geological materials. The pulverized sample is mixed with solid ammonium iodide, then heated in a test tube. The sublimate containing all the antimony as SbI_3 is dissolved in 2 M hydrochloric acid. This solution is used for atomic-absorption measurement at 2311.8 Å with air–acetylene flame. The limit of detection is 10 ppm of antimony in a 0.250 gram sample.

Flameless atomic absorption was used by Yanagesawa et al. (162b). The sample, in hydrochloric acid solution, is extracted into methyl isobutyl ketone. An aliquot of this solution is placed on a carbon rod and then dried and ashed at temperatures below 180°C. The sample is atomized for 3 sec at 2200°C using argon as an inert gas. The sensitivity for 1% absorption is 10^{-12} gram.

Dagnell et al. (20a) used atomic fluorescence for the determination of antimony. Fluorescence was generated in an air–propane flame by nebulizing aqueous solutions of antimony salts while irradiating the flame by means of a microwave excited, electrodeless discharge tube operating at 30 W. Strongest fluorescence was observed at 2311 A and a weaker fluorescence at 2068 and 2176. Antimony was determined in the range of 0.1–120 ppm. Kalihova and Synchra (58a) investigated various gas mixtures for the determination of antimony by atomic fluorescence. Most effective was a hydrogen–argon mixture burned in oxygen. Fluorescence was excited using a high-intensity hollow cathode lamp. Most sensitive measurements were made using the 217.6 nm line.

The theory and techniques of atomic absorption is discussed in Volume VI of the treatise and also in reference 125.

G.   NEUTRON ACTIVATION ANALYSIS

Natural antimony consists of two isotopes: ^{121}Sb with an abundance of 57.25% and ^{123}Sb, 42.75%. Both are converted to radioactive species upon bombardment with thermal neutrons. The 121 isotope has a cross-section of 6.8 barns for thermal neutron and is effectively converted into ^{122}Sb by an n-γ reaction. This isotope has a half-life of 2.8 days and emits betas with maximum energies of 0.74, 1.40, and 1.97 MeV, and gammas of 0.57, 0.69, 1.14, and 1.26 MeV. The ^{123}Sb has a cross-section of 2.5 barns and is converted to ^{124}Sb, having a half-life of 60.6 days, also decaying by β-γ emission. Because of its higher cross-section and shorter half-life of the product the ^{121}Sb $(n\gamma)$ ^{122}Sb reaction is preferred in activation analysis. The sensitivity has been determined as 9×10^{-6} μg for this reaction as compared with 1×10^{-4} μg for the ^{123}Sb (np) ^{124}Sb reaction (61).

Using this method traces of antimony have been determined in minerals, commercial products, a variety of metals and alloys (44), and in biological materials (54). In most procedures a known standard, antimony metal, or tartar emetic, is bombarded along with the unknown sample, and the amount of antimony is determined by comparison of the activities of the unknown and standard. Frequently a radiochemical separation is necessary to isolate the antimony from other activities. Both beta-counting techniques and gamma-ray spectrometry are suitable for measuring the activity. The major interference arises from ^{122}Te, which may be converted to ^{123}Sb by an np reaction. The theory, techniques, and applications of activation analysis are discussed in references 12 and 83. Sensitivities, interferences, and applications are given in reference 61. Various radiochemical methods are given in reference 84.

VII. PURITY TESTS FOR ANTIMONY AND ITS COMPOUNDS

Only a limited number of antimony compounds are used commercially or in the laboratory, and these include antimony metal, antimony trisulfide (stibnite), antimony potassium tartrate (tartar emetic), and antimony trichloride. Methods for the assay and/or tests for impurities in these substances are given in the following paragraphs.

A. ANTIMONY METAL

Antimony metal is best analyzed for impurities by emission spectroscopy. The ASTM recommends a point-to-plane spark technique (6). Cast disks or other antimony metal samples are used, having a flat surface area of at least 1 in. in diameter and large enough to prevent overheating under the action of the spark. A carbon counter electrode is used, and the concentrations of the impurities are determined by comparison of an analytical line of the desired element with the Sb 2858.04 line. Procedures are given by ASTM for the determination of Pb, Bi, As, Sn, Cu, Fe, Ni, and Ag in amounts ranging from 0.001 to 0.6%.

The ASTM (5) also gives chemical methods for the determination of impurities in antimony metal. To determine lead, the sample is dissolved in a mixture of hydrobromic acid and bromine. After dissolution the solution is fumed with sulfuric acid, then diluted, and the impure lead sulfate precipitate removed by filtration. The precipitate is dissolved in hot ammonium acetate, potassium dichromate is added, and the lead chromate is filtered, dried, and weighed. The method is suitable for samples containing 0.005–0.5% lead.

The filtrate from the lead sulfate separation is used for the determination of silver. This is treated with hydrogen sulfide to precipitate silver sulfide, which is filtered off, dissolved in nitric acid, and silver determined turbidimetrically as the chloride. The method is suitable for silver in the range, 0.001–0.2%.

Sulfur is determined on a separate sample by evolution of hydrogen sulfide, which is collected in ammoniacal cadmium chloride and then determined by titration with standard iodine. The method is useful in the range, 0.005–0.2%.

Arsenic, copper, iron, and nickel are determined photometrically in the range, 0.001–0.2%. For the arsenic determination the sample is dissolved by heating with a mixture of potassium hydrogen sulfate and sulfuric acid. After dissolution the mixture is cooled, diluted, and transferred to a distilling apparatus (Fig. 1). Hydrochloric acid, sodium chloride, and ferrous sulfate are added, and arsenic trichloride is distilled off at 105°C. The distillate is absorbed in water containing free bromine to oxidize As(III) to As(V). Ammonium molybdate is added, and the arsenomolybdate complex is reduced with hydrazine sulfate to the molybdenum blue whose color is measured at 660 nm.

Copper is determined as cupric bromide after dissolution of the sample in a mixture of hydrobromic acid and bromine. An aliquot is taken, and the red-violet cupric bromide complex is measured at 600 nm. Another aliquot is used for the determination of iron. Copper is removed by the addition of lead metal, and iron is determined in the filtrate using 1,10-phenanthroline.

Another aliquot is used for the determination of nickel. Copper is removed as in the iron determination, and nickel is then determined using the color produced by dimethylglyoxime. Detailed procedures are given in reference 6.

B.   STIBNITE

Antimony trisulfide (stibnite) is commonly used as a primer in explosives and as an additive to rubber. It is frequently analyzed for its sulfide content as well as for antimony. Sulfur may be determined by precipitation as barium sulfate after a preliminary oxidation with bromine (80). This procedure gives total sulfur. McNabb and Wagner (92) recommend an evolution method that determines only sulfide sulfur content. The sample is treated with concentrated hydrochloric acid and heated. The evolved hydrogen sulfide gas is swept out in a current of carbon dioxide, absorbed in ammoniacal cadmium chloride, and the cadmium sulfide precipitate is filtered off. This is dissolved in an excess of standard iodine solution, acidified, and the excess iodine is backtitrated with standard sodium thiosulfate. The solution remaining in the evolution flask may be used for antimony determinations.

Free sulfur is determined by extracting a dry sample with carbon tetrachloride and weighing the residue of sulfur after evaporation of the solvent. McNabb and Wagner found 0.05–0.10% free sulfur in various stibnite samples. Sulfate sulfur was determined by extracting another sample with water and precipitating barium sulfate, which is weighed. Most samples contained approximately 0.01% sulfate sulfur.

Iron is determined by boiling the stibnite sample with a solution of sodium hydroxide or sodium sulfide to remove most of the antimony sulfide. The residue is dissolved in a mixture of nitric and hydrochloric acids, diluted, then treated with hydrogen sulfide to remove any Group II sulfides. The precipitate is removed and discarded, and iron is determined in the filtrate by precipitation as $Fe(OH)_3$ followed by ignition of Fe_2O_3 or reduction and titration with permanganate. Samples contained 0.04–0.2% iron.

Norwitz et al. also studied the analysis of stibnite (108). Methods were developed for the determination of iron, lead, and arsenic. After dissolution of the sample in hydrochloric acid, tartaric acid is added to complex the antimony and iron determined colorimetrically with 1,10-phenanthroline. Another aliquot is used for the determination of lead. The sample is made basic with ammonia, and lead is determined colorimetrically after extraction of the dithizonate. Arsenic is determined on a separate sample that is dissolved in a mixture of sulfuric and nitric acids. This is fumed to expel nitric acid, then treated with a mixture of hydrobromic and hydrochloric acids. Arsenic is distilled at 93–95°C and determined colorimetrically by the molybdenum blue method in the distillate.

C. ANTIMONY POTASSIUM TARTRATE

The United States Pharmacopeia gives a procedure for the assay of antimony potassium tartrate (tartar emetic) and a test for its arsenic content (115). It is assayed by taking a known weight, dissolving it in water, buffering with sodium bicarbonate, followed by a titration with standard iodine using starch as indicator. The assay should lie between 99 and 103% $C_4H_4KO_7Sb \cdot \frac{1}{2}H_2O$.

The test for arsenic is made by dissolving a sample in hydrochloric acid and adding stannous chloride solution. After standing for 30 min, the solution is compared with a standard containing 200 ppm of As using Nessler tubes for comparison. To pass specifications, the sample should be no darker than the standard.

D. ANTIMONY TRICHLORIDE

Rosin (126) and Murray (101) give standards and tests for the assay and determination of impurities in reagent-grade antimony trichloride. The following tests are suggested.

It should dissolve in chloroform giving at most a slightly turbid solution. It may be assayed by dissolving in sodium potassium tartrate followed by the addition of sodium bicarbonate and iodimetric titration. It should contain no more than 0.002% arsenic, which is determined by dissolving the sample in hydrochloric acid, then adding stannous chloride. Only a slight darkening, arising from the formation of free arsenic, should be observed.

Sulfate is determined by the addition of barium chloride to a hydrochloric acid solution of the sample. It should contain no more than 0.01% sulfate. Iron is determined by the red thiocyanate complex and should be less than 0.010%.

To determine substances not precipitated by hydrogen sulfide a hydrochloric acid solution is treated with hydrogen sulfide gas, and the precipitate is filtered off. The residue remaining after evaporation of the filtrate should be less than 0.20%. The precipitate should dissolve completely in sodium sulfide solution leaving no dark residue of PbS or CuS.

VIII. DETERMINATION OF ANTIMONY IN SPECIFIC MATERIALS

A. METALS AND ALLOYS

A variety of alloys contain antimony. These range from certain brasses, bronzes, and steels containing only a few hundredths or tenths of a percent of antimony to bearing metals, pewter, and type metals that may contain as much as 50% antimony. The method of analysis must be suited to the amount of antimony in the sample. For trace and small amounts, spectrographic and colorimetric methods are recommended. With larger amounts of antimony titrimetric methods are suitable.

Nonferrous alloys composed largely of lead, tin, and antimony also containing small amounts of arsenic, copper, bismuth, iron, aluminum, and zinc are best analyzed by the following method. The sample is dissolved by heating with a mixture of concentrated sulfuric acid and potassium acid sulfate or potassium pyrosulfate. After dissolution the mixture is diluted with hydrochloric acid, sodium sulfite is added, and then it is boiled to volatilize any arsenic. The hot, acid solution is titrated with standard potassium bromate using methyl orange as indicator.

With alloys containing a lesser amount of antimony, 0.1–5.0% such as solders, separation of arsenic and antimony by distillation is recommended. The sample is dissolved as above then transferred to a distillation apparatus (Fig. 1). Sodium chloride and hydrochloric acid are added, and arsenic is distilled at 105°C in a current of carbon dioxide. Antimony may be distilled then at a temperature of 155–158°C in a current of carbon dioxide and with the dropwise addition of hydrochloric acid during the distillation.

Antimony may be determined in the distillate by bromate titration or iodimetrically. In the latter, tartaric acid is added to keep the antimony in solution, the mixture is neutralized with sodium hydroxide, sodium bicarbonate is added as buffer, and Sb(III) is titrated with standard iodine using starch as indicator.

The distillation-iodimetric method is especially recommended for the determination of arsenic and antimony in brasses and bronzes containing approximately 1% antimony. The sample is dissolved in a mixture of nitric and hydrochloric acid. After dissolution, ferric nitrate is added and the sample is made basic with ammonium hydroxide. Arsenic and antimony are collected in the ferric hydroxide precipitate, which is dissolved in dilute sulfuric acid. This is evaporated to SO_3 fumes, phosphoric and hydrochloric acids are added, and arsenic and antimony are then separated by distillation as discussed previously. Antimony is then determined by titration with iodine. Detailed procedures are given by the ASTM (5).

In alloys containing 0.1% antimony or less, concentration by co-precipitation with manganese dioxide is recommended. In this method the sample is dissolved in dilute nitric acid, and a solution of manganous nitrate is added. Potassium permanganate is then added, and the solution is boiled. The permanganate oxidizes Mn(II) ions to MnO_2, and in the process the permanganate ion is also reduced to MnO_2. The resulting manganese dioxide is insoluble in nitric acid, and arsenic, antimony, and tin are quantitatively carried down with it.

The precipitate is dissolved by fuming with sulfuric acid. Arsenic and antimony then may be separated by distillation and determined volumetrically as discussed previously. With small amounts of antimony a photometric determination is recommended, particularly the iodoantimonous acid method.

In this method, a sulfuric acid solution containing antimony is treated with an excess of potassium iodide to form the tetraiodoantimonate(III) complex, which has maximum absorbance at 420 and 330 nm. Sodium hypophosphite or ascorbic acid is added to remove oxidizing agents that may form free iodine. Bismuth interferes but is rarely encountered in copper-based alloys such as brasses and bronzes.

B.   ORES AND MINERALS

Stibnite, Sb_2S_3, is the most common mineral containing antimony, and its antimony content may vary from 1 or 2% to over 70%. Impurities commonly encountered in this ore are iron, a few tenths of a percent, lead, a few tenths of a percent, arsenic, a few hundredths of a percent, and silica which may range from a trace to 80–90%.

Most samples of stibnite will dissolve in concentrated hydrochloric acid. This should be performed in a covered beaker or flask with as little heating as possible to avoid losses of antimony trichloride. McNabb and Wagner recommend a reflux condenser and a current of carbon dioxide gas to avoid atmospheric oxidation and to help eliminate hydrogen sulfide gas. Antimony may be determined directly in the resulting solution by titration with standard bromate, or with standard iodine after adding tartrate and adjusting the acidity. In the bromate method any Fe(II) also will be titrated as well as As(III). These may be determined on separate samples and corrections applied. The arsenic content of most stibnite is usually so small that it does not add significantly to the antimony results. If more than a trace of iron is present the color of the $[FeCl_6]^{3-}$ complex will obscure the methyl orange endpoint and cause erratic results.

In the iodometric method As(III) will be titrated along with antimony. Iron(II) causes low results, and the error is directly proportional to the amount of Fe(II) present. Iron in the amount of 0.0004 gram caused an error of -0.0006 gram of antimony, and 0.0026 gram Fe gave an error of 0.0029 gram for Sb (93).

Next in order of abundance of antimony ores are the oxides, for example, valentenite, Sb_2O_3, and cervantite, Sb_2O_4. The antimony content and the impurities in these minerals are in the same range as those in stibnite, mentioned previously. The oxide ores are more difficult to dissolve than the sulfide ores, although an occasional sample may be soluble in concentrated hydrochloric acid. More drastic treatment usually is required. Heating the sample with a mixture of concentrated sulfuric acid, ammonium sulfate, and/or potassium sulfate in a Kjeldahl flask is recommended (38). Tartaric acid or filter paper also may be added as an "indicator." The sample is digested until the charred, carbonaceous material is oxidized, and the sample becomes clear. After cooling, tartaric acid and water are added to dissolve the salts. If a separation of arsenic is desired the mixture is made 8–10 M in hydrochloric acid, then treated with hydrogen sulfide gas to precipitate As_2S_3, which then is filtered off. Alternately arsenic may be separated from the mixture by distillation. Antimony may be determined in the solution by bromate titration or with standard iodine as discussed previously.

Certain refractory ores must be fused to attain solution. Sodium hydroxide alone or mixed with sodium peroxide is recommended. Iron, nickel, or zirconium crucibles must be used. Another effective fusion mixture is composed of equal volumes of anhydrous sodium carbonate and powdered sulfur. This is carried out in a covered porcelain crucible. The melts are taken up with water and acidified, and antimony is then determined by titration.

Other less common minerals such as antimonides and mixed sulfides frequently are soluble in mixtures of nitric and hydrochloric acids. If this treatment is not effective, solution may be effected by one of the techniques discussed above. The antimony content is then determined by titration with standard bromate or iodine.

C. ROCKS, SOILS, AND CERAMICS

Although antimony is not widely distributed in the earth's crust, it does occur in trace amounts in certain rocks and soils. It is also added to certain glasses, glazes, and ceramic material. All of these substances require drastic treatment to bring the antimony into solution; hence they are discussed together. Ward and Lakin (151) made a comprehensive study on the determination of antimony in rocks and soils. The sample is rendered soluble by fusion with sodium bisulfate. Although this fusion does not completely dissolve the sample, the authors believe that all the antimony is extracted in the process. Basic fusions as with sodium carbonate or sodium hydroxide are not recommended since they also solubilize the silica, which causes interference or occludes part of the antimony. After dissolution the antimony is determined colorimetrically with Rhodamine B.

Wise and Williams (162) dissolved glass samples by heating with hydrofluoric acid in the platinum dish. Silica was eliminated by evaporation of the acid solution to dryness. The sample was taken up in dilute hydrofluoric acid with the addition of oxalic acid. After evaporation and sublimation of excess oxalic acid at 150°C, antimony was determined by coulometric titration.

Many glazes and ceramic materials containing antimony oxide can be rendered soluble by heating with a mixture of sulfuric acid and potassium bisulfate as used for the dissolution of certain ores (80).

D. ORGANIC AND BIOLOGICAL MATERIALS

Antimony and its compounds may be associated with a variety of organic materials. It may be a constituent of the organic compound itself such as tartar emetic or metallo-organic compounds. It may be a filler or a pigment in materials such as rubber or plastics, or it may be associated with body fluids and wastes. Before analysis the antimony must be in the form of an inorganic cation. With many organic salts or metallo-organic compounds, dissolving in 1:1 hydrochloric acid will convert the antimony to a form suitable for analysis. In most situations, however, it is necessary to destroy the organic matter completely before analysis.

Dry-ashing procedures are not recommended because of possible losses caused by volatilization of oxides or halides. Wet oxidations usually are the methods of choice. The most effective method consists in using fuming sulfuric acid along with the addition of 30% hydrogen peroxide (143). If fuming sulfuric is not available, a mixture of concentrated nitric and sulfuric acid may be used. Potassium hydrogen sulfate is frequently added to the mixture to raise the boiling point and thus decrease the digestion time. The oxidation is best carried out in a Kjeldahl flask. If this is not available an Erlenmeyer flask or covered beaker may be substituted for it. The mixture is heated until a clear solution results. This is cooled and diluted, and antimony is then determined by an appropriate method.

IX. LABORATORY PROCEDURES

The procedures in the following section are divided into three major parts: first, procedures for the dissolution of the sample; second, procedures for separation; and finally, analytical procedures by titrimetric, colorimetric, and atomic absorption methods. A variety of detailed procedures also may be found in references 1, 5, 6, 38, 67, 80, and 128.

A. DISSOLUTION OF THE SAMPLE

1. Metals and Alloys

a. ALLOYS CONTAINING LARGELY ANTIMONY, TIN, AND LEAD (1–99% antimony)

(1) Dissolution in Sulfuric Acid–Potassium Hydrogen Sulfate

Procedure

Weigh out 0.1000–2.0000 grams of the alloy in the form of drillings or filings into a 500 ml Erlenmeyer flask. Add 10–15 ml conc. sulfuric acid and about 5–7 grams of potassium hydrogen sulfate (or potassium pyrosulfate). Heat slowly to decompose the sample, avoiding too high a temperature because the sample may melt and then be difficult to dissolve. When decomposition is complete, heat vigorously over an open flame to expel any free sulfur adhering to the walls of the flask. Let the mixture stand until cool, then determine antimony by titration with standard bromate.

b. MISCELLANEOUS ALLOYS CONTAINING 0.1–10% ANTIMONY

(1) Acid Mixtures Used to Dissolve Alloys

The following mixtures have been used to dissolve various antimony alloy and may prove useful. No detailed procedures are given. The general technique consists in treating the alloy first with the cold mixture then heating to increase the speed of reaction and finally heating to eliminate the excess oxidant.

a. Concentrated hydrochloric acid saturated with bromine.

b. Concentrated hydrochloric acid containing suspended solid iodine. This mixture oxidizes antimony to the Sb(III) state only.

c. Concentrated hydrochloric acid with the dropwise addition of 30% hydrogen peroxide.

d. Concentrated hydrochloric acid with the addition of 10–20 mg portions of solid potassium chlorate.

e. Aqua regia, that is, mixtures of hydrochloric acid and nitric acid, usually with an excess of hydrochloric acid.

After dissolution the acidity and so forth are adjusted, and the antimony is determined by bromate titration.

c. ALLOYS CONTAINING TRACES OF ANTIMONY

(1) Dissolution and Collection of Manganese Dioxide (76,111)

Alloys containing very small amounts of antimony and tin may be dissolved directly in 1:1 nitric acid and the antimony collected by coprecipitation with manganese dioxide. This method is especially useful for the determination of traces of antimony in lead.

Procedure

Weigh out a 1.0000–10.0000 gram sample into a 400 ml beaker. Add 50–100 ml of 1:1 nitric acid, warm and heat if necessary to initiate the reaction. After dissolution boil the solution 3–5 min to expel oxides of nitrogen. Use this solution for the collection of antimony on manganese dioxide (Section IX.B.2).

(2) Dissolution and Collection on Stannic Oxide

Alloys containing small amounts of tin such as brasses and bronzes also may be dissolved in 1:1 nitric acid. Tin is precipitated as the hydrated oxide, $SnO_2 \cdot xH_2O$, which carries down all of the antimony in the form of H_3SbO_3. The following procedure is recommended.

Procedure

Weigh out a 1.000 gram sample of the alloy into a 40 ml centrifuge tube. Add about 10–15 mg of pure tin metal to act as carrier for the antimony. Add about 10 ml 1:1 nitric acid, then warm gently if necessary to hasten the reaction. After all the alloy has reacted, heat and boil gently to expel oxides of nitrogen. Add about 20 ml of water, and digest for about 30 min with occasional stirring on a hot-water bath.

Centrifuge down the residue of $SnO_2 \cdot xH_2O$ and so forth, and decant and discard the supernate. Stir the residue with about 5 ml water. Centrifuge and discard washings. Add 5 ml conc. sulfuric acid, and heat to fumes of SO_3 to dissolve the residue. Use this solution for the iodoantimonite method.

2. Ores and Metallurgical Products

a. Dissolution in Hydrochloric Acid

High-grade stibnite ores usually will dissolve completely in concentrated hydrochloric acid.

Procedure

Weigh a sample of the ore into a beaker. Add 25–50 ml conc. hydrochloric acid, cover, and heat just at the boiling point until dissolved. Avoid prolonged boiling.

b. Dissolution in Hot Concentrated Sulfuric Acid

Oxides are best dissolved by fuming with a mixture of concentrated sulfuric acid and potassium hydrogen sulfate.

Procedure

Weigh a sample into a beaker or Erlenmeyer flask. Add 15 ml conc. sulfuric acid and 5–6 grams of solid potassium hydrogen sulfate. Cover, heat, and fume until dissolution is complete, and if necessary add an additional 5 ml of concentrated sulfuric acid during the process. Cool, and carefully rinse down the cover and sides of the container with 2–3 ml water. Add 0.5 gram solid hydrazine sulfate and fume again. Cool, and use this solution for the determination of antimony by bromate titration or colorimetrically.

3. Rocks, Soils, Ceramics, and Ores of High Silica Content

Two techniques are recommended: evaporation with hydrofluoric acid and sulfuric acid, and fusion with sodium hydroxide.

a. DISSOLUTION IN HYDROFLUORIC ACID–SULFURIC ACID

Procedure

Weigh the finely divided sample (80–100 mesh) into a platinum dish; then add 5 ml of 18 N (1:1) sulfuric acid and 10 ml conc. hydrofluoric acid. Cautiously evaporate to fumes of SO_3. Cool, add another 5 ml of 18 N sulfuric acid, and again evaporate to SO_3 fumes. The cool, acid solution then may be diluted with an appropriate solvent and used for the colorimetric or titrimetric determination of antimony. Frequently a residue of insoluble calcium sulfate will remain, but this usually contains no antimony.

b. DISSOLUTION BY FUSION WITH SODIUM HYDROXIDE

This technique is recommended only if everything else fails.

Procedure

Place about 5 grams of solid sodium hydroxide and about 0.1 gram solid KNO_3 in a zirconium, nickel, or iron crucible. Heat cautiously until it has melted and any water has been expelled. After it has cooled, weigh a finely divided sample (80–100 mesh) and place it on top of the fused sodium hydroxide. Spread out and cover with another 5 grams of solid sodium hydroxide. Fuse the mixture at low red heat for 10–15 min. Leach the melt from the crucible using about 50 ml of water. Acidify, and determine antimony by an appropriate method.

4. Organic Material

a. FUMING SULFURIC ACID–HYDROGEN PEROXIDE DECOMPOSITION

Procedure

Transfer a weighed sample (1–5 grams) to a 100 ml Kjeldahl flask, and cautiously add about 10 ml fuming (15% SO_3) sulfuric acid. Swirl and warm slightly if necessary to dissolve the substance. Add 30% hydrogen peroxide dropwise down the sides of the flask while agitating gently by hand. Continue the dropwise addition of H_2O_2 until the liquid is straw colored (usually a total of 1–5 ml is required); then warm until abundant fumes of SO_3 are evolved. If the liquid is not clear and colorless (or light straw colored) at this point, continue the dropwise addition of H_2O_2 and heating. Finally heat to decompose the excess hydrogen peroxide. Cool, dilute with an appropriate solvent, and determine antimony by the desired method. It is important to note, however, that all or part of the antimony may be in the 5+ oxidation state at this point. Reduction with sodium bisulfate or other suitable reductant will be necessary if the method is based on the reactions of Sb(III) ions.

b.   SULFURIC ACID–NITRIC ACID DECOMPOSITION

Procedure

Weigh out a sample of appropriate size (1–5 grams), and transfer it quantitatively to a 500 ml Kjeldahl flask (or 500 ml Erlenmeyer or 500 ml covered beaker). Cautiously add 30 ml of 1:1 sulfuric acid, 30 ml conc. nitric acid, and 8–10 grams of solid potassium hydrogen sulfate. Heat and boil (in hood) until all the nitric acid is expelled and fumes of SO_3 are evolved. Continue boiling until all of the organic matter is decomposed and the solution is clear. At this point the mixture should be colorless or at most a light straw color. If it is not, add an additional 10 ml conc. sulfuric acid and digest until all organic matter is destroyed. Cool, then cautiously dilute the mixture with a solvent suitable for the contemplated analysis. During this oxidation part of the antimony may be oxidized to the 5+ state. If the method is based on the reactions of Sb(III) ions, reduction with sodium bisulfate or other suitable reductant is required prior to the analysis.

B.   SEPARATIONS

1.   Precipitation of Antimony Trisulfide

This is best carried out in a 2–3 N hydrochloric acid solution using a 2% solution of thioacetamide as precipitant. A separation is achieved from Group III cations, that is, iron, aluminum, chromium, zinc, manganese, cobalt, and nickel, but not from arsenic, copper, tin, bismuth, and mercury. Lead and cadmium will precipitate partially.

Procedure

Adjust the solution to 2–3 N in hydrochloric acid, add 1 ml of 2% thioacetamide for each 10 mg Sb, cover, then heat and boil until the red precipitate of Sb_2S_3 is transformed into the black modification. Dilute the hot solution with approximately $\frac{1}{3}$ of its volume of distilled water, add an additional 5 ml of 2% thioacetamide, heat again to boiling, stir, let stand until lukewarm, and then filter and wash with cold water.

For a rough gravimetric determination the precipitate may be filtered into a tared, sintered glass crucible, washed with cold water, followed by alcohol, then dried at 180°C. Better results are obtained by drying at 280–300°C in a current of carbon dioxide.

For best results the sulfide precipitate should be filtered through filter paper and washed with cold water. The precipitate, filter paper and all, is then dissolved by heating with a mixture of nitric acid, sulfuric acid, and potassium sulfate as directed under the dissolution of the manganese dioxide precipitate followed by bromate titration.

If thioacetamide is not available, the hot solution may be treated with hydrogen sulfide gas.

2.   Collection by Coprecipitation with Manganese Dioxide

This procedure is recommended for samples containing 0.1% or less antimony. Tin, thallium, niobium, tungsten, and molybdenum also are carried down quantitatively in this precipitate. Arsenic, bismuth, iron, and traces of the major constituent also may be coprecipitated. Chlorides and other reducing agents must be absent.

Procedure

Dissolve the sample in nitric acid or sulfuric acid–potassium sulfate as directed previously. Transfer to a 500 ml beaker, then dilute with 1 M nitric acid to approximately 200–300 ml depending on the size of the original sample. Heat and boil for several minutes to remove oxides of nitrogen. Add 5 ml of 2% potassium permanganate to the boiling solution, then 5 ml of 10% manganous nitrate solution. Stir, heat, and boil for about 1 min; then add an additional 5 ml of 2% potassium permanganate. Continue to heat and boil for about 3 min, then remove the beaker from the heat and allow to settle for about 5 min. Filter while hot, and transfer the manganese dioxide quantitatively to the filter paper using hot water. Wash the precipitate two or three times with hot water; then discard the filtrate and washings.

Place the filter paper with the precipitate in the original beaker. Add 10 ml conc. nitric acid, 15 ml conc. sulfuric acid, and 5 grams of potassium hydrogen sulfate. Cover and heat cautiously until the filter paper is decomposed, then evaporate to fumes of SO_3. Cool, wash down the sides of the beaker with about 10 ml of water, add about 0.25 g of solid hydrazine sulfate, and evaporate again to SO_3 fumes. This solution may be used after appropriate dilution for the colorimetric determination of antimony by the iodoantimonite or Rhodamine B method. If larger amounts are present, antimony may be determined by bromate titration on this solution.

3.   Separation of Arsenic and Antimony by Distillation

Place 1.000–5.000 grams of the alloy or mineral in the distilling flask of the apparatus in Fig. 1. Add 15 ml conc. sulfuric acid and 5 grams of potassium hydrogen sulfate. Assemble the apparatus, and pass a stream of carbon dioxide through it at the rate of 3–5 bubbles/sec. Heat until fumes of SO_3 are evolved; then allow to cool.

Dip the end of the condenser into 50 ml of water in a 400 ml beaker; then add 20 ml of water through the acid bulb, mix, and then cool. Add 35 ml conc. hydrochloric acid, mixed with 1 ml 50% hypophosphous acid HPO_2, and mix with the solution in the flask. Place 75 ml conc. hydrochloric acid in the acid bulb, and pass a stream of CO_2 through the system at a rate of six to eight bubbles/sec. Heat the solution in the flask to 110–112°C, then drop hydrochloric acid from the bulb at a rate so the temperature is maintained at 110–112°C. Continue distillation until all the hydrochloric acid in the bulb has been added. Arsenic may then be determined in the distillate (see "Arsenic"). Place a new receiver containing 50 ml of water at the tip of the condenser, submerging the tip at least 5 min. Add 75 ml conc. hydrochloric acid to the acid bulb, pass a stream of CO_2 through at the rate of 6–8 bubbles/sec, heat the flask, and add the acid at a rate so the temperature is maintained at 155–158°C. Continue to distill until all acid has been added. All antimony now should be in the distillate.

Add 1 gram of solid sodium sulfite to the distillate, then evaporate down to 20 ml. Determine antimony by titration with iodine or with bromate.

C.   ATOMIC ABSORPTION

This method is perhaps the most versatile of all methods. It is applicable to alloys, minerals, and other commercial substances containing antimony in the range, 0.05–10%. No separations are required, and interference is encountered only when large amounts of other constituents are present. This interference can be eliminated by preparing standards containing approximately the same amounts of other major constituents as in the unknown sample, that is, when analyzing a brass sample, copper also should be added to the standards.

Samples are best dissolved in aqua regia containing an excess of hydrochloric acid. The final sample should be 2 N in hydrochloric acid as the absorption is influenced by the acid concentration. Sulfuric acid solutions should be avoided as the absorption is depressed by this acid.

Procedure

Weigh out a sample of suitable size into a beaker, add 10 ml of distilled water, 85 ml conc. hydrochloric acid, and 2 ml conc. nitric acid. Cover, heat, and warm until the sample has dissolved, then transfer to a 500 ml volumetric flask and dilute to volume with distilled water. Other mixtures containing hydrochloric acid plus an oxidizing agent also may be used for dissolving the sample. (See Section X.A.1.b.(1).) The excess oxidant then should be removed by boiling and the sample diluted to volume so that the final acidity is 2 N in hydrochloric acid. Measure the absorption under the operating conditions given in Table 4. Compare this with a series of standards prepared from aliquots of a solution containing 1000 ppm Sb prepared by dissolving 1.3343 grams of dry antimony potassium tartrate, $KSbOC_4H_4O_6$, in 2 N hydrochloric acid and diluting to 500 ml with 2 N hydrochloric acid. In preparing these standards, known amounts of the major constituent should also be added. For example, in analyzing an alloy containing 85% Pb, 10% Sb, and 5% Sn, 850 ppm of Pb^{2+} ion should be present in all standards.

D. BROMATE TITRATION

Procedure

Weigh out a sample containing 0.1000–0.2000 gram of antimony, and dissolve in a sulfuric acid–potassium hydrogen sulfate mixture (see Section IX.A.1.a.(1)). To the cold solution add 50 ml conc. hydrochloric acid and 150 ml water. Transfer to a 500 ml Erlenmeyer flask, add 25 ml of 6% sulfurous acid solution, then evaporate down to approximately 100 ml to expel sulfur dioxide and arsenic trichloride.

Add 100 ml of water, two drops of 0.1% methyl orange indicator, heat to boiling, and titrate rapidly with standard 0.1 N potassium bromate until the methyl orange is bleached. Reheat to boiling, add an additional two drops of methyl orange, and slowly titrate until the methyl orange changes from pink to colorless.

In place of methyl orange, a reversible indicator, α-naphthoflavone (8a), may be used. Add 0.5 ml of a 0.2% solution of α-naphthoflavone in ethanol. The color change is from pale straw color to deep brownish orange. The indicator is not destroyed by excess bromate.

Potassium bromate (0.1000 N) may be made by weighing out exactly 2.7836 grams of the analytical reagent and diluting to 1.000 liter with distilled water.

This solution may be standardized against arsenious oxide (see Arsenic Chapter) or against pure antimony metal dissolved and titrated as directed above.

E. IODINE TITRATION

Procedure

Dissolve the sample by a suitable method, cool, and dilute with 10 ml of water. Add about 4 grams of solid tartaric acid or 5 grams of sodium potassium tartrate (Rochelle salt). Neutralize most of the excess acid by slowly adding a 40% sodium hydroxide solution. At this point the solution should be only slightly acid. If it is basic add concentrated hydrochloric acid until just acid. Cool, then cautiously add saturated sodium bicarbonate solution until effervescence ceases, then about 15–20 ml in excess. Add about 5 ml of a 1% starch solution, and titrate with standard 0.1 N iodine solution to the blue-grey endpoint.

The preparation and standardization of 0.1 N iodine solution is discussed in Section IX of "**Arsenic**."

F. IODATE TITRATION

Procedure

Dissolve the sample by a suitable method, then transfer it or an aliquot to a glass-stoppered Erlenmeyer (iodine type) flask. Adjust acidity using ammonia or concentrated hydrochloric acid until the total acidity is 3.5 N in acid, add 5 ml chloroform (or carbon tetrachloride), then titrate with standard potassium iodate (0.1000 N), stoppering the flask and shaking vigorously after each addition of potassium iodate. The endpoint is reached when the violet color of iodine in the organic layer is bleached, owing to the formation of colorless ICl. The acidity should not be allowed to fall below 2.5 N during the titration. If necessary, add more concentrated hydrochloric acid.

G. IODOANTIMONITE METHOD

This method is useful for the determination of antimony in the range of 0.05–1.0 mg, measuring the absorbance of the $[SbI_4]^-$ complex at 420 nm. Bismuth, platinum, and palladium also form colored complexes and interfere. Ions that precipitate insoluble iodides such as Ag^+, Tl^+, and

Pd^{2+} cause low results by carrying down some antimony in the precipitate. Chlorides and fluorides also cause low results by bleaching the color. Oxidizing agents must be absent. By measuring the absorbance at 330 nm the sensitivity is increased 7.5 times. The solution must be 2.2–3.6 N in sulfuric acid at the time of measurement.

Procedure

Dissolve a weighed sample by an appropriate method, avoiding the use of hydrochloric acid, which may result in losses of antimony as volatile $SbCl_3$; then fume with sulfuric acid to remove other interfering anions. Or use the solution from the manganese dioxide collection method, the solution after the destruction of organic matter, or the residue after dissolving the sample in nitric acid (see Section IX.A).

Transfer the entire sample or an appropriate aliquot to a 50 ml volumetric flask, using small portions of water to effect the transfer if the solution has been evaporated with concentrated sulfuric acid. Add enough concentrated sulfuric acid so that the final solution after dilution will be about 3.0 ± 0.5 N in sulfuric acid.

Add 10 ml of a solution containing 100 grams KI + 20 grams of $NaH_2PO_2 \cdot H_2O$ dissolved in 100 ml of water (25.0 ml of a solution containing 140 grams KI + 10 grams ascorbic acid may be used in place of the 10 ml of this solution). Dilute to volume with distilled water, mix thoroughly, then let stand at least 10 min for development of color. Prepare a blank and a series of standards containing 0.05–1.0 mg of Sb. The standards may be prepared by fuming 0.1000 g of pure antimony metal with 10 ml conc. sulfuric acid and diluting this solution to 100 ml in a volumetric flask. The diluted solution contains 0.1 mg Sb/ml. Measure the absorbance at 420 nm, and compare with the standard curve, which should be linear in the range 0.05–0.50 mg.

H. RHODAMINE B METHOD

This is the most sensitive method for antimony and is useful in the range, 1–100 μg of antimony. Large amounts of gallium, gold, iron, tellurium, and thallium interfere. The order of addition of reagents as well as strict adherence to experimental details is important for a successful determination.

Procedure

Weigh out a sample containing 0.001–0.02 mg of antimony, and dissolve using one of the previous procedures so that the sample is finally obtained in the form of a solution in concentrated sulfuric acid and/or potassium hydrogen sulfate. The sample at this point should contain 5.0 ml or less of concentrated sulfuric acid.

To the cold, sulfuric acid solution of the sample, add 25 ml of distilled water and 5 ml conc. hydrochloric acid. Swirl and heat just to boiling. Cool to room temperature then filter off any residue such as lead sulfate, and wash the residue with two 5 ml portions of 1% sulfuric acid, catching the filtrate and washings in a 250 ml beaker. Discard the residue.

To the filtrate in the beaker add 10 ml conc. hydrochloric acid and 5 ml 6% sulfurous acid. Dilute to 60 ml, then heat and boil until the volume is down to 40 ml. Cool to room temperature, then transfer to a 150 ml separatory funnel, using two 10 ml portions of water to complete the transfer. Add 1 ml of ceric sulfate solution (0.5 gram ceric sulfate dissolved in 100 ml 1 N sulfuric acid) and immediately thereafter 3 ml of 0.2% aqueous solution of rhodamine B, then mix by swirling.

Add 15.0 ml of benzene, shake vigorously for 1 min, allow the layers to separate, then drain off and discard the aqueous layer. Shake once or twice, swirl, allow any suspended water droplets to settle, then drain these off and discard. Filter the benzene layer through glass wool into a clean, dry, absorption cell (2 cm), and measure the absorbance at 565 nm using benzene as a reference standard.

Antimony standards (1 μg/ml) are prepared by dissolving 0.0100 gram of pure antimony metal in 10 ml hot conc. sulfuric acid, then diluting to 100 ml. This solution contains 100 μg Sb per ml. A 1.0 ml aliquot is taken, and this is diluted to 100 ml with 2 N sulfuric acid giving a solution containing 1 μg Sb/ml. From this solution a series of standards is made using 0, 5.0, 10.0, 15.0, and 20.0 ml aliquots. These are transferred to beakers, and 5.0 ml conc. sulfuric acid and 5.0 gram of potassium hydrogen sulfate (if also present in the dissolved sample) are added. These are run through the procedure beginning with the second paragraph of the preceding discussion.

REFERENCES

1. Abel, G. A., "Antimony." in F. D. Snell and C. L. Hilton, Eds., *Encyclopedia of Industrial Chemical Analysis,* Vol. 6, Wiley-Interscience, New York, 1968, p. 39–70.

2. Ahrnes, L. H., *Spectrochemical Analysis,* Addison-Wesley, Cambridge, Mass., 1954.

3. Anderson, J. R. A., and H. Whitley, *Anal. Chim. Acta,* **6,** 517 (1952).

4. Andrews, L. W., *J. Amer. Chem. Soc.,* **25,** 756 (1903).

5. A.S.T.M., *Chemical Analysis of Metals, Sampling and Analysis of Metal Bearing Ores,* Part 32, May 1967, American Society for Testing Materials, Philadelphia, Pa., 1967.

6. A.S.T.M., *Methods for Emission Spectrochemical Analysis,* American Society for Testing Materials, Philadelphia, Pa., 1964.

7. Bachelder, M. C., and P. M. Sparrow, *Anal. Chem.,* **29,** 149 (1957).

8. Beard, H. C., and L. A. Lyerly, *Anal. Chem.,* **33,** 1781 (1961).

8a. Belcher, R., *Anal. Chim. Acta,* **3,** 588 (1949).

9. Bock, R., *Z. Anal. Chem.,* **133,** 110 (1951).

10. Bock, R., H. Kusche, and E. Bock, *Z. Anal. Chem.,* **138,** 167 (1953).

11. Bothorel, P., "Antimoine" in P. Pascal, Ed., *Nouveu Traite De Chemie Minerale,* Vol. XI, Masson et Cie., Paris, 1958.

12. Bowen, H. J. M., and D. Gibbons, *Radioactivation Analysis,* Oxford at the Clarenden, Oxford, England, 1963.

12a. Braman, R. C., L. L. Justen, and C. C. Forebock, *Anal. Chem.,* **44,** 2195 (1972).

13. Brennecke, E., *Schwefelwasserstoff als Reagens in der quantitativen Analyse,* Sluttgart, 1939.

14. Brown, R. A., and E. H. Swift, *J. Amer. Chem. Soc.,* **71,** 2717 (1949).

14a. Burke, K. E., *Anal. Chem.,* **42,** 1536 (1970).

14b. Burke, K. E., *Analyst (London),* **97,** 19 (1972).

15. Burke, R. W., and O. Menis, *Anal. Chem.,* **38,** 1719 (1966).

16. Burstall, F. H., G. R. Davies, R. P. Linstead, and R. A. Wells, *Nature,* **163,** 64 (1949).

17. Cheng, K. L., and B. L. Goydish, *Anal. Chem.,* **35,** 1965 (1963).

18. Christianson, Walter, G., *Organic Derivation of Antimony,* Chemical Catalog Co., New York, 1925.

19. Clark, S. G., *Analyst (London),* **53,** 373 (1928).

20. Cozzi, D., *Anal. Chim. Acta,* **4,** 204 (1950).

20a. Dagnell, R. M., K. C. Thompson, and T. S. West, *Talanta,* **14,** 1151 (1967).

21. Dawson, J., and R. J. Magee, *Mikrochim. Acta,* **3,** 325 (1958).

22. Dunlap, L. B., and W. D. Shults, *Anal. Chem.,* **34,** 499 (1962).

23. Duval, C., *Inorganic Thermogravimetric Analysis,* Elsevier, New York, 1953.

24. Eegriwe, E., *Z. Anal. Chem.,* **70,** 400 (1927).

25. Elkind, A., K. H. Gayer, and D. F. Boltz, *Anal. Chem.,* **25,** 1744 (1953).

26. Evans, B. S., *Analyst (London),* **56,** 171 (1931).

27. Eve, A. J., and E. T. Verdier, *Anal. Chem.,* **28,** 537 (1956).

28. Fairhall, L. T., and F. Hyslop, *Toxicology of Antimony,* U.S. Public Health Reports, Supplement 195, 1947.

29. Faris, J., *Anal. Chem.,* **32,** 520 (1960).

30. Feigl, F., *Z. Anal. Chem.,* **64,** 45 (1924).

30a. Filer, T. D., *Anal. Chem.,* **43,** 725 (1971).

31. Flaschka, H., and H. Jakobljevich, *Anal. Chim. Acta,* **4,** 247 (1951).

31a. Fogg, A. G., C. Burgess, and D. T. Burns, *Talanta,* **18,** 1175 (1971).

31b. Fogg, A. G., J. Jillings, R. D. Marriott, and D. T. Burns, *Analyst* (*London*), **94**, 768 (1969).

32. Fouckon, M. L., *J. Pharm. Chim.*, **25**, 537 (1937).

33. Frank, C. W., and C. E. Meloan, *Anal. Chem.*, **39**, 379 (1967).

34. Fredrick, W. G., *Anal. Chem.*, **13**, 922 (1941).

35. Friend, J. N., *Man and the Chemical Elements*, 2nd edit., Griffin and Co., London, 1961.

36. Furman, N. H., *Anal. Chem.*, **3**, 217 (1931).

37. Furman, N. H., *J. Amer. Chem. Soc.*, **54**, 4235 (1932).

38. Furman, N. H., *Standard Methods of Chemical Analysis*, Vol. I, 6th edit., Van Nostrand, Princeton, N. J., 1962.

39. Furman, N. H., W. B. Mason, and R. Pekola, *Anal. Chem.*, **21**, 1325 (1949).

40. Galliford, D. J., and J. T. Yardley, *Analyst* (*London*), **88**, 653 (1963).

40a. Gibert, T. G., and D. N. Hume, *Anal. Chim. Acta*, **65**, 451 (1973).

41. Goldstone, N. I., and M. B. Jacobs, *Anal. Chem.*, **16**, 206 (1944).

42. Goto, H., and Y. Kakita, *J. Chem. Soc.* (*Japan*), **64**, 515 (1943); through *Chem. Abstr.*, **41**, 3392 (1947).

42a. Gowin, J. U., J. L. Holt, and R. E. Miller, *Anal. Chem.*, **44**, 1042 (1972).

43. Grant, J., *Analyst* (*London*), **53**, 626 (1928).

44. Greendale, A. E., and D. C. Love, *Anal. Chem.*, **35**, 632 (1963).

45. Greenhalgh, R., and J. P. Riley, *Anal. Chim. Acta*, **27**, 305 (1962).

46. Gyory, S., *Z. Anal. Chem.*, **32**, 415 (1893).

47. Hammock, E. W., R. A. Brown, and E. H. Swift, *Anal. Chem.*, **20**, 1048 (1948).

48. Harrison, G. R., *M.I.T. Wavelength Tables*, Wiley, New York, 1960.

49. Hayakawa, H., *Bunseki Kagaku*, **7**, 296–300, 631–636 (1958); through *Chem. Abstr.*, **53**, 3998 (1959).

50. Hoffman, J. I., and G. E. F. Lundell, *J. Res. Nat. Bur. Stand.*, **22**, 465 (1939).

51. Holler, A. C., *Anal. Chem.*, **19**, 353 (1947).

52. Holness, H., and G. Cornish, *Analyst* (*London*), **67**, 221 (1942).

53. Hourigan, H. F., and J. W. Robinson, *Anal. Chim. Acta*, **10**, 281 (1954).

54. Howie, R. A., M. M. Molishia, and H. Smith, *Anal. Chem.*, **37**, 1059 (1965).

55. Inczedy, J., *Analytical Applications of Ion Exchangers*, Pergamon, London, 1966.

56. Irving, H. M., and F. J. C. Rossotti, *Analyst* (*London*), **77**, 801 (1952).

56a. Jacobsen, E., and T. Rojahn, *Anal. Chim. Acta*, **54**, 261 (1971).

57. Jamison, G. S., *J. Ind. Eng. Chem.*, **3**, 250 (1911).

58. Jander, G., and A. Brull, *Anal. Chem.*, **453**, 335 (1927).

58a. Kalihova, D., and V. Sychra, *Anal. Chim. Acta*, **59**, 477 (1972).

59. Kimura, K., and N. Saito, *J. Chem. Soc.* (*Japan*), *Pure Chem. Sec.*, **74**, 305 (1953); through *Chem. Abstr.*, **47**, 9850 (1953).

60. Klement, R., and A. Kuhn, *Z. Anal. Chem.*, **152**, 146 (1956).

61. Koch, R. C., *Activation Analysis Handbook*, Academic, New York, 1960.

62. Kolthoff, I. M., and J. J. Lingane, *Polarography*, Vol. II, 2nd edit., Wiley-Interscience, New York, 1952.

63. Kolthoff, I. M., and R. L. Probst, *Anal. Chem.*, **21**, 753 (1949).

64. Krause, K. A., and F. Nelson, *Proceedings of the First International Conference on the Peaceful Uses of Atomic Energy*, Vol. 7, United Nations, New York, 1956.

65. Kraus, R., and J. V. A. Novak, *Chemie*, **56**, 302 (1943).

66. Kristaleva, L. B., *Tr. Tomsk. Gos. Univ.*, **154**, 271 (1962).

67. Kroupa, E. K., and R. Klement, "Antimon" in W. Fresenius and G. Jander, Eds., *Handbuch der analytischen Chemie*, Vol. 5a, Springer Verlag, Berlin, 1951.

68. Kunin, R., and R. J. Myers, *Ion Exchange Resins*, Wiley, New York, 1950.

69. Kuznetsov, V. I., *Zh. Anal. Khim.*, **2**, 179 (1947).

69a. Leclercq, M., and S. Carlier, *Anal. Lett.*, **1**, 359 (1968).

70. Lederer, E., and M. Lederer, *Chromatography*, Elsevier, New York, 1957.

71. Lingane, J. J., *Anal. Chem.*, **15**, 583 (1943).

72. Lingane, J. J., *Electroanalytical Chemistry*, 2nd edit., Wiley-Interscience, New York, 1958.

73. Lingane, J. J., and C. Auerbach, *Anal. Chem.*, **23**, 986 (1951).

74. Lingane, J. J., and F. Nishida, *J. Amer. Chem. Soc.*, **69**, 530 (1947).

74a. Long, G. G., J. G. Stevens, L. H. Bowen, and S. L. Ruby, *Inorg. Nucl. Chem. Lett.*, **5**, 21 (1969).

75. Lowe, R. W., S. H. Prestwood, R. R. Richard, and E. I. Wyatt, *Anal. Chem.*, **33**, 874 (1961).

76. Luke, C. L., *Anal. Chem.*, **15**, 528 (1943).

77. Luke, C. L., *Anal. Chem.*, **16**, 448 (1944).

78. Luke, C. L., *Anal. Chem.*, **25**, 674 (1953).

79. Luke, C. L., *Anal. Chem.*, **31**, 1680 (1959).

80. Lundell, G. E. F., H. A. Bright, and J. I. Hoffman, *Applied Inorganic Analysis*, 2nd edit., Wiley, New York, 1953.

81. Lundell, G. E. F., and J. I. Hoffman, *Outlines of Methods of Chemical Analysis*, Wiley, New York, 1938.

82. Lure, Y. Y., and N. A. Fillippova, *Zavod. Lab.*, **14**, 159 (1948); through *Chem. Abstr.*, **42**, 8696.

83. Lyon, W. S., *Guide to Activation Analysis*, Van Nostrand, Princeton, N.J., 1964.

84. Maeck, W. J., *The Radiochemistry of Antimony*, Nuclear Science Series NAS-NS 3033, National Academy of Sciences—National Research Council, Office of Technical Services, Washington, D.C., 1961.

85. Maeck, W. J., G. L. Booman, M. E. Kussy, and J. E. Rein, *Anal. Chem.*, **33**, 1775 (1961).

86. Maren, T. H., *Anal. Chem.*, **19**, 487 (1947).

87. Mason, Brian, *Principles of Geochemistry*, Wiley, New York, 1952.

88. Matulis, R. M., and J. C. Guyon, *Anal. Chem.*, **37**, 1391 (1965).

89. McAlpine, R. K., and B. A. Soule, *Qualitative Chemical Analysis*, Van Nostrand, New York, 1933.

90. McCay, L. M., *Anal. Chem.*, **5**, 1 (1933).

91. McChesney, E. W., *Anal. Chem.*, **18**, 146 (1946).

92. McNabb, W. M., and E. C. Wagner, *Anal. Chem.*, **1**, 32 (1929).

93. McNabb, W. M., and E. C. Wagner, *Anal. Chem.*, **2**, 251 (1930).

94. Meites, L., *Polarographic Techniques*, 2nd edit., Wiley-Interscience, New York, 1965.

94a. Meranger, J. C., and E. Somers, *Analyst (London)*, **93**, 799 (1968).

95. Milner, G. W. C., *The Principles and Applications of Polarography*, Longmans Green, London, 1957.

96. Mogerman, W. D., *J. Res. Nat. Bur. Stand.*, **33**, 307 (1944).

97. Mohr, F., *Lehrbuch der chemisch-analytischen Titriermethoden*, R. Oppenheim Verlag Berlin, 1874.

98. Morrison, G. H., and H. Freiser, *Solvent Extraction in Analytical Chemistry*, Wiley, New York, 1957.

99. Mostyn, R. A., and A. F. Cunningham, *Anal. Chem.*, **39**, 433 (1967).

100. Moulds, D. E., "Antimony" in *Mineral Facts and Problems*, U. S. Bureau of Mines Bulletin 630, 1965 edit., U. S. Govt. Printing Office, Washington, D.C., 1965.

101. Murray, B. J., *Standards and Tests for Reagent and C. P. Chemicals*, Van Nostrand, New York, 1927.

101a. Nicolas, D. J., *Anal. Chim. Acta*, **55**, 59 (1971).

102. Neill, B. R., *Anal. Chem.*, **14**, 955 (1942).

103. Nielsch, W., and G. Boltz, *Mikrochim. Acta*, **1954**, 313 (1954).

104. Neumann, H. M., *J. Amer. Chem. Soc.*, **76**, 2611 (1954).

105. Neumann, H. M., *J. Amer. Chem. Soc.*, **78**, 1848 (1956).

106. Neuman, M., *Z. Phys. Chem.*, **14**, 218 (1914).

107. Norwitz, G., *Anal. Chem.*, **23**, 386 (1951).

108. Norwitz, G., J. Cohen, and M. E. Everett, *Anal. Chem.*, **32**, 1132 (1960).

109. Page, J. E., and F. A. Robinson, *J. Soc. Chem. Ind.*, **61**, 93 (1942).

110. Palmer, H. E., *Z. Anorg. Chem.*, **67**, 317 (1910).

111. Park, B., and E. J. Lewis, *Anal. Chem.*, **5**, 182 (1933).

112. Partington, J. R., *A History of Chemistry*, Vol. II, MacMillan, London, 1961.

113. Pascal, P., *Nouveau Traite de Chemie Minerale*, Vol. XI, *Arsenic Antimone Bismuth*, Masson, Paris, 1958.

114. Pfeil, E., G. Ploss, and H. Saran, *Z. Anal. Chem.*, **146**, 241 (1955).

115. Pharmacopeia of the United States of America XVII, 17th edit., Mack, Easton, Pa., 1965.

116. Pirtea, T. I., *Z. Anal. Chim.*, **118**, 26 (1939).

117. Pribil, R., *Proc. XIth Int. Congr. Pure Applied Chem. (London)* **5**, 893 (1953); through *Chem. Abstr.*, **48**, 5017 (1954).

118. Qureshi, M., and M. A. Khan, *Anal. Chem.*, **35**, 13 (1963).

119. Ramette, R. W., *Anal. Chem.*, **30**, 1158 (1958).

120. Ramette, R. W., and E. B. Sandell, *J. Amer. Chem. Soc.*, **78**, 4872 (1956).

121. Ray, H. N., *J. Indian Chem. Soc.*, **17**, 586 (1940); through *Chem. Abstr.*, **36**, 4772 (1942).

122. Remy, H., *Treatise on Inorganic Chemistry*, Vol. II, Elsevier, New York, 1956.

123. Reynolds, S. A., U. S. Atomic Energy Commission Report ORNL-1557, 1953.

124. Roberts, E. J., and F. Fenwick, *J. Amer. Chem. Soc.*, **50**, 2125 (1928).

125. Robinson, J. W., *Atomic Absorption Spectrophotometry*, Dekker, New York, 1966.

126. Rosin, J., *Reagent Chemicals and Standards*, 3rd edit., Van Nostrand, Princeton, N.J., 1961.

127. Samuelson, O., *Ion Exchangers in Analytical Chemistry*, Wiley, New York, 1963.

128. Sandell, E. B., *Colorimetric Determination of Traces of Metals*, 3rd edit., Wiley-Interscience, New York, 1959.

129. Sasaki, Y., *Bull. Chem. Soc. (Japan)*, **28**, 615 (1955); through *Chem. Abstr.*, **50**, 75656 (1956).

130. Scherrer, J. A., *J. Res. Nat. Bur. Stand.*, **21**, 95 (1938).

131. Schulek, E., *Z. Anal. Chem.*, **102**, 111 (1935).

132. Sen, B. N., *Aust. J. Sci.*, **13**, 49 (1950).

133. Sen, B. N., *Z. Anorg. Allg. Chem.*, **279**, 328 (1955).

134. Silaeva, E. V., and V. L. Kurbatova, *Zavodsk. Lab.*, **28**, 280 (1962).

135. Singh, B., and I. Ilahi, *J. Indian Chem. Soc.*, **13**, 717 (1936).

136. Smith, G. F., and R. L. May, *Anal. Chem.*, **13**, 460 (1941).

137. Smith, G. W., and S. A. Reynolds, *Anal. Chim. Acta*, **12**, 151 (1955).

138. Smith, O. C., *Inorganic Chromatography*, Van Nostrand, Princeton, N.J., 1953.

139. Spacu, G., and A. Pop, *Z. Anal. Chem.*, **111**, 254 (1937).

140. Stanton, R. E., and A. J. McDonald, *Analyst (London)*, **87**, 299 (1962).

141. Strain, H. H., *Chromatographic Adsorption Analysis*, Wiley-Interscience, New York, 1945.

142. Studlar, K., *Collect. Czech. Chem. Commun.*, **25**, 1965 (1960); through *Anal. Abstr., 8* (March 1961) p. 996.

143. Tabern, D. L., and E. F. Shelberg, *Anal. Chem.*, **4**, 401 (1932).

144. Takamoto, S., *J. Chem. Soc. Japan, Pure Chem. Sec.*, **76**, 1339 (1955); through *Chem. Abstr.*, **51**, 12732 (1957).

144a. Tamba, M. G., and N. Vantini, *Analyst (London)*, **97**, 542 (1972).

145. Thorneycroft, W. E., "Antimony" in J. Newton Friend, Ed., *A Textbook of Inorganic Chemistry*, Vol. VI, Part V, Charles Griffin, London, 1936.

146. Tomula, E. S., *Z. Anorg. Chem.*, **118**, 81 (1921).

146a. Toren, P. E., *Anal. Chem.*, **40**, 1152 (1968).

147. Treadwell, F. P., and W. T. Hall, *Analytical Chemistry*, Vol. II, Wiley, New York, 1930.

148. Van Aman, R. E., F. D. Hollibaugh, and J. H. Kanzelmeyer, *Anal. Chem.*, **31**, 1783 (1959).

148a. Van Dyck, G., and F. Verbeek, *Anal. Chim. Acta*, **66**, 251 (1973).

149. Wang, C. Y., *Antimony*, 3rd edit., Griffin, London, 1952.

150. Ward, F. N., and H. W. Lakin, *Anal. Chem.*, **21**, 1204 (1949).

151. Ward, F. N., and H. W. Lakin, *Anal. Chem.*, **26**, 1168 (1954).

152. Webster, S. H., and L. T. Fairhall, *J. Ind. Hyg. Toxicol.*, **27**, 184 (1945).

153. Weeks, Mary E., *Discovery of the Elements*, J. Chem. Educ., Easton, Pa., 1956.

154. Welcher, F. J., and R. B. Hahn, *Semimicro Qualitative Analysis*, Van Nostrand, New York, 1955.

155. West, P. W., and W. C. Hamilton, *Anal. Chem.*, **24**, 1025 (1952).

156. West, P. W., and A. J. Llacer, *Anal. Chem.*, **34**, 555 (1962).

157. West, T. S., *Metallurgia*, **53**, 292 (1956).

158. White, J. C., U. S. Atomic Energy Commission Report CF-56-9-18 (1956).

159. Willard, H. H., and L. H. Perkins, *Anal. Chem.*, **25**, 1634 (1953).

160. Willard, H. H., and P. Young, *J. Amer. Chem. Soc.*, **50**, 1372 (1928).

161. Willard, H. H., and P. Young, *J. Amer. Chem. Soc.*, **50**, 3260 (1928).

162. Wise, W. M., and J. P. Williams, *Anal. Chem.*, **36**, 1863 (1964).

162a. Yamamoto, Y., T. Kumamaru, and Y. Hayashi, *Anal. Lett.*, **5**, 419 (1972).

162b. Yamamoto, Y., T. Takeuchi, and M. Suzuki, *Anal. Chim. Acta*, **64**, 381 (1973).

162c. Yanagisawa, M., M. Suzuki, and T. Takeuchi, *Anal. Chim. Acta*, **47**, 121 (1969).

163. Zaikovski, F. V., *Z. Anal. Khim.* **9**, 155 (1954).

164. Zintl, E., and H. Wattenberg, *Berichte*, **56**, 472 (1922).

165. Zotta, M., *Gazz. Chim. Ital.*, **78**, 143 (1948); through *Analyst* (*London*), **74**, 75 (1949).

MOLYBDENUM

By Gordon A. Parker, *Department of Chemistry, University of Toledo, Toledo, Ohio*

Contents

I. INTRODUCTION

Additional information on the analytical chemistry of molybdenum is available in special monographs (133,235).

A. HISTORY

B. Quist and S. Rimman in 1754 first suggested that molybdenite, naturally occurring MoS_2, is a distinct substance (239), and C. W. Scheele in 1778 distinguished this material from graphite which has similar characteristics (1060). Scheele, too, succeeded in 1778 in preparing molybdenum oxide but it was P. J. Hjelm in 1782 who, at Scheele's request, first isolated metallic molybdenum by reduction of the oxide with carbon (876). Molybdenum gained prominence as an alloying element in steels during the early part of the twentieth century, and this and other uses have resulted in its continually increased demand.

B. OCCURRENCE

1. In Nature

Molybdenite, MoS_2, is the most common form and chief commercial ore of molybdenum. Other molybdenum-containing minerals include powellite, $CaMoO_4$, and wulfenite, $PbMoO_4$ (394). Molybdenum deposits are generally associated with igneous rocks (554), and the isotopic distribution of naturally occurring molybdenum has been studied by mass spectrometry (192). In the western world the United States is the leading producer of molybdenum, followed by Canada, Chile, and several other countries (683). Commercial molybdenite generally contains upwards of 0.5% molybdenum disulfide. Molybdenum ore is also obtained as a by-product in the mining of copper and tungsten.

Trace amounts of molybdenum are found throughout the earth's soil (766) and water (982). In addition it is found in plant species where it is an essential component for the proper utilization of nitrogen (246), in animals (318), and in humans (180). In animals, including man, it is associated with oxidation of xanthine (761), utilization of copper (911), and prevention of dental caries (434).

2. In Industrial Products

Molybdenum metal, molybdenum-based alloys, and molybdenum-containing alloys are widely used for applications where high-temperature strength is needed. This property coupled with favorable electrical conductance and coefficient of expansion makes molybdenum alloys suitable for electron tube filaments containing metal–glass seals. Molybdenum's low thermal neutron absorption allows it to be used with uranium in nuclear reactors. Molybdenum trioxide is the starting material for production of most molybdenum-containing products. These products find use as catalysts in the petroleum industry, as pigments for paints and ceramics, and in fertilizers. Molybdenum disulfide is a widely used solid lubricant with properties similar to those of graphite. Various summaries of the industrial applications of molybdenum are available (33,85,386,643,1073,1078).

C. TOXICOLOGY AND INDUSTRIAL HYGIENE

Molybdenum and its compounds are generally less toxic than the corresponding chromium analogs (673). Molybdenum oxides, chlorides, and ammonium molybdate are all more toxic than molybdenum itself (946). Molybdenosis can be a serious malady especially prevalent among those associated with the mining and manufacture of molybdenum unless special precautions are taken (238,672,1047). Most serious concern is breathing molybdenum containing particulates. Maximum permissible concentrations of molybdenum carbide and silicide of 6 mg/m^3 air and molybdenum boride of 4 mg/m^3 air have been suggested (674).

II. PROPERTIES OF MOLYBDENUM AND ITS COMPOUNDS

A. PHYSICAL PROPERTIES

Molybdenum is a silvery-white metal with a high melting point and excellent high-temperature strength. It exhibits good thermal and electrical conductivity and can be fabricated into a variety of shapes, for example, sheet, plate, wire, and rod. It has a low coefficient of thermal expansion. Table 1 lists some of the physical properties of molybdenum, and Table 2 lists some of the physical properties of representative molybdenum compounds.

B. ELECTROCHEMICAL PROPERTIES

The electrochemical properties of molybdenum have been reviewed (945). Any evaluation of these properties is complicated by the presence of various molybdenum isopoly and complex ion species, the exact nature of which depends upon specific solution conditions. This does not preclude inability to utilize various electrochemical transitions of molybdenum for analytical purposes but does require careful consideration in assigning observed behavior to specific molybdenum species. Latimer (582) tabulates the following electrode potentials, relative to the normal hydrogen electrode, for molybdenum(VI) transitions:

$$H_2MoO_4(aq) + 2\,H^+ + 1\,e^- = MoO_2^+ + 2\,H_2O \qquad E^0 = +0.4\ V$$
$$Mo(VI) + 1\,e^- = Mo(V)\ (in\ 2\,M\ HCl)\ E^0 = +0.53\ V$$

and for reduction of molybdenum(V):

$$MoO_2^+ + 4\,H^+ + 2\,e^- = Mo(III) + 2\,H_2O \qquad\qquad E^0 = 0.0\ V$$
$$Mo(V) + 2\,e^- = Mo(III)\ (green)\ (in\ 2\,M\ HCl) \qquad E^0 = -0.25\ V$$
$$Mo(V) + 2\,e^- = Mo(III)\ (red)\ (in\ 2\,M\ HCl) \qquad\ E^0 = +0.11\ V$$

TABLE 1
Physical Properties of Molybdenum (185,385,485,993)

Atomic number	42
Atomic weight	95.94
Melting point	2610°C
Boiling point	5560°C
Density (20°C)	10.22 gram/cm^3
Specific heat	0.059 cal/(gram)(°C)
Thermal expansion (0–20°C)	0.00000535/°C
Thermal conductivity	0.35 cal/(sec)(cm^2)(°C/cm)
Heat of fusion	70 cal/gram
Electrical resistivity (20°C)	5.2 $\mu\Omega$-cm
Crystal structure (body-centered cubic)	a = 3.1405 A at 25°C
Tensile strength	60,000–70,000 psi

Stable isotopes

molecular weight	abundance
92	15.8%
94	9.0%
95	15.7%
96	16.5%
97	9.5%
98	23.8%
100	9.6%

Radioactive isotopes

molecular weight	decay	half life
88?	β^+, γ	27 min
89?	β^+	7 min
90	EC, β^+, γ	5.7 hr
91m	IT, EC, β^+, γ	65 sec
91	EC, β^+	15.5 min
93m	IT, γ	6.9 hr
93	EC	3000 yr
99	β^-, γ	66.6 hr
101	β^-, γ	14.6 min
102	β^-	11.1 min
103m	β^-, γ	5.3 hr
103	β^-	66 sec
104	β^-, γ	1.3 min
105	β^-	41 sec
106		9.5 sec

TABLE 2

Physical Properties of Representative Molybdenum Compounds (485)

Compound	Formula	Appearance	Specific gravity	Melting point	Boiling point
Molybdenum trioxide	MoO_3	White-yellow	4.69	795°C	sublimes 1155°C
Molybdenum oxydichloride	MoO_2Cl_2	White-yellow	3.31	Sublimes	—
Molybdenum pentachloride	$MoCl_5$	Black	2.93	194°C	268°C
Molybdenum disulfide	MoS_2	Black	4.80	1185°C	Sublimes
Molybdenum trisulfide	MoS_3	Black	—	Decomposes	—
Sodium molybdenum	Na_2MoO_4	White	3.28	687°C	—
Ammonium paramolybdate	$(NH_4)_6Mo_7O_{24} \cdot 4\ H_2O$	White-yellow	2.50	—H_2O at 90°C, decomposes 190°C	—

345

The green and red forms of molybdenum(III) are assumed to result from respectively cationic and anionic molybdenum species. The electrode potential for reduction of molybdenum(III) is also given:

$$Mo(III) + 3\,e^- = Mo^0 \qquad E^0 = -0.2 \text{ V}$$

Another tabulation of electrode potentials (657) lists, in addition to those already cited, the following potentials relative to the normal hydrogen electrode:

$$Mo(VI) + 1\,e^- = Mo(V) \text{ (in } 1\,M \text{ HCl} + 2\,M \text{ KSCN)} \qquad E^0 = +0.5 \text{ V}$$
$$Mo(IV) + 1\,e^- = Mo(III) \text{ (in } 4.5\,M \text{ H}_2\text{SO}_4) \qquad E^0 = +0.1 \text{ V}$$
$$Mo(CN)_8^{3-} + 1\,e^- = Mo(CN)_8^{4-} \text{ (in } 0.25\,M \text{ KCl)} \qquad E^0 = +0.80 \text{ V}$$

Extrapolation to zero ionic strength of potentials measured for the Mo(VI)/Mo(V) couple in varying concentrations of HCl, up to 8 M, results in a standard electrode potential of +0.4826 V versus NHE at 30°C for the one-electron reduction (234). Representative formal potentials for the Mo(VI)/Mo(V) and Mo(V)/Mo(III) couples in the presence of varying complexing ligands are summarized in Table 3. The original reference should be consulted for more extensive listings.

Polarographic half-wave potentials of molybdenum in various supporting electrolytes are discussed in Section VI.D.

Potential-pH diagrams for aqueous molybdenum systems have been made (702).

C. OPTICAL PROPERTIES

Studies of ultraviolet absorption of molybdenum(VI) have been used to determine the exact nature of different molybdenum isopoly ions present in aqueous solutions of varying acidity (315). Absorbance values, in general, are weak, although measurement at 230 nm has been used for molybdenum determination (481). Molybdenum(V) chlorocomplex exhibits an absorbance maximum at 350 nm in 5 M HCl (1013), and studies have been made of the effects of varying temperature and acidity upon the absorbance of Mo(V) species in solution (664). The spectrum of molybdenum(IV) has been measured (155).

Infrared spectra are observed for various molybdenum species in the 900–1100 cm^{-1} region from metal–oxygen stretching vibrations (668).

Optical characteristics of various molybdenum complexes utilized for colorimetric determination are discussed in Section VI.C.

D. CHEMICAL PROPERTIES (235,643)

Molybdenum has the electron configuration $1s^2$, $2s^2$, $2p^6$, $3s^2$, $3p^6$, $3d^{10}$, $4s^2$, $4p^6$, $4d^5$, $5s^1$ and oxidation states of 0, 2+, 3+, 4+, 5+, and 6+. The

TABLE 3

Formal Electrode Potentials for Reduction of Molybdenum in Various Acid Media (460)

Medium	Electrode Potential (V vs NHE at 25°C)	
	Mo(VI)/Mo(V)	Mo(V)/Mo(III)
4 N HCl	0.501	0.129
4 N HCl + 10 N LiCl	0.624	0.223
4 N H$_2$SO$_4$	0.436	0.079
4 N H$_2$SO$_4$ + 10 N (NH$_4$)$_2$SO$_4$	0.459	0.115
2 N H$_2$SO$_4$ + 0.01 N F$^-$	0.439	0.109
4 N H$_3$PO$_4$	0.475	0.047
4 N H$_3$PO$_4$ + 10 N PO$_4^{3-}$ (total)	0.253	−0.233
1 M tartaric acid	0.285	0.024
1 M oxalic acid	0.511	—
1 M citric acid	0.356	—

6+ state is most common, and 3+, 5+, and 6+ forms are used for analytical purposes.

Molybdenum metal, while resisting oxidation at room temperature, is slowly oxidized at temperatures above 370°C and rapidly oxidized at temperatures above 650°C. It is resistant to attack by dilute hydrochloric or sulfuric acid and by alkaline hydroxides but dissolves in aqua regia, concentrated nitric or sulfuric acid, and fused oxidizing salts, for example, potassium chlorate. Molybdenum combines directly with boron, carbon, halogens, nitrogen, phosphorus, silicon, and sulfur. Different compounds can result from each of these reactions depending upon the reaction conditions.

Molybdenum trioxide, formed from calcination of molybdenum metal, is soluble in alkaline solution forming molybdate salts, MoO_4^{2-}, and in mineral acids. Acidification of molybdate solutions results in a precipitate that dissolves as the acid concentration is increased. Alkali metal molybdates are water soluble, while alkaline earth and heavy metal molybdates, generally, are not. Table 4 lists the solubility product constants of several insoluble molybdate salts.

1. Isopoly Compounds

Dilute solutions of molybdenum(VI), less than 0.001 M, in neutral or basic medium exist as the simple monomeric molybdate ion, MoO_4^{2-}. In acid solutions, pH values less than one, cationic species of molybdenum predominate. The simplest of these is the molybdenyl ion, MoO_2^{2+}. At the isoelectric point, occurring at pH 1–2 depending upon the choice of

TABLE 4

Solubility Product Constants of Representative Metal
Molybdates

Compound	Solubility product constant (K_{sp})	Reference
$CaMoO_4$	1.24×10^{-5}	809
$SrMoO_4$	2.58×10^{-7}	809
$BaMoO_4$	3.41×10^{-8}	809
$CdMoO_4$	6.05×10^{-8}	809
Ag_2MoO_4	2.8×10^{-12}	833
$PbMoO_4$	3.3×10^{-12}	924

mineral acid and the molybdenum concentration, molybdic acid, $MoO_3 \cdot x$ H_2O, precipitates only to redissolve as the pH either increases or decreases (50,1028). Various isopolymeric molybdenum species are found in dilute acid solution. There existence and interconversion as the acidity of a neutral molybdenum(VI) solution increases has been studied extensively. However, apparently conflicting results have often been obtained. The subject has been reviewed (862,1028).

Assuming no competing complexation reactions with auxiliary ligands, the protonization and polymerization of molybdate anion, MoO_4^{2-}, with increasing acidity has been shown, from potentiometric measurements in $3 M$ $NaClO_4$, to proceed according to the reactions (861):

$$H^+ + \quad MoO_4^{2-} = H\,MoO_4^- \qquad \log \beta_{1,1} = 3.89$$
$$2\,H^+ + \quad MoO_4^{2-} = H_2MoO_4 \qquad \log \beta_{2,1} = 7.50$$
$$8\,H^+ + 7\,MoO_4^{2-} = Mo_7O_{24}^{6-} + 4\,H_2O \qquad \log \beta_{8,7} = 57.74$$
$$\text{(paramolybdate)}$$

$$9\,H^+ + 7\,MoO_4^{2-} = H\,Mo_7O_{24}^{5-} + 4\,H_2O \qquad \log \beta_{9,7} = 62.14$$
$$10\,H^+ + 7\,MoO_4^{2-} = H_2Mo_7O_{24}^{4-} + 4\,H_2O \qquad \log \beta_{10,7} = 65.68$$
$$11\,H^+ + 7\,MoO_4^{2-} = H_3Mo_7O_{24}^{3-} + 4\,H_2O \qquad \log \beta_{11,7} = 68.21$$

β values are for the equilibrium formation constants of each equation written as shown. In addition to paramolybdate ion, protonated molybdate, and paramolybdate ions, other isopoly complexes have been identified in dilute acid solution. In $1 M$ NaCl with proton-to-molybdenum ratio of 1.5 the octamolybdate ion is present (42).

$$12\,H^+ + 8\,MoO_4^{2-} = Mo_8O_{26}^{4-} + 6\,H_2O \qquad \log \beta_{12,8} = 71.56$$

Numerous other anionic molybdenum species have been reported (177,- 351,564,828). Table 5 catalogs the names and formulas of the more common isopoly molybdate species.

With increasing acidity, below pH one, various cation forms of molybdenum are present in solution. These have been assigned the general formula $[(MoO_2)(MoO_3)_{x-1}]^{2+}$ with $x = 1$ representing the molybdenyl ion MoO_2^{2+}, $x = 2$ the dimer $Mo_2O_5^{2+}$, $x = 3$ the trimer $Mo_3O_8^{2+}$, and so forth (443). Proton and aquo forms of these complexes exist (1052). Formation of polymeric cation species is favored with increasing molybdenum concentration and decreasing acidity below pH 1. Formation constants for dimer and trimer forms in 1 M HClO$_4$ are known (1053).

$$2 \text{ monomer} = \text{dimer} \qquad \log \beta_1 = 2.72$$
$$3 \text{ monomer} = \text{trimer} \qquad \log \beta_2 = 5.96$$

The chemistry of other oxo containing molybdenum species, including those containing molybdenum(V), has been reviewed (668). Mixed-valence isopoly species, formed from partial reduction of molybdenum(VI) materials and known as molybdenum blue species, have also been reviewed (836).

2. Coordination Compounds

There are many metal–ligand complexes with molybdenum as the central metal ion bound to inorganic or organic ligands. Their chemistry has been reviewed (669,797). In addition to complexes of type Mo_xL_y where, especially if L represents a halide, several values of x and y are possible, complexes of type MoO_xL_y form. In dilute acid solution, in particular, molybdenum(VI) complexation involves the molybdenyl ion,

TABLE 5
Names and Formulas of Common Isopoly Molybdates (862,1028)

Name	Empirical Formula (M = univalent cation)	Formula[a]
Orthomolybdate	$M_2O \cdot MoO_3$	M_2MoO_4
Dimolybdate	$M_2O \cdot 2\ MoO_3$	$M_2Mo_2O_7$
Trimolybdate	$M_2O \cdot 3\ MoO_3$	$M_2Mo_3O_{10}$
Tetramolybdate (octamolybdate)	$M_2O \cdot 4\ MoO_3$	$M_4Mo_8O_{26}$
Hexamolybdate	$M_2O \cdot 6\ MoO_3$	$M_2Mo_6O_{19}$
Paramolybdate	$3M_2O \cdot 7\ MoO_3$	$M_6Mo_7O_{24}$
Decamolybdate	$M_2O \cdot 10\ MoO_3$	$M_2Mo_{10}O_{31}$
16-Molybdate	$M_2O \cdot 16\ MoO_3$	$M_2Mo_{16}O_{49}$

[a] Waters of hydration are omitted.

MoO_2^{2+}, and reactions of the type

$$MoO_2^{2+} + rL^{n-} = MoO_2L_r^{(2-rn)}$$

occur with L representing halide, sulfate, nitrate, or other complexing species. Complexation with dimeric molybdenum(VI) cation is also well known. An example of this is the tetraperoxo-1,2-dimolybdate(VI) ion $[(O_2)_2Mo(O)—O—Mo(O)(O_2)_2]^{2-}$ formed from reaction of hydrogen peroxide with acid molybdate solution (190).

Molybdenum(V) forms complexes similar to those of molybdenum(VI). The ion, MoO^{3+}, for example, complexes with many ligands. One of these is the orange-red molybdenum-thiocyanate complex that is widely used for colorimetric molybdenum determination. Several molybdenum(V)-thiocyanate complexes exist, and different formulas have been assigned to the complex of analytical interest (133,933). Among these are the oxopentathiocyanatomolybdate(V) ion, $[MoO(NCS)_5]^{2-}$ (1077), and a dimeric species, perhaps $[(MoO_2)_2(NCS)_6]^{4-}$ (670). In the dimeric species molybdenum(V) atoms are joined by two bridging oxygen atoms. Both molybdenum(VI) and molybdenum(V) form EDTA complexes with a 2:1 ratio of Mo:EDTA. The Mo(V)-EDTA complex has analytical significance. The Mo(VI)-EDTA complex contains two molybdenum groups attached separately at each end of the EDTA molecule, $Na_4[MoO_3EDTAMoO_3]\cdot 8\ H_2O$ (169,549), while the Mo(V)-EDTA complex consists of a single molybdenum dimer bound to an EDTA molecule, $Na_2[(MoO_2)_2EDTA]\cdot H_2O$ (668). Both molybdenum(V) and molybdenum(IV) form stable octacyanomolybdate complexes, $[Mo(CN)_8]^{3-}$ and $[Mo(CN)_8]^{4-}$. The oxidation-reduction potential of this couple is suitable for $[Mo(CN)_8]^{3-}$ to function as an oxidizing agent in basic medium (373). Aquotrihydroxotetracyanomolybdate(IV), as its tripotassium salt, is used as an analytical reagent for amperometric determination of various divalent heavy metal ions (447).

Complexes containing molybdenum(III) and molybdenum(II) are discussed in the reviews cited at the beginning of this section. Utilization of these complexes for analytical purposes has not been extensively explored.

Molybdenum in its several valence states forms many complexes with organic ligands. The molybdenum(VI) and molybdenum(V) species are widely used for analytical purposes. The formula for the 8-quinolinol complex of molybdenum(VI), $MoO_2(C_9H_6ON)_2$, with its 2:1 ratio of molybdenum(as molybdenyl cation):ligand is typical of the type of complex expected in weak acid solutions, although 1:1 complexes are also frequently encountered. Molybdenum(V) complexes are, generally,

TABLE 6

Formation Constants of Representative Molybdenum Complexes

Molybdenum	Ligand	Ratio Mo:L	Formation constant ($\log \beta$)	Reference
Mo(V)	EDTA	1:1	0.8	914
Mo(VI)	Alizarin red S	1:2	9.2	915
Mo(VI)	Ascorbic acid	1:1	4.6	965
Mo(VI)	Carminic acid	1:1	3.8	915
Mo(VI)	Catechol	1:2	4.6	914
Mo(VI)	EDTA	1:1	10.7	915
Mo(VI)	EDTA	2:1	19.5	915
Mo(VI)	Ferron	1:1	3.4	816
Mo(VI)	Pyrogallol	1:2	5.7	914
Mo(VI)	Rezarson	1:1	6.3	620
Mo(VI)	Toluene-3,4-dithiol	1:3	18.4	310
Mo(VI)	Xylenol orange	2:2	17.0	121

similar to the corresponding molybdenum(VI) species. Bonding between molybdenum and organic molecules can occur through oxygen, sulfur, and nitrogen linkages. Some organomolybdenum compounds are of biological interest (660,940). Table 6 lists some examples of organomolybdenum compounds used for analytical purposes.

3. Heteropoly Compounds

Heteropoly complexes contain a central metal element bonded to numerous oxygenated metal ions. Perhaps best known of these complexes is the 12-molybdophosphate ion, $[PMo_{12}O_{40}]^{3-}$, containing twelve molybdenum atoms about a central phosphorus(V) atom. Bonding occurs in a tetrahedral arrangement with four equivalent molybdenum groups, each of which contains molybdenum atoms joined to oxygen in an octahedral array (473). The 1:12 complexes are represented by the general formula, $[X^{+n}Mo_{12}O_{40}]^{-(8-n)}$, where X represents phosphorus(V), arsenic(V), silicon(IV), germanium(IV), titanium(IV), zirconium(IV), and other central metal ions (48); 1:12 heteropoly complexes of the type, $[X^{+n}Mo_{12}O_{42}]^{-(12-n)}$, occur when X represents cerium(IV) and thorium(IV). Heteropoly complexes also exist with a central metal-ion-to-molybdenum ratio of 1:9, and some of these are stable as dimers with a central metal ion to molybdenum ratio of 2:18. The 1:6 hetero ions exist and are represented by the general formula, $[X^{+n}Mo_6O_{24}]^{-(12-n)}$, with X representing tellurium(VI) and iodine(VII) and by the general formula, $[X^{+n}Mo_6O_{24}H_6]^{-(6-n)}$, with X representing cobalt(II), iron(III), aluminum(III), nickel(II), and other ions. Hetero ions containing other ratios of central metal ion to molybdenum are postulated (1026). Molybdenum is not unique in its ability to form hetero ions. Ions containing tungsten and other elements clustered about a central metal species are well characterized. Mixed heteropoly ions also occur. With mixed complexes, some of the molybdenum groups are replaced by those of a different element, for example, the $1:12$ complex, 11-molybdo-1-vanado-phosphate, $[PMo_{11}VO_{40}]^{5-}$. Many heteropoly complexes are readily soluble in water and oxygen-containing organic solvents such as ethers and alcohols. They have high molecular weights, are generally colored, and in the solid form contain large amounts of water of hydration. The absorption spectra of heteropoly molybdate complexes exhibit bands in the near-infrared, visible, and ultraviolet regions of the spectrum. Many of these bands are used for analytical determinations (429,430). Heteropoly complexes degrade in solutions of increasing basicity until finally only simple ions remain. 12-Molybdophosphate, for example, breaks up into molybdate ions and phosphate ions in basic solution.

Heteropoly molybdates are easily reduced (431,1027). In all cases the reduction product exhibits a deep blue color, molybdenum blue. The exact nature of the reduced form is uncertain and has been assigned to a complex containing both Mo(VI) and Mo(V) (62,934) and to a reduced species in which the added electrons reside with the complex as a whole and cannot be assigned to a particular ion within the complex (818).

Additional information on particular heteropoly molybdates is found in Section II.F.2, where applications of heteropoly species in analysis are described. Table 7 lists several representative heteropoly ions.

E. RADIOCHEMICAL AND NUCLEAR PROPERTIES (868)

Table 1 lists the naturally occurring and artifically produced isotopes of molybdenum. The radiochemical reaction of greatest analytical interest is the decomposition of ^{99}Mo formed from neutron bombardment of ^{98}Mo the most abundant natural isotope.

$$^{98}Mo + n \rightarrow {}^{99}Mo + \gamma$$
$$^{99}Mo \rightarrow {}^{99m}Tc + \beta^- \qquad t_{1/2} = 67 \text{ hr}$$
$$^{99m}Tc \rightarrow {}^{99}Tc + \gamma \qquad n_{1/2} = 6 \text{ hr}$$

Section VI.E contains additional information on the radiochemistry of molybdenum.

F. MOLYBDENUM AND ITS COMPOUNDS IN ANALYSIS

1. Molybdenum Electrodes

An electrode pair consisting of a molybdenum wire–saturated calomel electrode is responsive to changes in hydrogen ion activity. Satisfactory

TABLE 7

Photometric Determination of Various Elements Through Heteropoly Formation with Molybdenum

Element detected	Heteropoly species	Procedure	Wavelength (nm)	Reference
As(V)	12-Molybdoarsenate	Reduce with ascorbic acid	840	222
Bi(III)	18-Molybdobismuthophosphate	Reduce with ascorbic acid, measure A vs time	725	355
Ce(IV)	12-Molybdocerate	Precipitate, filter, dissolve, add Mo(VI)	380	883
Cs(I)	Cs compound of 12-molybdophosphate	Precipitate, filter, dissolve, reduce with hydrazine	805	399
Co(II)	8-Molybdo-4-tungstosilicate	Indirect, reduce with Co(II)	Red filter	512
Cu(II)	12-Molybdophosphate	Reduce with [Cu(CN)$_2$]$^{1-}$	725	1022
Fe(II)	6-Molybdo-6-tungstophosphate	Indirect, reduce with Fe(II)	725	985
Ge(IV)	12-Molybdogermanate	pH 1.5	315	424
Hf(IV)	12-Molybdohafnate	Reduce with Sn(II)	725	186
Nb	10-Molybdo-2-niobophosphate	pH 1, boil, cool extract with BuOH	413	902
P(V)	12-Molybdophosphate	Extract with BuOH, reduce with Sn(II)	725	617
P(V)	11-Molybdo-1-vanadophosphate	Extract with 1-penanol	308	425
Se(IV)	Several possible	0.5 M HNO$_3$	380	884
Si(IV)	12-Molybdosilicate	Reduce with Fe(II), extract with 2-pentanol	800	448
Te(IV)	11-Molybdo-1-tellurophosphate	Extract with 2-pentanol	360	545
Tl(I)	Tl Compound of 12-molybdophosphate	Precipitate, filter, dissolve, reduce with hydrazine	808	356
Th(IV)	Not specified	Reduce with Sn(II)	690	627
Ti(IV)	12-Molybdo-1-titanophosphate	Extract with cyclohexanol	313	903
V(V)	11-Molybdo-1-vanadophosphate	1 M HCl	323	423
Y(III)	Not specified	Reduce with Sn(II)	695	628
Zr(IV)	12-Molybdo-1-zirconophosphate	0.1 M HClO$_4$	350	700

353

titration curves are obtained for potentiometric titration of strong acids with sodium hydroxide (417) and organic acids in nonaqueous solvents (1050). A molybdenum-oxide-coated molybdenum wire, as indicating electrode, also responds to changes in hydrogen ion activity (40,778).

A molybdenum microelectrode is responsive to polarographic reduction of various metal species in both aqueous (853) and fused alkali halide (854) solvents. A silver wire indicating electrode–molybdenum wire reference electrode pair responds to polarographic reduction of silver in a 0.1 M sulfosalicylic acid supporting electrolyte (41). Amperometric titrations of strong oxidizing agents in 3 M sulfuric acid can be followed using a Mo–Pt electrode pair (265). Other applications of molybdenum wire electrodes to polarographic studies have been reviewed (208).

2. Molybdenum Heteropoly Compounds

Phosphorus is frequently determined as the 12-molybdophosphate complex either by weighing insoluble ammonium 12-molybdophosphate (518) or photometrically after reduction of the heteropoly complex to molybdenum blue (108). Silicon, too, is determined as 12-molybdosilicate either gravimetrically as the quinoline compounds (626) or colorimetrically after reduction to molybdenum blue (27). Molybdophosphate precipitates of cesium and rubidium (543), other metal ions (406), and various alkaloids (428) are used for quantitative determination of each of these materials. Qualitative identification of metal ions (609) aldehydes and ketones (88) and biological materials (1045) is possible through color formation with various heteropoly complexes. Polyaminocarboxylic acids, for example, are detected by molybdotungstophosphate spray reagent after paper chromatographic separation (570). Heteropoly complexes are also used for staining tissue samples prior to microscopic examination (276).

In addition to phosphorus and silicon, mentioned previously, other elements are determined quantitatively either as their heteropoly complex or by reaction with a previously formed heteropoly material (48,109,356,1046). Table 7 lists several inorganic elements that are determined photometrically by heteropoly reaction.

Ammonium molybdophosphate and other heteropoly complexes are used as ion-exchange materials for separation of alkali metal ions (187) and rare earth elements (984). Thin-layer chromatographic separation of various fission products occurs on coatings composed of heteropoly molybdenum compounds (119).

3. Molybdenum Compounds

Solutions of molybdenum(III) (248) and molybdenum(V) (262,1023) are suitable titrants for volumetric determination of oxidizing agents. Solutions of octacyanomolybdate(IV) ion are used for titrating various reducing agents (1043). Molybdenum(VI) solution is used for determining certain heavy metal ions by precipitation titrations (59,123) and, as a color-forming reagent, in the qualitative identification of various aliphatic and aromatic aldehydes and ketones (756).

III. DISSOLUTION OF MOLYBDENUM, MOLYBDENUM ALLOYS, AND MOLYBDENUM-CONTAINING MATERIALS

A. MOLYBDENUM, MOLYBDENUM ALLOYS AND COMPOUNDS

Molybdenum and molybdenum-based alloys are soluble in concentrated nitric or sulfuric acid (845), aqua regia (783), mixtures containing strong mineral acid and hydrofluoric acid (171,328), 30% hydrogen peroxide plus nitric acid (255), and by sodium peroxide–sodium carbonate fusion (467). Molybdenum trioxide is soluble in sodium hydroxide solution forming molybdate salts and in mineral acids (235). Molybdenum disulfide is removed from metal surfaces by combustion in a stream of moist oxygen (454). Most molybdoheteropoly species dissolve in water, alcohol, or in basic solution where they decompose forming simple, soluble ions (1026).

B. ORES

Molybdenite and other molybdenum-containing minerals are put into solution with concentrated nitric acid (1106) or mixtures of nitric acid and another strong mineral acid (667,733). Following this, sulfuric acid is added and the sample is evaporated to near dryness. Solution can also be achieved by fusing the molybdenum ore with potassium hydroxide (31), sodium peroxide (714), or sodium peroxide–sodium hydroxide (388) flux. The fused sample is dissolved in water or dilute acid.

C. MISCELLANEOUS MATERIALS

Soil samples for molybdenum analysis are treated with strong mineral acid (316), by basic fusion (1058), or by prolonged shaking with hot water (614) or organic acid (217,320). Combustion in a stream of oxygen is used for sample treatment when only a limited quantity of sample is available (686). Plant and animal samples are prepared by digesting them in strong mineral acid (708,843). Dry ashing of soil, plant, and animal samples is also possible (771,834,873). Molybdenum in natural waters is concentrated and collected by coprecipitation (168,486,559), extraction (174), or ion exchange (464).

IV. SEPARATION FROM OTHER ELEMENTS

Molybdenum, frequently determined by colorimetric or oxidation-reduction procedures, must first be separated from other ions that give similar responses to the complexing agents or reducing titrants generally employed. These separations often take the form of sulfide precipitation at 0.3 M hydrogen ion concentration, extraction of chloro complexes with oxygen-containing immiscible organic solvents, extraction of α-benzoin oxime or 8-quinolinol organomolybdenum complexes with chloroform, or ion exchange. Both cation- and anion-exchange resins are used for molybdenum, and molybdenum appears either as a positively or negatively charged species depending upon pH and formation of charged complexes with auxiliary ions present in solution. Elution occurs by appropriate choice of eluent to alter the complex form retained by the resin being used. Many different complexing agents, organic extracting solvents, and ion-exchange resin–eluent solution combinations have been developed to separate molybdenum from specific interferences. These are discussed in this section.

A. PRECIPITATION METHODS

Directions are available for precipitating the acid-insoluble sulfide group, including molybdenum, from solutions 0.3 M in hydrogen ion (269,383). Specific procedures using sulfide precipitation separate molybdenum from rhenium (694), vanadium (966), tungsten (1085), and titanium, zirconium, thorium, iron, nickel, and zinc (442). A series of reports describes sulfide separation procedures for molybdenum from the alkaline earth metals, aluminum, bismuth, cadmium, indium, lead, mercury, thorium, uranium, and zinc (1003). Sulfide precipitation is suggested for separation of molybdenum in silicate rocks (371) and steels (979).

Coprecipitation of molybdenum with insoluble metal hydroxides separates Mo(VI) from other ions (775,1009). In the presence of iron(III) hydroxide coprecipitation is nearly quantitative at low pH values but decreases with increasing pH; no coprecipitation is observed in the presence of excess strong base. Ammonia and dilute alkaline hydroxides are used as precipitating agents. Coprecipitation also occurs with zirconium(IV) hydroxide (773) and magnesium hydroxide (774). The latter reagent separates chromium(III) from molybdenum(VI) by coprecipitation of chromium at high pH. Hydroxide coprecipitation separates molybdenum from rhenium (736) and from various ions in natural waters (411). Coprecipitation of molybdenum with metastannic acid (731) and insoluble barium and lead salts (621) has been described.

Organic precipitating agents are used to concentrate molybdenum and other ions from a variety of materials. Precipitation of Mo, Cu, Fe, Mn, and Zn at pH 6.6 with 8-quinolinol is suggested for concentrating these ions from soil extracts prior to spectrographic analysis (571). α-Benzoin oxime (1063) and a mixed reagent containing 8-quinolinol, tannic acid, and thionalide (916) concentrate several ions, including molybdenum, from natural waters. Double precipitation separates molybdenum from solutions 0.2 M in hydrochloric acid first with a mixed reagent of tannic acid and methyl violet and then, after filtering and dissolving the initial precipitate, with ammonium thiocyanate and methyl violet (560). Nearly quantitative recovery of molybdenum is obtained, and addition of tartaric acid in the second step minimized tungsten interference.

B. EXTRACTION METHODS

Methods for extracting molybdenum have been summarized (527). Oxygen-containing organic solvents extract the molybdenum species, $MoOCl_3$ and MoO_2Cl_2, from aqueous solutions acidified with hydrochloric acid (142,143). A single extraction with diethyl ether from solutions 6 M in hydrochloric acid removes approximately 80% of the molybdenum present (21,1090). Ether extraction of molybdenum from solutions containing hydroiodic acid (496) or hydrofluoric acid (497) are not satisfactory; less than 10% of the molybdenum present is transferred to the organic phase. In contrast, the presence of hydrobromic acid enhances extraction into a variety of oxygen-containing organic solvents (719,1116). Use of β,β'-dichlorodiethyl ether is unsatisfactory for molybdenum extraction (37).

Various esters extract molybdenum from acidified aqueous solution. Butyl acetate is used both alone and as a 1:1 mixture with diethyl ether (1091). Amyl acetate separates molybdenum(VI) from rhenium(IV) (1093) and iron(III) (691), the later by a countercurrent technique. Molybdenum(V) is separated from vanadium(IV) by amyl acetate extraction (1094).

Methyl isobutyl ketone separates numerous metal species from hydrochloric acid solution (112,327), including molybdenum (887,1111). Data indicating the extent of extraction of metal ions with 0.1 M solutions of various β-diketones in benzene are reported (954). Acetylacetone in chloroform is satisfactory for removal of molybdenum from other metal ions in 3 M sulfuric acid solution (652) and has been applied to the separation of molybdenum from ashed biological samples (242) and fission products (284). Extraction with acetylacetone from solutions 0.5 M in sodium hydroxide separates rhenium from molybdenum (1092); only 0.02% of the molybdenum is transferred with rhenium to the organic phase.

Optimum conditions for extracting metal thiocyanate complexes, including molybdenum(V)-thiocyanate complex, are described using diethyl ether (99) and methyl isobutyl ketone (329). Selective separation of molybdenum(V)-thiocyanate into methyl isobutyl ketone occurs following reduction of molybdenum(VI) by thiosulfate ion at 10°C (1095). Molybdenum is returned to an aqueous phase by back extraction with ammonia in the presence of hydrogen peroxide. Molybdenum(V)-thiocyanate is also selectively extracted into chloroform-isoamyl alcohol in the presence of N-benzylaniline (483). The nature of the molybdenum(V)-thiocyanate complex is discussed in Section II.D.2.

Various heteropoly molybdate complexes are extracted from aqueous solution containing approximately 0.1 M hydrogen ion. Butanol (583,904), butanol–chloroform mixtures (1055), and butyl acetate (1034) are used for the organic phase. Reduced heteropoly species, molybdenum blue complexes, are also extracted into oxygen containing organic solvents, for example, butanol (502). Extraction of molybdophosphoric acid occurs in the presence of quinoline (565) and of reduced molybdophosphoric acid in the presence of di-n-octylamine (503). The nature of the molybdenum heteropoly species is discussed in Section II.D.3.

Molybdenum and other ions are readily extracted into organic, amine-containing, solvents (629,877,879,1076). Selectivity is achieved by proper choice of amine, solvent, and acidity. Rhenium is separated from molybdenum by extracting from 4–6 M NaOH solution with pyridine (661). Molybdenum remains in the aqueous phase. Molybdenum is extracted from aqueous solutions adjusted to pH 4.2 with hydrochloric acid using aniline as the organic solvent (266,561). Optimum conditions for the separation of vanadium, molybdenum, tungsten, and rhenium from low-grade ores by extraction with solutions of tri-n-octylamine are discussed (306). The liquid ion exchanger, Amberlite LA-1, dissolved in xylene extracts a variety of materials over a wide pH range (334,414). A 0.1 M solution of a similar liquid exchanger, Amberlite LA-2 in kerosene, has been applied to the separation of molybdenum(VI) (457). Solutions of quantary ammonium halides in kerosene extract molybdenum and vanadium from aqueous, alkaline solutions (304). Niobium is separated from molybdenum by extraction of its pyrogallol complex with a solution of tetra-*n*-butylammonium bromide in ethyl acetate (34). Molybdenum remains in the aqueous phase. Cupferron, ammonium salt of nitrosophenylhydroxylamine, extracts molybdenum (368) and various other ions (166). Nearly quantitative transfer of molybdenum into the organic phase occurs with a single extraction from 2 *M* hydrochloric acid solution by cupferron dissolved in chloroform. Extraction of the molybdenum(VI) complex of *N*-benzoyl-*N*-phenylhydroxylamine with chloroform at pH less than 6 separates molybdenum from iron (1117) and at pH 3 from tungsten (801). Separation of molybdenum from iron is possible by extracting molybdenum with 1:1 isoamyl alcohol–chloroform from an aqueous solution adjusted to pH 2.2 and containing diphenylguanidimium chloride and titron, 1,2-dihydroxybenzene-3,5-disulfonic acid disodium salt (150). Extraction of the molybdenum alizarine red S complex with 1,2-dichloroethane from solutions adjusted to pH 4.5–5.5 and containing tetradecyldimethylbenzylammonium chloride is reported (412).

Molybdenum is extracted from aqueous solutions containing 5% hydrochloric acid by α-benzoin oxime dissolved in chloroform (382). This procedure is commonly used to separate molybdenum from steel samples (319). Extraction with other oximes has also been discussed (277).

Various metal ions, including molybdenum, are extracted by chloroform solutions of 8-quinolinol (305,953) and other substituted quinolines (844). Extraction of molybdenum is best from acidic solution with pH values near one. Separation of rhenium(VII) from molybdenum(VI) is described by extracting the former with quinoline from solutions made alkaline with sodium hydroxide (1007).

Molybdenum(VI) and several other ions are separated from a variety of metal species by extraction from aqueous solution 0.5 *M* in hydrochloric acid and containing EDTA, ascorbic acid, and citric acid (495). The extraction is carried out with 0.5 *M* di-*n*-butyl phosphate in gasoline. The advantage of this procedure over other methods lies in the back extraction of the organic phase with aqueous 0.1 *M* hydrochloric acid solution containing 1.5% hydrogen peroxide. Only molybdenum is reported to pass into the aqueous phase, while the other metal ions present remain with the di-*n*-butyl phosphate in the organic phase. Tri-*n*-butyl phosphate, too, extracts various metals from aqueous solution (415). Molybdenum(VI) is separated from solutions containing hydrochloric acid by tri-*n*-butyl phosphate dissolved in carbon tetrachloride (209,436). If the acidity is lowered to 0.01 *M*, molybdenum is not extracted (1069). In the presence of sulfuric acid titanium (1118) and rhenium (408) are extracted with tri-*n*-butyl phosphate leaving molybdenum in the aqueous phase. Molybdenum(V) species (141) and peroxomolybdate species (493) are also extracted with tri-*n*-butyl phosphate. Bis(2-ethylhexyl) phosphoric acid in kerosene extracts molybdenum from solutions 0.01 M in nitric acid, while rhenium remains in the aqueous phase (487,1115). This reagent is used for extracting radioactive waste liquors (435); molybdenum, zirconium, and niobium are removed from aqueous solutions 2.5 M in nitric acid. Back extraction of molybdenum only, using aqueous nitric acid–ammonium fluoride, achieves separation from zirconium and niobium. Separation of molybdenum from tungsten is possible by extracting the former from hydrochloric acid solution at pH 2–3 with 0.1 M diisoamyl methylphosphonate in kerosene (573). Extraction of molybdenum with other substituted phosphonic acid esters is described (409). Various sulfur-containing organophosphorus compounds extract metal ions including molybdenum (145,354). For example, a 1 M solution of *O,O,S*-tripropyldithiophosphate in benzene separates molybdenum(V) from several other ions (147). The separation of molybdenum from solutions containing catechol and *n*-butyltriphenylphosphonium bromide with chloroform is also reported (1054).

Extraction of molybdenum(V) and molybdenum(VI) occurs with mercapto acids in the presence of amine salts (152). Various organic solvent systems for molybdenum extraction with thioglycolic acid in the presence of diphenylguanidinium chloride have been investigated (151). Molybdenum is separated from tungsten using thiolactic acid in isoamyl alcohol–benzene (1:1 vol./vol.) from solutions containing *o*-phenetidine salts (704). Table 8 lists various reagents for molybdenum extraction.

C. ION-EXCHANGE METHODS

Methods for the ion exchange of molybdenum have been discussed (527).

The ion-exchange properties of molybdenum are explainable if one considers the various cationic and anionic complex species formed by molybdenum in the presence of other reagents. Molybdenum(VI), as MoO_4^{2-}, easily passes through strong cation resins of the sulfonated polystyrene type from solutions 0.01 M in hydrochloric acid (241). This allows separation of molybdenum from cationic species which are retained to varying degrees upon the column (970). Addition of increasing amounts of ethanol to molybdenum(VI) solutions dissolved in 0.1 M hydrochloric acid has little effect upon this exchange (900,974). Molybdenum(V), too, has little affinity for sulfonated polystyrene resins when dissolved in 0.2 M hydrochloric acid (128,968). Hydrobromic acid

TABLE 8
Extracting Agents for Molybdenum

Aqueous phase	Complexing agent	Extractant	Separation from	Reference
3 M H_2SO_4	Acetylacetone	$CHCl_3$–acetylacetone, 1:1 vol./vol.	Fe	652
2 M HCl, citric acid	Acetylacetone	$CHCl_3$–acetylacetone, 1:1 vol./vol.	W	339
0.5 M H_2SO_4	Diethyldithiocarbamate	$CHCl_3$	Al, Mg	906,980
0.5 M HCl	N,N'-Diphenylthiocarbamo-hydroxamic acid	$CHCl_3$	Co, Ni	638
0.2 M HCl	Nitron	$CHCl_3$	Re	588
2 M $HClO_4$	Thenoyltrifluoroacetone	Amyl acetate	W	146
0.01 M HCl	$Bu_3NH^+Cl^-$, thioglycolic acid	CH_2Cl_2	V	1121
4 M HCl	Tributyl phosphate	CCl_4	Te	929

solutions (0.5 M) of molybdenum(VI) undergo little exchange upon passage through columns of strong acid sulfonated polystyrene resins (280), and addition of increasing amounts of various lower alcohols does not alter this condition (531). Molybdenum(VI) is also not retained when dissolved in solutions 0.1 M in hydrofluoric acid (281), hydrofluoric acid–sulfuric acid (196), hydrofluoric acid–nitric acid (196), or from hydrofluoric acid–organic solvent mixtures (530). The same is true of molybdenum solutions dissolved in comparable concentrations of sulfuric acid or nitric acid (973). Mixed solvents containing 0.6 M nitric acid and varying amounts of lower alcohols, however, cause reduction of molybdenum(VI), and the reduced species do tend to be retained, to varying degrees, by sulfonated polystyrene resin columns (529). Addition of certain complexing ligands to a weak acid solution of molybdenum(VI) assures formation of anionic species that pass through strong acid cation resin columns; citric acid (499,896), hydrogen peroxide (278), and EDTA solution (849) are suggested. Cationic molybdenum species are, however, reported to exist in the presence of 1–2 M acetic acid (198). Elution with citric acid solution has isolated molybdenum as part of a separation scheme for 35 metallic radioelements (95), from ashed plant material (888), and from steels (20). Molybdenum is separated from iron and steels with solutions containing hydrogen peroxide (19,956). It is separated from vanadium on a cation resin using 5% $NaNH_4HPO_4$ as eluent (650). Separation from beryllium is also described (71,969).

The presence of positively charged molybdenum species explains the separation of rhenium from molybdenum on KU-2 resin (294). Rhenium passes through the column, while molybdenum is retained from solutions containing 5% thiourea in 0.1 M acid. Cationic species, too, account for retention of molybdenum on columns of sulfonated coal (1014). In 0.1 M acid tungsten passes through a column of this material, while molybdenum is retained (895). Retention of both tungsten and molybdenum after their reduction with lead amalgam allows separation of phosphorus from these elements on sulfonated coal (510). Phosphorus is not retained. This same exchanger is used for separation of rhenium from molybdenum, where passage of solutions adjusted to pH 3–5 retains molybdenum (313). Some reduction of molybdenum is reported with this resin, and passage of the reduced species into the effluent occurs. To counter this, addition of bromine water to the effluent is suggested, and the sample is put through the column a second time (314) or passed through a column of alumina that also retains molybdenum (848).

Some cation-exchange studies of molybdenum from concentrated acid solutions are reported. Under conditions where MoO_2^{2+} is present exchange is possible, but the presence of MoO_4^{2-} and various isopoly anion species, in many cases, prevents quantitative retention. Greater retention of molybdenum(VI) on Dowex 50 is reported from solutions 9 M in $HClO_4$ than from solutions 9 M in HCl (721). Some retention of molybdenum(VI) also occurs in 9 M HBr solution on Dowex 50, and this condition is proposed for separation of molybdenum from tungsten (720). Tungsten appears first in the effluent.

Anion-exchange resins retain molybdenum to varying degree depending upon their basicity and upon the acid concentration of the molybdenum solution. Strong base quaternary amine resins of the type represented by Dowex 1 exhibit maximum retention for molybdenum(VI) in solutions approximately 4 M in hydrochloric acid (540). Molybdenum elution occurs in strongly basic solution (1114). Quaternary amine resins of the type represented by Dowex 2, however, exhibit a minimum retention for molybdenum(VI) in 1 M hydrochloric acid solution (125) and complete elution from 4 M sodium hydroxide solution (735). Retention of molybdenum and other ions from hydrochloric acid solution is also reported with moderately basic and weakly basic anion exchange resins (295,551,555) and from solutions containing dimethyl sulfoxide–methanol–hydrochloric acid (282). Generally, the extent to which molybdenum is retained decreases with decreasing resin basicity (886). A scheme for ion-exchange separation of the arsenic analytical group elements on Dowex 1 in the chloride form is available (790). Solvents of increasing pH are used and molybdenum elution, with 1 M ammonia–2 M ammonium nitrate, comes after most other elements have passed through the column.

Adsorption of molybdenum occurs on Dowex 1 resin from solutions containing concentrated hydrofluoric acid (249) and on various other anion resins in the presence of fluoride ion (240) due to formation of molybdenum oxyfluoride species. Other separations are reported in the presence of mixed hydrofluoric acid–hydrochloric acid solutions, for example, isolation of molybdenum as part of a separation scheme for 32 different ions using columns of Dowex 1 anion resin and Dowex 50 cation resin with varying concentrations of HF–HCl and other solvents (829). Separation of titanium, zirconium, tungsten, niobium, molybdenum, and tantalum collectively from iron and other elements is possible using 1 M hydrofluoric acid solution and the strong base anion resin De-Acidite FF (197,213,981). The ions named are retained on the resin, while iron passes into the effluent. For these six ions, separation is in the order listed with molybdenum elution occurring from a solution containing 3 M HF–3 M HCl. The other species are eluted with solutions containing different concentrations of these acids or, in the case of tantalum, with NH_3–NH_4Cl solution.

Retention of various ions on Dowex 1 resin in sulfuric acid medium is discussed (195). Molybdenum(VI) exhibits strong adsorption over a wide range of sulfuric acid concentrations. Addition of hydrogen peroxide decreases molybdenum retention for acid concentrations up to 1.5 M. A similar resin is used to separate molybdenum from any of the following mixtures: Y(III)–Th(IV)–U(VI), Th(IV)–Hf(IV)–Zr(IV), or Cr(III)–V(V)–W(VI). Separation depends upon sulfuric acid concentration, hydrogen peroxide concentration, and proper use of ammonia–ammonium nitrate solutions. All are employed as eluting solvents (972). Nitric acid medium is used to retain various species upon anion resins (250), for example, separation from uranium(VI) occurs with 8 M nitric acid eluent on a divided column containing both Dowex 1 and tributyl phosphate on Kel-F solid support (400). Uranium is retained on the column, while most other ions, including molybdenum, pass into the effluent. A solution containing 0.2 M phosphoric acid–0.6 M sodium sulfate passes rhenium from the anion resin EDE-10, while molybdenum and tungsten are retained (847).

Various separations involve molybdenum in the presence of a complexing agent to aid in recovery of separated species. Solutions containing potassium oxalate in place of sodium hydroxide separate molybdenum(VI) from rhenium(VII) on Amberlite IRA-400 resin (659). Molybdenum is separated from Zr, Nb, Ta, and W on Dowex 1 with solutions containing oxalic and citric acids (57). The presence of thiocyanate complexes and a Dowex 1 type resin permits separation of various metal species (104). The weak-base diethylaminoethyl (DEAE) cellulose anion-exchange material in the thiocyanate form retains Re(VII), Mo(VI), and W(VI) (413). Each of these is eluted separately and collected using 0.02 M NH$_4$SCN at pH 3 for rhenium, 0.1 M NH$_4$SCN at pH 5 for molybdenum, and 0.1 M NaOH–0.1 M NaCl for tungsten. Molybdenum is separated from uranium(VI) with solutions containing carbonate ion (482), from tungsten(VI) with solutions containing formic acid (897), and from a variety of transition metals with solutions containing tartrate ion (682). The chelating ion-exchange resin, Chelex 100, separates molybdenum(VI) and vanadium(V) from other metal ions in a solution adjusted to pH 5 with 0.2 M nitric acid (832); molybdenum and vanadium are then removed by passing 2 M ammonia through the resin column. Table 9 summarized various molybdenum separations using ion exchangers.

D.   OTHER CHROMATOGRAPHIC METHODS

Molybdenum(VI) is adsorbed into alumina columns from pH 4 to 6 in sulfuric acid medium (548); elution occurs with dilute ammonia. Alumina

TABLE 9
Ion-Exchange Separations of Molybdenum

Aqueous phase	Resin	Eluent	Separation from	Reference
H_2O	Amberlite IR-120 (H^+)	Citric acid, pH 2.5	V, U	762
0.3 M H^+	Bio-Rad AG-50W (H^+)	0.25 M H_2SO_4–1% H_2O_2	Ti	899,971
6 M HNO_3	Dowex 50 (H^+)	Tetrahydrofuran–6 M HNO_3, 9:1 vol./vol.	U	261
H_2SO_4–citric acid	KU-2 (H^+)	H_2SO_4–citric acid	Fe, W	476
0.05 M NaOH	Amberlite IR-400 (Cl^-)	2.5 M NaOH	Re	267,986
NH_3–NH_4OAc, pH 5.3	Dowex 1 (OAc^-)	NH_3–NH_4OAc, pH 5.3	W	405
1 M HCl	Dowex 1 (Cl^-)	1 M HCl	Re, Tc	401,509
4 M HCl	Dowex 1 (Cl^-)	1 M HCl	Co, Fe	662
MeOH–6 M HNO_3, 9:1 vol./vol.	Dowex 1 (NO_3^-)	MeOH–6 M HNO_3, 9:1 vol./vol.	Re	528
3 M HCl	Dowex 1 (SCN^-)	0.5 M NaOH–0.5 M NaCl	Sb, Sn	463
0.25 M HCl	EDP-10 (Cl^-)	0.25 M HCl	Fe, Al	962
0.1 M HCl	EDP-10P (Cl^-)	0.1 M HCl	V	898
0.5 M HF	EDP-10P (Cl^-)	1 M HCl	W	744
2 M HCl	TM (Cl^-)	2 M HCl	Pb	961

columns separate molybdenum and other components in fission product mixtures (1004), from iron and other alloying elements in the presence of citrate and tartrate complexing agents (480), and from rhenium (219). Columns containing activated carbon separate rhenium(VII) from molybdenum(VI) in dilute sulfuric acid solution (18); rhenium is retained while molybdenum passes through. Molybdenum is separated from ^{99}Tc (605) and other ions (562,824) on columns of hydrated iron(III) oxide. Molybdenum separation from germanium occurs with the sorbent SG (14).

Retention of molybdenum(VI) from acid solution by reverse-phase partition with methyl isobutyl ketone impregnated fluorocarbon polymer supports achieves separation from iron (283) and from Nb–Ta–W mixtures (279). With a tributyl phosphate stationary phase separation of tungsten, molybdenum, and rhenium is possible from solutions containing 0.2 M tartaric acid and varying concentrations of hydrochloric acid (687).

Many solvent systems are described for paper chromatographic separation of ions, including molybdenum; the subject has been thoroughly reviewed (592). Mixtures of butanol and hydrochloric acid are commonly used for separating molybdenum(VI) (709), for example, n-butanol saturated with a 1:1 (vol./vol.) mixture of 2 M HCl–2 M HNO$_3$ and containing 0.5% benzoylacetone separates molybdenum(VI) from other alloying elements in steels (781). Addition of hydrogen peroxide to butanol–hydrochloric acid solution achieves separation of titanium, vanadium, and molybdenum as their peroxo complexes (591). Mixed acetone–hydrochloric acid solvents separate molybdenum and other ions in ashed biological samples (243) and in oils (321). Trace metals in food stuffs are separated using methyl isobutyl ketone–12 M HCl–water (15:3:3 vol./vol.) (189). Several developing solvents containing chloroform, an alcohol, and hydrochloric acid successfully separate mixtures of various heavy metal ions, including molybdenum (84). Separation of molybdenum(VI) from various cations with ethanol–12 M HCl–water (4:1:5 vol./vol.) is discussed (23). Cr(VI)–Mo(VI)–W(VI)–V(V) mixtures are separated by paper chromatography on alumina-impregnated paper using a basic developing solvent at pH 9 (699). Molybdenum(VI) is separated from other metal ions on paper impregnated with tin(IV) tungstate using n-butanol–hydrochloric acid (7:3 vol./vol.) (803). Paper impregnated with a 0.5% solution of rutin in methanol separates iron, titanium, molybdenum, and tungsten using a methanol–1M hydrochloric acid (50:5 vol./vol.) solvent system (56,996). Paper treated with tributyl phosphate and developed with dilute nitric acid (398) or nitric acid–

ammonium thiocyanate solution (170) separates molybdenum from vanadium and tungsten. Separation of various molybdenum valence states is
achieved in the presence of thiocyanate (159) and oxinate (964) species.
Table 10 lists some paper-chromatographic separation conditions for
molybdenum.

Thin-layer chromatographic separations are described for Re–Mo–V–
W on alumina plates with various solvents (293) and on silica gel plates
for Re–Mo–W with methanol–3 M hydrochloric acid (7:3 vol./vol.)
(556,894). Separation of ^{99}Tc from ^{99}Mo occurs on silica gel plates using
acetone (637).

Table 11 lists spray reagents used to detect molybdenum after
chromatographic separation. Other color-identifying tests for molybdenum are listed in Section V.B and elsewhere (917).

Gas-chromatographic separation of volatile transition metal chlorides
(910) and fluorides (445,772) has been reported using Kel-F oil on
fluorocarbon polymer supports, and the procedure has been adapted for
quantitative determination of molybdenum in various Mo–W alloys
(446).

TABLE 10
Paper Chromatographic Separations of Molybdenum

Developing solvent	Separation from	Reference
AmOH–HCl–30% H_2O_2, 19:6:2 vol./vol.	V, W, Re, Fe	309
Aqueous saturated NaCl	Re	220
BuOH–anisole–6 M HCl, 4:1:1 vol./vol.	V, W	179
BuOH saturated with 3 M HCl	Sn, Fe, Ti	513
BuOH–12 M HCl, 1:1 vol./vol.	Cr, W, U	709
BuOH–12 M HCl–sym-collidine, 38:52:10 vol./vol.	Cr, V, W	1080
BuOH–30% H_2O_2–1 M HNO_3, 20:5:1 vol./vol.	V, W	178
12 M HCl–H_2O, 1:9 vol./vol.	Mo(VI), Mo(V), Fe	999
12 M HCl–H_2O–Et_2O–MeOH, 4:15:50:30 vol./vol.	Mo(VI), Mo(V)	779
MeOH–CCl_4–$CHCl_3$, 10:1:4 vol./vol.	Cr, W	978
16 M HNO_3–H_2O–dioxine–antipyrine, 1:2.5:100:1 vol./vol.	Fe, Pb, Mn, Co, Cu	229

TABLE 11

Spray Reagents for Detection of Molybdenum After Paper-Chromatographic Separation

Spray reagent	Observed color	Reference
0.5% Alizarine in ethanol, exposed solution to NH_3 vapors before spraying	Purple	513
0.1 gram Kojic acid + 0.5 gram oxine in 60 ml ethanol + 40 ml water	Brown	229
5% Pyrogallol in ethanol	Pale blue	1080
5% Aqueous $K_4Fe(CN)_6$	Brown	229
Sat'd aqueous KSCN + acetone (1:1 vol./ vol.), overspray with $SnCl_2$ in 12 M HCl	Red	229
0.025% rhodamine B + 10% KBr in 1 M HCl, overspray with 10% aqueous KI	Blue (fluorescence)	663
10% Aqueous tannic acid	Yellow	360
0.1% Toluene-3,4-dithiol in 0.25 M NaOH, overspray with 12 M HCl	Green	556

E. MISCELLANEOUS METHODS

Distillation of molybdenum chlorides and oxychlorides is possible. Volatilization occurs from a variety of molybdenum-containing salts in a stream of carbon tetrachloride–carbon dioxide at a dull red heat (427). Molybdenum chlorides are separated from tungsten chlorides by distillation (1112,1113). Volatilization of MoO_3 is not recommended for quantitative separation (1).

Paper electrophoresis has been used to separate various inorganic ions, including molybdenum. Separations are reported for molybdenum from rhenium in the presence of 0.3 M HCl at a potential of 13 V/cm (1099), from tungsten in the presence of 0.1 M lactic acid at a potential of 7 V/cm (98) and from technetium in the presence of 1.2 M NH_3 at a potential of 80 V/cm (54).

Separation of molybdenum and tungsten based upon the larger size of the tungsten complex with borate ion occurs during electrodialysis across an ion-exchange membrane (96) and with ion-sieve materials (97).

V. DETECTION

Molybdenum is found with arsenic as an acid-insoluble sulfide in the classical hydrogen sulfide scheme of qualitative chemical analysis (387,-788). Molybdenum(VI) sulfide, like arsenic sulfide, remains insoluble upon addition of hydrochloric acid and is separated from the sulfides of antimony and tin which dissolve. After dissolution of molybdenum and arsenic sulfides with ammonia and removal of arsenic, the presence of molybdenum is confirmed by a suitable color reaction. In the extended qualitative scheme of Noyes and Bray, molybdenum occurs in the tellurium group (25,665). The sulfides of this group are dissolved in aqua regia, tellurium is precipitated by passage of sulfur dioxide, the filtrate is evaporated to dryness, the residue is dissolved in hydrochloric acid, and molybdenum is detected by a suitable confirmatory test.

Various nonsulfide schemes of qualitative analysis also include molybdenum in their separation procedures, for example, in schemes based upon precipitation of ethyl xanthates (1067) and basic benzoates (616). In both procedures, however, molybdenum is separated subsequent to precipitation of other ions with the named reagents. A qualitative scheme based upon chloroform extraction of dithizone complexes at various pH values separates various ions, while Ti, W, V, and Mo remain in the aqueous phase (572). Another procedure, after removal of insoluble chlorides, calls for evaporation of sample to dryness followed by extraction of the residue with 1% nitric acid; tin, antimony, titanium, and molybdenum remain as insoluble materials (26). In another scheme, sodium carbonate is used to form insoluble precipitates with Mo, V, W and As; other ions are subsequently precipitated with other reagents (438). In all cases confirmatory tests are necessary to verify the presence of molybdenum after initial separation.

Several qualitative procedures utilizing paper and thin-layer chromatography include molybdenum; among these is one employing a 10% (wt./vol.) solution of benzoylacetone in butanol shaken with 0.1 M nitric acid and giving an R_f value of 0.25 for Mo(VI) (780). Chloroform–acetone–isopentanol–11 M hydrochloric acid (1:1:1:0.5 vol./vol.) is used in a circular chromatographic procedure to separate molybdenum and other ions; R_f values between 0.68 and 0.80 are found for Mo(VI) depending upon the other ions present (786). Both of these schemes use Whatman No. 1 paper. A procedure employing S & S No. 2043a paper and ethanol–12 M HCl–water (15:4:1 vol./vol.) results in an R_f of 0.7 for molybdenum(VI) (875). A thin-layer chromatographic separation on alumina gives an R_f value of 0.62 for molybdenum(VI) when acetone–4 M HCl–acetylacetone (45:3:2 vol./vol.) eluting solvent is used for samples separated from 7 M HCl solution by methyl isobutyl ketone–amyl acetate (2:1 vol./vol.) extraction (360). This same procedure is applied to a qualitative scheme based upon the ring oven technique (361). Other procedures for molybdenum using a ring oven (94) and a slotted oven (596) are reported. In all cases, suitable confirmatory tests are necessary.

A. PRECIPITATION REACTIONS

Several reagents form distinctly colored precipitates with molybdenum for confirmatory identification. A solution of tetramethyldiaminodiphenylmethane in dilute acetic acid yields a blue precipitate with molybdenum(VI) (747). The test will detect as little as 0.06 mg ammonium molybdate/ml. Tungsten(VI) yields a faint opalescence and vanadium(V) a green precipitate with this reagent. 1-Nitroso-2-naphthol in 95% alcohol forms a red precipitate with an acidified solution of Mo(VI) (891); under similar conditions, vanadium(V) yields a red-brown precipitate. The white precipitate of molybdenum(VI) in a 4 M HCl with 2% alcoholic α-benzoin oxime is used as a confirmatory test (173); W(VI) and V(V) also precipitate with this reagent. An acid solution of molybdenum(VI) forms a red-brown precipitate with alcoholic O,O'-dihydroxyazobenzene (164), zirconium(IV) being the only reported interference. The catalytic effect on the hydrogen peroxide reaction with thallium(I) to form insoluble brown thallium(III) hydroxide is used as a confirmatory test for molybdenum, tungsten, and vanadium (260).

B. COLOR REACTIONS

The red-colored species formed by molybdenum(VI) and thiocyanate ion in the presence of a reducing agent best characterizes molybdenum (259,541). Tin(II) chloride dissolved in 2.5 M HCl is added to an acidified Mo(VI) solution followed by addition of a 10% solution of ammonium or potassium thiocyanate. Generally, the red species formed is extracted into an organic solvent, for example, benzyl alcohol. Some caution is necessary to avoid prolonged contact with excessive amounts of tin(II) chloride (470), and a minimum acid concentration of 2 M is necessary for satisfactory results (808). The test is sensitive to 5 ppm of molybdenum. Further discussion of this color reaction is found in Section VI.C.1.a. A modification of this test, with increased sensitivity, utilizes the catalytic effect of molybdenum(VI) on the discolorization of the iron(III) thiocyanate complex (732) for detection of between 0.62 and 3 μg of Mo. Other species that catalyze this reaction interfere.

Dithiol, toluene-3,4-dithiol, yields a green precipitate with molybdenum(VI) soluble in a variety of organic solvents (182,183). The sample, containing Mo(VI), is made 2 M in hydrochloric acid, and organic solvent is added (n-butyl acetate or ethylene dichloride is satisfactory) followed by dithiol in 1 M sodium hydroxide. Alternately, solid zinc dithiol can be used. A green precipitate, soluble in the organic layer, indicates molybdenum. If ammonia is added to the green precipitate, the aqueous layer turns blue. Other ions also give position tests, especially rhenium(VII) and tungsten(VI) (666).

Acidified solutions of Mo(VI) form yellow peroxo species upon addition of hydrogen peroxide (82), and, if the solution is basic prior to addition of H_2O_2, a red color is observed (289).

A red species forms upon addition of saturated phenylhydrazine solution to a solution of molybdenum(VI) acidified with sulfuric acid (681,841). The compound formed can be extracted with amyl alcohol or cyclohexanol. The test is sensitive and subject to few interferences; lead, tin(II), and the cyanide complexes of iron do, however, interfere.

A 1% solution of pyrogallol produces coloration with solutions of molybdenum(VI) buffered at pH 4.4 (17). The color varies from yellow to orange-red depending upon the amount of molybdenum present. Optimum results are obtained for solutions containing between 0.005 and 0.1 mg Mo/ml. 2-Thiopyrogallol forms a brown-violet color with molybdenum(VI) in pH 5.2 acetate buffer (224). The same reagent yields a blue-violet colored molybdenum(VI) species in 0.1 M hydrochloric acid. Bismuth, silver, mercury(II), platinum(IV), and tungsten(VI), if present, produce other colors. The Pt, W, and Mo species can be extracted with isobutanol. Pyrocatechol forms an orange color with Mo(VI) in hydrochloric acid solution, while 4-nitropyrocatechol gives a yellow color (558). Titanium(IV), uranium(IV), and uranium(VI) also produce colored products. Other color forming reactions with phenols (874) and hydroxy azo compounds (369) are described.

Solid potassium ethyl xanthate added to a solution of molybdenum(VI), adjusted to pH 1.8–1.9 with hydrochloric acid, forms a red-violet color (523,757). Moderate amounts of tungsten(VI) do not interfere, and the test is sensitive to 0.04 μg of Mo. Increased sensitivity results if the colored species is allowed to form at the interface between aqueous and diethyl ether layers (525); after formation 2 M NaOH is added, the aqueous layer is withdrawn, filtered, and acidified with 2 M HCl producing a pink or red coloration that confirms the presence of molybdenum in the original sample. Additional increase in sensitivity is claimed if cetyl xanthate is substituted for ethyl xanthate in the conventional test (1008).

Various test procedures using filter paper strips detect molybdenum, for example, paper impregnated with zinc xanthate assumes a red color upon contact by a drop of molybdenum(VI) solution (184). One drop of o-hydroxyphenylfluorone placed on paper followed by the test solution, potassium fluoride, and dilute sulfuric acid yields a red color in the presence of molybdenum(VI) (311); cerium(IV), chromium(VI), and manganese(VII) interfere. Paper containing mercury(I) nitrate, ammonium thiocyanate, and dilute hydrochloric acid produces blue and red rings upon addition of one drop of Mo(VI) solution (15); the sensitivity is 0.4 mg/ml. Tungsten(VI) produces only a blue ring under these conditions. Filter paper itself, acting as a reducing agent, will, when in contact with a test solution for 1 hr at 90°C, produce a blue color along its edge if Mo(VI) is present (101). The test is sensitive to 50 μg of molybdenum; however, aluminum interferes.

Molybdenum(VI) solution, upon addition of an alcoholic solution of cochineal and ultraviolet excitation, produces a characteristic red fluorescence (325,998); tungsten(VI) and other ions give similar results.

Color-forming procedures for molybdenum using spray reagents after chromatographic separation are discussed in Section IV.D. Table 12 summarizes various color test reagents for molybdenum.

C. PHYSICAL METHODS

Numerous characteristic spectral lines serve to identify the presence of molybdenum, and compilations of atomic spectra for the elements, including molybdenum, are available (357,568,1109). From many observable lines the following are commonly used for molybdenum: 281.615, 317.035, 379.825, and 386.411 nm. A detection limit of 0.03 ppm of Mo is cited using the 379.825 nm line and flame emission with an air–acetylene flame (253); with nitrous oxide–acetylene flame and atomic absorption measurement the detection limit is 0.1 ppm of Mo for the same molybdenum line (767). Comprehensive listings of X-ray spectral lines are also available (8,66). A selection of these for molybdenum include:

	Å	KeV
$K\alpha_1$	0.709	17.479
$K\beta_1$	0.632	19.608
$L\alpha_1$	5.406	2.293
$L\beta_1$	5.177	2.395

Infrared detection of molybdenum is possible from observation of the infrared band at 11.0 μ with a shoulder at 10.6 μ for the nitron precipitate of molybdenum(VI) using a potassium bromide pellet (633). Peaks located at other positions are listed for tungsten(VI), chromium(VI), and other anion species precipitated by nitron.

Microscopic observation of characteristic crystal forms for molybdenum containing compounds is used for identification. Alkaline molybdenum(VI) solution forms, upon addition of thallium(I) nitrate, colorless or pale yellow hexagonal crystals of thallium molybdate (72). Red, slender

TABLE 12
Qualitative Identification of Molybdenum

Reagent	Condition	Color	Interference	Sensitivity (μg)	Reference
2,2'-Bipyridine	H^+, Sn(II)	Red-violet	W(VI)	0.4	519
Diethyldithiophosphonic acid	H^+	Red	Fe(III)	20	132
Diphenylcarbazide	H^+, SO_3^{2-}, Et_2O	Red	—	50	524
Dithiol	H^+, $C_2H_4Cl_2$	Green	Re(VII), W(VI)	20	182
Gossypol	H^+	Red	—	0.1	1048
Hydrazine sulfate	HOAc, boil	Blue	—		676
Methylene blue	$N_2H_4 \cdot H_2SO_4$, boil	Colorless	—	1	566
1,10-Phenanthroline	H^+, Sn(II), benzyl alcohol	Rose	W(VI), V(IV)	0.02	817
Phenylhydrazine	H^+	Red	Pb(II), Sn(II)	25	681
KSCN	H^+, Sn(II)	Red	Fe(III)	5	259
K Ethylxanthate	H^+	Red	Fe(III)	0.04	757
Xylenol orange	$NH_2OH \cdot HCl$, boil	Red-orange	EDTA	0.02	397

373

crystals form when molybdenum(VI) is reduced with acidic tin(II) chloride in the presence of potassium thiocyanate and quinoline (63), while yellow to yellow-green pleochromic crystals of molybdenum(VI) with 8-quinolinol are observed in 80% acetic acid (1037).

D.   DETECTION IN SPECIFIC MATERIALS

With regard to detection of molybdenum in metals, especially alloy steels, it is possible to perform a qualitative test directly upon the metal surface. A drop of suitable reagent is placed upon the clean, polished metallic surface, and after a few seconds the resulting solution is transferred to a spot plate or blotted onto filter paper. Bromine water (593), aqua regia (820), and nitric acid (713) are suitable sampling reagents. In an alternate procedure a hole is drilled directly into the metal surface, and this cavity is used to contain the test reagents (521). Once the sample is obtained any suitable test for molybdenum can be employed. Potassium ethyl xanthate (245) and gossypol (1049) are among the reagents specifically mentioned for detection of molybdenum in steels. Metal sample solution is also possible by an electrographic technique in which a sheet of filter paper moistened with potassium chloride or other suitable electrolyte is placed upon the clean sample surface (156,181,538). The sample itself serves as anode, and a graphite or copper cathode is placed in contact with the moistened paper. Brief passage of electric current at a potential of a few volts results in transfer of sample constituents to the filter paper. After this suitable confirmatory tests are applied.

Contact procedures are also used for detection of metals in ores and minerals, for example, a suitable reagent on cellophane- or gelatin-impregnated paper is placed in contact with the smooth polished sample surface (395). Pressure is applied, and after sufficient time the paper is removed and treated, if necessary, with additional color-forming reagents. Potassium hexacyanoferrate(II) is used to detect molybdenum by this method (1108). Grinding of sample with color forming reagent also serves for detection of metal ions. Sodium thiosulfate followed by potassium hydrogen sulfate (producing a red-brown color) (410) and ammonium thiosulfate followed by tin(II) chloride (producing a violet color) (606) are proposed for molybdenum. Dithiol, too, is claimed as a suitable reagent for detection of molybdenum in molybdenite ore after first leaching the ore with sodium hypochlorite (976).

Discussion of analytical procedures for molybdenum in a variety of other materials is found in Section VIII.

VI. DETERMINATION OF MOLYBDENUM

A. GRAVIMETRIC METHODS

Lead(II) salts, 8-quinolinol, and α-benzoin oxime are generally used to quantitatively precipitate molybdenum. These and other reagents are described in this section.

1. Insoluble Molybdates

a. ALKALINE EARTH MOLYBDATES

Alkaline earth metals form insoluble molybdate precipitates of decreasing solubility in the order barium, strontium, calcium (724,809). Quantitative precipitation of stoichiometric species is best from nearly neutral solutions. Barium molybdate is dried by ignition above 300°C and weighed (599); however, variations in ionic strength of the sample solution are reported to effect the composition of this precipitate (346). Precipitation of barium molybdate occurs in solutions containing 20% alcohol and a slight excess of ammonia (919). Precipitation from aqueous boiling solutions is reported for strontium and calcium molybdates (162,589).

b. HEAVY METAL MOLYBDATES

Lead molybdate is precipitated quantitatively from boiling molybdate solution containing acetic acid–acetate buffer upon dropwise addition of lead(II) solution (1062); an excess of lead should be avoided (594). The precipitate is filtered, washed with dilute ammonium acetate solution, and ignited at 600–650°C to $PbMoO_4$. Homogeneous precipitation of lead molybdate is described based upon the slow release of lead(II) ion from $PbEDTA^{2-}$ upon dropwise addition of chromium(III) ion (726). Silver molybdate, Ag_2MoO_4, is precipitated by addition of silver(I) ion to neutral molybdate solution (651) and after filtering dried at 250°C and weighed (462). Procedures for formation of insoluble molybdate precipitates with thallium(I) (1102), mercury(II), (1104) and cadmium (437) are available.

2. Sulfide Precipitates

Molybdenum sulfide, MoS_3, precipitates from slightly acid solutions containing Mo(VI) upon passage of hydrogen sulfide. Precautions are necessary to minimize reduction of molybdenum, formation of thiomolybdates, and formation of colloidal particles (967,1025). Precipitation in the presence of potassium triiodide (258) and with solutions containing formic acid (963) are suggested to improve quantitative results. Ammonium sulfide (1001) and sodium sulfide (1002) serve as alternate sources of sulfide ion, while homogeneous sulfide precipitation is possible using trithiocarbonic acid (292), ammonium thiosulfate–sodium hypophosphite (819), and thioacetamide (270). With thioacetamide optimum results for molybdenum are reported when precipitation proceeds from solutions 0.7 M in perchloric acid (130). Addition of tartaric acid prior to sulfide precipitation prevents tungsten interference (131,655). Molybdenum sulfide is converted to molybdenum oxide, MoO_3, for weighing. Ignition below 600°C is commonly suggested (115), but considerably lower temperatures have been proposed to further minimize loss by volatilization (758,1081). Treatment of the ignited molybdenum oxide with nitric acid is claimed to improve quantitative results (90).

3. Miscellaneous Inorganic Precipitates

Successful quantitative precipitation of $[Cr(NH_3)_5Cl][MoS_4]$ from ammonical solution is reported in the presence of appropriate masking agents that prevent interference from several species commonly associated with molybdenum (935,936). Ammonium phosphomolybdate precipitate is also utilized for molybdenum determination (498).

4. 8-Quinolinol

Molybdenum(VI) forms a precipitate with 8-quinolinol from solutions within the pH range 3.3–7.6 (324). Quantitative procedures generally employ acetic acid–acetate buffer and precipitation from hot solution. The presence of ammonium tartrate prevents coprecipitation of the corresponding tungsten species (937), while addition of EDTA eliminates precipitation of iron(III) and other ions. Titanium, vanadium, and tungsten, however, interfere even in the presence of EDTA (791). Molybdenum(VI) oxinate precipitate, $MoO_2(C_9H_6ON)_2$, is dried above 140°C (111).

5. α-Benzoin Oxime

A precipitate of molybdenum(VI) and α-benzoin oxime forms in solutions containing up to 5% sulfuric acid (507). Addition of bromine water is necessary during precipitation to prevent reduction of molybdenum(VI). The precipitate is dried to constant weight at 105°C as $MoO_2(C_{14}H_{12}O_2N)_2$ (382) or ignited above 500°C to MoO_3 (416). Homogeneous precipitation of molybdenum is possible with this reagent (199).

6. Miscellaneous Organic Precipitates

Applications of organic reagents for gravimetric determination of molybdenum have been reviewed (30). Table 13 lists various organic precipitating reagents for molybdenum.

B. TITRIMETRIC METHODS

From the numerous titration procedures described in this section oxidation of a reduced form of molybdenum with one of the common oxidizing titrants is generally used for molybdenum determination.

1. Acid-Base Titrations

Two inflection points are observed with a glass calomel electrode pair and sodium hydroxide titrant during the titration of molybdic acid prepared by passing sodium molybdate through a strong cation-resin ion-exchange column in the hydrogen form (690). Quantitative potentiometric titration of molybdenum(VI) in solutions initially adjusted to pH 6.7 is possible using hydrochloric acid titrant and an antimony–Ag/AgCl electrode pair (584), while visual titration of molybdenum solution, initially adjusted to pH 4.6, with standard citric acid titrant employs a mixed diphenylcarbazone–brilliant green indicator (547).

2. Precipitation Titrations

a. LEAD SALTS

Direct titration of Mo(VI) is achieved with a standard solution of lead(II) when either 1-(2-pyridylazo)-2-resorcinol, PAR, or 1-(2-pyridylazo)-2-naphthol, PAN, and a trace of copper salt is used as indicator. The reaction is performed upon molybdate solutions buffered at pH 6 and heated to 80°C with slow titrant addition (578,799). Besides anions that react with lead, the presence of acetate, tartrate, citrate, fluoride, and sulfate cause interference (579). Dithizone is also a suitable indicator for direct titration of Mo(VI); reaction occurs at pH 4.8–5.5 using an acetic acid–acetate buffer (1030). The applicability of various other indicators has been reviewed (370).

Addition of a known excess of lead salt to a molybdenum solution and titration of the excess lead with standard Mo(VI) titrant is described using diphenylcarbazone (206) and other indicators (158,557). Titration of the excess lead with EDTA is also proposed for the indirect determination of molybdenum (1033).

Molybdenum–lead titrations have been followed conductometrically (840) and amperometrically using both one (44,52) and two indicating electrodes (176).

TABLE 13
Organic Precipitating Agents for Molybdenum

Reagent	Condition	Weighing form	Interference	Reference
Acridine 0.5%, in 5% HOAc	Aqueous solution	Ignite to MoO_3	V, Cr, W	264
N-Benzoyl-N-phenylhydroxylamine 2.5%, in EtOH	1 M HCl, heat	110°C $MoO_2(C_{13}H_{10}O_2N)_2$	Cr, W	920
α-Benzoin oxime 0.05 M, in acetone–water 1:1 vol./vol.	5% H_2SO_4	105°C $MoO_2(C_{14}H_{12}O_2N)_2$	V, W	382
4-Amino-4'-chlorobiphenyl 0.36%, in alcohol	pH 1.8–2.8, boil	Ignite to MoO_3	W	601
Cupferron 6%, aqueous	10% HCl	Ignite to MoO_3	Fe	892
8-Quinolinol 3%, in acetate buffer	HOAc–NH$_4$OAc, boil	140°C $MoO_2(C_9H_6ON)_2$	Ti, V, W	791
β-Naphthoquinone 2% + tannin 2%, aqueous	pH 1–2	Ignite to MoO_3	W	893
1-Nitroso-2-naphthol 2%, in EtOH	pH 1–2.3	Ignite to MoO_3	W	753
Pyridoin 1%, in EtOH	pH 2–9.5	Ignite to MoO_3	W	64
Purpurogallin 1%, in EtOH	pH 3.5, boil	Ignite to MoO_3	Ti, Zr	221
Vanillylidine–benzidine 3%, in 50% HOAc	Aqueous solution, boil	Ignite to MoO_3	W	396

b. MISCELLANEOUS METAL SALTS

Molybdenum(VI) is titrated with standard silver nitrate solution using
xylenol orange indicator (175) or potentiometric end point detection with
a silver–calomel electrode pair (901). Conductometric (157) and amperometric (932) end point detection is also reported for this reaction. Other
metal titrants for molybdenum(VI) include barium (116), calcium (581),
mercury(I) (642), and thorium (866).

c. ORGANIC REAGENTS

Direct amperometric titration of molybdenum(VI) with 8-quinolinol is
possible from solutions buffered at pH 3.7–5.5 with acetic acid–acetate
buffer (645). Alternately the molybdenum precipitate can be filtered,
dissolved, and the released 8-quinolinol titrated (680,938). If a known
excess of 8-quinolinol is used to precipitate molybdenum an aliquot of
the excess from the filtrate can be titrated directly avoiding the need to
dissolve the molybdenum precipitate (419). Substituted 8-quinolinol
(712), 8-mercaptoquinoline (760,948), N-benzoyl-N-phenylhydroxylam-
ine (930), diantipyrylmethane (1120), and tetramethyldiaminodiphenylmethane (748) have been used for titration of molybdenum(VI), generally with amperometric end point detection.

3. Complexometric Titrations

a. INDIRECT TITRATION WITH EDTA

A known excess of standard EDTA solution is added to an acid
solution of molybdenum(VI), hydroxylamine hydrochloride (580,1086) or
hydrazine sulfate (576,1012) is added, and the mixture is boiled to reduce
molybdenum to Mo(V). After reduction the solution is adjusted to pH
4.5, diluted with ethanol, heated, and the excess EDTA titrated with
standard Cu(II) titrant using 1-(2-pyridylazo)-2-naphthol, PAN, indicator. Addition of tartrate masks Ti, Nb, Ta, and W, while addition of
NH_4F, after reduction, masks Th, Al, Ce, La, and U (577). Titration of
the excess EDTA with zinc(II)–chromogene black ET–00 indicator (136),
lead(II)–xylenol orange indicator (237), or iron(III)–sulfosalicylic acid
indicator (607) are described. Titration of the remaining EDTA has been
followed photometrically (363,506,784).

b. DIRECT TITRATION WITH EDTA

Direct titration of molybdenum(VI) solution, adjusted to pH 4.5 with
acetic acid–acetate buffer, is possible with EDTA using a mixed
indicator of pyrocatechol–indigo carmine (135). A diphenylcarbazone-

methylene blue–vanadium(V) indicator mixture also serves for direct molybdenum(VI)–EDTA titration (855). Amperometric detection of the molybdenum(VI)–EDTA end point using a rotating electrode is also possible (823,1119).

4. Oxidation-Reduction Titrations

a. PRELIMINARY REDUCTION OF MOLYBDENUM(VI)

Reduction of molybdenum(VI) is necessary before titration with standard oxidizing agents. The effects of various reducing agents and of various acid media and concentrations have been discussed (67,389,944). Formation of molybdenum(V) is readily achieved, but quantitative formation of molybdenum(III) requires exclusion of air. Shaking a Mo(VI) solution, previously adjusted to an acid concentration of 2–3.5 M HCl, with mercury for several minutes results in quantitative conversion to the characteristic red-brown molybdenum(V) species (290,952). The solution is then filtered to remove mercury(I) chloride and unreacted merucy. Quantitative conversion to molybdenum(V) also results upon passage of molybdenum(VI) solution, adjusted to 1.5 M HCl, through a reductor tube containing bismuth (29). Molybdenum(VI) solutions in 2 M HCl are reduced to the five state with a silver reductor at 70°C (338,379).

Reduction of molybdenum(VI) to molybdenum(III) and subsequent quantitative titration is accomplished under an inert atmosphere of carbon dioxide or nitrogen. The green molybdenum(III) species can also be determined indirectly by collecting it in a solution of standard oxidizing reagent, frequently iron(III) ammonium sulfate, and titrating the reduced species formed with standard oxidizing titrant. At low heat reduction of Mo(VI) to Mo(III) occurs in the presence of powdered aluminum from solutions containing 2.5 M H_2SO_4 (830). Reduction also occurs with cadmium amalgam in 6 M HCl (943), with cobalt in 0.9–1.4 M H_2SO_4 (782), with Na–Pb alloy in 8 M HCl at 95–100°C (227), with zinc in dilute sulfuric acid (163), and with zinc amalgam in 5% H_2SO_4 (662,1103).

b. PERMANGANATE

Molybdenum(III) formed by an appropriate reduction procedure is titrated directly with a standard permanganate solution (205,272,693). Frequently the reduced molybdenum is passed directly into a solution of iron(III) ammonium sulfate, and the resulting iron(II) salt is titrated with potassium permanganate (218,291). Phosphate is generally added to mask the color of iron(III) (988), while the addition of ammonium fluoride prevents tungsten from interfering (942).

c. CERIUM

Molybdenum(V) is titrated directly with a standard solution of cerium(IV) sulfate (952). Heating, carbon dioxide purge, and the addition of manganese(II) sulfate have all been suggested to improve the reaction (957,958). Addition of phosphoric acid is also recommended (1101). Vanadium does not interfere if the titrate solution contains 4–5 M H_2SO_4 (212). Ferroin, diphenylaminesulfonic acid, or N-phenylanthranilic acid are frequently used as indicators for this reaction (36,813), although other indicators have been suggested (811,814,851,864).

d. CHROMIUM

Potentiometric titration curves for the reaction of molybdenum(III) in acid solution with standard dichromate titrant exhibit two breaks indicating oxidation to Mo(V) followed by oxidation to Mo(VI) (835); diphenylamine (1061) or other indicators (949) are used for visual end point detection. Selective preliminary reduction allows dichromate titration to be used for determination of both molybdenum and iron in the same solution (366,698).

Potentiometric titration of molybdenum(VI) is possible with a standard chromium(II) solution provided the deaerated molybdenum solution is kept near its boiling point. Two breaks are observed in the titration curve indicating reduction to Mo(V) followed by reaction to Mo(III). Platinum–calomel (117) and bimetallic electrode pairs (144) follow the course of this titration, and addition of oxalic acid masks interference of tungsten (233,323). Prior titration at ambient temperature allows determination of iron and molybdenum in the same solution (926). Amperometric titration of molybdenum(V) with chromium(II) has been reported (297,729).

e. VANADIUM

Molybdenum(V) is titrated directly with a standard solution of ammonium vanadate (567,960); N-phenylanthranilic acid (994), copper(II) phthalocyaninetetrasulfonic acid (812), or chloropromazine hydrochloride (332) are used as indicators. Direct titration of molybdenum(VI) with vanadium(II) titrant under a nitrogen atmosphere is also discussed (342).

f. MISCELLANEOUS INORGANIC TITRANTS

Molybdenum(V) is titrated directly with a standard solution of iron(III) salt in heated, acidified, and deaerated solutions using methylene blue indicator (852), while direct potentiometric titration of molybdenum(VI) with standard iron(II) titrant is possible in solutions containing 12 *M* phosphoric acid (810). This later reaction serves as the basis for a coulometric titration procedure with internally generated iron(II) titrant (11). Reduced molybdenum species are titrated in the presence of strong base with a standard solution of potassium ferricyanide titrant and indigo carmine indicator; addition of pyrophosphate (484) or EDTA (362) prevents precipitate formation. Molybdenum(VI) in acid solution is titrated with a standard solution of titanium(III) titrant and monoindigo-sulfonic acid indicator for visual end point detection (511), although potentiometric and amperometric detection are more common (730,-1072). The reaction proceeds slowly (89). Coulometric generation of titanium(III) titrant is also possible (730,1098). Other inorganic titrants for molybdenum include tin(II) chloride (10,372) for reduction of molybdenum(VI), and lead(IV) acetate (76) for oxidation of molybdenum(III).

g. ORGANIC TITRANTS

Ascorbic acid titrant is discussed for reduction of molybdenum(VI) (918) with an amperometric titration procedure while potentiometric titration with chloramine T (1087) and visual titration with methylene blue (215) are used for oxidation of molybdenum(III).

C. OPTICAL METHODS

Colorimetric absorption procedures for molybdenum are widely used. The thiocyanate complex of molybdenum(V) in particular is employed for determination of molybdenum in a variety of materials. Toluene-3,4-dithiol also finds extensive use as a complexing ligand for colorimetric molybdenum determination. Various reduced molybdenum heteropoly species, although applied to determination of phosphorus and other elements, have not been extensively used for molybdenum. Many other color-forming agents for molybdenum have been tested. In general, prior separation and/or masking is necessary for satisfactory determination in the presence of other ions commonly associated with molybdenum.

Emission, X-ray, and atomic absorption spectroscopic procedures, too, are widely used for determining small amounts of molybdenum in various materials.

1. Colorimetric

Compilations of procedures are available for colorimetric determinations of the elements, including molybdenum (172,858,927).

a. THIOCYANATE

Molybdenum(V) forms an orange-red complex with thiocyanate ion in acid medium with an absorbance maximum near 470 nm depending upon solution conditions (47,1032). A second more intense peak occurs near 320 nm in the ultraviolet region of the spectrum (647). Complex formation is achieved by reduction of molybdenum in solutions containing Mo(VI) and potassium or ammonium thiocyanate. Tin(II) chloride generally serves as the reducing agent, although evidence indicates formation of molybdenum(III) and other reduced species in addition to molybdenum(V) (75,193). Addition of trace amounts of iron and copper salts enhances development of maximum color intensity (211,703). A 5% HCl solution is proposed as optimum for color development (402), although other acid conditions are satisfactory and in some cases preferred depending upon the choice of reducing agent. Ascorbic acid reduces molybdenum exclusively to the five state in 1.4 M H_2SO_4 solution (5,28,586). Alternate reducing agents, employed for reduction of molybdenum prior to absorbance measurement of the thiocyanate complex, include hydrazine (326), titanium(III) chloride (127), thiourea (649), and others (231,312,850). Photochemical reduction is an alternate means of generating molybdenum(V) (532,723), and direct reduction of molybdenum(VI) by thiocyanate ion itself occurs in the presence of copper(II), which catalyzes the reaction (68,825). The molybdenum(V)-thiocyanate complex is stabilized by the presence of acetone (600). Other reagents, forming stable ternary complexes with molybdenum(V)-thiocyanate include 2,2'-bipyridine (83), butyl cellosolve (458), diantipyrylmethane (1011), 3,4-dichlorobenzyltriphenylphosphonium chloride (717), ferroin (1010), nitron (46), tricapyrylmethylammonium chloride (1077), and tetraphenylarsonium chloride (514). The ternary complexes are generally extracted into water-immiscible solvents prior to absorbance measurement. Extraction of the molybdenum(V)-thiocyanate complex itself is preferred after its formation in aqueous solution. Extracting solvents include butanol (326), butyl acetate (426), cyclohexanol (403), methyl isobutyl ketone (378), isopropyl ether (763), and tributyl phosphate (842). Reduction of molybdenum(VI) and formation of the thiocyanate complex can occur after molybdenum extraction with thenoyltrifluoroacetone in amyl acetate (146).

Effects of various diverse ions upon the accuracy of the molybdenum(V)-thiocyanate absorbance measurement have been studied (126,-734), and specific procedures are available for eliminating the interference of aluminum (776), iron (869), rhenium (384), tungsten (49,789), and nitrate ion (931).

Procedures for approximate determination of molybdenum as its thiocyanate complex are discussed using visual comparison of the molybdenum(V)-thiocyanate color with permanent color standards prepared from potassium chromate and cobalt(II) nitrate solutions (701,-792).

b. Diethyldithiocarbamate

Diethyldithiocarbamate complexes with various metal species (563) including molybdenum(VI) (550). The yellow molybdenum complex, formed in the presence of hydrochloric acid, is extracted by chloroform (980). Complexation also occurs with molybdenum(V) (860). Procedures are available for determining molybdenum using diethyldithiocarbamate in the presence of uranium (344) and in the presence of tungsten using 5-phenylpyrazoline-1-dithiocarbamate (134). 3-Oxapentamethylenedithiocarbamate (603) and pyrrolidinedithiocarbamate (455) also complex with molybdenum.

c. Heteropoly Methods

Solutions containing 10 M acetic acid, molybdenum(VI), and KH_2PO_4 form, upon addition of phenylthiosemicarbazide and heating, a reduced molybdenum blue species. After adjustment to 1 M acid with H_2SO_4, the molybdenum blue is extracted into amyl alcohol–diethyl ether (1:1 vol./vol.), and absorbance measurements are made in the near-infrared region of the spectrum (520). Ascorbic acid reduction of various tungstomolybdophosphate heteropoly species produce reduced forms suitable for determination of molybdenum (826,827).

d. Peroxomolybdate

A yellow peroxomolybdate complex forms from reaction of molybdenum(VI) in perchloric acid medium upon addition of hydrogen peroxide (1016). Hydrogen peroxide has been used for simultaneous determination of titanium, vanadium, and molybdenum, all of which form peroxo complexes (1064). Extraction of the peroxomolybdate complex is possible with tributyl phosphate (1040).

e. Phenylhydrazine

A stable red complex forms upon reaction of molybdenum(VI) and phenylhydrazine in solutions containing 50% acetic acid (45). The procedure is applied to determining molybdenum in steels following reduction of iron (113).

f. Polyhydric Compounds

Various polyhydric phenols form colored complexes with molybdenum(VI) (120,997). A stable orange pyrocatechol complex forms in either 0.4% NaOH solution in the presence of sodium metabisulfite (348,878) or in acetate buffer (755). The tiron, disodium-1,2-dihydroxybenzene-3,5-disulfonate, complex of molybdenum is stable in acetate buffer (859), and prior separation of molybdenum by α-benzoin oxime extraction removes many interfering ions (1070). Extraction of ternary molybdenum-pyrocatechol-amine complexes isolates molybdenum from the effects of interfering ions (149,1006). Pyrogallol red complexes with molybdenum have been discussed (392,1005).

Quercetin, 3,3',4',5,7-pentahydroxyflavone (319), and dihydroquercetin (167) form yellow complexes with molybdenum after its extraction from aqueous solution with α-benzoin oxime in chloroform. The procedures are used to determine molybdenum in steels. The morin, 2',3,4',5,7-pentahydroxyflavone, complex of molybdenum, formed in aqueous 0.1 M HCl solution, is extracted into butanol prior to measurement of its absorbance (24).

Applications of trihydroxyfluorone derivatives for the colorimetric determination of molybdenum are reviewed (908).

g. 8-QUINOLINOL

8-Quinolinol forms a complex with molybdenum(VI) at pH 3–3.5, which is extracted into chloroform prior to absorbance measurement (226,689). Substituted oximes are also used (122,228,816) as is complex formation with molybdenum(V) (137). The yellow 8-mercaptoquinoline complex of molybdenum(VI) is extracted into toluene before proceeding with absorbance measurement (58,634); moderate amounts of tungsten do not interfere (12).

h. THIO SUBSTITUTED ACIDS

Mercaptoacetic (138), mercaptosuccinic (139), 2,3-dimercaptopro-prionic, (140) and dithiovanillic (846) acids form yellow or yellow-green complexes with molybdenum(VI). The mercaptoacetic and mercaptosuccinic acid complexes are best formed at pH 3.5 in acetate buffer, the 2,3-dimercaptoproprionic acid complex in 0.2 M HCl and the dithiovanillic acid complex in 0.10–0.15 M HCl. Reduction of molybdenum(VI) to molybdenum(V) can occur with thio acids at other hydrogen ion concentrations and formation of the corresponding molybdenum(V) complexes. The presence of potassium chlorate stabilizes the color of the molybdenum-mercaptoacetic acid complex (1071), and addition of suitable masking agents prevents interference from several other metal ions (77,522,743). Extraction of a ternary molybdenum complex containing diphenylguanidine and mercaptosuccinic acid is discussed (148). Thio acid complexation is widely used for determination of molybdenum is steels.

i. TOLUENE-3,4-DITHIOL

Toluene-3,4-dithiol forms a green complex with molybdenum(VI) in strong acid solution (310). The insoluble complex is extracted into a variety of immiscible solvents, for example, amyl acetate, before absorbance measurement (13). Both molybdenum and tungsten react with this reagent, the molybdenum complex being preferentially extracted from acid solution (22,452,722). Dithiol has been used for determination of molybdenum in various materials. Some precaution is necessary, however, in the preparation and storage of the reagent as it is easily air oxidized (335).

j. MISCELLANEOUS COLOR FORMING REAGENTS

Numerous additional color forming reagents are available for determination of molybdenum; for example, applications of azo dyes (526) and pyridine derivatives (359) are described. The monographs cited at the beginning of this chapter and current reviews (110) should be consulted for further details. Table 14 summarizes conditions for various colorimetric molybdenum procedures.

2. Infrared

Variations in intensity of certain infrared bands of 8-quinolinol complexes of molybdenum, vanadium, and tungsten in the 10–12 μ region are related to the concentration of each metal ion (635). The metal is precipitated with 8-quinolinol, and the precipitate is collected, dried, weighed, and pressed into a KBr pellet. Increasing error results with this procedure if more than one of the metals is present in the 8-quinolinol precipitate.

3. Fluorimetric

Relatively few fluorimetric procedures appear in the chemical literature for determination of molybdenum. Addition of carminic acid to molybdenum(VI) solution buffered at pH 5.2 yields maximum fluorescent emission at 590 nm after excitation at 560 nm and is used to determine molybdenum in mild steels (491). Fluorescent measurement of 8,8-diphenylboroxyquinoline, formed upon addition of sodium tetraphenylborate to 8-quinolinol displaced from the molybdenum-oxine complex, serves as an indirect measure of molybdenum (1021).

4. Polarimetric

Linear variations in the optical rotation of D-tartaric acid (343) and L-malic acid (422) occur upon addition of increasing amounts of molybdenum(VI). The procedure has found application in determining molybdenum in steel after its separation from the other alloying elements present (759).

5.   Emission Spectroscopy

a.   ARC-SPARK EMISSION

Emission spectroscopy is widely used for qualitative and quantitative analysis of materials containing varying amounts of molybdenum in the presence of other chemical species. Sensitivity in the parts per million range allows for accurate determination of molybdenum in, for example, alloys, ores, soils, and biological samples. Various analytical lines are selected for monitoring the amount of molybdenum present, although overlapping iron lines limit the choice to some extent (653,991). The 317.035 nm line is frequently used for molybdenum determination, although other lines in the ultraviolet and visible spectrum are also suitable. The intensity of molybdenum spectral lines is effected by matrix composition, and studies have been made upon molybdenum spectra in the presence of various cationic (129,675,928) and anionic (2,537) species. Excitation in the presence of halide ions, for example, results in increased sensitivity for molybdenum because of the increased volatility of molybdenum halide (420,785,794). Numerous internal standards have been suggested for determination of molybdenum. The following line pairs are representative examples: Mo 317.04/Fe 317.14 (871), Mo 281.62/Fe 282.86 (60), and Mo 281.62/Co 269.47 (1056). A compilation of specific procedures for determination of molybdenum by emission spectroscopy in various materials is available (235).

b.   FLAME EMISSION

Special conditions are necessary for flame excitation of molybdenum to prevent formation of refractory oxides and to achieve adequate sensitivity from samples containing only minute amounts of molybdenum. Determination at the parts per million level is possible in oxyhydrogen flames with molybdenum solutions containing 99% ethanol (925), in fuel-rich oxy–acetylene flames with solutions containing 75% ethanol (254), and in fuel-rich nitrous oxide–acetylene flames (490). Separated air–acetylene (381) and nitrous oxide–acetylene (489,1068) flames are recommended for excitation of molybdenum. The emission lines at 379.83, 386.41, and 390.30 nm are frequently cited for monitoring the molybdenum content of sample solutions.

6.   Atomic-Absorption Spectroscopy

Many auxiliary ions, if present with molybdenum, effect the magnitude of absorbance values observed in determining molybdenum by

TABLE 14

Color-Forming Agents for Photometric Molybdenum Determination

Reagent	Condition	Wavelength (nm)	Interference	Reference
2-Amino-4-chlorobenzenethiol hydrochloride 4%, in EtOH	pH 2, extract with CHCl	720	Ti, Zr	492
0.0006 M catechol violet + 0.004 M cetyltrimethylammonium bromide, aqueous	pH 3.85	675	W, Sb, EDTA	53
5,7-Dibromo-8-hydroxyquinoline 0.04%, in EtOH	pH 1.09, 80% EtOH	386	Ti, Fe, V	228
Dihydroquercetin 0.3%, in EtOH	pH 5	410	W	167
0.001 M ferron, aqueous	0.2 M HCl, extract with cyclohexanol	400	Cu, V, W	816
NH$_2$OH 0.1% + 0.001 M PAR, aqueous	pH 6, boil	530	Fe	574
Hydrogen peroxide 3%, aqueous	2 M HClO$_4$	330	Fe, V, F$^-$	1016
0.001 M magneson, in EtOH	9 M HCl, extract with AmOH, add reagent	560	Fe, V, Cr	223
Maleonitrile dithiolate 0.1%, aqueous	3 M HCl–acetonitrile 1:1 vol./vol.	665	V, W	165
Mercaptoacetic acid 0.4%, aqueous	pH 4, 0.4% KClO$_3$	365	Nb, Ta, Zr	743

Reagent	Conditions			
9-Methyl-2,3,7-trihydroxy-6-fluorone 0.025%, in EtOH + 3 M H$_2$SO$_4$, 20:1 vol./vol.	pH 1.5–2.3, 0.02% gelatin	510	Ti, Zr, Cr, W	636
N-1,1'-(3,3'-Disulfonato-4,4'-biphenylene)-bis-(3-N-hydroxy-3-N-phenyltriazene) 0.5%, aqueous	pH 2.5–3.5	420	Pd, Fe, Cu	542
Phenylfluorone 0.015%, in MeOH	pH 2, 0.10% gum arabic 50% acetic acid, boil	550	Ge, V, W	93
Phenylhydrazine hydrochloride, solid		505	V, W	45
KSCN 25% + SnCl$_2$ 25%, in 12 M HCl	HCl, Fe(II), Cu(II), extract with BuCOMe	500	Ti, Re, Pt	378
Pyrazine-2,3-dicarboxylic acid 2%, aqueous	HCl, Sn(II), heat	550	Fe, NO$_3^-$	358
Pyrogallol red 0.005% + dodecyltrimethylammonium bromide 3%, aqueous	pH 5.0–5.3	587	V, W, Sn	1005
Pyrrolidinedithiocarbamate 2%, aqueous	pH 4.5, extract with CHCl$_3$	388	Fe, Co, Bi	455
Rezarson 0.02%, in EtOH	0.5 M HCl–EtOH, 1:1 vol./vol.	510	Cu, Ti, Cr	620
Sulfochlorophenol S 0.05%, aqueous	0.12 M H$_2$SO$_4$–EtOH, 4:1 vol./vol.	656	Cu, Nb, F$^-$	230
Toluene-3,4-dithiol 0.5% + thioglycolic acid 1%, in 0.5 M NaOH	4 M HCl, HF, NH$_2$OH, extract with CCl$_4$	680	Cu, W	380

atomic-absorption spectroscopy (800,806). Aspiration of molybdenum samples containing 2% ammonium chloride plus 1000 μg/ml of aluminum ion removes the interference of several species when using an air–acetylene flame (200). In addition, the presence of aluminum ion enhances the observed absorbance reading for molybdenum over that expected from samples containing only molybdate ions. This enhancement is more pronounced in a nitrous oxide–acetylene flame and is also observed with certain other ions present in the molybdenum solution. Solutions containing 5% HCl, 2% NH_4Cl, and 3000 μg/ml of Na_2SO_4 are claimed to give optimum results for molybdenum in a nitrous oxide–acetylene flame (474). Use of a flameless atomic-absorption burner, consisting of a high-temperature graphite cell, and no suppressing or enhancing agents is satisfactory for molybdenum determination (100); only titanium and niobium, of several elements generally alloyed with molybdenum, interfere. Amyl methyl ketone extraction of the 8-quinolinol complex of molybdenum from aqueous solution at pH 1 concentrates trace amounts of molybdenum while simultaneously removing some interfering ions (154); the extract is aspirated directly into an air–acetylene flame.

The molybdenum line at 313.26 nm is frequently chosen for determination of molybdenum by atomic-absorption spectroscopy. The less sensitive line at 379.83 nm is also used. A compilation of detection limits for various elements reports for the 313.26 nm molybdenum line, experimental limits of 0.2 and 0.03 ppm of Mo using air–acetylene and nitrous oxide–acetylene flames, respectively (751).

7. X-Ray Spectroscopy

X-Ray fluorescence spectroscopy generally utilizes the intensities of the molybdenum K lines ($K\alpha_1$ at 0.709 Å and $K\beta_1$ at 0.632 Å) for analysis of molybdenum-containing materials. Measurements are effected by matrix composition, and various means have been devised to correct for this effect in ores (478), steels (679,983), and other materials (468,612). In addition to measurement of solid samples, borax melts (618) and aqueous solutions (623,752,770) are used for molybdenum determination. Various internal standards are employed including niobium for determination of molybdenum in the presence of zirconium (349); titanium (1039), copper (575), and tungsten (247) for determination of molybdenum in the presence of tungsten; and measurement of constant background intensity for determination of molybdenum in the presence of various alloying elements (350). Applications of nondispersive X-ray spectrometry for determination of molybdenum have been reviewed (569,1089).

X-Ray absorption spectroscopy, utilizing the absorption-edge technique, is available for molybdenum determination (61,79,477).

D. ELECTRICAL METHODS

Reduction of molybdenum(VI) at the dropping mercury electrode generally yields well-defined polarographic waves depending upon solution conditions. Diffusion currents are proportional to molybdenum concentration and allow determination of molybdenum in various matrixes.

1. Direct-Current Polarography

Comparative investigations have been made of the effects of various acids upon the polarographic reduction of molybdenum (232,390). Three polarographic waves are observed for molybdenum in a supporting electrolyte containing 0.1 M H_2SO_4–0.2 M Na_2SO_4 (440) with $E_{1/2}$ values occurring at +0.06, −0.29, and −0.60 V versus SCE. Diffusion current measurements at the first plateau, indicating reduction to molybdenum(V), are proportional to molybdenum concentration in the millimolar range, and the procedure is used to determine molybdenum in the presence of vanadium (340). Up to four polarographic waves result for molybdenum depending upon the particular complex present at different sulfuric acid concentrations. Molybdenum waves are measured in solutions containing 1 M H_2SO_4 (515), 4 M H_2SO_4 (598), and 9 M H_2SO_4 (307). A supporting electrolyte of 0.6 M H_2SO_4–1.0 M $HClO_4$ serves for determining molybdenum in steels (553), and solutions containing 0.5 M H_2SO_4–0.5 M HF are used for determining molybdenum in niobium-based alloys (364). Two or three polarographic waves occur with molybdenum in phosphoric acid medium depending upon the phosphoric acid concentration. In 6 M H_3PO_4 $E_{1/2}$ values of +0.17 and −0.36 V versus SCE are used for determining molybdenum in steels (465). Molybdenum is determined in the presence of uranium following reduction to Mo(V) with hydroxylamine hydrochloride in 3–4 M HCl solution (207). The solution is then adjusted to pH 1 and the polarogram recorded. With this procedure an $E_{1/2}$ value for molybdenum occurs at −0.54 V versus SCE. Millimolar amounts of Mo(VI) are determined in the presence of trace amounts of tungsten from solutions containing 4–6 M HCl–0.54 M NH_4F (941).

Molybdenum exhibits two well-defined polarographic waves in citric acid medium adjusted to pH 2. $E_{1/2}$ values occur at −0.22 and −0.65 V versus SCE (333). Prior precipitation removes iron as insoluble hydroxide, and other metals that complex with citrate are, generally, reduced at more negative potentials. The procedure is applied to molybdenum in steels. Molybdenum can be determined in the presence of tungsten from solutions 1 M in sodium citrate and buffered at pH 7 (656). Polarographic determination of molybdenum has been made in solutions containing citric acid–disodium hydrogen phosphate (107) and citric acid–sulfuric acid (365,1082). $E_{1/2}$ values of -0.22 and -0.52 V versus SCE are observed for molybdenum in 0.1 M tartaric acid solution at pH 2.0. Purging with sulfur dioxide prior to diffusion current measurement on the plateau of the second wave removes the interfering effects of Fe(III) and Cr(III) (750). Diffusion-current measurements of molybdenum at potentials beyond the $E_{1/2}$ potential of -1.27 V versus SCE, when present in 0.25 M ammonium tartrate, are used for determining molybdenum in ores and alloys (391).

Polarographic reduction of the molybdenum(V)-thiocyanate complex extracted into diethyl ether occurs with an $E_{1/2}$ value of 0.10 V versus Hg-pool electrode (9). Lithium chloride dissolved in methanol is added as supporting electrolyte, and the procedure is applied for determination of molybdenum in steels. Acetylacetone extracts of molybdenum from dilute sulfuric acid also yield well-defined polarographic waves (746).

In addition to diffusion-controlled polarographic waves molybdenum also exhibits kinetic polarographic waves in acid media. Kinetic waves are characterized by their independent relation to mercury column height and by a large temperature coefficient. Reduced molybdenum species from polarographic reduction are reoxidized by ions in the solution. The overall effect is the reduction of the auxiliary ion and an enhancement of the current observed from the polarographic wave of the ion of interest. Polarographic reduction of molybdenum(V) in the presence of perchlorate ion exhibits this effect, and chloride ions are formed in solution (821,922). Quantitative molybdenum determination is carried out in 1.0 M $HClO_4$–0.75 M H_2SO_4 (347) or 1.0 M $NaClO_4$–0.75 M H_2SO_4 (921). A catalytic molybdenum wave is observed in 0.1 M H_2SO_4–0.2 M Na_2SO_4–0.5 M $NaNO_3$ due to the effect of nitrate ion (440). Even greater sensitivity is claimed for molybdenum if catalytic current measurements in 1.0 M H_2SO_4–0.5 M KNO_3 are made at 45°C rather than 25°C (86). Chloric acid (516) and sulfuric acid–hydrogen peroxide solutions (517) are also employed as media for catalytic polarographic determination of molybdenum.

Table 15 summarizes various polarographic procedures for molybdenum.

2.   Other Polarographic Techniques

Molybdenum undergoes four reduction steps in a supporting electrolyte containing 0.5 M nitrilotriacetic acid at pH 3.0, and from these the

TABLE 15
Complexing Agents for Polarographic Molybdenum Determination

Electrolyte solution	$E_{1/2}$ vs SCE (V)	Interference	Reference
0.016 M 8-hydroxyquinoline-5-sulfonic acid + HOAc pH 4.0	−0.63	V, W	80
0.5 M $H_3C_6H_5O_7$ + 0.025 M H_2SO_4 pH 3.5 + 0.05 M $Th(NO_3)_4$	−0.27 −0.64	Fe, Cu	365,1082
0.05 M malic acid	−0.23 −0.51	W	341
0.1 M mandelic acid + thymol 0.005% + camphor 0.1%	−0.26 −0.63	Ti, V, U	671
0.1 M tartaric acid pH 2.0	−0.22 −0.52	V, NO_3^-	750

$E_{1/2}$ value at −0.48 V versus SCE is used to monitor molybdenum concentrations in thorium-uranium oxides by derivative polarography (644). Molybdenum is determined in the presence of tungsten and other ions by alternating current polarography in solutions containing 0.5 M citric acid–0.25% hydrochloric acid (1038). Molybdenum in zirconium alloys is determined by square-wave polarography (1083), and oscillographic polarography is used for determination of uranium, molybdenum, and vanadium in solutions containing ferron and acetate buffer (81). Determination of molybdenum is described using anodic stripping (308).

3. Coulometric

Quantitative controlled-potential coulometric reduction of molybdenum occurs at −0.25 V versus SCE in an electrolyte solution of 0.2 M $(NH_4)_2C_2O_4$–1.3 M H_2SO_4 at pH 2.1. The procedure is adapted to determining molybdenum in alloys containing tungsten and rhenium (831). A 0.3 M HCl–acetate buffer at pH 1.5 is used for coulometric reduction of molybdenum in the presence of chromium(III) (404).

References to coulometric titration of molybdenum with electrochemically generated lead(II), iron(II), tin(II), and titanium(III) are cited in Section VI.B.

E. RADIOCHEMICAL METHODS

Direct monitoring of beta or gamma emissions from ^{99}Mo, $t_{1/2} = 67$ hr, is used for quantitative molybdenum determination. Molybdenum in fission products is isolated prior to counting by extraction with α-benzoin oxime in ethyl acetate from acidified aqueous solutions (630).

Ammonium molybdenum is added to the sample as carrier, and the presence of potassium permanganate assures Mo(VI). Potassium iodide and iron(III) are present as holdback carrier and scavenger, respectively, and addition of ammonia removes manganese(II) and iron(III) prior to acidification and extraction. After extraction molybdenum is stripped from the organic phase with ammonia. The ammonical solution is acidified, and lead(II) is added to precipitate lead molybdate, which is collected and counted. Other methods for separation of molybdenum prior to counting include precipitation with α-benzoin oxime (867) or nitron (802).

Addition of ^{99}Mo to molybdenum-containing sample solutions with utilization of an isotopic-dilution technique (303) and indirect measurement, through formation of an insoluble Ag_2MoO_4 precipitate containing ^{110}Ag (331), are both used for determining molybdenum.

A widely used alternate to direct monitoring of the ^{99}Mo X-ray peak at 0.78 MeV is measurement of the 0.14 MeV X-ray peak of ^{99m}Tc daughter as it decays to ^{99}Tc, $t_{1/2}$ = 6 hr (615). Various procedures isolate technetium from molybdenum (535,868), including ion exchange upon Dowex-1 in HCl (401), solvent extraction with tetraphenylarsonium chloride in $CHCl_3$ (608), and a ring-oven procedure (504), Rhenium is frequently used as a carrier for technetium in its separation from molybdenum.

Excitation of molybdenum-containing samples by neutron activation results in conversion of ^{98}Mo, the most abundant isotope (natural abundance, 24%), to ^{99}Mo. After sufficient time, generally 24 hr (55), equilibrium is established between ^{99}Mo/^{99m}Tc, and the technetium is separated and counted. The ^{101}Mo isotope formed from ^{100}Mo (natural abundance 9.6%) and its daughter ^{101}Tc is sometimes used in molybdenum determination following activation (441). Other isotopes have also been measured, but their decay is rapid. Both fast and slow neutrons are used as radiation sources, and irradiation times have varied from a few minutes to several days. After activation chemical separations are frequently necessary to remove interfering effects from other elements present in the sample. Detection of microgram and nanogram quantities of molybdenum is possible depending upon the energy and duration of the neutron flux. Molybdenum has been determined by neutron activation in the presence of aluminum (716), germanium dioxide (55,955), selenium (885), uranium (533,959), vanadium pentoxide (6), and tungsten (1019).

F.   METHODS BASED ON CATALYTIC ACTIVITY

Several metals, including molybdenum, catalyze the oxidation of iodide ion to iodine with hydrogen peroxide in acid solution, and it is possible to determine the concentration of metal species present, in the parts per billion range, by following the rate of iodine formation (301,1096). The slope of the curve obtained from monitoring the iodine concentration as a function of time, after a fixed time interval and for different known concentrations of molybdenum, is plotted as a function of molybdenum concentration. Iodine formation is followed by absorbance measurement upon the iodine-starch complex (32) or amperometrically (456,881), and both procedures have been automated (345,1065). Use of specially purified starch and 35°C temperature is claimed to yield increased sensitivity for molybdenum from this reaction (51). Tungsten interference is removed by addition of oxalate or citrate ion (124).

The effect of molybdenum upon the hydrogen peroxide-iodine reaction has been followed using a Landolt-type reaction (989,990). Here a small amount of ascorbic acid is added to consume the iodine formed from oxidation of iodide. After all ascorbic acid has reacted the color of the iodine-starch complex suddenly appears, and the time interval from initial mixing to color appearance is proportional to the molybdenum concentration. Plots of reciprocal time as a function of molybdenum concentration using standard solutions serve as calibration curve for this procedure. Addition of variamine blue indicator enhances color formation, and EDTA masks iron(III).

The oxidation of rubeanic acid with hydrogen peroxide in 0.1 M HCl (745) and of iodide ion with sodium perborate in acetate buffer (1020,-1075) are also catalyzed by molybdenum, and both reactions are used for molybdenum determination. Other reactions employed for molybdenum based upon its catalytic action include oxidation of 1-naphthylamine by potassium bromate (1097), reduction of malachite green by titanium(III) (741), reduction of methylene blue by hydrazine at elevated temperature (102), and reduction of selenate ion with tin(II) (585). In the later reaction free selenium formed from reduction is monitored turbidmetrically.

G. BIOLOGICAL METHODS

Growth of the fungus *Aspergillus niger* in a solution containing fixed amounts of all but one of its essential micronutrients is proportional to the amount of the varied constituent (728). By following this growth molybdenum, among the essential micronutrients Fe, Cu, Zn, Mn, and Mo, has been determined quantitatively (375,692). Growth is monitored after a fixed incubation period, the duration of which has been varied from 2.5 days at 35°C (738) to 6 days at 25°C (727), and the dry weight of fungus material resulting from this growth is measured. The procedure is sensitive for molybdenum in the parts per billion range and compares with chemical procedures suitable for this level of molybdenum. Optimum acidity for molybdenum is pH 2, and special precautions are necessary to avoid contamination from trace amounts of molybdenum contained within the metal salts used in the nutrient solution. The procedure has found use in determination of molybdenum in soils, plants, fertilizers, and related materials. Molybdenum determination based upon growth of other substances is also described (590,839).

VII.   ANALYSIS OF MOLYBDENUM, ITS ALLOYS AND COMPOUNDS

A.   HIGH-PURITY MATERIALS

1.   Molybdenum Metal

Metallic and nonmetallic trace constituents in high-purity molybdenum metal are determined by a variety of analytical procedures including emission (685), atomic-absorption (697), and X-ray spectrometry (939); neutron activation (255); and spark-source mass spectrometry (1044). Determination of gaseous impurities in high-purity molybdenum has been reviewed (453,640,768). Table 16 lists procedures for determining specific components in molybdenum metal.

2.   Molybdenum Compounds

Fifteen trace impurities in molybdenum trioxide, MoO_3, are determined in amounts less than 0.1% by spectroscopic measurement using graphite and with NaCl carrier to surpress molybdenum lines (677). Addition of radioactive tracers aids in ascertaining impurities in molybdenum(V) chloride (742). Impurities in high purity ammonium molybdate are removed by scavenging with tin(IV) hydroxide (749), while photometric procedures are described for determining traces of iron (299) and silicon (804) in molybdenum compounds. Water and ammonia content of ammonium molybdate is measured by thermal decomposition and infrared spectrometry (494).

B.   MOLYBDENUM ALLOYS

Titanium–zirconium–molybdenum alloy, containing approximately 0.5% Ti, 0.07% Zr, and 0.01% C, is a substitute for pure molybdenum in most applications and, in addition, possesses greater high-temperature strength than unalloyed molybdenum metal. After dissolving a sample of this alloy in aqua regia and fuming with sulfuric acid, titanium content is measured colorimetrically with diantipyrylmethane. Hydroxylamine is

TABLE 16
Methods for Determination of Trace Elements in Pure Molybdenum

Element	Method	Reference
Al	Neutron activation	256
Sb	Spectrophotometric	587
As	Spectrophotometric	889
B	Spectroscopic	695
C	Titrimetric	38
Co	Spectrophotometric	798
Cu	Spectrophotometric	764
H	Manometric	765
Fe	Polarographic	287
Mn	Spectrophotometric	216
Ni	Spectrophotometric	796
Nb	Spectrophotometric	1084
N	Spectroscopic	1079
O	Neutron activation	3
Pd	Neutron activation	257
Si	Spectrophotometric	288
Na	Neutron activation	214
S	Conductometric	1107
Th	Radiometric	16
W	Spectrophotometric	822
Zr	Spectroscopic	275

first added to reduce iron(III) and vanadium(V), which would otherwise interfere (783). Zirconium in a separate aliquot is determined colorimetrically with Arsenazo III. Anion-exchange separation of titanium, zirconium, and molybdenum occurs with eluents containing varying amounts of sulfuric and oxalic acid. Each isolated component is then determined by a suitable colorimetric procedure (882). Spectroscopic analysis of this alloy is also discussed (696).

Molybdenum–tungsten alloys, after being rendered soluble by sodium carbonate fusion, are treated with hydrazine to reduce molybdenum to Mo(V), a known excess of EDTA is added, and the excess EDTA is titrated with standard zinc titrant at pH 5 using xylenol orange indicator (787). Both tungsten and molybdenum in a second aliquot precipitate upon adding a known excess of lead(II) solution. The excess lead(II) is titrated with EDTA, and from the combined data tungsten content is found by difference. Various wet-chemical procedures (539) and spectroscopic examinations (595) are used to measure trace constituents in molybdenum–tungsten alloys.

Ferromolybdenum is used as a source of molybdenum in the manufacture of molybdenum-containing steels. Ferromolybdenum contains from

TABLE 17

Methods for Determination of Trace Elements in Ferromolybdenum

Element	Method	Reference
Al	Gravimetric	449
Sb	Spectrophotometric	913
As	Titrimetric	91
Bi	Spectrophotometric	160
Cr	Spectrophotometric	407
Cu	Titrimetric	479
Pb	Spectroscopic	905
P	Spectrophotometric	631
Si	Spectrophotometric	552
S	Gravimetric	449
Sn	Spectrophotometric	912
Ti	Spectrophotometric	328
W	Spectrophotometric	451

50 to 70% molybdenum, iron, and trace amounts of other elements and is soluble in aqua regia or by fusion with sodium peroxide. Molybdenum content is found by sulfide precipitation followed by ignition to the oxide and weighing (508). Precipitation of hydrous iron(III) oxide with ammonia traps some molybdenum. Cation exchange separates iron and molybdenum. Molybdenum is then determined, after reduction, by titrating with standard $KMnO_4$ titrant (1035). Titration of excess EDTA with standard copper(II) titrant, after addition of a known amount of EDTA to a ferromolybdenum sample solution and reduction of the molybdenum present with hydrazine sulfate, is an alternate procedure for molybdenum (467). Direct spectrophotometric measurement of the colored molybdenum(V) species formed by reaction with hydrazine sulfate is not subject to iron interference if iron(III) is first reduced to iron(II) by addition of sodium sulfite (39). Simultaneous determination of trace components of ferromolybdenum is done using emission spectroscopy (905) and X-ray fluorescence (1029). Table 17 summarizes procedures for determining specific trace components in ferromolybdenum.

Procedures for determining gases in molybdenum alloys are similar to those for measuring gases in pure molybdenum and are cited in Section VII.A.1.

VIII.   DETERMINATION IN SPECIFIC MATERIALS

A.   METALLURGICAL SAMPLES

1.   Iron and Steel

Molybdenum in cast iron is determined simultaneously with other minor constituents by emission spectroscopy (461) or X-ray spectroscopy (544). It is determined uniquely by precipitation as lead molybdate (1015) or colorimetric measurement (1074).

Molybdenum in steels, added to increase high-temperature strength, is determined by any of several analytical techniques. Various wet-chemical procedures have been reviewed (500). Among these are precipitation and weighing of lead molybdate either directly (336) or following prior separation of molybdenum sulfide after conversion of this precipitate to the oxide and redissolving before adding lead(II) ion (501). Molybdenum oxinate precipitates from steel samples in the presence of EDTA. The precipitate is filtered, washed with 2 M HCl to remove vanadium and tungsten, and dried at 140°C before weighing (807). Molybdenum(VI) remains in the filtrate after precipitation of insoluble hydrous oxides from a sample solution. Following separation it is reduced to a lower valence state and titrated with standard potassium permanganate (1088) or vanadium(V) (271) titrant. Best results are obtained with steels having 1% or more molybdenum. Colorimetric measurement of molybdenum as the Mo(V)-thiocyanate complex is suitable for steels containing upwards of 0.01% molybdenum (613). Ascorbic acid reduction and perchloric acid solutions are recommended (632), while extraction of the colored complex aids in removing interferences (65,1031). An automated procedure for molybdenum in steels uses the thiocyanate reaction (114). Thio-substituted acids (439) and hydrogen peroxide (274) are also used as color-forming agents for determining molybdenum in steel samples.

Selection of suitable analytical lines for the emission spectroscopic examination of molybdenum in steels has been discussed (737), and various means of improving molybdenum emissions in the far ultraviolet are available (330,987). Flame excitation of the methyl isobutyl ketone extract of molybdenum with α-benzoin oxime into a fuel-rich oxygen–acetylene flame is an alternate emission procedure (995). Satisfactory atomic absorption measurement of the 379.8 nm Mo line occurs from a solution containing ammonium chloride and using an air–acetylene flame (688). Nitrous oxide–acetylene mixtures and monitoring of the 313.3 nm line is also used for molybdenum. With perchloric acid solutions and, if necessary, added iron salts for signal enhancement, as little as 0.002% molybdenum in steel can be determined (236,1018). X-Ray fluorescence is used for determining various alloying elements in steel samples including molybdenum when present in amounts greater than 0.001% (469).

Diffusion current measurement of the second polarographic plateau of Mo(VI), $E_{1/2}$ potential of -0.57 V versus SCE, is used for determining molybdenum in steels (754). The supporting electrolyte is citric acid–sodium dihydrogen phosphate–hydrochloric acid at pH 1.6. Iron interferes unless removed by hydrous oxide precipitation before polarographic measurement of the molybdenum wave. Prior reduction of molybdenum with hydrazine sulfate before polarographic measurement is an alternate means of determining molybdenum in steels (466). Titanium and iron are first removed by hydrous oxide precipitation, while small amounts of tungsten do not interfere. It is possible to titrate molybdenum(VI) from steel samples amperometrically with 8-mercaptoquinoline. Anode oxidation of excess titrant after the equivalence point occurs at a rotating platinum anode with impressed potential of $+1.0$ V versus SCE (103). Prior addition of sodium hydroxide is necessary to remove iron, vanadium, and chromium as their hydrous oxides, and phosphoric acid is added to prevent tungsten from interfering.

Procedures are available for determining several minor constituents in steels, including molybdenum, by neutron-activation analysis (118,705) and mass spectrometry (472).

2. Nonferrous Alloys

Molybdenum is used in many different alloys. These include, for example, heat-resistant cobalt- or nickel-based alloys and uranium-based alloys used as nuclear fuel elements. A compilation of properties and uses of many nonferrous molybdenum containing alloys is available (646). Table 18 lists methods for determining molybdenum in several molybdenum-containing alloys.

TABLE 18

Methods for Determination of Molybdenum in Various Nonferrous Alloys

Principle alloying element	Approximate molybdenum content (%)	Method	Reference
Al	0.3	Spectrophotometric, thiocyanate	975
Co	1–7	Spectroscopic	856
Cu	20–85	Spectrophotometric, 2,2'-bicinchonic acid	35
Ni	1.5–9	Atomic absorption	1066
Nb	2–10	Spectrophotometric, thiocyanate	263
Ti	11–40	X-Ray fluorescence	1039
U	0.5–10	Gravimetric, α-benzoin oxime	302
U	2.5	Atomic absorption	870
Zr	0.01	Spectroscopic	252

TABLE 19

Methods for Determination of Trace Molybdenum in Various High Purity
Elements

Element	Method	Reference
Al	Neutron activation	597
Be	Spectroscopic	161
Bi	Spectroscopic	273
B	Spectroscopic	201
Cd	Polarographic	624
Cu	Spectroscopic	795
Fe	Spectrophotometric, diethyldithiocarbamate	909
Ga	Spectroscopic	740
Pb	Spectroscopic	459
Nb	Atomic absorption	488
Pt	Spectroscopic	604
Pu	Spectroscopic	210
Re	Spectrophotometric, thioxene	602
Si	Spectroscopic	1051
Th	Spectroscopic	203
Sn	Spectroscopic	639
Ti	Spectrophotometric, thiocyanate	907
W	Spectrophotometric, dithiol	153
U	Spectrophotometric, thiocyanate	1100
V	Spectrophotometric thiocyanate	505
Zr	Spectroscopic	202

B. ELEMENTS AND COMPOUNDS

Molybdenum appears as a trace impurity in many pure chemical
elements. Spectrophotometric procedures, for example, are available for
determining between 10^{-5} and $10^{-2}\%$ Mo in refractory metals (711) and
for microgram amounts of Mo in Fe, Cr, Ni, and Mn (909). Table 19
summarizes procedures for determining trace amounts of molybdenum
in high-purity elements.

Molybdenum impurity in alkali halide salts is determined by emission
spectroscopy in the parts per million range (251,1110) and in tin(II)
chloride by cathode-ray polarography (710). Trace amounts of molybdenum in uranium tetrafluoride are found by spectrophotometric (546) and
spectroscopic (777,863) measurement. Spectrophotometric measurement

of the molybdenum-morin complex is used to determine microgram amounts of molybdenum in alkali metal hydroxides (317).

Methods for determining molybdenum as an impurity in various metal oxides are summarized in Table 20.

C. ROCKS AND MINERALS

Spectrographic (421), colorimetric (648,857), and polarographic (388) procedures are used to determine molybdenum in rock samples. Various standard rock samples have been analyzed for their molybdenum content by an automated colorimetric procedure (285) and by atomic absorption spectrometry (815). On-site determination of molybdenum is possible with a portable X-ray fluorescence analyzer (296).

Molybdenum content of ores and their concentrates is obtained by many methods. After preliminary sample treatment, involving alkali hydroxide or alkali peroxide fusion, a molybdenum-containing aliquot can be treated with lead(II) to precipitate $PbMoO_4$ (105) or, following reduction with bismuth to produce Mo(V), titrated with standard vanadate titrant (31). Colorimetric measurement of the Mo(V)-thiocyanate complex after sample digestion in strong mineral acid is an alternate method of determining molybdenum in ores (733). Polarographic (475), emission-spectroscopic (1024) and X-ray spectroscopic (977) techniques are also used. Molybdenum, in several molybdate and tungstate min-

TABLE 20

Methods for Determination of Trace Molybdenum in Various Metal Oxides

Oxide	Method	Reference
BeO	Spectrophotometric, thiocyanate	377
Fe_2O_3	Spectrophotometric, thioglycolic acid	536
Nb_2O_5	Spectroscopic	947
SiO_2	Neutron activation	450
Ta_2O_5	Mass spectrometry	641
ThO_2	Spectroscopic	707
TiO_2	Neutron activation	718
U_3O_8	Spectroscopic	43
V_2O_5	Radiometric	300
WO_3	Spectroscopic	684
Y_2O_3	Spectrophotometric, thiocyanate	376
ZrO_2	Spectroscopic	706

erals, is successfully determined by spectrophotometric measurement of the dissolved 8-quinolinol precipitate of molybdenum(VI) (225).

Complete analysis of molybdenum ores is possible by wet-chemical (191) and spectroscopic (715) procedures.

A procedure is available for spectroscopic determination of minor components in molybdenite, MoS_2, one of the principle molybdenum ores (610). In addition, other methods are used to determine individual elements in molybdenite. These include procedures for copper (890), germanium (204), iron (805), lead (471), niobium (667), rhenium (714), and zinc (890).

Molybdenum is present as a trace constituent in many ores. Procedures are described for its determination in ores mined chiefly for copper (4), niobium (880), tungsten (432), and uranium (194).

D. SOILS AND PLANTS

Molybdenum is an essential micronutrient in plants, necessary for proper nitrogen fixation. Its presence in plants and the soil in which they grow has been studied extensively (7,106,838,865).

In soils molybdenum, typically present to the extent of a few parts per million, is isolated by treating a soil sample with strong mineral acid (316), basic fusion (1058), prolonged extraction with ammonium oxalate buffered at pH 3.3 (322,337), or treatment with other extracting agents (353). Molybdenum content obtained from a mild extracting procedure correlates best with the molybdenum content of plants grown in the soil being tested (418,658). After isolation molybdenum is determined colorimetrically with thiocyanate (625) or dithiol (950), by polarographic measurement (87), emission spectroscopy (872), atomic absorption (611), or neutron-activation analysis (1122).

In plants molybdenum, present in the parts per million range, is determined following either wet or dry ashing by any of several procedures including colorimetric measurement (73,92), polarography (708), emission spectroscopy (739), X-ray spectroscopy (1017), and neutron-activation analysis (725).

E. ANIMAL MATTER

Neutron activation is used for determining microgram amounts of molybdenum in marine organisms (286), rabbit plasma (837), sheep wool (367), and cow's milk (1105).

In humans, variations from 1 to 40 μg of molybdenum occur in a 24-hr urine sample with a mean value of 15.6 μg/24-hr sample. Determination is by colorimetric measurement of the dithiol complex after concentrations of molybdenum with cupferron (298). Other colorimetric procedures are used for finding the molybdenum content of blood (1042) and liver (843). Atomic-absorption spectrometry is used to find the molybdenum content of blood and urine (769), and a neutron-activation procedure is used for determination of molybdenum in teeth (608).

F.   WATER AND AIR

Molybdenum is present in natural waters at a concentration of several parts per billion. It is collected from water samples by coprecipitation (168,486,559), extraction (174), or ion exchange (464) and determined by colorimetric measurement or other procedures. A comparison of various methods has been made (268).

Molybdenum has been found in city air in amounts of several micrograms per 1000 m^3 (374) and to a lesser extent in rural locations (654). Air samples are collected from filters, ashed, and dissolved in dilute acid prior to analysis for molybdenum by colorimetric measurement of the thiocyanate (1000) or phenylhydrazine complex (1041). Molybdenum has been measured in the vicinity of electric welding operations (534) and in particular matter collected from high-altitude nuclear explosions (74).

TABLE 21
Methods for Determination of Molybdenum in Various Materials

Material	Approximate molybdenum content (%)	Method	Reference
Alcoholic beverages	10^{-6}	Spectrophotometric, thiocyanate	244,793
Ceramics	10^{-3}	Polarographic	70
Cermets	10	X-Ray fluorescence	1057
Coal	10^{-3}	Spectrophotometric, rhodanide + thiocarbamide	69
Drugs	—	Polarographic	678
Fertilizer	10^{-4}	Spectrophotometric, thiocyanate	393
Foodstuffs	10^{-5}	Spectrophotometric, thiocyanate	78,951
Glass	1	X-Ray fluorescence	433
Graphite	10^{-2}	Spectroscopic	1059
Paint	—	Titrimetric, $KMnO_4$	992
Petroleum products	10^{-4}	Neutron activation	188,1036
Soap	—	Gravimetric, MoO_3	923
Solid lubricants	8	Atomic absorption	444

G. MISCELLANEOUS MATERIALS

Molybdenum plays an important role in the composition of many materials. Molybdenum disulfide, for example, is used as a solid lubricant and molybdenum compounds in fertilizers enhance plant growth. Table 21 lists procedures for determining molybdenum in a variety of materials.

IX. RECOMMENDED PROCEDURES

A. GRAVIMETRIC DETERMINATION WITH 8-QUINOLINOL (791)

Precipitation of molybdenum(VI) oxinate in the presence of a sixfold excess of ethylenediaminetetraacetic acid eliminates interference from Pb, Bi, Hg(II), Cd, Cu, Fe, Al, Cr(III), Be, U, Co, Zn, Mn, and Ni. Titanium(IV) is removed before molybdenum precipitation by addition of sodium hydroxide. Vanadium and tungsten interfere.

Reagents

8-Quinolinol, 3%. Dissolve 30 grams of 8-quinolinol in a small amount of glacial acetic acid. Neutralize the solution with 4 M NH_3 until a precipitate just begins to form. Redissolve the precipitate by adding 4 M HOAc. Dilute to 1 liter with 4 M HOAc.

Ethylenediaminetetraacetate, 5%. Dissolve 50 grams of $Na_2H_2EDTA \cdot 2H_2O$ in distilled water, and dilute to 1 liter with additional distilled water.

Acetate buffer. Mix three parts of a 50% (wt./vol.) solution of ammonium acetate with four parts of a 50% (vol./vol.) solution of acetic acid.

Procedure

Neutralize the sample solution containing molybdenum(VI) to the methyl red color transition. Add 15 ml of EDTA solution and 5 ml of acetate buffer. Dilute to approximately 80 ml with distilled water. Heat to boiling, and slowly add 8-quinolinol solution until precipitate formation is complete and the supernatant liquid is slightly yellow. Digest for 2–3 min, and filter through a previously weighed fine porosity scintered-glass crucible. Wash the precipitate thoroughly with hot distilled water until the washings are colorless. Dry for 1 hr and then to constant weight at 130–140°C. The precipitate is $MoO_2(C_9H_6NO)_2$, molecular weight 416.26, and contains 23.05% Mo.

Interference

The presence of EDTA and proper buffering conditions prevent interference of Pb, Bi, Hg(II), Cd, Cu, Fe, Al, Cr(III), Be, U, Co, Zn, Mn, and Ni. Removal of Ti as the insoluble hydrous oxide prior to molybdenum precipitation eliminates its interference. Vanadium and tungsten, if present with molybdenum in the sample, interfere with this procedure.

B. TITRIMETRIC DETERMINATION WITH CERIUM(IV) (67)

In the absence of other species that undergo oxidation-reduction reactions, molybdenum(VI) can be determined by reduction to Mo(V) and titration using standard cerium(IV) sulfate titrant. Quantitative preliminary reduction of molybdenum with metallic mercury occurs in solutions containing 2–4 M HCl.

Reagents

Cerium(IV) sulfate, 0.1 N. Dissolve approximately 63 grams of $Ce(SO_4)_2 \cdot 2(NH_4)_2SO_4 \cdot 2H_2O$ in 500 ml of distilled water containing 30 ml of 18 M sulfuric acid. Stir, cool, and filter if necessary. Dilute to 1 liter with distilled water. Standardize the solution against a suitable primary standard, for example, arsenic(III) oxide.

N-Phenylanthranilic acid, 0.2%. Dissolve 1.0 gram of n-phenylanthranilic acid in 20 ml of 5% (wt./vol.) $NaHCO_3$. Dilute to 500 ml with distilled water.

Hydrochloric acid, 6 M.

Mercury.

Procedure

Dilute the neutral sample solution, containing molybdenum(VI), with an equal volume of 6 M HCl. Place the solution in a separatory funnel, and add 25 ml of metallic mercury. Shake for 10 min. Draw off the mercury, and filter the aqueous solution through paper to remove mercury(I) chloride formed during molybdenum reduction. Wash the separated mercury twice with 50 ml portions of 3 M HCl, and add the washings to the sample solutions. Wash the insoluble residue on the filter paper three times with 30 ml portions of 3 M HCl, and add these to the sample solution. Add five drops of indicator, and titrate the sample with standard cerium(IV) solution. The indicator changes to purple after the end point of the titration. Alternately 0.025 M ferroin can be used as indicator for this titration. With ferroin the solution is pale blue after the end point.

Interference

All elements that are reduced by metallic mercury and subsequently reoxidized by Ce(IV) interfere with this procedure unless previously removed. Iron, chromium, vanadium, and others are reduced by metallic mercury.

C. COLORIMETRIC DETERMINATION IN STEELS WITH AMMONIUM THIOCYANATE (619)

In this procedure molybdenum reduction with tin(II), beyond the desired Mo(V) necessary for color formation, is prevented by proper control of acidity. A stable Mo(V)-SCN complex is maintained by adding hydroquinone to prevent molybdenum oxidation. Preliminary extraction of molybdenum(VI) with α-benzoin oxime separates it from iron(III), and methyl isobutyl ketone extraction of the Mo(V)-SCN complex improves sensitivity with colorimetric measurement. If large amounts of tungsten are present, an additional prior extraction of molybdenum with cupferron is necessary to separate it from tungsten. The procedure is useful for steels containing 0.001–0.1% molybdenum.

Reagents

Tin(II) chloride, 1.8 M. Dissolve 40 grams of $SnCl_2 \cdot 2H_2O$ in 25 ml of 6 M HCl. Heat to dissolve. Cool and dilute to 100 ml with distilled water. Prepare fresh every few days.

Ammonium thiocyanate, 5%. Dissolve 25 grams of NH_4SCN in distilled water, and dilute to 500 ml with additional distilled water. Store in a polyethylene bottle.

α-Benzoil oxime, 1%. Dissolve 2 grams of α-benzoin oxime in 20 ml of warm ethanol. Dilute to 200 ml with chloroform.

Citric acid, 10%. Dissolve 10 grams of citric acid in distilled water, and dilute to 100 ml with additional distilled water.

Hydroquinone.

Methyl isobutyl ketone.

Sulfuric acid, 18 M.

Hydrochloric acid, 12 M.

Nitric acid, 16 M.

Perchloric acid, 70%.

Hydrofluoric acid, 48%.

Molybdenum standard, 10 μg Mo/ml. Dissolve 0.252 gram of $Na_2MoO_4 \cdot 2H_2O$ in distilled water, and dilute to 1 liter in a volumetric flask with additional distilled water. Dilute 100 ml of this solution to 1 liter in a volumetric flask with distilled water.

Procedure

Dissolve by heating 25–500 mg of steel, containing up to 40 μg of molybdenum, in 2–5 ml of HCl–HNO_3 (4:1 vol./vol.) plus 2–15 drops of HF. Add 2–5 ml $HClO_4$, and evaporate carefully over a flame. Continue evaporation until 0.5–1.0 ml of liquid remains. Cool, and add 5 ml HCl to dissolve the residue. Heat until just boiling, add 35 ml water, and cool. Transfer the solution to a separatory funnel, and dilute to 50 ml with additional distilled water. Add 10 ml of α-benzoin oxime solution, and shake for 1 min. Separate and remove the lower chloroform layer. Wash the aqueous layer with 10 ml of chloroform and separate again. Combine the $CHCl_3$ extracts, and add to them 15 ml of 0.6 M HCl. Shake and collect the $CHCl_3$ layer. Wash the aqueous layer with 3 ml of $CHCl_3$. Collect the chloroform layer, and combine it with the main chloroform extract. Evaporate the organic solution to remove $CHCl_3$. Add 5 ml of H_2SO_4, and heat to white fumes to carbonize organic matter. Add 1 ml of HNO_3 and 0.5 ml $HClO_4$. Heat with slowly increasing temperature to destroy the organic residue. Evaporate over a flame until only 4 ml of liquid remain. Cool and add 20 ml of distilled water. Cool the solution again, and add 10 ml of citric acid solution. Transfer to a separatory funnel using an additional 10 ml of water. Dilute to approximately 70 ml with distilled water.

To form the molybdenum(V)-thiocyanate complex add 5 ml of $SnCl_2$ solution, and wait 5 min. Add 10 ml of NH_4SCN solution, 10 ml of methyl isobutyl ketone, and shake vigorously for 30 sec. Drain off the lower, aqueous layer and discard. Immediately transfer the organic layer to a 50 ml conical flask containing 100 mg hydroquinone. Dissolve the hydroquinone, transfer a portion of the solution to a 1.00 cm absorption cell, wait 1 min, and measure the absorbance at 470 nm against a methyl isobutyl ketone blank.

A calibration curve is prepared by transferring 1.0–4.0 ml of molybdenum standard solution to separate separatory funnels each containing 10 ml of 10% citric acid solution and 20 ml of 3.6 M H_2SO_4. Dilute each sample to approximately 70 ml with distilled water, and proceed as previously described beginning with the steps necessary to form the molybdenum(V)-thiocyanate complex.

Interference

Interference from most of the elements associated with molybdenum in steels is eliminated by the double-extraction procedure. Interference does occur in the presence of rhenium and large amounts of tungsten.

D. DETERMINATION BY ATOMIC ABSORPTION IN LAKE WATER (174)

Molybdenum in natural waters, in the parts per billion range, is extracted with 8-quinolinol in methyl isobutyl ketone and aspirated directly into a nitrous oxide–acetylene flame. Addition of ascorbic acid prior to extraction eliminates the interfering effects of V(V), Fe(III), Cr(VI), and W(VI) at the concentration levels normally found in natural waters.

Reagents

8-Quinolinol, 1%. Dissolve 1 gram of 8-quinolinol in redistilled reagent-grade methyl isobutyl ketone. Dilute to 100 ml with additional solvent. Prepare fresh each week.

Ascorbic acid, 1%. Dissolve 1 gram of ascorbic acid in distilled water. Dilute to 100 ml with additional distilled water.

Nitric acid, 6 M.

Molybdenum standard, 100 μg Mo/ml. Dissolve 0.0750 gram of dried MoO_3 in 5–10 ml of 0.1 M NaOH. Add approximately 90 ml of distilled water and sufficient 6 M HCl to make the solution slightly acid. Dilute with distilled water to 500 ml in a volumetric flask. A working standard, 1 μg Mo/ml, consists of 1.00 ml of this stock solution diluted to 100 ml with distilled water.

Molybdenum-free lake water. Adjust a sample of filtered lake water to pH 5 ± 0.2, and pass it through a 120 × 15 mm ion-exchange column, containing Chelex-100 resin, at a flow rate of 5 ml/min. The resin is prepared by washing with 2 M HNO_3 and then washing with distilled water until the effluent has a pH of 4–5.

Operating Conditions

Wavelength	313.3 nm
Entrance slit	100 μ
Exit slit	150 μ
Lamp current	15 mA
Nitrous oxide	30 psi, 11 scfh flow rate

Acetylene   15 psi. Adjust flow until flame is just about to
                   become luminous.
Burner height   Adjust for optimum reading.
Scale expansion   Tenfold

Procedure

The water sample is filtered immediately after collection through a 0.5 μ millipore membrane and then acidified with nitric acid to pH 2. To 100 ml of sample in a separatory funnel add 5 ml of ascorbic acid solution. Adjust pH to the range 2.0–2.4. Add 5 ml of 8-quinolinol solution, and mix on a mechanical shaker for 15 min. Separate the organic phase, and aspirate it directly into the flame. Blank and calibration samples are processed with the same procedure using 100 ml of molybdenum-free lake water for the blank and 100 ml of molybdenum-free lake water containing 0.5–1 ml of dilute Mo standard. For samples with less than 3 parts per billion of Mo larger initial samples, up to 200 ml, should be used and the amount of 8-quinolinol solution increased to 7 ml.

Interference

Interference from moderate amounts of Fe(III), V(V), Cr(VI), and W(VI) are prevented by addition of ascorbic acid prior to 8-quinolinol extraction; otherwise they are extracted with molybdenum and cause errors in the atomic absorption measurement. Other ions in amounts commonly found in natural waters do not interfere with this procedure.

E. POLAROGRAPHIC DETERMINATION IN NIOBIUM ALLOYS (365)

Molybdenum and titanium are determined simultaneously in the presence of Nb, Ta, W, and Zr in a supporting electrolyte containing citric acid–sulfuric acid–thorium nitrate. Neither nickel nor chromium(III) interfere with molybdenum determination, but iron(III) must be absent. The $E_{1/2}$ value for reduction of Mo(VI) to Mo(V) under the experimental conditions employed is -0.273 V versus SCE, and the diffusion current constant for molybdenum is 1.21. Diffusion current values for constructing a calibration curve are measured on the first reduction plateau at -0.48 V versus SCE. The second molybdenum wave, reduction of Mo(V) to Mo(III), occurs at -0.635 V versus SCE and overlaps the titanium reduction wave. Titanium, in this procedure, is determined by subtracting the amount of molybdenum present from the combined concentration measured from the second plateau.

Reagents

Nitric acid, 16 M.
Sulfuric acid, 18 M.
Hydrofluoric acid, 48%
Sodium hydroxide, 10 M.
Citric acid monohydrate, solid.
Thorium nitrate hexahydrate, solid.
Molybdenum standard, 3.00 $\times$ 10^{-3} M. Dissolve 0.442 gram of reagent-grade MoO_3 in boiling 15 M ammonia. Evaporate to dryness. Dissolve the residue in a solution containing 1 M citric acid–0.1 M sulfuric acid, and dilute with additional solvent to 1 liter in a volumetric flask.

Procedure

For alloys containing approximately 5% Mo, dissolve about 200 mg of alloy, accurately weighed, in 5 ml 16 M HNO_3 plus 10 ml 48% HF in a platinum crucible. Add 2.77 ml of 18 M H_2SO_4 from a buret, and evaporate the mixture, heating until sulfur trioxide fumes appear. Dissolve the residue in approximately 50 ml of distilled water containing 21.02 grams citric acid monohydrate. Transfer the sample to a 100 ml volumetric flask, and dilute to the mark with distilled water. Place 5.00 ml of this solution in a beaker, and add 45 ml of 1 M citric acid and 30 ml of distilled water. Adjust the pH to 2.0–3.0 with 10 M NaOH. Add 2.94 gram of $Th(NO_3)_4 \cdot 6H_2O$ crystals. Dissolve these, and add additional 10 M NaOH until the pH is exactly 3.5. Transfer to a 100 ml volumetric flask, and dilute to volume with distilled water. Approximately 40 ml of diluted sample is deaerated 10 min with oxygen-free nitrogen, and a polarogram is recorded from 0 to -1.0 V versus SCE. Diffusion currents for molybdenum are taken from the first reduction plateau at -0.48 V versus SCE. A calibration curve of diffusion current versus molybdenum concentration is prepared by treating between 0.50 and 5.00 ml of molybdenum standard as described for sample preparation starting with addition of 1 M citric acid.

Interferences

Iron is the most serious interfering element in determining molybdenum by this procedure and must be absent from the sample solution. Ti, Nb, Ta, W, and Zr all give well-defined polarographic waves under the experimental conditions, but their reductions all occur at more negative potentials and do not overlap the plateau of the first molybdenum wave.

ACKNOWLEDGMENT

Assistance from Ms. Nancy Parker and Ms. Mary J. Reger for collecting reference material and Ms. Sandra K. Gunsberg and Ms. Joan H. Kent for typing the manuscript is gratefully acknowledged.

REFERENCES

1. Aaremae, A., and G. Assarsson, *Z. Anal. Chem.*, **144**, 412 (1955).

2. Abashidze, N. F., *Tr. Metrol. Inst. SSSR*, No. 96, 165 (1968); through *Chem. Abstr.*, **70**, 34036b (1969).

3. Abdurakhmanova, S. R., V. A. Kireev, L. V. Novalikhin, and Yu. N. Talanin, *Zh. Anal. Khim.*, **23**, 1188 (1968).

4. Adamiec, I., *Chem. Anal. (Warsaw)*, **11**, 1175 (1966).

5. Adamiec, I., *Chem. Anal. (Warsaw)*, **11**, 1183 (1966).

6. Adams, F., and J. Hosh, *Radiochem. Radioanal. Lett.*, **3**, 31 (1970).

7. Adhate, S., *Indian J. Agron.*, **5**, 57 (1960).

8. Adler, I., in Meites, L., Ed., *Handbook of Analytical Chemistry*, McGraw-Hill, New York, 1963.

9. Afghan, B. K., and R. M. Dagnall, *Talanta*, **14**, 239 (1967).

10. Agasyan, P. K., E. R. Nikolaeva, and K. K. Tarenova, *Zavod. Lab.*, **35**, 1034 (1969).

11. Agasyan, P. K., K. K. Tarenova, E. R. Nikolaeva, and R. M. Katina, *Zavod. Lab.*, **33**, 547 (1967).

12. Agrinskaya, N. A., and V. I. Petrashen, *Tr. Novocherk. Politekh. Inst.*, **143**, 27 (1963); through *Chem. Abstr.*, **61**, 1259g (1964).

13. Agrinskaya, N. A., and V. I. Petrashen, *Tr. Novocherk. Politekh. Inst.*, **143**, 35 (1963); through *Chem. Abstr.*, **61**, 28b (1964).

14. Akerman, K., D. Wiater, Z. Kozak, K. Jurkiewicz, and T. Krol, *Przem. Chem.*, **43**, 442 (1964); through *Chem. Abstr.*, **61**, 12986g (1964).

15. Aleksandrov, A., and P. Vasileva-Aleksandrova, *Mikrochim. Ichnoanal. Acta*, **1963**, 23. (1963).

16. Aleksandrov, L. N., *Zavod. Lab.*, **26**, 975 (1960).

17. Alekseev, R. I., *Zavod. Lab.*, **7**, 863 (1938).

18. Alexander, G. B., *J. Amer. Chem. Soc.*, **71**, 3043 (1949).

19. Alimarin, I. P., and A. M. Medvedeva, *Zavod. Lab.*, **21**, 1416 (1955).

20. Alimarin, I. P., and A. M. Medvedeva, *Tr. Kom. Anal. Khim., Akad. Nauk SSSR, Inst. Geokhim. Anal. Khim.*, **6**, 351 (1955); through *Chem. Abstr.*, **50**, 12741a (1956).

21. Alimarin, I. P., and V. N. Polyanskii, *Zh. Anal. Khim.*, **8**, 266 (1953).

22. Allen, S. H., and M. B. Hamilton, *Anal. Chim. Acta*, **7**, 483 (1952).

23. Almassy, G., and J. Straub, *Magy. Kem. Foly.*, **60**, 104 (1954); through *Chem. Abst.*, **52**, 6053a (1958).

24. Almassy, G., and M. Vigvari, *Acta Chim. Acad. Sci. Hung.*, **20**, 243 (1959).

25. Alstodt, B. S., and A. A. Benedetti-Pichler, *Ind. Eng. Chem. Anal. Ed.*, **11**, 294 (1939).

26. Alvarez Querol, M. C., and C. L. Wilson, *Mikrochem. Mikrochim. Acta*, **36/37**, 224 (1951).

27. Andersson, L. H., *Ark. Kemi*, **19**, 223 (1962).

28. Andreeva, Z. F., and L. P. Timofeeva, *Dokl. TSKHA (Timiryazev. Sel'skokhoz. Akad.)*, **124**, 319 (1967); through *Chem. Abstr.*, **69**, 39072z (1968).

29. Ankudimova, E. V., *Tr. Kom. Anal. Khim., Akad. Nauk SSSR, Otdel. Khim. Nauk*, **5**, 197 (1954); through *Chem. Abstr.*, **49**, 12187d (1955).

30. Ankudimova, E. V., *Tr. Novocherk. Politekh. Inst.*, **143**, 17 (1963); through *Chem. Abstr.*, **61**, 4931e (1964).

31. Ankudimova, E. V., and V. I. Petrashen, *Tr. Novocherk. Politekh. Inst.*, **31**, 73 (1955); through *Chem. Abstr.*, **53**, 12944g (1959).

32. Anokhina, L. G., Yu. A. Sopin, and N. A. Agrinskaya, *Tr. Novocherk. Politekh. Inst.*, **143**, 63 (1963); through *Chem. Abstr.*, **61**, 27g (1964).

33. Anonymous, *Chem. Eng. News*, **49**, 14 (1971).

34. Arecasis, S. M., and B. Scott, *J. Inorg. Nucl. Chem.*, **27**, 2665 (1965).

35. Arkhangel'-skaya, A. S., and G. P. Batalina, *Zavod. Lab.*, **34**, 408 (1968).

36. Arrington, C. E., and A. C. Rice, *US Bur. Mines Rep. Invest.*, No. 3441, 1 (1939).

37. Artyukin, P. I., A. A. Bezzubenko, E. N. Gil'bert, B. I. Peshchevitskii, V. A. Pronin, and A. V. Nikolaev, *Dokl. Akad. Nauk SSR*, **169**, 98 (1966).

38. Aruina, A. S., *Zavad. Lab.*, **12**, 411 (1946).

39. Ashy, M. A., and J. B. Headridge, *Anal. Chim. Acta*, **59**, 217 (1972).

40. Athavale, V. T., V. P. Apte, M. R. Dhaneshwar, and R. G. Dhaneshwar, *Indian J. Chem.*, **6**, 656 (1968).

41. Athavale, V. T., M. R. Dhaneshwar, and R. G. Dhaneshwar, *J. Electroanal. Chem.*, **14**, 31 (1967).

42. Aveston, J., E. W. Anacker, and J. S. Johnson, *Inorg. Chem.*, **3**, 735 (1964).

43. Avni, R., *Spectrochim. Acta*, **23B**, 619 (1968).

44. Aylward, G. H., *Anal. Chim. Acta*, **14**, 386 (1956).

45. Ayres, G. H., and B. L. Tuffly, *Anal. Chem.*, **23**, 304 (1951).

46. Babenko, A. S., and V. N. Tolmachev, *Ukr. Khim. Zh.*, **27**, 732 (1961).

47. Babko, A. K., *J. Gen. Chem. USSR*, **17**, 642 (1947).

48. Babko, A. K., *Talanta*, **15**, 721 (1968).

49. Babko, A. K., and O. F. Drako, *Zh. Anal. Khim.*, **12**, 342 (1957).

50. Babko, A. K., and G. I. Gridchina, *Zh. Neorg. Khim.*, **13**, 123 (1968).

51. Babko, A. K., G. S. Lisetskaya, and G. F. Tsarenko, *Zh. Anal. Khim.*, **23**, 1342 (1968).

52. Babko, A. K., and A. I. Volkova, *Zavod. Lab.*, **24**, 135 (1958).

53. Bailey, B. W., J. E. Chester, R. M. Dagnall, and T. S. West, *Talanta*, **15**, 1359 (1968).

54. Bailey, R. A., and L. Yaffe, *Can. J. Chem.*, **38**, 1871 (1960).

55. Baishya, N. K., and R. B. Heslop, *Anal. Chim. Acta*, **50**, 209 (1970).

56. Balogh-Kelemen, I., *Acta Phys. Chem. Debrecina*, **10**, 133 (1964); through *Chem. Abstr.*, **63**, 14036d (1965).

57. Bandi, W. R., E. G. Buyok, L. L. Lewis, and L. M. Melnick, *Anal. Chem.*, **33**, 1275 (1961).

58. Bankovskis, J., E. Svarcs, and A. Ievms, *Zh. Anal. Khim.*, **14**, 313 (1959).

59. Banks, C. V., and H. Diehl, *Anal. Chem.*, **19**, 222 (1947).

60. Bardocz, A., and F. Varsanyi, *Acta Tech. Acad. Sci. Hung.*, **13**, 409 (1955); through *Chem. Abstr.*, **50**, 7662a (1956).

61. Barieau, P. E., *Anal. Chem.*, **29**, 348 (1957).

62. Barkovskii, V. F., N. A. Alikina, and V. S. Shvarev, *Zh. Anal. Khim.*, **25**, 341 (1970).

63. Baro-Graf, J. C., *Publ. Inst. Invest. Microquim. Univ. Nac. Litoral (Rosario, Argent.)*, **14**, 147 (1950); through *Chem. Abstr.*, **46**, 5483h (1952).

64. Batt. A, N., and B. D. Jain, *Z. Anal. Chem.*, **195**, 424 (1963).

65. Bauer, G. A., *Anal. Chem.*, **37**, 155 (1965).

66. Bearden, J. A., *Rev. Mod. Phys.*, **39**, 78 (1967).

67. Becker, J., and C. J. Coetzee, *Analyst (London)*, **92**, 166 (1967).

68. Bednyak, N. A., *Nauch. Zap. Dnepropetr. Gos. Univ.*, **77**, 144 (1962); through *Chem. Abstr.*, **61**, 13917d (1964).

69. Bekyarova, E., E. Khristova, and D. Ruschev, *Godishnit Khim.-Tekhnol. Inst.*, **11**, 105 (1964); through *Chem. Abstr.*, **65**, 16717h (1966).

70. Beleuta, I. L., and C. Bratu, *Radiochem. Radioanal. Lett.*, **2**, 233 (1969).

71. Belyavshaya, T. A., and V. I. Fadeeva, *Vestn. Mosk. Univ., Ser. Fiz.-Mat. Estestven. Nauk*, **4**, 73 (1956); through *Chem. Abstr.*, **51**, 11162i (1957).

72. Benedetti-Pichler, A. A., *Identification of Materials*, Springer-Verlag, Wein, 1964.

73. Benne, E. J., and E. I. Linden, *J. Ass. Offic. Agr. Chem.*, **43**, 510 (1960).

74. Benson, P., C. E. Gleit, and L. Leventhal, *US At. Energy Comm. Symp. Ser.*, **5**, 108 (1965).

75. Bergh, A. A., and G. P. Haught, *Inorg. Chem.*, **8**, 189 (1969).

76. Berka, A., J. Dolezal, I. Nemec, and J. Zyka, *J. Electroanal. Chem.*, **3**, 278 (1962).

77. Bermejo Martinez, F., and M. Meijon Mourino, *Inform. Quim. Anal. (Madrid)*, **18**, 7 (1964).

78. Bernstein, I., *Rocz. Panstw. Zakl. Hig.*, **10**, 263 (1959); through *Chem. Abstr.*, **54**, 2621b (1960).

79. Bertin, E. P., R. J. Lingobucco, and R. J. Carves, *Anal. Chem.*, **36**, 641 (1964).

80. Bertoglio Riolo, C., T. Fulle Soldi, and C. Occhipinti, *Ann. Chim. (Rome)*, **57**, 1344 (1967).

81. Bertoglio Riolo, C., T. Fulle Soldi, and G. Spini, *Ann. Chim. (Rome)*, **58**, 3 (1968).

82. Bettel, *Chem. News*, **97**, 40 (1908).

83. Bhadra, A. K., and S. Banerjee, *Talanta*, **20**, 342 (1973).

84. Bhatnagar, R. P., and K. D. Sharma, *Indian J. Appl. Chem.*, **29**, 133 (1966).

85. Bigeon, J., *Ind. Chim. (Paris)*, **53**, 271 (1966).

86. Bikbulatova, R. U., and S. I. Sinyakova, *Zh. Anal. Khim.*, **19**, 134 (1964).

87. Bikbulatova, R. U., and S. I. Sinyakova, *Agrokhimiya*, **1967**, 123 (1967).

88. Billman, J. H., D. B. Borders, J. A. Buehler, and A. W. Seiling, *Anal. Chem.*, **37,** 264 (1965).

89. Binder, E., G. Goldstein, P. Lagrange, and J. P. Schwing, *Bull. Soc. Chim. Fr.,* **1965,** 2807 (1965).

90. Binder, O., *Chem.-Ztg.*, **42,** 255 (1918).

91. Binder, O., *Chem.-Ztg.*, **42,** 619 (1918).

92. Bingley, J. B., *J. Agr. Food Chem.*, **7,** 269 (1959).

93. Black, A. H., and J. D. Bonfiglio, *Anal. Chem.*, **33,** 431 (1961).

94. Blackman, L. C. F., *Mikrochim. Acta,* **1956,** 1366 (1956).

95. Blaedel, W. J., E. D. Olsen, and R. F. Buchanan, *Anal. Chem.,* **32,** 1866 (1960).

96. Blasius, E., and G. Lange, *Z. Anal. Chem.,* **160,** 169 (1958).

97. Blasius, E., and H. Pittack, *Angew. Chem.*, **71,** 445 (1959).

98. Blum, L., *Rev. Chim. (Bucharest),* **9,** 28 (1958).

99. Bock, R., *Z. Anal. Chem.,* **133,** 110 (1951).

100. Bodrov, N. V., and G. I. Nikolaev, *Zh. Anal. Khim.,* **24,** 1314 (1969).

101. Bogatskii, V. D., L. Ya. Dzyubanna, and N. L. Olenovich, *Zavod. Lab.,* **9,** 473 (1940).

102. Bognar, J., *Nehezip. Musz. Egyet. Idegennyelvu Kozlem. Bamyaszat, Kohaszat, Gepeszet,* **23,** 233 (1964); through *Chem. Abstr.,* **62,** 1070b (1965).

103. Bogovina, V. I., V. P. Novak, and V. F. Mal'tsev, *Zh. Anal. Khim.,* **20,** 951 (1965).

104. Bok, L. D. C., and V. C. O. Schueler, *J. S. Afr. Chem. Inst.,* **14,** 45 (1961).

105. Boldina, S. M., and N. T. Kazantseva, *Zavod. Lab.,* **25,** 661 (1959).

106. Bol'shakov, V: A., Yu. I. Dobritskaya, D. N. Ivanov, and L. P. Orlova, *Agrokhimiya,* **1967,** 142 (1967).

107. Boltz, D. F., T. DeVries, and M. G. Mellon, *Anal. Chem.,* **21,** 563 (1949).

108. Boltz, D. F., and C. H. Lueck, Phosphorus, in Boltz, D. F., Ed., *Colorimetric Determination of Non-metals,* Wiley-Interscience, New York, 1958, Chap. II.

109. Boltz, D. F., and M. G. Mellon, *Anal. Chem.,* **19,** 873 (1947).

110. Boltz, D. F., and M. G. Mellon, *Anal. Chem.,* **44,** 300R (1972).

111. Borrel, M., and R. Paris, *Anal. Chim. Acta,* **4,** 267 (1950).

112. Boswell, C. R., and R. R. Brooks, *Mikrochim. Ichnoanal. Acta,* **1965,** 814 (1965).

113. Bozsai, I., *Talanta,* **10,** 543 (1963).

114. Braithwaite, K., and J. D. Hobson, *Analyst (London),* **93,** 633 (1968).

115. Brinton, P. H. M. P., and A. E. Stoppel, *J. Amer. Chem. Soc.,* **46,** 2454 (1924).

116. Brintzinger, H., and E. Jahn, *Z. Anal. Chem.,* **94,** 396 (1933).

117. Brintzinger, H., and F. Oschatz, *Z. Anorg. Allg. Chem.,* **165,** 221 (1927).

118. Brune, D., and K. Jirlow, *Nukleonik,* **6,** 242 (1964).

119. Buchtela, K., and M. Lesegang, *Mikrochim. Ichnoanal. Acta,* **1965,** 67.

120. Buchwald, H., and E. Richardson, *Talanta,* **9,** 631 (1962).

121. Budesinsky, B., *Zh. Anal. Khim.,* **18,** 1071 (1963).

122. Buganov, Kh. G., in Tishchenko, V. A., Ed., *Mater. Nauch Konf. Aspir., Rostov.-Na-Donu Gos. Univ., 7th, 8th, 1967,* Izd. Rostov. Univ., Rostov-on-Don, USSR, 1968; through *Chem. Abstr.,* **71,** 35691v (1969).

123. Bukowska, A., *Chem. Anal. (Warsaw)*, **12**, 501 (1967).

124. Bulgakova, A. M., A. P. Mirnaya, and N. V. Obodynskaya, *Metody Anal. Khim. Reaktivov Prep.*, No. 13, 151 (1966); through *Chem. Abstr.*, **68**, 9062h (1968).

125. Bunney, L. R., N. E. Ballou, J. Pascual, and S. Foti, *Anal. Chem.*, **31**, 324 (1959).

126. Burriel-Marti, F., and A. P. Bouza, *Quim. Ind. (Bilbao)*, **3**, 168 (1956); through *Chem. Abstr.*, **53**, 21414i (1959).

127. Burriel-Marti, F., and G. A. Cobo, *An. Quim.*, **64**, 973 (1968); through *Chem. Abstr.*, **70**, 81522s (1969).

128. Burriel-Marti, F., and C. A. Herrero, *An. Edafol. Agrobiol. (Madrid)*, **25**,325 (1966); through *Chem. Abstr.*, **66**, 5984h (1967).

129. Burriel-Marti, F., and S. Jimenez-Gomez, *An. Real Soc. Espan. Fis. Quim.*, Ser. B, **48**, 783 (1952); through *Chem. Abstr.*, **47**, 4243c (1953).

130. Burriel-Marti, F., and A. M. Vidan, *An. Real Soc. Espan. Fis. Quim.*, Ser. B, **61**, 867 (1965); through *Chem. Abstr.*, **64**, 10381g (1966).

131. Burriel-Marti, F., and A. M. Vidan, *An. Real Soc. Espan. Fis. Quim.*, Ser. B, **62**, 139 (1966); through *Chem. Abstr.*, **65**, 4632b (1966).

132. Busev, A. I., *Zh. Anal. Khim.*, **4**, 234 (1949).

133. Busev, A. I., *Analytical Chemistry of Molybdenum*, Ann Arbor-Humphrey Science, Ann Arbor, Mich., 1969.

134. Busev, A. I., V. M. Byr'ko, and I. I. Grandberg, *Vestn. Mosk. Univ., Ser. II*, **15**, 76 (1960); through *Chem. Abstr.*, **55**, 1289e (1961).

135. Busev, A. I., and F. Chang, *Vestn. Mosk. Univ., Ser. Mat. Mekh. Astron. Fiz. Khim.*, **14**, 203 (1959); through *Chem. Abstr.*, **54**, 8457d (1960).

136. Busev, A. I., and F. Chang, *Zh. Anal. Khim.*, **14**, 445 (1959).

137. Busev, A. I., and F. Chang, *Vestn. Mosk. Univ., Ser. II*, **16**, 36 (1961); through *Chem. Abstr.*, **55**, 25596a (1961).

138. Busev, A. I., and F. Chang, *Zh. Anal. Khim.*, **16**, 39 (1961).

139. Busev, A. I., and F. Chang, *Zh. Anal. Khim.*, **16**, 171 (1961).

140. Busev, A. I., F. Chang, and Z. P. Kuzyaeva, *Zh. Anal. Khim.*, **16**, 695 (1961).

141. Busev, A. I., and V. A. Frolkina, *Zh. Neorg. Khim.*, **9**, 2481 (1964).

142. Busev, A. I., and V. A. Frolkina, *Zh. Neorg. Khim.*, **14**, 1289 (1969).

143. Busev, A. I., V. A. Frolkina, and M. Ya. Koroleva, *Zh. Anal. Khim.*, **24**, 205 (1969).

144. Busev, A. I., and L. Gyn, *Zh. Anal. Khim.*, **13**, 519 (1958).

145. Busev, A. I., and T. V. Rodionova, *Vestn. Mosk. Univ. Khim.*, **23**, 63 (1966); through *Chem. Abstr.*, **70**, 23540k (1969).

146. Busev, A. I., and T. V. Rodionova, *Zh. Anal. Khim.*, **23**, 877 (1968).

147. Busev, A. I., and T. V. Rodionova, *Anal. Lett.*, **2**, 9 (1969).

148. Busev, A. I., and G. P. Rudzit, *Zh. Anal. Khim.*, **18**, 840 (1963).

149. Busev, A. I., and G. P. Rudzit, *Zh. Anal. Khim.*, **19**, 102 (1964).

150. Busev, A. I., and G. P. Rudzit, *Zh. Anal. Khim.*, **19**, 569 (1964).

151. Busev, A. I., G. P. Rudzit, and M. Dzintarnieks, *Zh. Anal. Khim.*, **21**, 176 (1966).

152. Busev, A. I., G. P. Rudzit, and C. A. Naku, *Khim. Osnovy Ekstraktsion. Metoda Razdeleneiya Elementov, Akad. Nauk SSSR, Inst. Geokhim. Anal. Khim.*, **1966**, 81; through *Chem. Abstr.*, **65**, 11322d (1966).

153. Buss, H., H. W. Kohlschutter, and S. Miedtank, *Z. Anal. Chem.*, **178**, 1 (1960).

154. Butler, L. R. P., and P. M. Mathews, *Anal. Chim. Acta*, **36**, 319 (1966).

155. Cadiot, M., and M. Lamache-Duhameaux, *C.R.H. Acad. Sci., Ser. C*, **264**, 1282 (1967).

156. Calamari, J. A., R. Hubata, and P. B. Roth, *Ind. Eng. Chem. Anal. Ed.*, **15**, 71 (1943).

157. Candea, C., and I. G. Murgulescu, *Bull. Soc. Chim. Romania*, **17**, 103 (1935).

158. Candea, C., and I. G. Murgulescu, *Ann. Chim. Anal. Chim. Appl.*, **18**, 33 (1936).

159. Candela, M. I., E. J. Hewitt, and H. M. Stevens, *Anal. Chim. Acta*, **14**, 66 (1956).

160. Carlstrom, C. G., and V. Palvarinne, *Jernkontoret Ann.*, **146**, 403 (1962); through *Chem. Abstr.*, **57**, 15790h (1962).

161. Carpenter, L., R. W. Lewis, and K. A. Hazen, *Appl. Spectrosc.*, **20**, 44 (1966).

162. Carriere, E., and A. Dautheville, *Bull. Soc. Chim. Fr.*, **10**, 264 (1943).

163. Carriere, E., and A. Dantheville, *Bull. Soc. Chim. Fr.*, **12**, 923 (1945).

164. Casassas, E., and L. Esk, *An. Real Soc. Espan. Fis. Quim., Ser. B*, **61**, 577 (1965); through *Chem. Abstr.*, **63**, 12295a (1965).

165. Chakrabarti, A. K., and S. P. Bag, *Talanta*, **19**, 1187 (1972).

166. Chalmers, R. A., and G. Svehla, *Solvent Extr. Chem. Proc. Int. Conf., Goteborg*, **1966**, 600 (1967).

167. Chan, F. L., and R. W. Moshier, *Talanta*, **3**, 272 (1960).

168. Chan, K. M., and J. P. Riley, *Anal. Chim. Acta*, **36**, 220 (1966).

169. Chan, S. I., R. J. Kula, and D. T. Sawyer, *J. Amer. Chem. Soc.*, **86**, 377 (1964).

170. Chang, C. C., and H. H. Yang, *Hua Hsueh Tung Pao*, **1965**, 45 (1965); through *Chem. Abstr.*, **63**, 7634d (1969).

171. Chapman, F. W., G. G. Marvin, and S. Y. Tyree, *Anal. Chem.*, **21**, 700 (1949).

172. Charlot, G., *Colorimetric Determination of Elements*, Elsevier, Amsterdam, 1964.

173. Charlot, G., D. Bezier, and R. Gauguin, *Rapid Detection of Cations*, Chemical, New York, 1954.

174. Chau, Y. K., and K. Lum-Shue-Chan, *Anal. Chim. Acta*, **48**, 205 (1969).

175. Cheng, K. L., and B. L. Goydish, *Microchem. J.*, **13**, 35 (1968).

176. Chirkov, S. K., and L. S. Studenskaya, *Byull. Nauch.-Tekh. Inform. Ural. Nauch.-Issled, Inst., Chern. Metal.*, **1959**, 113; through *Chem. Abstr.*, **55**, 4247c (1961).

177. Chojnacka, J., *Omagiu Raluca Ripan*, **1966**, 191; through *Chem. Abstr.*, **69**, 24300y (1968).

178. Chow, S. F., and S. C. Liang, *Sci. Sinica (Peking)*, **8**, 196 (1959).

179. Chow, S. F., and S. C. Liang, *Sci. Sinica (Peking)*, **11**, 207 (1962).

180. Christian, G. D., *Anal. Chem.*, **41**, 24A (1969).

181. Clark, G. C., and E. E. Hale, *Analyst (London)*, **78**, 145 (1953).

182. Clark, R. E. D., *Analyst (London)*, **83**, 396 (1958).

183. Clark, R. E. D., and R. G. Neville, *J. Chem. Educ.*, **36**, 390 (1959).

184. Clarke, B. L., and H. W. Hermance, *Ind. Eng. Chem. Anal. Ed.*, **9**, 292 (1937).

185. Climax Molybdenum Company, *Molybdenum Metal*, Climax Molybdenum, New York, 1961.

186. Clowers, C. C., and J. C. Guvon, *Anal. Chem.*, **41**, 1140 (1969).

187. Coetzee, C. J., and E. F. C. H. Rohwer, *Anal. Chim. Acta*, **44**, 293 (1969).

188. Colombo, U., and G. Sironi, *Anal. Chem.*, **36**, 802 (1964).

189. Connolley, J. F., and M. F. Maguire, *Analyst (London)*, **88**, 125 (1963).

190. Connor, J. A., and E. A. V. Ebsworth, "Peroxy compounds of transition metals," in H. J. Emeleus and A. G. Sharpe, Eds., *Advances in Inorganic Chemistry and Radiochemistry*, Vol. 6, Academic, New York, 1964.

191. Crook, W. J., and L. Crook, *Mining Sci. Press*, **117**, 313 (1918).

192. Crouch, E. A. C., and T. A. Tuplin, *Nature*, **202**, 1282 (1964).

193. Crouthamel, C. E., and C. E. Johnson, *Anal. Chem.*, **26**, 1284 (1954).

194. Crowther, A. B., *UK At. Energy Authority, Ind. Group Hdq.*, No. SCS-R-91 (1959).

195. Danielsson, L., *Acta Chem. Scand.*, **19**, 670 (1965).

196. Danielsson, L., *Acta Chem. Scand.*, **19**, 1859 (1965).

197. Danielsson, L., *Ark. Kemi*, **27**, 467 (1967).

198. Darbinyan, M. V., and D. S. Gaibakyan, *Izv. Akad. Nauk Arm. SSR, Khim. Nauk*, **16**, 335 (1963); through *Chem. Abstr.*, **60**, 11356c (1964).

199. Das, M., *Diss. Abstr. B*, **29**, 1950 (1968).

200. David, D. J., *Analyst (London)*, **93**, 79 (1968).

201. Degtyareva, O. F., and M. F. Ostrovskaya, *Zh. Anal. Khim.*, **22**, 1863 (1967).

202. Degtyareva, O. F., and L. G. Sinitsyna, *Zh. Anal. Khim.*, **20**, 603 (1965).

203. Degtyareva, O. F., and L. G. Sinitsyna, *Zh. Anal. Khim.*, **22**, 1500 (1967).

204. Dekhtrikyan, S. A., *Dokl. Akad. Nauk Arm. SSR*, **28**, 213 (1959).

205. Deniges, G., *Bull. Soc. Pharm. Bordeaux*, **65**, 50 (1927).

206. Deshmukh, G. S., *Bull. Chem. Soc. Jap.*, **29**, 27 (1956).

207. Deshmukh, G. S., and J. P. Srivastava, *Z. Anal. Chem.*, **176**, 28 (1960).

208. Dhaneshwar, R. G., "Voltammetric Studies Using Different Electrode Systems," in Kapoor, R. C., Ed., *Proc. Symp. Electrode Processes*, pp. 55–74, Univ. Jodhpur, Jodhpur, India, 1966; through *Chem. Abstr.*, **70**, 43362g (1969).

209. Dhara, S. C., and S. M. Khopkar, *Indian J. Chem.*, **5**, 12 (1967).

210. Dhumwad, R. K., M. V. Joshi, and A. B. Patwardham, *Anal. Chim. Acta*, **42**, 334, (1968).

211. Dick, A. T., and J. B. Bingley, *Nature*, **158**, 516 (1946).

212. Dikshitulu, L. S. A., and G. G. Rao, *Z. Anal. Chem.*, **202**, 344 (1964).

213. Dixon, E. J., and J. B. Headridge, *Analyst (London)*, **89**, 185 (1956).

214. Doege, H. G., *Anal. Chim. Acta*, **38**, 207 (1967).

215. Dolezal, J., B. Moldan, and J. Zyka, *Collect. Czech., Chem. Commun.*, **24**, 3769 (1959).

216. Donaldson, E. M., and W. R. Inman, *Talanta*, **13**, 489 (1966).

217. Doran, J. W., and D. C. Martens, *J. Ass. Offic. Anal. Chem.*, **53**, 228 (1970).

218. Doring, T., *Z. Anal. Chem.*, **82**, 193 (1930).

219. Duca, A., D. Stanescu, and M. Puscasu, *Rev. Roum. Chim.*, **11**, 833 (1966).

220. Duca, A., D. Stanescu, and M. Puscasu, *Rev. Roum. Chim.*, **11**, 839 (1966).

221. Dutt, Y., and R. R. Singh, *Curr. Sci.*, **35**, 122 (1966).

222. Duval, L., *Chim. Anal.*, **51**, 415 (1969).

223. D'yachenko, S. S., N. A. Agrinskaya, and V. I. Petrashen, *Zavod. Lab.*, **36**, 23 (1970).

224. Dziomko, V. M., and A. I. Cherepakhin, *Vses. Zaochnyi Politekh. Inst., Sb. Statei*, No. 9, 65 (1955); through *Chem. Abstr.*, **51**, 4196e (1957).

225. Easton, A. J., and A. A. Moss, *Mineral. Mag.*, **35**, 995 (1966).

226. Eberle, A. R., and M. W. Lerner, *Anal. Chem.*, **34**, 627 (1962).

227. Edge, R. A., and G. W. A. Fowles, *Anal. Chim. Acta*, **32**, 191 (1965).

228. Elbeih, I. I. M., and M. A. Abon-Elnaga, *Can. J. Chem.*, **46**, 1379 (1968).

229. Elbeih, I. I. M., A. Elbakry, and M. M. Aly, *Chem.-Anal.*, **54**, 39 (1965).

230. Elinson, S. V., S. B. Savvin, and T. I. Nezhnova, *Zh. Anal. Khim.*, **22**, 531 (1967).

231. Elles, R., and R. V. Olson, *Anal. Chem.*, **22**, 328 (1950).

232. El-Shamy, H. K., and M. F. Barakat, *Egypt. J. Chem.*, **2**, 101 (1959).

233. El-Shamy, H. K., and M. F. Barakat, *Egypt. J. Chem.*, **2**, 191 (1959).

234. El-Shamy, H. K., and A. M. El-Aggan, *J. Amer. Chem. Soc.*, **75**, 1187 (1953).

235. Elwell, W. T., and D. F. Wood, *Analytical Chemistry of Molybdenum and Tungsten*, Pergamon, Oxford, 1971.

236. Endo, Y., T. Hata, and Y. Nakahara, *Bunseki Kagaku*, **18**, 878 (1969).

237. Endo, Y., and T. Tomori, *Bunseki Kagaku*, **11**, 1310 (1963).

238. Eolyan, S. L., *Zh. Eksp. Klim. Med.*, **5**, 70 (1965); through *Chem. Abstr.*, **66**, 118575q (1967).

239. Erametsa, O., *Suomen Kemi.*, **A40**, 121 (1967); through *Chem. Abstr.*, **67**, 113566r (1967).

240. Eristavi, D. I., F. I. Brouchek, and T. G. Macharashvili, *Tr. Gruz. Politekh. Inst.*, No. 3, 53 (1968); through *Chem. Abstr.*, **71**, 45346z (1969).

241. Erkelens, P. C. van, *Anal. Chim. Acta*, **25**, 42 (1961).

242. Erkelens, P. C. van, *Anal. Chim. Acta*, **25**, 129 (1961).

243. Erkelans, P. C. van, *Anal. Chim. Acta*, **25**, 226 (1961).

244. Eschnauer, H., and R. Neeb, *Z. Lebensm.-Unters. Forsch.*, **116**, 1 (1961).

245. Evans, B. S., and D. G. Higgs, *Analyst (London)*, **70**, 75 (1945).

246. Evans, H. J., *Soil Sci.*, **81**, 199 (1956).

247. Fagel, J. E., H. A. Liebhafsky, and P. D. Zemany, *Anal. Chem.*, **30**, 1918 (1958).

248. Farah, M., and S. Z. Mikhail, *Z. Anal. Chem.*, **166**, 24 (1959).

249. Faris, J. P., *Anal. Chem.*, **32**, 520 (1960).

250. Faris, J. P., and R. Buchanan, *Anal. Chem.*, **36**, 1157 (1964).

251. Farquhar, M. C., J. A. Hill, and M. M. English, *Anal. Chem.*, **38**, 208 (1966).

252. Farrell, R. F., G. J. Harter, and R. M. Jacobs, *Anal. Chem.*, **31**, 1550 (1959).

253. Fassel, V. A., and D. W. Golightly, *Anal. Chem.*, **39**, 466 (1967).

254. Fassel, V. A., R. B. Myers, and R. N. Kniseley, *Spectrochim. Acta*, **19**, 1187 (1963).

255. Fedoroff, M., *Bull. Soc. Chim. Fr.*, **1968**, 3451 (1968).

256. Fedoroff, M., *C.R.H. Acad. Sci., Paris, Ser. C*, **267**, 1227 (1968).

257. Fedoroff, M., *C.R.H. Acad. Sci., Paris, Ser. C,* **270,** 486 (1970).
258. Feigl, F., *Z. Anal. Chem.,* **65,** 24 (1924).
259. Feigl, F., *Qualitative Analysis by Spot Tests,* Elsevier, New York, 1946.
260. Feigl, F., and C. Costa Neto, *Monatsh. Chem.,* **86,** 336 (1955).
261. Feik, F., and J. Korkisch, *Mikrochim. Acta,* **1967,** 900 (1967).
262. Feldman, F. J., and G. D. Christian, *J. Electroanal. Chem.,* **12,** 199 (1966).
263. Ferraro, T. A., *Talanta,* **15,** 923 (1968).
264. Fidler, J., *Collect. Czech. Chem. Commun.,* **14,** 645 (1949).
265. Filenko, A. I., *Zh. Anal. Khim.,* **22,** 161 (1967).
266. Fischer, C., P. Muehl, and G. Guenzler, *Z. Chem.,* **8,** 235 (1968).
267. Fisher, S. A., and V. W. Meloche, *Anal. Chem.,* **24,** 1100 (1952).
268. Fishman, M. J., and E. C. Mallory, *J. Water Pollut. Contr. Fed.,* **40,** R67 (1968).
269. Flaschka, H., *Z. Anal. Chem.,* **137,** 107 (1952).
270. Flaschla, H., and H. Jakobljevich, *Anal. Chim. Acta,* **4,** 356 (1950).
271. Fogel'son, E. I., and N. V. Kalmykova, *Zavod. Lab.,* **11,** 31 (1945).
272. Fontes, G., and L. Thivolle, *Bull. Soc. Chim. Fr.,* **33,** 835 (1923).
273. Forrest, J., and H. L. Finston, *Appl. Spectrosc.,* **14,** 127 (1960).
274. Freegarde, M., and B. Jones, *Analyst (London),* **84,** 393 (1959).
275. Frishberg, A. A., *Zh. Prikl. Spektrosk.,* **5,** 12 (1966).
276. Fritz, H., *Mikroskopie,* **19,** 110 (1964).
277. Fritz, J. S., and D. R. Beuerman, *Anal. Chem.,* **44,** 692 (1972).
278. Fritz, J. S., and L. H. Dahmer, *Anal. Chem.,* **37,** 1272 (1965).
279. Fritz, J. S., and L. H. Dahmer, *Anal. Chem.,* **40,** 20 (1968).
280. Fritz, J. S., and B. B. Garralda, *Anal. Chem.,* **34,** 102 (1962).
281. Fritz, J. S., B. B. Garralda, and S. K. Karraker, *Anal. Chem.,* **33,** 882 (1961).
282. Fritz, J. S., and M. L. Gillette, *Talanta,* **15,** 287 (1968).
283. Fritz, J. S., and C. E. Hedrick, *Anal. Chem.,* **36,** 1324 (1964).
284. Fritze, K., *Radiochim. Acta,* **5,** 164 (1966).
285. Fuge, R., *Analyst (London),* **95,** 171 (1970).
286. Fukai, R., and W. W. Meinke, *Limnol. Oceanogr.,* **7,** 186 (1962).
287. Fukker, K., and A. J. Hegedus, *Mikrochim. Acta,* **1961,** 92 (1961).
288. Fukker, K., and A. J. Hegedus, *Mikrochim. Acta,* **1961,** 226 (1961).
289. Funck, A. D., *Z. Anal. Chem.,* **68,** 283 (1926).
290. Furman, N. H., and W. M. Murry, *J. Amer. Chem. Soc.,* **58,** 1689 (1936).
291. Gagliardi, E., and W. Pilz, *Monatsh. Chem.,* **82,** 1012 (1951).
292. Gagliardi, E., and W. Pilz, *Z. Anal. Chem.,* **136,** 103 (1952).
293. Gaibakyan, D. S., *Arm. Khim. Zh.,* **23,** 91 (1970).
294. Gaibakyan, D. S., and M. V. Darbinyan, *Izv. Akad. Nauk Arm. SSR, Khim. Nauk,* **15,** 321 (1962); through *Chem. Abstr.,* **58,** 8395d (1963).
295. Gaibakyan, D. S., and M. V. Darbinyan, *Izv. Akad. Nauk Arm. SSR, Khim. Nauk,* **17,** 631 (1964); through *Chem. Abstr.,* **63,** 2418h (1965).
296. Gallagher, M. J., *Inst. Mining Met. Trans., Sec. B,* **77,** 129 (1968).

297. Gallai, Z. A., *Nauch. Dokl. Vyssh. Shkolv, Khim. Khim. Tekhol.*, No. 3, 498, (1958); through *Chem. Abstr.*, **53**, 981a (1959).

298. Galli, A., *Ann. Biol. Clin. Paris*, **24**, 165 (1966).

299. Galliford, D. J. B., and E. J. Neuman, *Analyst (London)*, **87**, 68 (1962).

300. Ganiev, A. G., M. K. Nazmitdinov, and S. S. Sabirov, *Izv. Akad. Nauk Uzb. SSR, Ser. Fiz.-Mat. Nauk*, **12**, 40 (1968); through *Chem. Abstr.*, **69**, 15710e (1968).

301. Garcia, C., and G. L. J. Feliz, *Rev. Soc. Quim. Mex.*, **13**, 222A (1969).

302. Gardner, R. D., C. H. Ward, and W. H. Ashley, *NASA Accession Rep.*, No. N65-36212, (1964).

303. Geldhof, M. L., J. Eeckhaut, and P. Cornand, *Bull. Soc. Chim. Belges*, **65**, 706 (1956).

304. General Mills Inc., Brit. Pat. 907,465, Nov. 18, 1959; through *Chem. Abstr.*, **57**, 16225f (1962).

305. Gentry, C. H. R., and L. G. Sherrington, *Analyst (London)*, **75**, 17 (1950).

306. Gerisch, S., and S. Ziegenbolg, *Freiberg Forschung. B*, **128**, 67 (1968); through *Chem. Abstr.*, **69**, 45378v (1968).

307. Geyer, R., *Z. Anorg. Allg. Chem.*, **271** 93 (1952).

308. Geyer, R., G. Henze, J. Henze, and E. G. Mueller, *Proc. Conf. Appl. Phy.-Chem. Methods Chem. Anal.*, Budapest, **1**, 211 (1966).

309. Geyer, R., W. Hermann, and B. Bogatzki, *Wiss. Z. Tech. Hochsch. Chem. "Carl Schorlemner," Leuna-Mersburg*, **9**, 1 (1967); through *Chem. Abstr.*, **67**, 121950b (1967).

310. Gilbert, T. W., and E. B. Sandell, *J. Amer. Chem. Soc.*, **82**, 1087 (1960).

311. Gillis, J., A. Claeys, and J. Hoste, *Anal. Chim. Acta*, **1**, 421 (1947).

312. Ginzburg, L. B., and Yu. Yu. Lur'e, *Zavod. Lab.*, **14**, 538 (1948).

313. Ginzburg, L. B., and E. P. Shkrobot, *Analiz Rud Tsvetnykh Metal. Productov Ikt Pererabotki*, **1956**, 89; through *Chem. Abstr.*, **51**, 11141e (1957).

314. Ginzburg, L. B., and E. P. Shkrobot, *Sb. Nauch. Tr. Gos. Nauch.-Issled. Inst. Tsvet. Metal.*, **1956**, 89; through *Chem. Abstr.*, **54**, 7429i (1960).

315. Glemser, O., W. Holznagel, and S. I. Ali, *Naturforscher*, **20B**, 192 (1965).

316. Glushchenko, A. V., and I. N. Lyubimova, *Agrokhimiya*, **1967**, 134 (1967).

317. Godneva, M. M., and R. D. Vodyannekova, *Zh. Anal. Khim.*, **20**, 831 (1965).

318. Goinatskii, M. N., *Mikroelem. Med.*, **1968**, 103; through *Chem. Abstr.*, **71**, 47204a (1969).

319. Goldstein, G., D. L. Manning, and O. Menis, *Anal. Chem.*, **30**, 539 (1958).

320. Golubev, I. M., *Agrokhimiya*, **1966**, 135 (1966).

321. Gorbach, G., and K. Kvisler, *Fette Seifen, Anstrichm.*, **70**, 659 (1968); through *Chem. Abstr.*, **70**, 12821b (1969).

322. Gorlach, E., *Rocz. Gleboznawcze*, **14**, 15 (1964); through *Chem. Abstr.*, **64**, 1344c (1966).

323. Goryushina, V. G., and T. V. Cherkashina, *Zavod. Lab.*, **14**, 255 (1948).

324. Goto, H., *J. Chem. Soc. Jap.*, **56**, 314 (1935).

325. Goto, H., *Sci. Rep. Res. Inst., Tohoku Imp. Univ., Ser. A*, **29**, 204 (1940); through *Chem. Abstr.*, **35**, 1721[7] (1941).

326. Goto, H., and S. Ikeda, *Nippon Kagaku Zasshi,* **77,** 82 (1956).

327. Goto, H., and Y. Kakita, *Sci. Rep. Res. Inst., Tohoku Univ., Ser. A,* **11,** 1 (1959); through *Chem. Abstr.,* **54,** 18173b (1960).

328. Goto, H., Y. Kakita, and K. Hirokawa, *Nippon Kagaku Zasshi,* **78,** 1343 (1957).

329. Goto, H., Y. Kakita, and M. Namiki, *Nippon Kagaku Zasshi,* **82,** 580 (1961).

330. Goto, H., and A. Saito, *Bunseki Kagaku,* **17,** 194 (1968).

331. Govaerts, J., and C. Barcia-Goyanes, *Nature,* **168,** 198 (1951).

332. Gowda, H. S., and R. Shakunthala, *Talanta,* **13,** 1375 (1966).

333. Grasshoff, K., and H. Hahn, *Z. Anal. Chem.,* **186,** 132 (1962).

334. Green, H., *Talanta,* **11,** 1561, (1964).

335. Greenberg, P., *Anal. Chem.,* **29,** 896 (1957).

336. Gregory, E., R. B. Foulston, and F. W. Gray, *Analyst (London),* **66,** 444 (1941).

337. Grigg, J. L., *Analyst (London),* **78,** 470 (1953).

338. Grubitsch, H., K. Halvorsen, and G. Schindler, *Z. Anal. Chem.,* **173,** 414 (1960).

339. Grubitsch, H., and T. Heggeboe, *Monatsh. Chem.,* **93,** 274 (1962).

340. Gupta, C. M., and J. K. Gupta, *J. Indian Chem. Soc.,* **44,** 526 (1967).

341. Gupta, J. K., and C. M. Gupta, *Inorg. Chim. Acta,* **3,** 358 (1969).

342. Gusev, S. I., and E. M. Nikolaeva, *Zh. Anal. Khim.,* **19,** 715 (1964).

343. Haas, W., and H. Faber, *Z. Anal. Chem.,* **193,** 89 (1963).

344. Haas, W., and T. Schwarz, *Mikrochim. Ichnoanal. Acta,* **1963,** 253 (1963).

345. Hadjiioannou, T. P., *Anal. Chim. Acta,* **35,** 360 (1966).

346. Haeringer, M., G. Goldstein, P. Lagrange, and J. P. Schwing, *Bull. Soc. Chim. Fr.,* **1967,** 723 (1967).

347. Haight, G. P., *Anal. Chem.,* **23,** 1505 (1951).

348. Haight, G. P., and V. Paragamian, *Anal. Chem.,* **32,** 642 (1960).

349. Hakkila, E. A., R. G. Hurley, and G. R. Waterbury, *US At. Energy Comm.,* CONF-285-3, 1 (1961).

350. Hakkila, E. A., and G. R. Waterbury, *Talanta,* **6,** 46 (1960).

351. Halasz, A., and E. Pungor, *Magy. Kem. Foly.,* **74,** 545 (1968).

352. Halasz, A., and E. Pungor, *Talanta,* **18,** 557 (1971).

353. Haley, L. E., and S. W. Melsted, *Soil Sci. Soc. Amer. Proc.,* **21,** 316 (1957).

354. Handley, T. H., *Nucl. Sci. Eng.,* **16,** 440 (1963).

355. Hargis, L. G., *Anal. Chem.,* **41,** 597 (1969).

356. Hargis, L. G., and D. F. Boltz, *Anal. Chem.,* **37,** 240 (1965).

357. Harrison, G. R., *Massachusetts Institute of Technology Wavelength Tables,* MIT, Cambridge, Mass., 1969.

358. Hartkamp, H., *Z. Anal. Chem.,* **231,** 161 (1967).

359. Hartkamp, H., *Z. Anal. Chem.,* **241,** 66 (1968).

360. Hashmi, M. H., M. A. Shahid, A. A. Ayaz, F. R. Chughtai, N. Hassan, and A. S. Adil, *Anal. Chem.,* **38,** 1554 (1966).

361. Hashmi, M. H., M. A. Shahid, N. A. Chughtai, and M. A. Rana, *Mikrochim. Acta,* **1968,** 1143 (1968).

362. Hayashi, S., *Bunseki Kagaku,* **11,** 438 (1962).

363. Headridge, J. B., *Analyst (London)*, **85,** 379 (1960).

364. Headridge, J. B., and D. P. Hubbard, *Analyst (London)*, **90,** 173 (1965).

365. Headridge, J. B., and D. P. Hubbard, *Anal. Chim. Acta*, **35,** 85 (1966).

366. Headridge, J. B., and M. S. Taylor, *Analyst (London)*, **88,** 590 (1963).

367. Healy, W. B., and L. C. Bates, *Anal. Chim. Acta*, **33,** 443 (1965).

368. Healy, W. B., and W. J. McCabe, *Anal. Chem.*, **35,** 2117 (1963).

369. Hemmeler, A., and E. Boni, *Studi. Urbinati, Fac. Farm.*, **30,** 18 (1956); through *Chem. Abstr.*, **54,** 22144e (1960).

370. Henkel, H., *Z. Anal. Chem.*, **119,** 326 (1940).

371. Henrickson, R. B., and E. B. Sandell, *Anal. Chim. Acta*, **7,** 57 (1952).

372. Henze, G., R. Geyer, and W. Lahl, *Neue Huette*, **14,** 54 (1969).

373. Hernandez Mendez, J., and F. Lucena Conder, *An. Quim.*, **64,** 71 (1968).

374. Hettche, H. O., *Air Water Pollut.*, **8,** 185 (1964).

375. Hewitt, E. J., and D. G. Hallas, *Plant Soil*, **3,** 366 (1951).

376. Hibbits, J. O., W. F. Davis, and M. R. Menke, *US At. Energy Comm.*, APEX-519, 1 (1959).

377. Hibbits, J. O., W. F. Davis, and M. R. Menke, *Talanta*, **4,** 104 (1960).

378. Hibbits, J. O., and R. T. Williams, *Anal. Chim. Acta*, **26,** 363 (1962).

379. Hiskey, C. F., V. F. Springer, and V. W. Meloche, *J. Amer. Chem. Soc.*, **61,** 3125 (1939).

380. Hobart, E. W., and E. P. Hurley, *Anal. Chim. Acta*, **27,** 144 (1962).

381. Hobbs, R. S., G. F. Kirkbright, and T. S. West, *Analyst (London)*, **94,** 554 (1969).

382. Hoenes, H. J., and K. G. Stone, *Talanta*, **4,** 250 (1960).

383. Hoffman, J. I., *Chem.-Anal.*, **50,** 30 (1961).

384. Hoffman, J. I., and G. E. F. Lundell, *J. Res. Nat. Bur. Stand.*, **23,** 497 (1939).

385. Holden, N. E., and F. W. Walker, *Chart of the Nuclides*, 10th edit., US At. Energy Comm., Washington, D.C., 1969..

386. Holliday, R. W., "Molybdenum," in *Mineral Facts and Problems*, US Bureau of Mines, Bulletin 630, US Dept. of Interior, Washington, D.C., 1965.

387. Holness, H., and K. R. Lawrence, *Analyst (London)*, **78,** 356 (1953).

388. Holten, C. H., *Acta Chem. Scand.*, **15,** 943 (1961).

389. Holtje, R., and R. Geyer, *Z. Anorg. Allg. Chem.*, **246,** 243 (1941).

390. Holtje, R., and R. Geyer, *Z. Anorg. Allg. Chem.*, **246,** 258 (1941).

391. Holzapfel, H., O. Guertler, and B. Tempel, *Z. Anal. Chem.*, **235,** 413 (1968).

392. Honsa, I., and V. Suk, *Collect. Czech. Chem. Commun.*, **35,** 1283 (1970).

393. Hoover, W. L., and S. C. Duren, *J. Ass. Offic. Anal. Chem.*, **50,** 1269 (1967).

394. Horton, F. W., *Molybdenum: Its Ores and Their Concentration*, US Bureau of Mines, Bulletin 111, US Dept. Interior, Washington, D.C., 1916.

395. Hosking, K. F. G., *Mining Mag. (London)*, **109,** 22 (1963).

396. Hovorka, V., *Collect. Czech. Chem. Commun.*, **10,** 527 (1938).

397. Hsu, P. Y., and C. Y. Wang, *Hua Hsueh Hsueh Pao*, **31,** 264 (1965); through *Chem. Abstr.*, **63,** 15531c (1965).

398. Hu, Z. T., and S. C. Shi, *Hua Hsueh Tung Pao*, **1963,** 312; through *Chem. Abstr.*, **60,** 6d (1964).

399. Huey, F., and L. G. Hargis, *Anal. Chem.,* **39,** 125 (1967).

400. Huff, E. A., *Anal. Chem.,* **37,** 533 (1965).

401. Huffman, E. H., R. L. Oswalt, and L. A. Williams, *J. Inorg. Nucl. Chem.,* **3,** 49 (1956).

402. Hurd, L. C., and H. O. Allen, *Ind. Eng. Chem. Anal. Ed.,* **7,** 396 (1935).

403. Hurd, L. C., and F. Reynolds, *Ind. Eng. Chem. Anal. Ed.,* **6,** 477 (1934).

404. Ibrahim, S. H., and A. P. M. Nair, *J. Madras Univ.,* **26B,** 521 (1956); through *Chem. Abstr.,* **52,** 12668b (1958).

405. Iguchi, A., *Sci. Papers Coll. Gen. Educ., Univ. Tokyo,* **6,** 153 (1957); through *Chem. Abstr.,* **51,** 9255e (1957).

406. Illingworth, J. W., and J. A. Santos, *Nature,* **134,** 971 (1934).

407. Imai, T., and S. Nagumo, *Kogyo Kagaku Zasshi,* **61,** 53 (1958).

408. Iordanov, N., and S. Mareva, *C.R. Acad. Bulg. Sci.,* **19,** 913 (1966).

409. Iordanov, N., S. Mareva, G. Borisov, and B. Iordanov, *Talanta,* **15,** 221 (1968).

410. Isakov, P. M., *Nauch. Byull. Leningrad. Gos. Univ., A. A. Zhdanova,* No. 33, 31 (1955); through *Chem. Abstr.,* **54,** 11844c (1960).

411. Ishibashi, M., T. Fujinaga, and T. Kuwamoto, *Nippon Kagaku Zasshi,* **79,** 1496 (1958).

412. Ishibashi, N., H. Kohara, and K. Abe, *Bunseki Kagaku,* **17,** 154 (1968).

413. Ishida, K., and R. Kuroda, *Anal. Chem.,* **39,** 212 (1967).

414. Ishimori, T., H. M. Sammour, K. Kimura, and H. Murakami, *Nippon Genshiryoku Gakkaishi,* **3,** 698 (1961); through *Chem. Abstr.,* **58,** 3955g (1963).

415. Ishimori, T., and K. Watanabe, *Bull. Chem. Soc. Jap.,* **33,** 1443 (1960).

416. Isibasi, S., *J. Chem. Soc. Jap.,* **61,** 125 (1940).

417. Issa, I. M., and H. Khalifa, *Anal. Chim. Acta,* **10,** 567 (1954).

418. Ivanov, D. N., and V. A. Bol'shakov, *Khim. Sel. Khoz.,* **7,** 299 (1969); through *Chem. Abstr.,* **71,** 29593x (1969).

419. Ivanov, N., and B. Karadakov, *Godishnik Khim.-Tekhnol. Inst.,* **11,** 1 (1964); through *Chem. Abstr.,* **65,** 12855c (1966).

420. Ivanova, G. F., *Zh. Anal. Khim.,* **20,** 82 (1965).

421. Ivanova, G. F., *Zh. Anal. Khim.,* **21,** 1307 (1966).

422. Jacobsohn, K., and M. D. Azevedo, *Rev. Port. Quim.,* **5,** 99 (1963).

423. Jakubiec, R. J., and D. F. Boltz, *Anal. Chim. Acta,* **43,** 137 (1968).

424. Jakubiec, R. J., and D. F. Boltz, *Anal. Chem.,* **41,** 78 (1969).

425. Jakubiec, R. J., and D. F. Boltz, *Mikrochim. Acta,* **1969,** 181 (1969).

426. James, L. H., *Ind. Eng. Chem. Anal. Ed.,* **4,** 89 (1932).

427. Jannasch, P., and O. Laubi, *J. Prakt. Chem.,* **97,** 154 (1918).

428. Jean, M., *Ann. Chim. (Paris),* **3,** 470 (1948).

429. Jean, M., *Chim. Anal. (Paris),* **44,** 195 (1962).

430. Jean, M., *Anal. Chim. Acta,* **42,** 543 (1968).

431. Jean, M., *Chim. Anal. (Paris),* **51,** 396 (1969).

432. Jeffrey, P. G., *Analyst (London),* **82,** 558 (1957).

433. Jelen, K., K. Ostrowski, E. Rulikowska, P. Schleifer, and E. Rykiert, *Szklo Ceram.,* **19,** 36 (1968); through *Chem. Abstr.,* **69,** 99005q (1968).

434. Jenkins, G. N., *Brit. Dent. J.*, **122**, 435 (1967).

435. Jenkins, I. L., and A. G. Wain, *J. Appl. Chem.*, **14**, 449 (1964).

436. Jezowska-Trzebiatowska, B., S. Kopacz, and A. Bartecki, *Zh. Neorg. Khim.*, **13**, 1899 (1968).

437. Jilek, A., and B. Bieber, *Acad. Tchneque Sci. Bull. Intern., Caisse,* **52**, 447 (1951); through *Chem. Abstr.*, **49**, 9443e (1955).

438. Jimeno, S., *Inform. Quim. Anal. (Madrid)*, **15**, 161 (1961).

439. Jogdeo, S. M., and L. M. Mahajan, *Indian J. Chem.*, **3**, 468 (1965).

440. Johnson, M. G., and R. J. Robinson, *Anal. Chem.*, **24**, 366 (1952).

441. Johnson, R. A., *Radiochim. Acta*, **5**, 231 (1966).

442. Johri, K. N., N. K. Kaushik, and K. Singh, *Talanta*, **16**, 432 (1969).

443. Jones, M. M., *J. Amer. Chem. Soc.*, **76**, 4233 (1954).

444. Julietti, R. J., and J. A. Wilkinson, *Analyst (London)*, **93**, 797 (1968).

445. Juvet, R. S., and R. L. Fisher, *Anal. Chem.*, **37**, 1752 (1965).

446. Juvet, R. S., and R. L. Fisher, *Anal. Chem.*, **38**, 1860 (1966).

447. Kabir-ud-Din, H. N., A. A. Khan, and M. A. Beg, *J. Electroanal. Chem.*, **19** 175 (1968).

448. Kakita, Y., and H. Goto, *Talanta*, **14**, 543 (1967).

449. Kakita, Y., E. Sudo, and M. Namiki, *Sci. Rep. Res. Inst. Tohoku Univ., Ser. A,* **13**, 199 (1961); through *Chem. Abstr.*, **56**, 4082i (1962).

450. Kalinin, A. I., R. A. Kuznekov, V. V. Moiseev, and V. E. Cepurnieks, *Radiokhim. Metody Opred Microelem., Akad. Nauk SSSR, Sb. Statei,* **1965**, 161; through *Chem. Abstr.*, **63**, 10658f (1965).

451. Kalinskii, Ya. M., E. G. Naruta, N. I. Chabanenko, and K. I. Nikiforova, *Sb. Tr. Chelyabinsk. Elektromet. Komb.*, **1**, 182 (1968); through *Chem. Abstr.*, **71**, 56358f (1969).

452. Kallmann, S., E. W. Hobart, and H. K. Oberthin, *Talanta*, **15**, 982 (1968).

453. Kallmann, S., R. Liu, H. Oberthin, R. Stevenson, and F. Lux, *US Dept. Comm.,* AD 622275, 1 (1965).

454. Kalnin, I. L., *Anal. Chem.*, **36**, 886 (1964).

455. Kalt, M. B., and D. F. Boltz, *Anal. Chem.*, **40**, 1086 (1968).

456. Kambara, T., S. Tanaka, and K. Fukada, *Bunseki Kagaku*, **17**, 1144 (1968).

457. Kamiya, S., M. Tokutomi, and Y. Matsuda, *Bull. Chem. Soc. Jap.*, **40**, 407 (1967).

458. Kapron, M., and P. L. Hehman, *Ind. Eng. Chem. Anal. Ed.*, **17**, 573 (1945).

459. Karabash, A. G., L. S. Bondarenko, G. G. Morozova, and S. I. Peizulaev, *Zh. Anal. Khim.*, **15**, 623 (1960).

460. Karpova, L. A., E. F. Speranskaya, and M. T. Kozlovskii, *Zh. Anal. Khim.*, **23**, 13 (1968).

461. Kashima, J., M. Kubota, and T. Kawasaki, *Bunko Kenkyu*, **17**, 215 (1969); through *Chem. Abstr.*, **71**, 45418z (1969).

462. Kato, H., and H. Hosimiya, *J. Chem. Soc. Jap.*, **60**, 1115 (1939).

463. Kawabuchi, K., *Nippon Kagaku Zasshi*, **87**, 262 (1966).

464. Kawabuchi, K., and R. Kuroda, *Anal. Chim. Acta*, **46**, 23 (1969).

465. Kawahata, M., H. Mochizuki, and R. Kajiyama, *Bunseki Kagaku*, **8**, 25 (1959).

466. Kawahata, M., H. Mochizuki, R. Kajiyama, and K. Irikura, *Bunseki Kagaku,* **11,** 317 (1962).

467. Kawahata, M., H. Mochizuki, R. Kajiyama, and M. Ishii, *Bunseki Kagaku,* **11,** 748 (1962).

468. Kawashima, I., *Bunko Kenkyu,* **17,** 22 (1968); through *Chem. Abstr.,* **70,** 43760k (1969).

469. Kawashima, I., and K. Tokiwa, *Nippon Kinzoku Gakkaishi,* **29,** 1201 (1965).

470. Kedesdy, E., *Mitt. Kgl. Materialprufungsant.,* **31,** 173 (1913); through *Chem. Abstr.,* **7,** 3940[1] (1913).

471. Kedrova, Yu. K., *Analiz Rud Tsvetnykh Metal. Produktov Ikh Pererabothi, Sb. Nauch. Trudov,* **14,** 21 (1958); through *Chem. Abstr.,* **53,** 21410h (1959).

472. Keene, B. J., *Talanta,* **13,** 1443 (1966).

473. Kepert, D. L., *Inorg. Chem.,* **8,** 1556 (1969).

474. Kerbyson, J. D., and C. Ratzkowski, *Can. Spectrosc.,* **15,** 43 (1970).

475. Kessler, F. M., *Prace Ustavu Pro Vyzkum Vyuziti Paliv,* No. 1, 140 (1955); through *Chem. Abstr.,* **50,** 3150f (1956).

476. Khalizova, V. A., L. P. Volkova, and E. P. Smirnova, *Mineral Syr'e Moscow Sb.,* **1,** 307 (1960); through *Chem. Abstr.,* **55,** 26843a (1961).

477. Khan, G. A., and S. F. Abdullin, *Tsvet. Metal.,* **37,** 10 (1964); through *Chem. Abstr.,* **62,** 5884g (1965).

478. Khan, G. A., A. V. Titovskii, and G. S. Mansurov, *Izv. Vyssh. Ucheb. Zaved., Tsvet. Met.,* **11,** 146 (1968); through *Chem. Abstr.,* **71,** 27198y (1969).

479. Khanna, V. B., and B. M. Sood, *ISI (Indian Stand. Inst.) Bull.,* **21,** 133 (1969).

480. Kharlamov, I. P., and P. Ya. Yakovlev, *Zavod. Lab.,* **23,** 535 (1957).

481. Kharlamov, I. P., P. Ya. Yakovlev, and M. I. Lykova, *Zavod. Lab.,* **26,** 933 (1960).

482. Khophar, S. M., and A. K. De, *Anal. Chim. Acta,* **23,** 147 (1960).

483. Khosla, M. M., and S. P. Rao, *Anal. Chim. Acta,* **57,** 323 (1971).

484. Kiboku, M., *Bunseki Kagaku,* **6,** 356 (1957).

485. Killeffer, D. H., and A. Lenz, *Molybdenum Compounds, Wiley-Interscience, New York, 1952.*

486. Kim, Y. S., and H. Zeitlin, *Anal. Chim. Acta,* **51,** 516 (1970).

487. Kimura, K., *Bull. Chem. Soc. Jap.,* **34,** 63 (1961).

488. Kirkbright, G. F., M. K. Peters, and T. S. West, *Analyst (London),* **91,** 705 (1966).

489. Kirkbright, G. F., M. K. Peters, and T. S. West, *Talanta,* **14,** 789 (1967).

490. Kirkbright, G. F., A. Semb, and T. S. West, *Spectrosc. Lett.,* **1,** 7 (1968).

491. Kirkbright, G. F., T. S. West, and C. Woodward, *Talanta,* **13,** 1645 (1966).

492. Kirkbright, G. F., and J. H. Yoe, *Talanta,* **11,** 415 (1964).

493. Kiss, A., *Magy. Kem. Foly.,* **69,** 524 (1963); through *Chem. Abstr.,* **61,** 85e (1964).

494. Kiss, A., *Magy. Kem. Foly.,* **75,** 302 (1969); through *Chem. Abstr.,* **71,** 108722s (1969).

495. Kiss, A., and A. J. Hegedus, *Mikrochim. Acta,* **1966,** 771 (1966).

496. Kitahara, S., *Rep. Sci. Res. Inst. (Japan),* **24,** 454 (1948).

497. Kitahara, S., *Rep. Sci. Res. Inst. (Japan),* **25,** 165 (1949).

498. Kitajima, S., *Sci. Pap. Inst. Phys. Chem. Res. (Tokyo)*, **16,** 285 (1931).

499. Klement, R., *Z. Anal. Chem.*, **136,** 17 (1952).

500. Klinger, P., *Arch. Eisenhuettenw.*, **14,** 157 (1940).

501. Klinkmann, A., *Metall (Berlin)*, **15,** 1198 (1961).

502. Klitina, V. I., F. P. Sudakov, and I. P. Alimarin, *Zh. Anal. Khim.*, **20,** 1145 (1965).

503. Klitina, V. I., F. P. Sudakov, and I. P. Alimarin, *Zh. Anal. Khim.*, **21,** 338 (1966).

504. Klockow, D., *Talanta*, **14,** 817 (1967).

505. Klug, O. N., and S. Metlenko, *Chem. Anal. (Warsaw)*, **13,** 7 (1968).

506. Klygin, A. E., N. S. Kolyada, and D. M. Zavorazhnova, *Zh. Anal. Khim.*, **16,** 442 (1961).

507. Knowles, H. B., *J. Res. Nat. Bur. Stand.*, **9,** 1 (1932).

508. Koch, W., and H. Brockmann, *Arch. Eisenhuettenw.*, **34,** 441 (1963).

509. Kojima, M., and T. Okubo, *Tokyo Kogyo Shikensho Hokoku*, **61,** 372 (1966); through *Chem. Abstr.*, **66,** 72181v (1967).

510. Kokorin, A. I., *Uch. Zap. Kishinev. Gos. Univ.*, **14,** 111 (1954); through *Chem. Abstr.*, **52,** 5204c (1958).

511. Kokorin, A. I., and N. A. Polotebnova, *Uch. Zap. Kishinev. Gos. Univ.*, **14,** 115 (1954); through *Chem. Abstr.*, **52,** 5203g (1958).

512. Kokorin, A. I., and S. K. Radenko, *Uch. Zap. Kishinev. Gos. Univ.*, **68,** 45 (1963); through *Chem. Abstr.*, **63,** 2375c (1965).

513. Kolier, I., and C. Ribando, *Anal. Chem.*, **26,** 1546 (1954).

514. Kolling, O. W., *Trans. Kansas Acad. Sci.*, **58,** 430 (1955).

515. Kolthoff, I. M., and I. Hodara, *J. Electroanal. Chem.*, **4,** 369 (1962).

516. Kolthoff, I. M., and I. Hodara, *J. Electroanal. Chem.*, **5,** 165 (1963).

517. Kolthoff, I. M., and E. P. Parry, *J. Amer. Chem. Soc.*, **73,** 5315 (1951).

518. Kolthoff, I. M., E. B. Sandell, E. J. Meehan, and S. Bruckenstein, *Quantitative Chemical Analysis*, Macmillan, New York, 1969.

519. Komarovskii, A. S., and N. S. Poluektov, *J. Appl. Chem. USSR*, **10,** 565 (1937).

520. Komatsu, S., *Nippon Kagaku Zasshi*, **82,** 265 (1961).

521. Kon'kov, A. G., *Zh. Anal. Khim.*, **4,** 158 (1949).

522. Kono, N., *Tokyo Ritsu Kogyo Shoreikan Hokoku*, **15,** 59 (1963); through *Chem. Abstr.*, **61,** 12622c (1964).

523. Koppel, J., *Chem.-Ztg.*, **43,** 777 (1919).

524. Korenman, I. M., and F. S. Frum, *J. Appl. Chem. USSR*, **16,** 416 (1943).

525. Korenman, I. M., and N. P. Tardov, *J. Appl. Chem. USSR*, **14,** 666 (1941).

526. Korkisch, J., *Mikrochim. Acta*, **1961,** 564 (1961).

527. Korkisch, J., *Modern Methods for the Separation of Rarer Metal Ions*, Pergamon, Oxford, 1969.

528. Korkisch, J., and F. Feik, *Anal. Chim. Acta*, **37,** 364 (1967).

529. Korkisch, J., F. Feik, and S. S. Ahluwalia, *Talanta*, **14,** 1069 (1967).

530. Korkisch, J., and A. Huber, *Talanta*, **15,** 119 (1968).

531. Korkisch, J., and E. Klakl, *Talanta*, **16,** 377 (1969).

532. Korobova, Z. P., and I. P. Kharlamov, *Sovrem. Metody Khim. Tekhnol. Kontr. Proizvod*, **1968,** 110; through *Chem. Abstr.*, **72,** 8953x (1970).

533. Kosta, L., and G. B. Cook, *Talanta,* **12,** 977 (1965).

534. Kosternaya, A. F., and N. A. Zavorovskaya, *Vses. Nauch.-Issled. Inst. Okhr. Tr.,* **1967,** 297; through *Chem. Abstr.,* **70,** 92904u (1969).

535. Kotegov, K. V., O. N. Pavlov, and V. P. Shvedov, "Technetium," in H. J. Emelius, and A. G. Sharpe, Eds., *Advances in Inorganic Chemistry and Radiochemistry,* Vol. 11, Academic, New York, 1968.

536. Kovaleva, A. G., and L. D. Sidel'nikova, *Peredovye Metody Khim. Tekhnol. Kontolya Proizv. Sb.,* **1964,** 200; through *Chem. Abstr.,* **62,** 15417f (1965).

537. Kozhevnikov, L. A., and A. M. Shavrin, *Uch. Zap. Perm. Gos. Univ.,* **19,** 129 (1960); through *Chem. Abstr.,* **58,** 5029h (1963).

538. Krajovan-Marjanovic, V., *Kem. Ind.,* **7,** 33 (1958); through *Chem. Abstr.,* **53,** 21393d (1959).

539. Kral, S., *Hutn. Listy,* **22,** 199 (1967).

540. Kraus, K. A., and F. Nelson, *Proc. Intern. Conf. Peaceful Uses At. Energy, Geneva, 1955,* **7,** 113 (1956).

541. Krauskopf, F. C., and C. E. Swartz, *J. Amer. Chem. Soc.,* **48,** 3021 (1926).

542. Krishnaswamy, N., and H. D. Bhargawa, *Z. Anal. Chem.,* **249,** 241 (1970).

543. Krivy, I., and J. Krtil, *Collect. Czech. Chem. Commun.,* **29,** 587 (1964).

544. Kruithof, R., *Int. Foundry Congr., 31st,* Amsterdam, 1964.

545. K'u, T. L., F. P. Sudakov, and Z. F. Shakhova, *Zh. Anal. Khim.,* **19,** 968 (1964).

546. Kuehn, P. R., O. H. Howard, and C. W. Weber, *Anal. Chem.,* **33,** 740 (1961).

547. Kugai, L. N., and T. N. Nazarchuk, *Zn. Anal. Khim.,* **17,** 1082 (1962).

548. Kuhn, K., *Z. Anal. Chem.,* **130,** 210 (1950).

549. Kula, R. J., *Anal. Chem.,* **38,** 1581 (1966).

550. Kul'berg, L. M., and A. G. Kovaleva, *Dokl. Akad. Nauk SSSR,* **98,** 79 (1954).

551. Kunin, R., and R. J. Myers, *J. Amer. Chem. Soc.,* **69,** 2874 (1947).

552. Kurbatova, V. I., and E. V. Silaeva, *Tr. Vses. Nauchn.-Issled. Inst. Standartn. Obraztsov Spektrosk. Etalanov.,* **1,** 79 (1964); through *Chem. Abstr.,* **64,** 1347b (1966).

553. Kurobe, M., H. Terada, and N. Tajima, *Bunseki Kagaku,* **11,** 767 (1962).

554. Kuroda, P. K., and E. B. Sandell, *Geochim. Cosmochim. Acta,* **6,** 35 (1954).

555. Kuroda, R., K. Ishida, and T. Kirirjama, *Anal. Chem.,* **40,** 1502 (1968).

556. Kuroda, R., K. Kawabuchi, and T. Ito, *Talanta,* **15,** 1486 (1968).

557. Kutzelnigg, A., *Z. Anal. Chem.,* **129,** 382 (1949).

558. Kuznetsov, V. I., *Dokl. Akad. Nauk SSSR,* **50,** 233 (1945).

559. Kuznetsov, V. I., and L. G. Loginova, *Zh. Anal. Khim.,* **13,** 453 (1958).

560. Kuznetsov, V. I., and G. A. Myasoedova, *Tr. Kom. Anal. Khim. Akad. Nauk SSSR, Inst. Geokhim. Anal. Khim.,* **9,** 89 (1958); through *Chem. Abstr.,* **53,** 3981b (1959).

561. Kuznetsov, V. I., and P. D. Tetov, *Izv. Sib. Otd. Akad. Nauk SSSR,* No. 3, 58 (1960); through *Chem. Abstr.,* **54,** 18029c (1960).

562. Kyriacon, D., *Surface Sci.,* **8,** 370 (1967).

563. Lacoste, R. J., M. H. Earing, and S. E. Wiberley, *Anal. Chem.,* **23,** 871 (1951).

564. Lagrange, P., and J. P. Schwing, *Bull. Soc. Chim. Fr.,* **1967,** 718 (1967).

565. Lakshmanan, V. I., and B. C. Haldor, *Proc. Nucl. Rad. Chem. Symp.*, **1967**, 398 (1967).

566. Lang, R., *Z. Anal. Chem.*, **128**, 165 (1948).

567. Lang, R., and S. Gottlieb, *Z. Anal. Chem.*, **104**, 1 (1936).

568. Lange, N. A., Ed., *Handbook of Chemistry*, 10th edit., McGraw-Hill, New York, 1961.

569. Langheinrich, A. P., J. W. Foster, *Advan. X-Ray Anal.*, **11**, 275 (1967).

570. L'Annunziata, M. F., and W. H. Fuller, *J. Chromatogr.*, **34**, 270 (1968).

571. Laqunas-Gil, R., and M. Dean-Guelbenzu, *An. Real Soc. Espan. Fis. Quim.*, *Ser. B*, **62**, 935 (1966); through *Chem. Abstr.*, **66**, 62550a (1967).

572. Lardera, M. R., and A. Mori, *Ann. Chim. (Rome)*, **45**, 869 (1955).

573. Laskorin, B. B., V. S. Ul'yanova, and R. A. Sviridova, *Zh. Prikl. Khim.*, **35**, 2409 (1962).

574. Lassner, E., R. Pueschel, K. Katzengruber, and H. Schedle, *Mikrochim. Acta*, **1969**, 134 (1969).

575. Lassner, E., R. Pueschel, and H. Schedle, *Metall*, **20**, 724 (1966).

576. Lassner, E., and R. Scharf, *Z. Anal. Chem.*, **167**, 114 (1959).

577. Lassner, E., and R. Scharf, *Z. Anal. Chem.*, **168**, 429 (1959).

578. Lassner, E., R. Scharf, and R. Pueschel, *Z. Anal. Chem.*, **165**, 29 (1959).

579. Lassner, E., and H. Schedle, *Talanta*, **13**, 326 (1966).

580. Lassner, E., and H. Schedle, *Talanta*, **15**, 623 (1968).

581. Lassner, E., and H. Schlesinger, *Z. Anal. Chem.*, **158**, 195 (1957).

582. Latimer, W. M., *The Oxidation State of the Elements and Their Potentials in Aqueous Solutions*, Prentice-Hall, New York, 1952.

583. Lavrukhina, A. K., and A. I. Zaitseva, *Tr. Kom. Anal. Khim. Akad. Nauk SSSR, Inst. Geokhim. Anal. Khim.*, **16**, 210 (1968); through *Chem. Abstr.*, **69**, 24169n (1968).

584. Lazarev, A. I., *Zavod. Lab.*, **26**, 935 (1960).

585. Lazarev, A. I., *Zh. Anal. Khim.*, **22**, 1836 (1967).

586. Lazarev, A. I., and V. I. Lazareva, *Zavod. Lab.*, **24**, 798 (1958).

587. Lazarev, A. I., and V. I. Lazareva, *Zavod. Lab.*, **25**, 405 (1959).

588. Lazarev, A. I., V. I. Lazareva, S. Sh. Zak, and T. M. Ustenko, *Zavod. Lab.*, **28**, 1316 (1962).

589. Lebedev, K. B., *Izv. Akad. Nauk Kazakh SSR, Ser. Met. Obogashehen. Ogneuporov*, No. 1, 56, 1959; through *Chem. Abstr.*, **54**, 16118c (1960).

590. Lebughe, P., and W. van Pee, *Bull. Agr. Congo Belge*, **51**, 739 (1960); through *Chem. Abstr.*, **60**, 7455e (1964).

591. Lederer, M., *Anal. Chim. Acta*, **8**, 259 (1953).

592. Lederer, M., and C. Majani, *Chromatogr. Rev.*, **12**, 239 (1970).

593. Leiba, S. P., and M. M. Shapiro, *Zavod. Lab.*, **3**, 503 (1934).

594. Lelubre, R., *Ing. Chim. (Brussels)*, **25**, 101 (1941).

595. Lenskaya, K. K., O. F. Tikhomirova, V. N. Golubeva, N. N. Sorokina, and L. M. Suchelenkova, *Sb. Tr. Tsentr. Nauchn.-Issled. Inst. Chern. Met.*, **49**, 48 (1966); through *Chem. Abstr.*, **66**, 25846f (1967).

596. Leonard, M. A., S. A. E. F. Shahine, and C. L. Wilson, *Mikrochim. Ichnoanal. Acta,* **1965,** 160 (1965).

597. Lesbats, A., *Ann. Chim. (Paris),* **3,** 293 (1968).

598. Li, P. C., T. H. Chao, and Y. H. Li, *Chi-Lin Ta Hsueh Tzu Jan K'o Hsueh Pao,* No. 1, 99 (1959); through *Chem. Abstr.,* **56,** 10903i (1962).

599. Liang, S. C., and H. H. P. Hsu, *J. Chin. Chem. Soc.,* **17,** 90 (1950).

600. Liang, S. C., and T. H. Shen, *Sci. Record (China),* **3,** 209 (1950); through *Chem. Abstr.,* **47,** 1002b (1953).

601. Liang, S. C., and S. J. Wang, *Hua Hsueh Hsueh Pao,* **24,** 117 (1958); through *Chem. Abstr.,* **53,** 970c (1959).

602. Licis, J., *Latv. PSR Zinat. Akad. Vestia,* **3,** 374 (1968); through *Chem. Abstr.,* **69,** 102766w (1969).

603. Likussar, W., W. Beyer, and O. Wawschinck, *Mikrochim. Acta,* **1968** 735 (1968).

604. Lincoln, A. J., and J. C. Kohler, *Anal. Chem.,* **34,** 1247 (1962).

605. Lindner, L., *US At. Energy Comm.,* KFK-216, (1964).

606. Lepchinski, A., and M. Kr'steva, *Godishnit Khim.-Tekhnol. Inst.* **4,** 13 (1957); through *Chem. Abstr.,* **55,** 14170f (1961).

607. Liteanu, C., I. Crisan, and F. Gheorghe, *Studia Univ. Babes-Bolyai, Ser. Chem.,* **8,** 107 (1963); through *Chem. Abstr.,* **61,** 11333c (1964).

608. Livingston, H. D., and H. Smith, *Anal. Chem.,* **39,** 538 (1967).

609. Lo, C. P., and L. J. Y. Chu, *Ind. Eng. Chem. Anal. Ed.,* **16,** 637 (1944).

610. Lontsikh, S. V., and Ya. D. Raikhbaum, *Nauch. Tr., Irkutsk. Gos. Nauchn.-Issled. Inst. Redk. Tsvet. Metal.,* **17,** 74 (1968); through *Chem. Abstr.,* **71,** 45468r (1969).

611. Loon, J. C. van, *At. Absorption Newslett.,* **11,** 60 (1972).

612. Losev, N. F., A. N. Smagunova, R. A. Belova, and Yu. A. Studennikov, *Zavod. Lab.,* **32,** 154 (1966).

613. Lounamaa, N., *Anal. Chim. Acta,* **33,** 21 (1965).

614. Lowe, R. H., and H. F. Massey, *Soil Sci.,* **100,** 238 (1965).

615. Lowenthal, G. C., J. Robson, and R. Deshpande, *Aust. At. Energy Comm.,* AAEC/TM-374, (1967).

616. Luca, C., and S. Armeanu, *An. Univ. Bucuresti, Ser. Stiint. Natur.,* **14,** 121 (1965); through *Chem. Abstr.,* **67,** 78642p (1967).

617. Lueck, C. H., and D. F. Boltz, *Anal. Chem.,* **28,** 1168 (1956).

618. Luke, C. L., *Anal. Chem.,* **35,** 56 (1963).

619. Luke, C. L., *Anal. Chim. Acta,* **34,** 302 (1966).

620. Lukin, A. M., G. S. Petrova, and N. A. Kaslina, *Zh. Anal. Khim.,* **24,** 39 (1969).

621. Lukovnikov, A. F., and A. N. Ponomarev, *Radiokhimiya,* **4,** 19 (1962).

622. Lundell, G. E. F., and H. B. Knowles, *Ind. Eng. Chem.,* **16,** 723 (1924).

623. Lux, F., F. Ammentorp-Schmidt, and W. Opavsky, *Z. Anorg. Allg. Chem.,* **341,** 172 (1965).

624. Lysenko, V. I., *Metody Analiza Veshchestv Vysokoi Chistoty, Akad. Nauk SSSR, Inst. Geokhim. Anal. Khim.,* **1965,** 382; through *Chem. Abstr.,* **65,** 4642c (1966).

625. Lyubimova, I. N., and A. V. Glushchenko, *Agrokhimiya,* **1969,** 137 (1969).

626. MacDonald, A. M. G., and F. H. Van der Voort, *Analyst (London)*, **93**, 65 (1968).

627. Madison, B. L., and J. C. Guyon, *Anal. Chem.*, **39**, 1706 (1967).

628. Madison, B. L., and J. C. Guyon, *Anal. Chim. Acta*, **42**, 415 (1968).

629. Maeck, W. J., G. L. Booman, M. E. Kussy, and J. E. Rein, *Anal. Chem.*, **33**, 1775 (1961).

630. Maeck, W. J., M. E. Kussy, and J. E. Rein, *Anal. Chem.*, **33**, 237 (1961).

631. Maekawa, S., and M. Ebihara, *Bunseki Kagaku*, **9**, 731 (1960).

632. Maekawa, S., and K. Kato, *Bunseki Kagaku*, **14**, 433 (1965).

633. Magee, R. J., S. A. F. Shahine, and C. L. Wilson, *Mikrochim. Ichnoanal. Acta*, **1964**, 479 (1964).

634. Magee, R. J., and A. S. Witwit, *Anal. Chim. Acta*, **29**, 27 (1963).

635. Magee, R. J., and A. S. Witwit, *Anal. Chim. Acta*, **29**, 517 (1963).

636. Majumdar, A. K., and C. P. Savariar, *Anal. Chim. Acta*, **22**, 158 (1960).

637. Maki, Y., K. Tanaka, and Y. Murakami, *Kanagawa-Ken Kogyo Shikensho Kenkyo Hokoku*, **1968**, 40; through *Chem. Abstr.*, **70**, 16309p (1969).

638. Maklakova, V. P., and I. P. Ryazanov, *Zavod. Lab.*, **34**, 1049 (1968).

639. Malakhov, V. V., N. P. Protopopova, V. A. Trukhacheva, and I. G. Yudelevich, *Tr. Kom. Anal. Khim. Akad. Nauk SSSR, Inst. Geokhim. Anal. Khim.*, **16**, 89 (1968); through *Chem. Abstr.*, **69**, 24279y (1968).

640. Malamand, F., *Office Nat. Etud. Rech. Aerospatiales, Note Tech.*, No. 105, 1967; through *Chem. Abstr.*, **69**, 15865j (1968).

641. Malm, L. L., *Appl. Spectrosc.*, **22**, 318 (1968).

642. Mambetkaziev, E. A., O. A. Songina, and V. A. Zakharov, *Sb. Statei Aspir. Soiskatelei, Min. Vyssh. Sredn. Spets. Obrazov. Kaz. SSR, Khim. Khim. Tekhnol.*, **1967**, 1301; through *Chem. Abstr.*, **70**, 43773s (1969).

643. Mann, J. W., "Molybdenum," in F. D. Snell and C. L. Hilton, Eds., *Encyclopaedia of Industrial Chemical Analysis*, Vol. 16, Wiley-Interscience, New York, 1972.

644. Manning, D. C., R. G. Ball, and O. Menis, *Anal. Chem.*, **32**, 1247 (1960).

645. Manok, F., and C. Kovacs, *Studia Univ. Babes-Bolyai, Ser. Chem.*, **9**, 85 (1964); through *Chem. Abstr.*, **61**, 12623g (1964).

646. Manzone, M. G., and J. Z. Briggs, *Mo: Less Common Alloys of Molybdenum*, Climax Molybdenum, New York, 1962.

647. Markle, G. E., and D. F. Boltz, *Anal. Chem.*, **25**, 1261 (1953).

648. Marshall, N. J., *Econ. Geol.*, **63**, 291 (1968).

649. Mashezov, V. Kh., and S. D. Beskov, *Uch. Zap. Kabardino-Balkar. Gos. Univ. Ser. Sel.-Khoz. Khim.-Biol.*, No. 29, 263 (1966); through *Chem. Abstr.*, **68**, 18383r (1968).

650. Matsuo, T., and A. Iwase, *Bunseki Kagaku*, **4**, 148 (1955).

651. McCay, L. W., *J. Amer. Chem. Soc.*, **56**, 2548 (1934).

652. McKaveney, J. P., and H. Freiser, *Anal. Chem.*, **29**, 290 (1957).

653. McKenzie, R. M., *Spectrochim. Acta*, **18**, 1009 (1962).

654. McMullen, T. B., R. B. Faoro, and G. B. Morgan, *J. Air Pollut. Contr. Ass.*, **20**, 369 (1970).

655. McNerney, W. N., and W. F. Wagner, *Anal. Chem.*, **29**, 1177 (1957).

656. Meites, L., *Anal. Chem., 25,* 1752 (1953).

657. Meites, L., *Handbook of Analytical Chemistry,* McGraw-Hill, New York, 1963.

658. Mekenzie, R., *Aust. J. Exp. Agr. Animal Husbandry,* **6,** 21 (1966).

659. Meloche, V. W., and A. F. Preuss, *Anal. Chem.,* **26,** 1911 (1954).

660. Meriwether, L. S., W. F. Marzluff, and W. G. Hodgson, *Nature* **212,** 465 (1966).

661. Meslri, D., and B. Haedar, *J. Sci. Ind. Res.,* **20B,** 551 (1961).

662. Michaelis, C., N. S. Tarlano, J. Clune, and R. Yolles, *Anal. Chem.,* **34,** 1425 (1962).

663. Miketukova, V., *J. Chromatogr.,* **24,** 302 (1966).

664. Mikhail, S. Z., and D. F. Paddleford, *Anal. Chem.,* **35,** 24 (1963).

665. Miller, C. C., *J. Chem. Soc.,* **1941,** 786 (1941).

666. Miller, C. C., *J. Chem. Soc.,* **1941,** 792 (1941).

667. Minczewski, J., and C. Rozycki, *Chem. Anal. (Warsaw),* **10,** 965 (1965).

668. Mitchell, P. C. H., *Quart. Rev. (London),* **20,** 103 (1966).

669. Mitchell, P. C. H., *Coord. Chem. Rev.,* **1,** 315 (1966).

670. Mitchell, P. C. H., and R. J. P. Williams, *J. Chem. Soc.,* **1962,** 4570 (1962).

671. Mittal, M. L., *Z. Naturforsch., B,* **24,** 1053 (1969).

672. Mogilevskaya, O. Ya., *Gig. Sanit.,* **1950,** 18 (1950).

673. Mogilevskaya, O. Ya., *Toksikol. Redk. Metal.,* **1963,** 26; through *Chem. Abstr.,* **60,** 2247d (1964).

674. Mogilevskaya, O. Ya., *Gig. Truda Prof. Zabolevaniya,* **9,** 40 (1965); through *Chem. Abstr.,* **63,** 12218c (1965).

675. Mohamed, M., and N. I. Tarasevich, *Vestn. Mosk. Univ., Ser. II,* **21,** 92 (1966); through *Chem. Abstr.,* **66,** 24095y (1967).

676. Moir, J., *J. Chem. Met. Soc. S. Africa,* **16,** 191 (1916).

677. Molenda, K., *Pr. Inst. Mech. Precyz.,* **15,** 47 (1967); through *Chem. Abstr.,* **69,** 24181k (1968).

678. Molle, L., G. Patriarche, and A. A. Gerbaux, *J. Pharm. Belg.,* **20,** 263 (1965).

679. Momoki, K., *Bunseki Kagaku,* **10,** 330 (1961).

680. Montequi, R., and A. Doadrio, *An. Fis. Quim. (Madrid),* **43,** 1141 (1947).

681. Montignie, E., *Bull. Soc. Chim.,* **47,** 128 (1930).

682. Morie, G. P., and T. R. Sweet, *J. Chromatogr.,* **16,** 201 (1964).

683. Morning, J. L., Molybdenum, in US Bureau Mines, *Minerals Yearbook 1969,* US Dept. Interior, Washington, D.C., 1971.

684. Moroshkina, T. M., and Yu. V. Abramychev, *Vestn. Leningrad. Univ. Ser. Fiz. Khim.,* **2,** 161 (1960); through *Chem. Abstr.,* **54,** 18189d (1960).

685. Morris, W. F., and E. F. Worden, *Appl. Spectrosc.,* **25,** 305 (1971).

686. Morsches, B., and G. Toelg, *Z. Anal. Chem.,* **219,** 61, (1966).

687. Moskvin, L. N., L. I. Shur, L. G. Tsaritsyna, and G. T. Martysh, *Radiokhimiya,* **9,** 377 (1967).

688. Mostyn, R. A., and A. F. Cunninghon, *Anal. Chem.,* **38,** 121 (1966).

689. Motojina, K., H. Hashitani, K. Izawa, and H. Yoshidu, *Bunseki Kagaku,* **11,** 47 (1962).

690. Moulik, S. P., *Sci. Cult. (Calcutta)*, **27**, 45 (1961).

691. Muehl, P., C. Fischer, and G. Guenzler, *Solvent Extr. Chem., Proc. Int. Conf., Goteborg*, **1966**, 625 (1967).

692. Mulder, E. G., *Anal. Chim. Acta*, **2**, 793 (1948).

693. Muller, E., *Z. Elektrochem. Angew. Phy. Chem.*, **33**, 182 (1927).

694. Muller, J. H., and W. A. LaLande, *J. Amer. Chem. Soc.*, **55**, 2376 (1933).

695. Muntz, J. H., *Develop. Appl. Spectrosc.*, **3**, 298 (1963).

696. Muntz, J. H., *US At. Energy Comm.*, ML-TDR-64-36, 1 (1964).

697. Muntz, J. H., and L. L. Roush, *US At. Energy Comm.*, AED-CONF-66-172-9, 1 (1966).

698. Muralikrishna, U., and G. G. Rao, *Talanta*, **15**, 143 (1968).

699. Murata, A., *Nippon Kagaku Zasshi*, **78**, 395 (1957).

700. Murata, K., Y. Yokoyama, and S. Ikeda, *Anal. Chim. Acta*, **48**, 349 (1969).

701. Murty, G. V. L. N., and S. J. Rao, *Metallurgia*, **45**, 212 (1952).

702. Muylder, J. van, and M. Pourbaix, *Proc. Int. Comm. Electrochem. Thermodynam. Kinet.*, **1958**, 218 (1958).

703. Nabivanets, B. I., *Ukr. Khim. Zh.*, **24**, 775 (1958).

704. Nacu, A., D. Nacu, and R. Mocanu, *An. Stiint. Univ. "Al. I. Cuza," Iasi, Sect. Ic*, **12**, 27 (1966)); through *Chem. Abstr.*, **67**, 6192u (1967).

705. Nadkarni, R. A., and B. C. Haldar, *Talanta*, **16**, 116 (1969).

706. Nakajima, T., and H. Fukushima, *Bunseki Kagaku*, **9**, 81 (1960).

707. Nakajima, T., and H. Fukushima, *Bunseki Kagaku*, **9**, 830 (1960).

708. Nangniot, P., *Collect. Czech. Chem. Commun.*, **30**, 40 70 (1965).

709. Nascutiu, T., *Acad. Rep. Populare Romine, Studii Cercetari Chim.*, **9**, 719 (1961); through *Chem. Abstr.*, **57**, 11d (1962).

710. Naumann, R., and W. Schmidt, *Z. Anal. Chem.*, **239**, 1 (1968).

711. Nazarenko, V. A., and M. B. Shustova, *Spektr. Khim. Metody Analiza Mat., Sb. Metodik*, **1964**, 150; through *Chem. Abstr.*, **62**, 4610b (1965).

712. Nazarenko, V. A., and S. Ya. Vinkovetskaya, *Zh. Anal. Khim.*, **11**, 572 (1956).

713. Neace, J. C., *Chem.-Anal.*, **50**, 77 (1961).

714. Nebesar, B., *Anal. Chim. Acta*, **39**, 309 (1967).

715. Nedler, V. V., *Zavod. Lab.*, **7**, 795 (1938).

716. Neeb, K. H., H. Stoeckert, and W. Gebauhr, *Z. Anal. Chem.*, **219**, 69 (1966).

717. Neeb, R., *Z. Anal. Chem.*, **182**, 10 (1961).

718. Neirinckx, R., F. Adams, and J. Hoste, *Anal. Chim. Acta*, **46**, 165 (1969).

719. Nelidow, I., and R. M. Diamond, *J. Phys. Chem.*, **59**, 710 (1955).

720. Nelson, F., and D. C. Michelson, *J. Chromatogr.*, **25**, 414 (1966).

721. Nelson, F., T. Murase, and K. A. Kraus, *J. Chromatogr.*, **13**, 503 (1964).

722. Nelson, G. B., and G. R. Waterbury, *US At. Energy Comm.*, TID-7629, 62 (1961).

723. Nemodruk, A. A., and E. V. Bezrogova, *Zh. Anal. Khim.*, **23**, 884 (1968).

724. Nesmeyanov, A. N., I. A. Savich, M. F. El'kind, and V. A. Koryazhkin, *Vestn. Mosk. Univ., Ser. Mat. Mekh., Astron. Fiz. Khim.*, No. 1, 221 (1956); through *Chem. Abstr.*, **51**, 11816i (1957).

725. Neuberger, M., and A. Fourcy, *J. Radioanal. Chem.*, **1**, 289 (1968).

726. Newcomb, G., and J. J. Markham, *Anal. Chim. Acta*, **35**, 261 (1966).

727. Nicholas, D. J. D., *Analyst (London)*, **77**, 629 (1952).

728. Nicholas, D. J. D., *Ann. N.Y. Acad. Sci.*, **137**, 217 (1966).

729. Nikolaeva, E. R., P. K. Agasyan, and K. K. Tarenova, *Vestn. Mosk. Univ., Ser. II*, **23**, 135 (1968); through *Chem. Abstr.*, **69**, 113254d (1968).

730. Nikolaeva, E. R., P. K. Agasyan, K. K. Tarenova, and S. I. Boikova, *Vestn. Mosk. Univ., Ser. II*, **23**, 73 (1968); through *Chem. Abstr.*, **70**, 43771q (1969).

731. Nishida, H., *Bunseki Kagaku*, **12**, 57 (1963).

732. Nishida, H., *Bunseki Kagaku*, **12**, 567 (1963).

733. Nishida, H., *Bunseki Kagaku*, **12**, 732 (1963).

734. Nishida, H., *Bunseki Kagaku*, **13**, 140 (1964).

735. Nomitsu, T., and H. Fujinaka, *Yamaguchi Daigaku Rika Hokoku*, **10**, 107 (1959); through *Chem. Abstr.*, **54**, 14868e (1960).

736. Novikov, A. I., *Zh. Anal. Khim.*, **16**, 588 (1961).

737. Novotny, M., and K. Oulehla, *Hutn. Listy*, **24**, 535 (1969).

738. Nowosielski, O., *Rocz. Gleboznawcze*, **10**, 151 (1961); through *Chem. Abstr.*, **56**, 3764g (1962).

739. Oertel, A. C., *Aust. J. Appl. Sci.*, **1**, 259 (1950).

740. Oldfield, J. H., and E. P. Bridge, *Analyst (London)*, **86**, 267 (1961).

741. Omarova, E. S., E. F. Speranskaya, and M. T. Kozlovskii, *Izv. Akad. Nauk. Kaz. SSR, Ser. Khim.*, **18**, 32 (1968); through *Chem. Abstr.*, **69**, 32763s (1968).

742. Oppermann, H., and H. Liebschler, *Int. Symp. "Reinstoffe Wiss. Tech.," Tagunsber*, **1965**, 1303 (1966); through *Chem. Abstr.*, **70**, 25266z (1969).

743. Otterson, D. A., and J. W. Graab, *Anal. Chem.*, **30**, 1282 (1958).

744. Pakholkov, V. S., and V. E. Panikarovskikh, *Izv. Vyssh, Ucheb. Zavel., Khim. Khim. Tekhnol.*, **10**, 168 (1967); through *Chem. Abstr.*, **67**, 50044w (1967).

745. Pantaler, R. P., *Zh. Anal. Khim.*, **18**, 603 (1963).

746. Pantani, F., *Ric. Sci.*, **1**, 12 (1961).

747. Papafil, M., and R. Cernatesco, *Ann. Sci. Univ. Jassy*, **15**, 384 (1929); through *Chem. Abstr.*, **23**, 3187[9] (1929).

748. Papafil, M., and M. Furnica, *An. Stiint. Univ. "Al. I. Cuza," Iasi, Sect. Ic*, **10**, 17 (1964); through *Chem. Abstr.*, **63**, 15547b (1965).

749. Papageorgios, P., W. Walczyk, and M. Plonka, *Chem. Stosow., Ser. A*, **12**, 469 (1968); through *Chem. Abstr.*, **70**, 92708h (1969).

750. Parry, E. P., and M. B. Yakubik, *Anal. Chem.*, **26**, 1294 (1954).

751. Parsons, M. L., and P. M. McElfresh, *Appl. Spectrosc.*, **26**, 472 (1972).

752. Partley, H., *Z. Anal. Chem.*, **209**, 398 (1965).

753. Patil, S. V., *Indian J. Chem.*, **2**, 317 (1964).

754. Patriarche, G., A. A. Gerbaux, and L. Molle, *Bull. Soc. Chim. Belges*, **74**, 240 (1965).

755. Patrovsky, V., *Chem. Listy*, **49**, 854 (1955).

756. Paul, J., *Microchem. J.*, **13**, 594 (1968).

757. Pavelka, F., and A. Laghi, *Mikrochem. Ver. Mikrochim. Acta*, **31**, 138 (1943).

758. Pavelka, F., and A. Zucchelli, *Mikrochem. Ver. Mikrochim. Acta,* **31,** 69 (1943).

759. Pavlinova, A. V., A. A. Fedorov, and V. V. Yakovlev, *Zh. Anal. Khim.,* **24,** 561 (1969).

760. Pavlova, I. M., and O. A. Songina, *Sb. Statei Aspir. Soiskatelei, Min. Vyssh. Sredn. Spets. Obrazov. Kaz. SSR, Khim. Khim. Tekhnol.,* **3–4,** 218 (1965); through *Chem. Abstr.,* **66,** 101394z (1967).

761. Peive, J., *Agrokhimiya,* **1969,** 61 (1969).

762. Pekarek, V., and S. Maryska, *Collect. Czech. Chem. Commun.,* **33,** 1612 (1968).

763. Peng, P. Y., and E. B. Sandell, *Anal. Chim. Acta,* **29,** 325 (1963).

764. Penner, E. M., and W. R. Inman, *Talanta,* **10,** 407 (1963).

765. Petushkov, E. E., A. A. Tserfas, and T. M. Maksumov, *Metody Opred. Issled. Sostoyaniya Gazov Metal.,* **1968,** 266; through *Chem. Abstr.,* **71,** 27186t (1969).

766. Pflugmacher, A., and H. Beck, *Z. Pflanzenernahr Dung Bodenk,* **77,** 218 (1957); through *Chem. Abstr.,* **53,** 3570g (1959).

767. Pickett, E. E., and S. R. Koirtyohann, *Spectrochim. Acta,* **23B,** 235 (1968).

768. Picklo, G. A., *Rep. NRL (Naval Res. Lab. US) Prog.,* **1961,** 1 (1961).

769. Pierce, J. O., and J. Cholak, *Arch. Environ. Health,* **13,** 208 (1966).

770. Pierron, E., and R. Munch, *Develop. Appl. Spectrosc.,* **2,** 360 (1963).

771. Pijck, J., *Verhandel. Koninkl. Vlaam. Acad. Wetenschap. Belg. Kl. Wetenschap.,* No. 67, 1961; through *Chem. Abstr.,* **57,** 2531d (1962).

772. Pitak, O., *Chromatographia,* **1,** 29 (1970).

773. Plotnikov, V. I., and V. L. Kochetkov, *Izv. Akad. Nauk Kaz. SSR, Ser. Fiz.-Mat.,* **5,** 53 (1967); through *Chem. Abstr.,* **69,** 90509z (1968).

774. Plotnikov, V. I., and V. L. Kochetkov, *Izv. Akad. Nauk Kaz. SSR, Ser. Fiz.-Mat.,* **5,** 10 (1967); through *Chem. Abstr.,* **69,** 73616q (1968).

775. Plotnikov, V. I., and V. L. Kochetkov, *Zh. Anal. Khim.,* **23,** 377 (1968).

776. Podberezskaya, N. K., *Tr. Kaz. Nauchn.-Issled. Inst. Mimeral. Syr'ya,* No. 2, 232 (1960); through *Chem. Abstr.,* **60,** 2321d (1964).

777. Podobnik, B., and M. Spenko, *Anal. Chim. Acta,* **34,** 294 (1966).

778. Podurovskaya, O. M., and V. E. Petrakovich, *Zavod. Lab.,* **27,** 157 (1961).

779. Pollard, F. H., J. F. W. McOmie, and A. J. Banister, *Chem. Ind. (London),* 1955, 1958.

780. Pollard, F. H., J. F. W. McOmie, and H. M. Stevens, *J. Chem. Soc. (London),* **1951,** 1863 (1951).

781. Pollard, F. H., J. F. W. McOmie, H. M. Stevens, and J. G. Maddock, *J. Chem. Soc.,* **1953,** 1338 (1953).

782. Polotebnova, N. A., and L. M. Branover, *Uch. Zap. Kishinev. Gos. Univ.,* **14,** 119 (1954); through *Chem. Abstr.,* **52,** 5203h (1958).

783. Polyak, L. Ya., *Zavod. Lab.,* **32,** 1317 (1966).

784. Polyak, L. Ya., and I. S. Bashkirova, *Zh. Anal. Khim.,* **21,** 682 (1966).

785. Pometun, E. A., and V. G. Sinyakova, *Zh. Anal. Khim.,* **22,** 440 (1967).

786. Poonia, N. S., *J. Chem. Educ.,* **44,** 477 (1967).

787. Popova, O. I., and O. G. Seroya, *Zh. Anal. Khim.,* **23,** 791 (1968).

788. Porter, L. E., *Ind. Eng. Chem. Anal. Ed.,* **6,** 138 (1934).

789. Potrokhov, V. P., and L. I. Lebedeva, *Zh. Anal. Khim.*, **21**, 182 (1966).

790. Preobrazhenskii, B. K., and L. N. Moskvin, *Radiokhimiya*, **3**, 309 (1961).

791. Pribl, R., and M. Malat, *Collect. Czech. Chem. Commun.*, **15**, 120 (1950).

792. Prints, B. S., *Tr. Sverdl. Sel'skokhoz. Inst.*, **11**, 513 (1964); through *Chem. Abstr.*, **64**, 4251b (1966).

793. Pro, M. J., and R. A. Nelson, *J. Ass. Offic. Agr. Chem.*, **39**, 945 (1956).

794. Pszonicki, L., *Acta Chim. Acad. Sci. Hung.*, **30**, 351 (1962).

795. Publicover, W., *Anal. Chem.*, **37**, 1680 (1965).

796. Pueschel, R., and E. Lassner, *Mikrochim. Ichnoanal. Acta*, **1**, 17 (1965).

797. Pueschel, R., and E. Lassner, "Chelates and Chelating Agents in the Analytical Chemistry of Molybdenum and Tungsten," in H. A. Flaschka and A. J. Barnard, Eds., *Chelates in Analytical Chemistry*, Vol. 1, Dekker, New York, 1967.

798. Pueschel, R., E. Lassner, and A. Illaszewicz, *Chem.-Anal.*, **52**, 40 (1966).

799. Pueschel, R., E. Lassner, and R. Scharf, *Z. Anal. Chem.*, **163**, 104 (1958).

800. Purushottam, A., P. P. Naider, and S. S. Lal, *Talanta*, **19**, 1193 (1972).

801. Pyatnitskii, I. V., and L. F. Kravtsova, *Ukr. Khim. Zh.*, **35**, 77 (1969).

802. Qaim, S. M., and F. D. S. Butement, *Anal. Chim. Acta*, **28**, 591 (1963).

803. Qureshi, M., K. N. Mathur, and A. H. Israili, *Talanta*, **16**, 503 (1969).

804. Radovskaya, T. L., and V. F. Barkovskii, *Zavod. Lab.*, **34**, 411 (1968).

805. Rakhmilevich, N. M., *Zavod. Lab.*, **30**, 507 (1964).

806. Ramakrishna, T. V., P. W. West, and J. W. Robinson, *Anal. Chim. Acta*, **44**, 437 (1969).

807. Rao, A. L. N., and M. M. Tillu, *Indian J. Chem.*, **3**, 320 (1965).

808. Rao, D. V. R., *Curr. Sci.*, **21**, 257 (1952).

809. Rao, D. V. R., *J. Sci. Ind. Res.*, **13B**, 309 (1954).

810. Rao, G. G., and L. S. A. Dikshitulu, *Talanta*, **10**, 1023 (1963).

811. Rao, G. G., and N. V. Rao, *Z. Anal. Chem.*, **190**, 213 (1962).

812. Rao, G. G., and T. P. Sastri, *Z. Anal. Chem.*, **167**, 1 (1959).

813. Rao, G. G., and M. Suryanarayana, *Z. Anal. Chem.*, **169**, 161 (1959).

814. Rao, K. B., *Rec. Trav. Chim. Pays-Bas*, **84**, 71 (1965).

815. Rao, P. D., *At. Absorption Newslett.*, **10**, 118 (1971).

816. Rao, V. P. R., and K. V. Rao, *J. Inorg. Nucl. Chem.*, **30**, 2445 (1968).

817. Rao, V. P. R., K. V. Rao, and P. V. R. B. Sarma, *Mikrochim. Acta*, **1968**, 89 (1968).

818. Rasmussen, P. G., *J. Chem. Educ.*, **44**, 277 (1967).

819. Ray, H. N., *Analyst (London)*, **78**, 217 (1953).

820. Reboul, P., *Chim. Ind.*, **64**, 574 (1950).

821. Rechnitz, G. A., and H. A. Laitinen, *Anal. Chem.*, **33**, 1473 (1961).

822. Reef, B., and H. G. Doge, *Talanta*, **14**, 967 (1967).

823. Reishakhrit, L. S., M. P. Pustoshkina, and Z. I. Tikhonova, *Vestn. Leningrad. Univ., Ser. Fiz. Khim.*, No. 1, 122 (1964); through *Chem. Abstr.*, **60**, 13871h (1964).

824. Reyes, E. D., and J. J. Jurinak, *Soil Sci. Soc. Amer. Proc.*, **31**, 637 (1967).

825. Reznik, B. E., and G. M. Ganzburg, *Ukr. Khim. Zh.*, **28**, 115 (1962).

826. Reznik, B. E., G. M. Ganzburg, and G. V. Mal'tseva, *Zh. Anal. Khim.*, **23**, 1848 (1968).

827. Reznik, B. E., G. M. Ganzburg, and V. F. Milovanova, *Zavod. Lab.*, **33**, 18 (1967).

828. Reznikov, A. A., and A. A. Nechaeva, *Inform. Sb., Vses. Nauchn.-Issled. Geol. Inst.*, **56**, 127 (1962); through *Chem. Abstr.*, **60**, 11426a (1964).

829. Ricq, J. C., *J. Radioanal. Chem.*, **1**, 443 (1968).

830. Riegel, E. R., and R. D. Schwartz, *Anal. Chem.*, **26**, 410 (1954).

831. Rigdon, L. P., and J. E. Harrar, *Anal. Chem.*, **40**, 1641 (1968).

832. Riley, J. P., and D. Taylor, *Anal. Chim. Acta*, **41**, 175 (1968).

833. Ringbom, A., *Complexation in Analytical Chemistry*, Wiley-Interscience, New York, 1963.

834. Rinkis, G., *Mikroelem. Sel. Khoz. Med.* (Kiev: Gos. Izd. Sel. Lit. Ukr. SSR) *Sb.*, **1963**, 351; through *Chem. Abstr.*, **62**, 15059a (1965).

835. Rius, A., and J. M. Coronas, *An. Fis. Quim. (Madrid)*, **40**, 206 (1944).

836. Robin, M. B., and P. Day, "Mixed Valence Chemistry—A Survey and Classification," in H. J. Emeleus and A. G. Sharpe, Eds., *Advances in Inorganic Chemistry and Radiochemistry*, Vol. 10, Academic, New York, 1967.

837. Robinson, G. A., and A. McCarter, *Can. J. Biochem.*, **45**, 141 (1967).

838. Robinson, W. O., *J. Ass. Offic. Agr. Chem.*, **38**, 246 (1955).

839. Roey, G. van, and R. Bastin, *Rev. Ferment. Ind. Aliment.*, **9**, 121 (1964); through *Chem. Abstr.*, **63**, 913a (1965).

840. Rother, E., and G. Jander, *Z. Angew. Chem.*, **43**, 930 (1930).

841. Rovira, L., *Rev. Fac. Cienc. Quim. Univ. Nac. La Plata*, **16**, 235 (1941); through *Chem. Abstr.*, **36**, 6108[6] (1942).

842. Rozycki, C., *Chem. Anal. (Warsaw)*, **12**, 735 (1967).

843. Rubtsov, A. F., *Vop. Sud. Med. Ekspertizy, Sb.* (Ashkhabad: Turkmenistan), **1**, 127 (1965); through *Chem. Abstr.*, **65**, 17349d (1966).

844. Rudenko, N. P., K. K. Awad, V. I. Kuznetsov, and L. S. Gudym, *Vestn. Mosk. Univ., Ser. Khim.*, **23**, 36 (1968); through *Chem. Abstr.*, **70**, 14899g (1969).

845. Ruder, W. E., *J. Amer. Chem. Soc.*, **34**, 387 (1912).

846. Rudzitis, G., A. Kauke, and E. Jansons, *Uch. Zap. Latv. Gos. Univ.*, **1967**, 88; through *Chem. Abstr.*, **69**, 92662t (1968).

847. Ryabchikov, D. I., L. V. Bousova, and Yu. B. Gerlit, *Zh. Anal. Khim.*, **17**, 890 (1962).

848. Ryabchikov, D. I., and A. I. Lazarev, *Tr. Kom. Anal. Khim. Akad. Nauk SSSR, Inst. Geokhim. Anal. Khim.*, **7**, 41 (1956); through *Chem. Abstr.*, **50**, 15333i (1956).

849. Ryabchikov, D. I., P. N. Palei, and Z. K. Mikhailova, *Zh. Anal. Khim.*, **15**, 88 (1960).

850. Sadzhaya, N. D., and L. Petruzashvili, *Tr. Tbilissk. Gos. Pedagog. Inst.*, No. 11, 673 (1957); through *Chem. Abstr.*, **53**, 8934g (1959).

851. Sagi, S., and G. G. Rao, *Z. Anal. Chem.*, **189**, 229 (1962).

852. Sagi, S., and G. G. Rao, *Acta Chim. Acad. Sci. Hung.*, **38**, 89 (1963).

853. Saito, M., *Rev. Polarogr.*, **12**, 76 (1964).

854. Saito, M., *Rev. Polarogr.*, **12**, 122 (1964).

855. Sajo, I., *Z. Anal. Chem.*, **199**, 16 (1963).

856. Sakai, S., and K. Yoshino, *Bunko Kenkyu*, **10**, 23 (1961); through *Chem. Abstr.*, **57**, 15787d (1962).

857. Sandell, E. B., *Ind. Eng. Chem. Anal. Ed.*, **8**, 336 (1936).

858. Sandell, E. B., *Colorimetric Determination of Traces of Metals*, Wiley-Interscience, New York, 1959.

859. Sarma, B., *J. Sci. Ind. Res.*, **16B**, 478 (1957).

860. Sarma, V. B., and M. Suryanarayana, *Z. Anal. Chem.*, **240**, 6 (1968).

861. Sasaki, Y., and L. G. Sillen, *Acta Chem. Scand.*, **18**, 1014 (1964).

862. Sasaki, Y., and L. G. Sillen, *Ark. Kemi*, **29**, 253 (1968).

863. Sasonlas, R., *Spectrochim. Acta*, **23B**, 227 (1968).

864. Sastry, T. P., P. S. Sastry, E. L. R. Dayanand, and K. A. N. Reddy, *Chem.-Anal.*, **56**, 66 (1967).

865. Sauchelli, V., *Trace Elements in Agriculture*, Van Nostrand-Reinhold, New York, 1969.

866. Saxena, R. S., and M. L. Mital, *Z. Phys. Chem.*, **34**, 319 (1962).

867. Scadden, E. M., *Nucleonics*, **15**, 102 (1957).

868. Scadden, E. M., and N. E. Ballou, *US At. Energy Comm.*, NAS-NS-3009, 1 (1960).

869. Scaife, J. F., *Anal. Chem.*, **28**, 1636 (1956).

870. Scarborough, J. M., *Anal. Chem.*, **41**, 250 (1969).

871. Scharrer, K., and G. K. Judel, *Z. Pflanzenernahr Dung Bodenk*, **73**, 107 (1956); through *Chem. Abstr.*, **51**, 2457b (1957).

872. Scharrer, K., and G. K. Judel, *Z. Anal. Chem.*, **156**, 340 (1957).

873. Scharrer, K., and H. Munk, *Agrochimica*, **1**, 44 (1956).

874. Schnaiderman, S. Ya., and I. B. Roberova, *Izv. Kiev Politekh. Inst., Khim.*, **20**, 108 (1957); through *Chem. Abstr.*, **52**, 19720i (1958).

875. Schneer-Erdey, A., *Talanta*, **10**, 591 (1963).

876. Schofield, M., *Metallurgia*, **68**, 31 (1963).

877. Seeley, F. G., and D. J. Crouse, *J. Chem. Eng. Data*, **11**, 424 (1966).

878. Seifter, S., and B. Novic, *Anal. Chem.*, **23**, 188 (1951).

879. Selmer-Olsen, A. R., *Acta Chem. Scand.*, **20**, 1621 (1966).

880. Sen, S., *Sci. Cult.* (Calcutta), **23**, 318 (1957).

881. Shafran, I. G., V. P. Rozenblyum, and G. A. Shteenberg, *Metody Anal. Khim. Reaktivov Prep.*, No. 13, 120, (1966); through *Chem. Abstr.*, **68**, 26588b (1968).

882. Shakashiro, M., and H. Freund, *Anal. Chim. Acta*, **33**, 597 (1965).

883. Shakhova, Z. F., and S. A. Gavrilova, *Zh. Neorg. Khim.*, **3**, 1370 (1958).

884. Shakhova, Z. F., S. A. Gavrilova, and V. F. Zakharova, *Vestn. Mosk. Univ., Ser. II*, **21**, 64 (1966); through *Chem. Abstr.*, **66**, 82118v (1967).

885. Shamaev, V. I., *Radiokhimiya*, **2**, 624 (1960).

886. Shamsiev, S. M., and M. M. Senyavin, *Fiz.-Khim. Tekhnol. Issled. Miner. Syr'ya (Tashkent: Nauka)*, **1965**, 105; through *Chem. Abstr.*, **66**, 22585j (1967).

887. Shapiro, K. Ya., I. V. Volk-Karachevskaya, V. V. Kulakova, and Yu. N. Yurkevich, *Zh. Neorg. Khim.*, **12**, 2767 (1967).

888. Sharrer, K., and W. Hofner, *Z. Pflanznenernahr Dung Bodenk*, **86**, 49 (1959); through *Chem. Abstr.*, **55**, 11729c (1961).

889. Shcherbakov, V. G., and G. V. Onuchina, *Khim. Svoistva Metody Anal. Tugoplavkikh Soedin.*, **1969**, 128; through *Chem. Abstr.*, **72**, 38513u (1970).

890. Shcherbakov, V. G., Yu. N. Yurkevich, and R. A. Antonova, *Sb. Tr. Vses. Nauchn.-Issled. Inst. Tverdykh Splavov*, **3**, 31 (1960); through *Chem. Abstr.*, **57**, 1533a (1962).

891. Shemyakin, F. M., and A. N. Belokon, *C.R. Acad. Sci. USSR*, **18**, 277 (1938); through *Chem. Abstr.*, **32**, 4467[5] (1938).

892. Sheskolskaya, A. Ya., *Zh. Anal. Khim.*, **21**, 1138 (1966).

893. Sheskolskaya, A. Ya., *Zh. Anal. Khim.*, **22**, 812 (1967).

894. Shiobara, Y., *Bunseki Kagaku*, **19**, 243 (1970).

895. Shishkov, D. A., *Godishnik Minnv.-Geol. Inst. Sofiya*, **5**, 131 (1958); through *Chem. Abstr.*, **54**, 24112b (1960).

896. Shishkov, D. A., and E. G. Koleva, *C.R. Acad. Bulg. Sci.*, **17**, 909 (1964).

897. Shishkov, D. A., and E. G. Koleva, *Talanta*, **12**, 865 (1965).

898. Shishkov, D. A., and L. G. Shishkova, *Khim. Ind. (Sofia)*, **35**, 210 (1963).

899. Shishkov, D. A., and L. G. Shishkova, *C.R. Acad. Bulg. Sci.*, **17**, 137 (1964).

900. Shishkov, D. A., and L. G. Shishkova, *Talanta*, **12**, 857 (1965).

901. Shivahare, G. C., *Z. Anal. Chem.*, **219**, 187 (1966).

902. Shkaravskii, Yu. F., *Zh. Anal. Khim.*, **18**, 196 (1963).

903. Shkaravskii, Yu. F., *Zh. Anal. Khim.*, **19**, 320 (1964).

904. Shkaravskii, Yu. F., *Ukr. Khim. Zh.*, **30**, 670 (1964).

905. Shubina, S. B., A. B. Shaevich, and E. P. Basova, *Zavod. Lab.*, **26**, 1364 (1960).

906. Shustova, M. B., *Tr. Kom. Anal. Khim., Akad. Nauk SSSR, Inst. Geokhim. Anal. Khim.*, **15**, 111 (1965); through *Chem. Abstr.*, **63**, 4936h (1965).

907. Shustova, M. B., and V. A. Nazarenko, *Khim. Prom.*, No. 4, 78 (1962); through *Chem. Abstr.*, **59**, 5762b (1963).

908. Shustova, M. B., and V. A. Nazarenko, *Zh. Anal. Khim.*, **18**, 964 (1963).

909. Shustova, M. B., and E. I. Shelikhina, *Zavod. Lab.*, **33**, 810 (1967).

910. Sie, S. T., J. P. A. Bleumer, and G. W. A. Rijnders, *Separ. Sci.*, **1**, 41 (1966).

911. Siegel, L. M., and K. J. Monty, *J. Nutr.*, **74**, 167 (1961).

912. Silaeva, E. V., and V. I. Kurbatova, *Zavod. Lab.*, **27**, 1462 (1961).

913. Silaeva, E. V., and V. I. Kurbatova, *Zavod. Lab.*, **28**, 280 (1962).

914. Sillen, L. G., and A. E. Martell, *Stability Constants of Metal-Ion Complexes*, Special Publication No. 17, The Chemical Society, London, 1964.

915. Sillen, L. G., and A. E. Martell, *Stability Constants of Metal-Ion Complexes, Supplement No. 1*, Special Publication No. 25, The Chemical Society, London, 1971.

916. Silvey, W. D., and R. Brennan, *Anal. Chem.*, **34**, 784 (1962).

917. Singh, C., *J. Sci. Ind. Res.*, **19C**, 306 (1960).

918. Singh, D., and A. Varma, *J. Sci. Res. Banama Hendi Univ.*, **11**, 202 (1960–1961); through *Chem. Abstr.*, **56**, 9400g (1962).

919. Singh, H., and P. K. Bhattacharya, *Indian J. Appl. Chem.*, **29**, 213 (1967).

920. Sinha, S. K., and S. C. Shome, *Anal. Chim. Acta*, **24**, 33 (1961).

921. Sinyakova, S. I., and M. I. Glinkina, *Zh. Anal. Khim.*, **11**, 544 (1956).

922. Sinyakova, S. I., and M. I. Glinkina, *Tr. Chetvertogo Soveshch. Elektrokhim.*, **1956**, 201; through *Chem. Abstr.*, **54**, 9557g (1960).

923. Skellon, J. H., and J. W. Spence, *J. Soc. Chem. Ind. (London)*, **67**, 365 (1948).

924. Skobets, E. M., D. S. Turova, and A. I. Karnauk'hov, *Ukr. Khim. Zh.*, **36**, 33 (1970).

925. Skogerboe, R. K., A. T. Heybey, and G. H. Morrison, *Anal. Chem.*, **38**, 1821 (1966).

926. Slavik, J., *Chem. Tech. (Berlin)*, **6**, 528 (1954).

927. Snell, F. D., and C. T. Snell, *Colorimetric Methods of Analysis*, Volume IIA, Van Nostrand, New York, 1959.

928. Snopov, N. G., *Zh. Prikl. Spektrosk.*, **9**, 919 (1968).

929. Sokolov, V. A., and V. I. Levin, *Zh. Neorg. Khim.*, **9**, 742 (1964).

930. Sokolova, A. A., and Z. S. Mukhina, *Zavod. Lab.*, **34**, 1060 (1968).

931. Songina, O. A., and M. T. Kozlovskii, *Zavod. Lab.*, **13**, 677 (1947).

932. Songina, O. A., V. A. Zakharov, and E. A. Mambetkaziev, *Zh. Anal. Khim.*, **23**, 453 (1968).

933. Souchay, P., M. Cadiot, and M. Duhameaux, *C. R. Acad. Sci.*, **260**, 186 (1965).

934. Souchay, P., R. Massart, and M. Biquard, *Proc. Int. Conf. Coord. Chem.*, *Vienna*, **1964**, 394 (1964).

935. Spacu, P., and C. Gheorghiu, *Rev. Chim. Acad. Rep. Populaire Roumaine*, **2**, 21 (1957).

936. Spacu, P., and C. Gheorghiu, *Z. Anal. Chem.*, **159**, 209 (1958).

937. Spacu, P., C. Gheorghiu, and I. Paralescu, *Acad. Repub. Pop. Rom., Stud. Cercet. Chim.*, **10**, 157 (1962); through *Chem. Abstr.*, **58**, 3881f (1963).

938. Spacu, P., C. Gheorghiu, and I. Paralescu, *Z. Anal. Chem.*, **195**, 321 (1963).

939. Spano, E. F., and T. E. Green, *Anal. Chem.*, **38**, 1341 (1966).

940. Spence, J. T., *Coord. Chem. Rev.*, **4**, 475 (1969).

941. Speranskaya, E. F., and M. T. Kozlovskii, *Zavod. Lab.*, **30**, 403 (1964).

942. Speranskaya, E. F., and V. E. Mertsalova, *Khim. Khim. Tekhnol., Almaata Sb.*, **2**, 89 (1964); through *Chem. Abstr.*, **64**, 1344d (1966).

943. Speranskaya, E. F., and V. E. Mertsalova, *Izv. Vyssh. Ucheb. Zaved. Khim. Khim. Tekhnol.*, **8**, 893 (1965); through *Chem. Abstr.*, **64**, 13742g (1966).

944. Speranskaya, E. F., and V. E. Mertsalova, *Sb. Statei Aspir. Soiskatelei, Min. Vyssh. Sredn. Spets. Obrazov. Kaz. SSR, Khim. Khim. Tekhnol.*, **3–4**, 109 (1965); through *Chem. Abstr.*, **66**, 79993v (1966).

945. Speranskaya, E. F., V. E. Mertsalova, and I. I. Kulev, *Usp. Khim.*, **35**, 2129 (1966).

946. Spiridonova, V. S., and S. V. Surorov, *Gig. Sanit.*, **32**, 79 (1967).

947. Spitz, J., J. J. Chazee, and Tran Van Danh, *Method Phys. Anal.*, **4**, 375 (1968).

948. Suprunovich, V. I., and Yu. I. Usatenko, *Khim. Tekhnol., Respub. Mezhvedom. Nauch.-Tekh. Sb.*, No. 4, 27 (1966); through *Chem. Abstr.*, **68**, 18272d (1968).

949. Suryanarayana, M., and G. G. Rao, *Z. Anal. Chem.*, **173**, 358 (1960).

950. Stanton, R. E., and A. J. Hardwick, *Analyst (London)*, **92**, 387 (1967).

951. Stanton, R. E., and A. J. Hardwick, *Analyst (London)*, **93**, 193 (1968).

952. Stark, J. G., *J. Chem. Educ.*, **46**, 505 (1969).

953. Stary, J., *Anal. Chim. Acta*, **28**, 132 (1963).

954. Stary, J., and D. Hladsky, *Anal. Chim. Acta*, **28**, 277 (1963).

955. Stary, J., J. Ruzicka, and A. Zerna, *Anal. Chim. Acta*, **29**, 103 (1963).

956. Stefkin, F. S., *Uch. Zap., Mord. Gos. Univ.*, **1967**, 42; through *Chem. Abstr.*, **70**, 83877k (1969).

957. Stehlik, B., *Chem. Listy*, **26**, 533 (1932).

958. Stehlik, B., *Chem. Listy*, **38**, 1 (1944).

959. Steinnes, E., *Anal. Chim. Acta*, **57**, 249 (1971).

960. Stepin, V. V., *Zavod. Lab.*, **8**, 799 (1939).

961. Stepin, V. V., and V. I. Ponosov, *Zavod. Lab.*, **25**, 806 (1959).

962. Stepin, V. V., and V. I. Ponosov, *Tr. Ural. Nauchn.-Issled. Inst. Chern. Metal.*, **2**, 272 (1963); through *Chem. Abstr.*, **61**, 3665g (1964).

963. Sterba-Bohm, J., and J. Vostrebal, *Z. Anorg. Allg. Chem.*, **110**, 81 (1920).

964. Stevens, H. M., *Anal. Chim. Acta*, **14**, 126 (1956).

965. Stolyarov, K. P., and I. A. Amantova, *Talanta*, **14**, 1237 (1967).

966. Stoppel, A. E., C. F. Sidener, and P. H. M. P. Brinton, *Chem. News*, **130**, 353 (1925).

967. Straumanis, M., and B. Ogrins, *Z. Anal. Chem.*, **117**, 30 (1939).

968. Strelow, F. W. E., *Anal. Chem.*, **32**, 1185 (1960).

969. Strelow, F. W. E., *Anal. Chem.*, **33**, 542 (1961).

970. Strelow, F. W. E., *J.S. Afr. Chem. Inst.*, **14**, 51 (1961).

971. Strelow, F. W. E., *Anal. Chem.*, **35**, 1279 (1963).

972. Strelow, F. W. E., and C. J. C. Bothma, *Anal. Chem.*, **39**, 595 (1967).

973. Strelow, F. W. E., R. Rethemeyer, and C. J. C. Bothma, *Anal. Chem.*, **37**, 106 (1965).

974. Strelow, F. W. E., C. R. Van Zyl, and C. J. C. Bothma, *Anal. Chim. Acta*, **45**, 81 (1969).

975. Stross, W., and J. Clark, *Metallurgia*, **67**, 47 (1963).

976. Stubbs, M. F., *Analyst (London)*, **93**, 59 (1968).

977. Studennikov, Yu. A., R. A. Belova, and N. F. Losev, *Zavod. Lab.*, **33**, 1504 (1967).

978. Suchy, K., and E. Buchar, *Sb. Vys. Sk. Pedagog. (Prague), Prir, Vedy*, **1**, 111 (1957); through *Chem. Abstr.*, **52**, 14422b (1958).

979. Sudo, E., *Sci. Rep. Res. Inst., Tohoku Univ., Ser. A*, **2**, 618 (1950); through *Chem. Abstr.*, **46**, 4424d (1952).

980. Sudo, E., *Sci. Rep. Res. Inst., Tohoku Univ., Ser. A*, **8**, 380 (1956); through *Chem. Abstr.*, **51**, 4201d (1957).

981. Sugawara, K. F., *Anal. Chem.*, **36**, 1373 (1964).

982. Sugawara, K. F., and S. Okabe, *J. Earth Sci. Nagoyu Univ.*, **8**, 86 (1960); through *Chem. Abstr.*, **55**, 21707b (1961).

983. Sugimoto, M., *Bunseki Kagaku*, **12**, 539 (1963).

984. Surgutskii, V. P., and V. V. Serebrennikov, *Tr. Tomskogo Gos. Univ., Ser. Khim.*, **157**, 29, (1963); through *Chem. Abstr.*, **61**, 11308e (1964).

985. Suteu, A., and C. Maniu, *Rev. Roum. Chim., 13,* 1201 (1968).

986. Suvorovskaya, N. A., and S. D. Karavaeva, *Ionoobmen. Ekstraktsion. Metody Khim.-Obogat. Protsessaleh, Akad. Nauk SSSR, Inst. Gorn. Dela,* **1965**, 39; through *Chem. Abstr., 63,* 15527f (1965).

987. Suzuki, M., *Bunseki Kagaku, 18,* 176 (1969).

988. Suzuki, S., K. Harimaya, M. Ueno, N. Tsuji, and N. Yamaoka, *Bull. Chem. Soc. Jap., 30,* 766 (1957).

989. Svehla, G., *Analyst (London), 94,* 513 (1969).

990. Svehla, G., and L. Erdy, *Microchem. J., 7,* 221 (1963).

991. Swaine, D. J., *Spectrochim. Acta, 19,* 841 (1963).

992. Sward, G. G., "Chemical Analysis of Pigments," in G. G. Sward, Ed., *Paint Testing Manual,* ASTM Special Technical Publication 500, 13th edit., ASTM, Philadelphia, Pa., 1972.

993. Syre, R., *AGARDograph,* No. 94, 1965.

994. Syrokomskii, V. S., and Yu. V. Klimenko, *Zavod. Lab., 7,* 1093 (1938).

995. Syty, A., and J. A. Dean, *Anal. Lett., 1,* 241 (1968).

996. Szarvas, P., I. Balogh-Kelemen, and Z. Jarabin, *Acta. Univ. Debrecen.,* No. 2, 119 (1961); through *Chem. Abstr., 57,* 14417b (1962).

997. Szarvas, P., Z. Jarabin, and M. V. Braun, *Proc. Conf. Appl. Phys.-Chem. Methods Chem. Anal.,* Budapest, **3,** 155 (1966).

998. Szebelledy, L., and J. Jonas, *Mikrochim. Acta, 1,* 46 (1937).

999. Szilagyi, M., *ATOMKI (Atommag Kut. Intez.) Kozlem., 8,* 97 (1966); through *Chem. Abstr., 66,* 51864e (1967).

1000. Tabor, E. C., *Health Lab. Sci., 7,* 149 (1970).

1001. Taimni, I. K., and R. P. Agarwal, *Anal. Chim. Acta, 9,* 203 (1953).

1002. Taimni, I. K., and S. N. Tandon, *Anal. Chim. Acta, 22,* 34 (1960).

1003. Taimni, I. K., and S. N. Tandon, *Anal. Chim. Acta, 22,* 437 (1960).

1004. Takahashi, S., and E. Shikata, *J. Nucl. Sci. Technol., 7,* 130 (1970).

1005. Takeuchi, T., *Bunseki Kagaku, 15,* 473 (1966).

1006. Talipov. Sh. T., G. P. Gor'kovaya, and R. Kh. Dzhiyanbaeve, *Nauch. Tr. Tashkent Gos. Univ., 284,* 83 (1967); through *Chem. Abstr., 69,* 73687p (1968).

1007. Talwar, U. B., and B. C. Haldar, *Indian J. Chem., 3,* 452 (1965).

1008. Tamchyna, J. V., *Chem. Listy, 24,* 465 (1930).

1009. Tanaka, M., *Mikrochim. Acta,* **1958,** 204 (1958).

1010. Tanaka, T., K. Hiiro, and Y. Yamamoto, *Nippon Kagaku Zasshi, 91,* 354 (1970).

1011. Tananaiko, M. M., and L. A. Blukke, *Ukr. Khim. Zh., 29,* 974 (1963).

1012. T'ao, T. C., and W. Yank, *Hua Hsueh Tung Pao,* **1966,** 57; through *Chem. Abstr., 65,* 19302e (1966).

1013. Tarayan, V. M., and A. N. Pogosyan, *Arm. Khim. Zh., 19,* 586 (1966).

1014. Tarkovskaya, I. A., S. I. Shevchenko, and A. N. Chernenko, *Ukr. Khim. Zh., 35,* 1160 (1969).

1015. Taylor-Austin, E., *Analyst (London), 62,* 107 (1937).

1016. Telep, G., and D. F. Boltz, *Anal. Chem., 22,* 1030 (1950).

1017. Theisen, A. A., and A. Pinkerton, *Soil Sci. Soc. Amer. Proc., 32,* 440 (1968).

1018. Thomerson, D. R., and W. J. Price, *Analyst (London)*, **96**, 321 (1971).

1019. Thomson, B. A., and P. D. LaFieur, *Anal. Chem.*, **41**, 1888 (1969).

1020. Thompson, H., and G. Svehla, *Z. Anal. Chem.*, **247**, 244 (1969).

1021. Titkov, Yu. B., *Ukr. Khim. Zh.*, **36**, 613 (1970).

1022. Tobia, S. K., Y. A. Gawargious, and M. F. El-Shahat, *Anal. Chim. Acta*, **39**, 392 (1967).

1023. Tourky, A. R., M. Y. Farah, and H. K. El-Shamy, *Analyst (London)*, **73**, 258 (1948).

1024. Trapitsyn, N. F., and F. I. Mullayanov, *Zavod. Lab.*, **28**, 950 (1962).

1025. Tridot, G., and J. C. Bernard, *Acta Chim. Acad. Sci. Hung.*, **34**, 179 (1962).

1026. Tsigdinos, G. A., *Heteropoly Compounds of Molybdenum and Tungsten*, Bulletin Cdb-12a, Climax Molybdenum, Ann Arbor, Mich., 1969.

1027. Tsigdinos, G. A., *Electrochemical Properties of Heteropoly Molybdates*, Bulletin Cdb-15, Climax Molybdenum, Ann Arbor, Mich., 1971.

1028. Tsigdinos, G. A., and C. J. Hallada, *Isopoly Compounds of Molybdenum Tungsten and Vanadium*, Bulletin Cdb-14, Climax Molybdenum, Ann Arbor, Mich., 1969.

1029. Tsukamoto, A., and I. Shimizu, *Nippon Kinzoku Gakkaishi*, **32**, 473 (1968).

1030. Ueda, S., Y. Yamamoto, and H. Takenouchi, *Nippon Kagaku Zasshi*, **88**, 1299 (1967).

1031. Uehara, F., and A. Furuno, *Bunseki Kagaku*, **19**, 183 (1970).

1032. Ul'ko, N. V., *Ukr. Khim. Zh.*, **31**, 887 (1965).

1033. Umeda, M., *Bunseki Kagaku*, **9**, 172 (1960).

1034. Umland, F., and G. Wvensch, *Z. Anal. Chem.*, **225**, 362 (1967).

1035. Usatenko, Yu. I., and O. V. Datsenko, *Zavod. Lab.*, **15**, 779 (1949).

1036. Vahemets, H., *Uch. Zap. Tartu Gos. Univ.*, **193**, 121 (1966); through *Chem. Abstr.*, **69**, 40971y (1968).

1037. Vasilenko, V. D., B. E. Reznik, and L. A. Nakonechnaya, *Tr. Kom. Anal. Khim. Akad. Nauk SSSR, Otdel. Khim. Nauk*, **5**, 112 (1954); through *Chem. Abstr.*, **49**, 12181f (1955).

1038. Vasil'eva, L. N., and A. A. Pozdnyakova, *Sb. Nauch. Tr. Gos. Nauchn.-Issled. Inst. Tsvet. Metal.*, No. 27, 44 (1967); through *Chem. Abstr.*, **70**, 25430y (1969).

1039. Vasillaros, G. L., and J. P. McKavaney, *Talanta*, **16**, 195 (1969).

1040. Venegas, H. Z., *An. Fac. Quim. Farm., Univ. Chile*, **15**, 175 (1963); through *Chem. Abstr.*, **62**, 15417g (1965).

1041. Vengerskaya, Kh. Ya., *Nov. Obl. Prom.-Sanit. Khim.*, **1969**, 195; through *Chem. Abstr.*, **71**, 116274n (1969).

1042. Vengerskaya, Kh. Ya., and S. S. Salikhodzhaev, *Nov. Obl. Prom.-Sanit. Khim.*, **1969**, 237; through *Chem. Abstr.*, **72**, 619v (1970).

1043. Vicente-Perez, S., and R. Cordova-Orellana, *Inform. Quim. Anal. (Madrid)*, **22**, 38 (1968).

1044. Vidal, G., P. Galmard, and P. Lanusse, *Rech. Aerosp.* No. 122, 29 (1968); through *Chem. Abstr.*, **69**, 15642j (1968).

1045. Vignoli, L., B. Cristau, and A. Pfister, *Bull. Soc. Pharm. Marseilles*, **4**, 71 (1955).

1046. Vignoli, L., B. Cristau, and A. Pfister, *Chim. Anal. (Paris)*, **38**, 392 (1956).

1047. Vignoli, L., and J. P. Defretin, *Biol. Med. (Paris)*, **52**, 319 (1963).

1048.   Vioque-Pizarro, A., *Anal. Chim. Acta,* **6,** 105 (1951).

1049.   Vioque-Pizarro, A., and H. Malissa, *Mikrochem. Ver. Mikrochim. Acta,* **40,** 396 (1953).

1050.   Virasoro, E., *Rev. Fac. Ing. Quim.,* **20,** 83 (1951); through *Chem. Abstr.,* **49,** 5165g (1955).

1051.   Vivarat-Perrin, J., and E. Bonnier, *Chim. Anal. (Paris),* **48,** 511 (1966).

1052.   Vorob'ev, S. P., I. P. Davydov, and I. V. Shilin, *Zh. Neorg. Khim.,* **12,** 2142 (1967).

1053.   Vorob'ev, S. P., I. P. Davydov, and I. V. Shilin, *Zh. Neorg. Khim.,* **12,** 2665 (1967).

1054.   Vrchlabsky, M., and L. Sommer, *Talanta,* **15,** 887 (1968).

1055.   Wadelin, C., and M. G. Mellon, *Anal. Chem.,* **25,** 1668 (1953).

1056.   Waggoner, C. A., *Appl. Spectrosc.,* **13,** 31 (1959).

1057.   Wagner, J. C., and E. J. Violante, *Appl. Spectrosc.,* **19,** 195 (1965).

1058.   Ward, F. N., *Anal. Chem.,* **23,** 788 (1951).

1059.   Webb, M. S. W., J. C. Cotterill, and T. W. Jones, *Anal. Chim. Acta,* **26,** 548 (1962).

1060.   Weeks, M. E., and H. M. Leicester, *Discovery of the Elements,* 7th edit., Chemical Education, Easton, Pa., 1968.

1061.   Weiner, R., and P. Boriss, *Z. Anal. Chem.,* **160,** 343 (1958).

1062.   Weiser, H. B., *J. Phys. Chem.,* **20,** 640 (1916).

1063.   Weiss, H. V., and M. G. Lai, *Talanta,* **8,** 72 (1961).

1064.   Weissler, A., *Ind. Eng. Chem., Anal. Ed.,* **17,** 695 (1945).

1065.   Weisz, H., D. Klockow, and H. Ludwig, *Talanta,* **16,** 921 (1969).

1066.   Welcher, G. G., and H. Kriege, *At. Absorption Newslett.,* **8,** 97 (1969).

1067.   Wenger, P., R. Duckert, and E. Ankajdi, *Helv. Chim. Acta,* **28,** 1592 (1945).

1068.   West, A. C., R. N. Kniseley, and V. A. Fassel, *Anal. Chem.,* **45,** 815 (1973).

1069.   West, T. S., *Ind. Quim.* (Buenos Aires), **17,** 83 (1955); through *Chem. Abstr.,* **49,** 15609b (1955).

1070.   Will, F., and J. H. Yoe, *Anal. Chim. Acta,* **8,** 546 (1953).

1071.   Will, F., and J. H. Yoe, *Anal. Chem.,* **25,** 1363 (1953).

1072.   Willard, H. H., and F. Fenwick, *J. Amer. Chem. Soc.,* **45,** 928 (1923).

1073.   Williams, W. W., and J. W. Conley, *Ind. Eng. Chem.,* **47,** 1507 (1955).

1074.   Willmer, T. K., *Arch. Eisenhuettenw.,* **29,** 1959 (1958).

1075.   Wilson, A. M., *Anal. Chem.,* **38,** 1784 (1966).

1076.   Wilson, A. M., L. Churchill, K. Kiluk, and P. Hovsepian, *Anal. Chem.,* **34,** 203 (1962).

1077.   Wilson, A. M., and O. K. McFarland, *Anal. Chem.,* **36,** 2488 (1964).

1078.   Winer, W. O., *Wear,* **10,** 422 (1967).

1079.   Winge, J. K., and V. A. Fassel, *Anal. Chem.,* **37,** 67 (1965).

1080.   Witwit, A. S., R. J. Magee, and C. L. Wilson, *Talanta,* **9,** 495 (1962).

1081.   Wolf, K., *Z. Angew. Chem.,* **31,** 140 (1918).

1082.   Wolfson, H., *Nature,* **153,** 375 (1944).

1083.   Wood, D. F., and R. T. Clark, *Analyst (London)*, **87**, 342 (1962).

1084.   Wood, D. F., and J. T. Jones, *Analyst (London)*, **93**, 131 (1968).

1085.   Yagoda, H., and H. A. Fales, *J. Amer. Chem. Soc.*, **58**, 1494 (1936).

1086.   Yaguchi, H., and T. Kajiwara, *Bunseki Kagaku*, **14**, 785 (1965).

1087.   Yakovlev, P. Ya., *Zavod. Lab.*, **14**, 1132 (1948).

1088.   Yakovlev, P. Ya., and E. F. Pen'kova, *Zavod. Lab.*, **15**, 34 (1949).

1089.   Yamamoto, S., *Anal. Chem.*, **41**, 337 (1969).

1090.   Yamamoto, S., *Bull. Shimane Univ.*, No. 3, 61 (1951); through *Chem. Abstr.*, **48**, 1880i (1954).

1091.   Yamamoto, S., *Nippon Kagaku Zasshi*, **76**, 417 (1955).

1092.   Yatirajam, V., and L. Kakkar, *Anal. Chim. Acta*, **44**, 468 (1969).

1093.   Yatirajam, V., and R. Prosad, *Indian J. Chem.*, **3**, 345 (1965).

1094.   Yatirajam, V., and R. Prosad, *Indian J. Chem.*, **3**, 544 (1965).

1095.   Yatirajam, V., and J. Ram, *Mikrochim. Acta*, **1973**, 77 (1973).

1096.   Yatsimirskii, K. B., *Catalysis Chem. Kinet.*, **1964**, 201 (1964).

1097.   Yatsimirskii, K. B., and A. P. Filippov, *Zh. Anal. Khim.*, **20**, 815 (1965).

1098.   Yen, H. Y., and Y. H. Liu, K'o *Hsueh T'ung Pao*, **17**, 279 (1966); through *Chem. Abstr.*, **66**, 25801n (1967).

1099.   Ying, P. H., and T. T. Cheng, *Hua Hsueh Tung Pao*, **1964**, 375; through *Chem. Abstr.*, **62**, 2229c (1965).

1100.   Yoshimori, T., and T. Takeuchi, *Bunseki Kagaku*, **9**, 689 (1960).

1101.   Yoshimura, C., *Nippon Kagaku Zasshi*, **76**, 883 (1955).

1102.   Yoshimura, C., S. Uno, and H. Noguchi, *Bunseki Kagaku*, **12**, 42 (1963).

1103.   Yoshimura, T., *J. Chem. Soc. Jap.*, **73**, 122 (1952).

1104.   Yosida, Y., *J. Chem. Soc. Jap.*, **61**, 130 (1940).

1105.   Yule, H. P., *Anal. Chem.*, **38**, 818 (1966).

1106.   Yurkevich, Yu. N., and K. Ya. Shapiro, *Met. Vol'frama Molibdena Niobiya*, **1967**, 53; through *Chem. Abstr.*, **69**, 98546e (1968).

1107.   Yurkevich, Yu. N., and V. G. Shchorbakov, *Zh. Anal. Khim.*, **16**, 617 (1961).

1108.   Yushko, S. A., and L. D. Goslavskaya, *Mineral. Mikrovklyuchenjua, Akad. Nauk SSR, Inst. Minerolog. Geokhim. Kristallokhim. Redk. Elem.*, **1965**, 42; through *Chem. Abstr.*, **64**, 1331c (1966).

1109.   Zaidel', A. N., V. K. Prokof'ev, S. M. Raiskii, V. A. Slavnyi, and E. Ya. Shreider, *Tables of Spectral Lines*, IFI/Plenum, New York, 1970.

1110.   Zaremba, J., *Chem. Anal. (Warsaw)*, **12**, 305 (1967).

1111.   Zelikman, A. N., I. G. Kalinina, and R. A. Smol'nikova, *Zh. Neorg. Khim.*, **13**, 2778 (1968).

1112.   Zelikman, A. N., O. E. Krein, L. A. Nisel'son, N. N. Gorovits, and Z. I. Ivanova, *Razdelenie Blizikh Po Svoistvam Redk. Metal.*, **1962**, 186; through *Chem. Abstr.*, **58**, 2223c (1963).

1113.   Zelikman, A. N., O. E. Krein, L. A. Nisel'son, and Z. I. Ivanova, *Zh. Prikl. Khim.*, **35**, 1467 (1962).

1114.   Zelikman, A. N., and G. Meisner, *Zh. Prikl. Khim.*, **39**, 992 (1966).

1115.   Zelikman, A. N., and V. M. Nerezov, *Zh. Neorg. Khim.*, **14**, 1307 (1969).

1116. Zharovskii, F. G., and A. G. Sakhno, *Ukr. Khim. Zh.,* **28,** 145 (1962).

1117. Zharovskii, F. G., E. A. Shpak, and E. V. Piskunova, *Ukr. Khim. Zh.,* **29,** 102 (1963).

1118. Zharovskii, F. G., L. M. Vyazovskaya, and R. V. Kostova, *Ukr. Khim. Zh.,* **34,** 181 (1968).

1119. Zhdanov, A. K., and I. A. Umurzakova, *Nekotorye Vopr. Khim. Tekhnol. Fiz.-Khim. Analiza Neorgan. Sistem, Akad. Nauk Uz. SSR, Otd. Khim. Nauk,* **1963,** 163; through *Chem. Abstr.,* **61,** 1259e (1964).

1120. Zhivopistsev, V. P., and G. A. Sadakov, *Uch. Zap. Perm. Gos. Univ.,* No. 141,174 (1966); through *Chem. Abstr.,* **68,** 56321b (1968).

1121. Ziegler, M., and H. G. Horn, *Z. Anal. Chem.,* **166,** 362 (1959).

1122. Zmijewska, W., and J. Minczewski, *Chem. Anal. (Warsaw),* **14,** 23 (1969).

TUNGSTEN

By Gordon A. Parker, *Department of Chemistry, University of Toledo, Toledo, Ohio*

Contents

447

I. INTRODUCTION

Tungsten, wolfram, finds use in a variety of applications ranging from lamp filaments to hardened steels. The nature of its chemistry, particularly its resistance to dissolving in concentrated mineral acids, and the variety of elements with which it is alloyed create a challenge for the analyst interested in its determination. Reviews are available describing the properties of tungsten (561,813,907) and, in particular, the analytical chemistry of tungsten (114,215,574,620).

A. HISTORY

In 1781 the Swedish chemist, C. W. Scheele, reported the discovery of a new material, tungsten oxide, associated with calcium in the mineral scheelite. It was, however, the Spanish brothers, J. J. and F. de Elhuyar, one of whom had studied for a time in Sweden, who first reported the isolation of tungsten metal in 1783 by reduction of the oxide with carbon (852,1033). Tungsten was first used as a constituent of alloyed steels in the middle 1800s, tungsten lamp filaments were first employed in the early 1900s, and tungsten-carbide cutting tools were first used in the late 1920s. Since then the application of tungsten and other refractories for high-temperature alloys in the aerospace industry has further enhanced its importance.

B. OCCURRENCE

Chief tungsten ores are scheelite ($CaWO_4$) and wolframite, a mixture of ferberite ($FeWO_4$) and hübnerite ($MnWO_4$). Numerous other tungsten-containing ores and minerals exist (561). Major deposits of tungsten are found in China, the United States, and various other countries throughout the world. The geochemistry of tungsten has been studied (21,401,-633,942), and its distribution in coal (117), plants (864), natural waters (133), and the atmosphere (851) has been reported.

C.   TOXICOLOGY AND INDUSTRIAL HYGIENE

The toxicity of tungsten and tungsten compounds is of concern. Studies with rats (40,452) and other animals (439,740) indicate that, although much of the administered tungsten is excreted within a day, the portion retained causes altered physiological activity (124,338,936). The effects upon rabbits of eating vegetables grown in tungsten-rich soil is reported (203,859), and toxic levels of 0.1 mg/liter of tungsten in water (657) and 6 mg/m^3 of tungsten refractory compounds in air (623) are proposed. In humans tungsten is concentrated in the blood and various internal organs (202,272,933). Certain tungsten compounds are more toxic than tungsten itself (941). Tungsten(VI) chloride, for example, hydrolyzes, forming hydrochloric acid (928). Of special concern are the hazards involving workers in plants where tungsten is refined (307), fabricated (1009), and utilized, for example, in welding and applications employing tungsten electrodes (952). Consequences of inhaling radioactive tungsten in nuclear facilities are reported (647), as are the hazards of working with tungsten carbide and resulting lung disorders (43,177,468,-583).

II.   PROPERTIES OF TUNGSTEN AND ITS COMPOUNDS

A.   PHYSICAL PROPERTIES

Tungsten, after carbon, has the highest melting point of any known element. In addition it is one of the most heavy metals, with a density comparable to gold. Its low thermal expansion and electrical resistivity make it admirably suited for glass-metal seals as filaments in electron tubes. It is widely used as the filament in incandescent lamps and for high-temperature thermocouples, especially when alloyed with rhenium (29). Alone and as an alloying element in steels and nonferrous alloys, its hardness and high-temperature strength find extensive industrial application (45,255,862).

The physical properties of tungsten are compiled in several reviews (27,813,846,956). Table 1 lists some of the more common physical properties of tungsten.

Table 2 lists some physical properties of representative tungsten compounds.

B.   ELECTROCHEMICAL PROPERTIES

The electrochemistry of tungsten has been summarized (1003). Electrochemical measurements of tungsten-containing solutions are compli-

TABLE 1
Physical Properties of Tungsten (368,378,824,1045)

Atomic number	74
Atomic weight	183.85
Melting point	3410°C
Boiling point	5930°C
Density (20°C)	19.3 gram/cm^3
Specific heat	0.032 cal/(gram)(°C)
Thermal expansion (20–100°C)	0.0000044/°C
Thermal conductivity	0.40 cal/(sec)(cm^2)(°C/cm)
Heat of fusion	44 cal/gram
Heat of vaporization	1150 cal/gram
Electrical resistivity (20°C)	0.0000056(ohm)(cm)
Spectral emissivity (2000°K, 600 nm)	0.44
Crystal structure (body centered cubic)	a = 3.1652 Å at 25°C
Tensible strength (drawn wire)	250,000–600,000 psi
Stable isotopes	
180	0.14%
182	26.4%
183	14.4%
184	30.6%
186	28.4%

Radioactive isotopes		
173	EC	16 min
174	EC	29 min
175	EC, γ	34 min
176	EC, γ	2.5 hr
177	EC, γ	2.2 hr
178	EC	21.5 day
179 m	IT, γ	6 min
179	EC, γ	38 min
180 m	IT, γ	0.005 sec
181 m	IT, γ	14.6 μsec
181	EC, γ	130 day
183 m	IT, γ	5.2 sec
185 m	IT, γ	1.6 min
185	β^-, γ	75 day
187	β^-, γ	23.9 hr
188	β^-, γ	69 day
189	β^-, γ	11.5 min

cated by the presence of isopoly tungstates and complex ion formation with other ions present in the system. Studies of tungsten reduction in HCl indicate that the potential is pH dependent and can result in different reduced forms of tungsten (288,924). Electrode potentials for tungsten in a variety of other media are reported (427,477). Various compilations of electrochemical potentials list the following values for

TABLE 2

Physical Properties of Representative Tungsten Compounds (586,720,822,1045)

Compound	Formula	Appearance	Specific gravity	Melting point (°C)	Boiling point (°C)
Tungsten trioxide	WO_3	Yellow	7.16	sub. >1100	—
Tungsten hexachloride	WCl_6	Blue-black	3.52	275	346
Tungsten pentachloride	WCl_5	Green	—	248	276
Tungsten oxychloride	$WOCl_4$	Scarlet	11.9	211	228
Tungsten dioxychloride	WO_2Cl_2	Yellow	—	265	—
Ditungsten carbide	W_2C	Gray	17.2	2850	—
Tungsten carbide	WC	Gray	15.6	2900	—
Sodium tungsten	$Na_2WO_4 \cdot 2\,H_2O$	White	3.25	692	—

reductions of tungsten (113,141,545,566,617). Potential values are given relative to the normal hydrogen electrode.

$$2\,WO_3(s) + 2\,H^+ + 2\,e^- \rightarrow W_2O_5(s) + H_2O \qquad E° = -0.03\ V$$
$$WO_3(s) + 6\,H^+ + 6\,e^- \rightarrow W(s) + 3\,H_2O \qquad E° = -0.09\ V$$
$$WO_4^{2-} + 4\,H_2O + 6\,e^- \rightarrow W(s) + 8\,OH^- \qquad E° = -1.01\ V$$
$$W_2O_5(s) + 2\,H^+ + 2\,e^- \rightarrow 2\,WO_2(s) + H_2O \qquad E° = -0.04\ V$$
$$WO_2(s) + 4\,H^+ + 4\,e^- \rightarrow W(s) + 2\,H_2O \qquad E° = -0.12\ V$$
$$[W(CN)_8]^{3-} + e^- \rightarrow [W(CN)_8]^{4-} \qquad E° = +0.46\ V$$

Some formal potentials for tungsten in 12 M HCl are:

$$W(VI) + e^- \rightarrow W(V) \qquad E = +0.26\ V$$
$$W(V) + 2\,e^- \rightarrow W(III)\ red \qquad E = -0.2\ V$$
$$W(V) + 2\,e^- \rightarrow W(III)\ green \qquad E = +0.1\ V$$
$$W(V) + e^- \rightarrow W(IV) \qquad E = -0.3\ V$$

Tungsten can be placed in solution by anodic dissolution in the presence of strong base (28,362,370). The electrochemical behavior of tungsten has also been investigated in buffered (361) and acidic (22,105) media and in fused solvents at elevated temperatures (735).

Polarographic reactions of tungsten(VI) are discussed in Section VI.D. In general, the polarographic waves are affected by competing isopoly, heteropoly, and complexation reactions (365,397,482,548). Kinetic rather than diffusion-controlled waves are frequently observed for tungsten-containing solutions in the presence of reducible ions. Waves in the presence of chlorate (366,481) and hydrogen peroxide (484) especially have been extensively studied.

The use of tungsten and various tungsten bronzes for electrodes in electrochemical studies of other species is discussed in Section II.F.

C. OPTICAL PROPERTIES

The emission spectrum of tungsten has been extensively studied, and over 2000 separate lines have been recorded (353,547,1002). Extensive studies, too, have been made on the emissivity of tungsten (8,254,534). Infrared studies of various tungsten compounds are reported. These include refractory compounds (434) and tungstates (120). Raman spectra of various tungsten compounds have also been made (34,328,558,1021). The X-ray spectrum of tungsten has been compiled (51), and studies have been made of the applicability of the tungsten $K\alpha_1$ line as a standard for X-ray wavelength measurement (106).

D. CHEMICAL PROPERTIES

Tungsten metal, atomic number 74, has the electronic configuration, $1s^2\,2s^2\,2p^6\,3s^2\,3p^6\,3d^{10}\,4s^2\,4p^6\,4d^{10}\,4f^{14}\,5s^2\,5p^6\,5d^4\,6s^2$. It exhibits oxidation states of 2+, 3+, 4+, 5+, 6+, and with certain organometallic compounds 2−, 1−, 0, and 1+ (720,822). Presently only the 6+ state and various reduced tungsten species are of analytical interest. Tungsten is resistant to attack by the common mineral acids. It does, however, dissolve readily in a mixture of $HF-HNO_3$. It is attacked by fused alkali hydroxides, especially in the presence of an oxidizing agent, for example, KNO_2 or $KClO_3$ (487). Tungsten is not affected by air or water at ambient temperature but begins to tarnish at 400–500°C, forming WO_3. At temperatures between 600 and 700°C, this coating converts to a nonprotective form and severe corrosion begins (1045). Finely powdered tungsten metal burns readily in air (672). Tungsten is attacked by fluorine at room temperature and by other halogens, carbon, boron, silicon, and so forth at elevated temperatures.

Tungsten trioxide, WO_3, is the final product from ignition of acid-insoluble hydrous tungsten(VI) oxide. Dehydration is incomplete below 500°C, and volatilization occurs at temperatures greater than 700–800°C (126). The oxide is insoluble in mineral acid, although quantitative removal from a sample may require several repeated precipitations. The oxide dissolves readily in basic solution. The hydrous oxide, tungstic acid, from which WO_3 is obtained itself exists in alternate forms. There is an amorphous yellow tungstic acid, $WO_3 \cdot H_2O$, formed upon addition of strong mineral acid to a hot solution of alkaline tungstate and a white gelatinous, often colloidal, hydrated tungstic acid, $WO_3 \cdot 2H_2O$, formed during precipitation from cold solution (813). Other modifications have also been identified (425). The stoichiometric dioxide, WO_2, brown in color is also a stable tungsten oxide. Intermediate between WO_3 and WO_2 are several mixed valence oxides observed mainly at elevated temperatures; these include $WO_{2.72}(W_{18}O_{49})$ and $WO_{2.90}(W_{20}O_{58})$ (797,-821).

Tungsten oxides and soluble tungstates, WO_4^{2-}, upon acidification and in the presence of a reducing agent, for example, Sn(II), form mixed-valence tungsten oxides with a characteristic blue color, tungsten blue. The exact composition of these species is uncertain (298,924).

Tungsten forms a series of binary compounds with boron, carbon, nitrogen, and silicon (855). Typical of these are W_2B, WB, W_2C, WC, W_2N, WN, WSi_2, and W_2Si_3. Because of their refractory properties these materials are finding extensive use in the fabrication of high-temperature components, especially in the aerospace industry.

Tungsten(VI) forms a series of halide compounds of type WX_6 and a series of oxohalide compounds of type WOX_4 and WO_2X_2, where X represents a halide ion. Also common are mixed halides, substituted compounds where X is partially or wholly replaced by a nonhalide species, and anionic forms, for example, WF_8^{2-} and $WO_2F_4^{2-}$ (166,720).

Somewhat similar halides exist for the lower oxidation states of tungsten.

Normal tungstates exist corresponding to M_2WO_4, where M is a univalent cation, and MWO_4, where M is a divalent cation. Table 3 lists the solubility-product constants of several insoluble metal tungstates. In addition to the normal tungstates there is another series of compounds, the tungsten bronzes, with the general formula, M_xWO_3 (191,263,795). M represents an alkali metal, a rare earth ion, or other metallic cation and x is variable from near zero to one. Tungsten bronzes are highly colored, metallic in appearance, and exhibit electrical conductance or semiconductor properties depending upon the value of x (869).

The isopoly, heteropoly, and complexation chemistry of tungsten are discussed in subsequent sections.

1. Isopoly Compounds

When alkaline solutions containing simple tungstate ion, WO_4^{2-}, are acidified condensed tungstate species, isopoly ions, begin to form as the pH is lowered below approximately seven. Attempts have been made to identify specific isopoly species as they predominate at a unique pH value (467,663,817), but the complexity of the reactions have lead to divergent views regarding the exact nature of various isopoly tungstates. Ultimately, of course, as the acidity increases, hydrous tungstic oxide precipitates. The study of isopoly tungstates is the subject of several reviews (406,454,820,984).

It is generally accepted that the initial condensation product, formed about pH 6, be designated paratungstate A. This species slowly changes, over several days, to an alternate form. Possible intermediate species are designated by some as paratungstates B and the final condensation product as paratungstate Z. Others refer to the final condensation

TABLE 3
Solubility Product Constants of Representative Metal
Tungstates

Compound	Solubility product constant, K_{sp}	Reference
Ag_2WO_4	5.5×10^{-12}	816
$CaWO_4$	2.1×10^{-9}	813
$CdWO_4$	1.9×10^{-6}	586
$HgWO_4$	1.0×10^{-17}	816
$PbWO_4$	4.3×10^{-7}	586
$SrWO_4$	1.7×10^{-5}	586

product at this pH as paratungstate B. As the pH is lowered, to approximately pH 4, other isopoly species predominate. These are termed the metatungstates. First a pseudo (Ψ) metatungstate forms, probably from conversion of a paratungstate. This is transformed, through one or more intermediates, into "true" metatungstate.

Paratungstates form with a hydrogen ion to tungsten(VI) ratio of 1.16, that is, 7 H^+ to 6 W(VI):

$$7 H^+ + 6 WO_4^{2-} \rightarrow [HW_6O_{21}]^{5-} \quad + 3 H_2O$$
paratungstate A
$$14 H^+ + 12 WO_4^{2-} \rightarrow [W_{12}O_{41}]^{10-} \quad + 7 H_2O$$
paratungstate B

Metatungstates form with a hydrogen ion to tungsten(VI) ratio of 1.5, that is, 18 H^+ to 12 W(VI):

$$18 H^+ + 12 WO_4^{2-} \rightarrow \quad [W_{12}O_{39}]^{6-} \quad + 9 H_2O$$
metatungstate

Equilibrium constant values for these reactions are, respectively: log $K_{7,6}=54$, log $K_{14,12} = 110$, and log $K_{18,12} = 133$ (34). Other similar values for these constants are reported based upon measurements utilizing a variety of experimental techniques (302,454,990). The various isopoly species are capable of being protonated (328,854), and salts containing these ions are generally highly hydrated (223). Isopoly species containing other ratios of H^+ to W(VI) are reported (709) including a third identifiable metatungstate, tungsten X (84,149,555), and a tungsten species with a hydrogen ion to tungsten(VI) ratio of 1.58, that is, 19 H^+ to 12 W(VI), tungsten Y (85).

The structure of various isopoly tungstates is widely studied; Table 4 lists some alternate forms for various isopoly tungstates that have been proposed.

2. Coordination Compounds

The chemistry of inorganic tungsten complexes is reviewed periodically (275). Numerous complexes exist with most of the various oxidation states of tungsten. Halide and oxohalide complexes (720), sulfur complexes (720), carbonyl complexes (822), and phosphine and arsine complexes (644,774,1068) are all well characterized. Various tungsten-peroxo complexes with tungsten to peroxide ratios of 1:4, $[W(O_2)_4[^{2-}$; 1:3, $[WO(O_2)_3]^{2-}$; 1:2, $[W(O)_2(O_2)_2]^{2-}$; and 1:1, $[WO_3(O_2)]^{2-}$ are known. Either the 1:4 or 1:2 complex is most likely to occur in aqueous solution, depending upon pH (72,163,181,497). The 1:2 form is dimeric, that is, $[(O_2)_2W(O)OW(O)(O_2)_2]^{2-}$ with a bridging oxygen joining the two

TABLE 4
Names and Formulas of Common Isopoly Tungstates

Empirical formula (M = univalent cation)	Name	Molecular formula[a]	Reference
5 M$_2$O·12 WO$_3$	Paratungstate A	[HW$_6$O$_{21}$]$^{5-}$	454
		[W$_6$O$_{19}$(OH)$_3$]$^{5-}$	991
	Paratungstate B (or Z)	[W$_{12}$O$_{41}$]$^{10-}$	223
		[H$_2$W$_{12}$O$_{42}$]$^{10-}$	223
		[H$_{10}$W$_{12}$O$_{46}$]$^{10-}$	454
		[W$_{12}$O$_{36}$(OH)$_{10}$]$^{10-}$	702
3 M$_2$O·12 WO$_3$	Pseudo(Ψ)metatungstate	[H$_3$W$_6$O$_{21}$]$^{3-}$	454
		[W$_{12}$O$_{39}$]$^{6-}$	85
		[W$_{24}$O$_{72}$(OH)$_{12}$]$^{12-}$	702
	Metatungstate	[W$_{12}$O$_{39}$]$^{6-}$	223
		[H$_2$W$_{12}$O$_{40}$]$^{6-}$	297
		[H$_6$W$_{12}$O$_{42}$]$^{6-}$	223
		[W$_{12}$O$_{38}$(OH)$_2$]$^{6-}$	702

[a] Contains various structural water molecules, but omits waters of hydration.

tungsten atoms (327). Mixed complexes containing peroxo groups and inorganic ions, for example, chloride (1038,1074), are known as are mixed complexes containing peroxo groups and organic ligands. Among the organic ligands studied are oxalate (903), tartrate (96), pyridine (54), bipyridine (53), and others (455,631). Of major analytical interest is the W(V)-thiocyanate complex, which, with its characteristic yellow color, is frequently used for colorimetric determination of tungsten. The exact nature of this complex is unknown, and several formulas are reported for W(V)-thiocyanate depending upon solution acidity (720). [WO(NCS)$_5$]$^{2-}$ and a dimer, [W$_2$O$_3$(NCS)$_8$]$^{4-}$, are included among several possible formulas. Several complexes of W(VI)-thiocyanate are also known (994). The octacyanotungstate(IV) and octacyanotungstate(V) complexes form an oxidation-reduction couple, and [W(CN)$_8$]$^{3-}$ is sometimes used as an oxidizing agent for the titration of unknowns (1010).

Organic complexes of tungsten are reviewed (194,779). Complexes in which tungsten is part of a five- or six-membered ring are preferred when developing chromophoric reagents for colorimetric tungsten determination (490). Phenolic reagents in particular are suggested. Many complexes of tungsten(VI) form with the tungstyl ion, WO$_2^{2+}$ (664). Typical of these is the W(VI)-8-quinolinol complex, WO$_2$(C$_9$H$_6$ON)$_2$. 8-Quinolinol complexes of W(V) and W(IV) are reported to be of the form, [WQ$_4$]$^+$ and [WO$_4$], respectively (76), where Q represents 8-quinolinol. Carboxylic acids complex readily with tungsten(VI) (44,969), generally forming anions of type [WO$_3$L]$^{2-}$ or [WO$_2$L$_2$]$^{2-}$, where L^{2-} represents

TABLE 5

Formation Constants of Representative Tungsten(VI) Complexes

Ligand	Ratio W:L	Formation constant (log β)	Reference
Alizarin red S	1:2	7.8	893
Carmimic acid	1:1	5.5	894
Catechol	1:2	6.53	349
Citric acid	1:2	3.8	888
Malonic acid	1:1	3.09	894
Pyrogallol	1:2	6.98	349
Oxalic acid	1:1	4.85	894
8-Quinalizarin	1:2	9.72	796
Tartaric acid	1:1	3.93	894

the carboxylate anion. Tartrate (509), oxalate (870,1066), and citrate (782,888) complexes have all been specifically studied. A tungsten(V) oxalate complex, $[WO_2(C_2O_4)_2]^{3-}$, is also known (729). Table 5 lists the formation constants of some representative tungsten complexes.

3. Heteropoly Compounds

Like molybdenum and certain other transition metal ions, tungsten forms mixed complexes containing oxygen-bound tungsten groups clustered about a central species, the heteroatom. Heteropoly chemistry of tungsten is reviewed (223,441,799,983,1027). Complexes containing various ratios of tungsten to central metal atom are known, and of these the 12:1 complexes—12-tungstophosphate, $[PW_{12}O_{40}]^{3-}$, is typical—are perhaps best known. The 9:1 complexes containing either phosphorous or arsenic as heteroatom are dimeric, for example, $[P_2W_{18}O_{62}]^{6-}$. There are two types of 6:1 complexes. Those with either tellurium or iodine have the general formula, $[X^{n+}W_6O_{24}]^{(12-n)-}$, where X represents either Te(VI) or I(VII), and those with gallium or nickel have the general formula, $[X^{n+}W_6O_{24}H_6]^{(6-n)-}$, where X represents either Ga(III) or Ni(II). Other heteropoly series are also known having tungsten to heteroatom ratios of 10:1, 17:2, and $(6:1)_m$, where the value of m varies. Numerous mixed heteropoly complexes exist, for example, 11-tungstovanadophosphate, $[PVW_{11}O_{40}]^{4-}$ (977,985).

The heteropoly anions are represented structurally by groups of WO_6 octahedrons surrounding a XO_n polyhedron. Corners, edges, or faces can be shared depending upon the heteropoly series under consideration (1027). In general (441), the heteropoly acids are of high molecular weight, water soluble, highly hydrated, easily reduced forming the heteropoly blue color, and decomposed in strongly basic solution. Optimum pH ranges for maximum stability of several heteropoly tungsten complexes are reported (645,819,1028). Considerable interest is shown in the reduction of the tungsten heteropoly complexes (360,768,-938,999), but the exact nature of the heteropoly blues produced is still uncertain.

Some applications of heteropoly tungsten complexes for analysis of other materials are cited in Section II.F.

E. RADIOCHEMISTRY AND NUCLEAR PROPERTIES

Table 1 lists the naturally occurring and artificially produced isotopes of tungsten. ^{181}W ($t_{1/2}$, 130 days) and ^{185}W ($t_{1/2}$, 75 days) are generally used when radioactive tungsten isotopes are to be employed. With neutron activation, emissions from the artificially produced ^{187}W ($t_{1/2}$, 23.9 hr) or ^{183m}W ($t_{1/2}$, 5.2 sec) are generally monitored. Additional information on the radiochemistry of tungsten is given in Section VI.E and in a published monograph (649).

F. TUNGSTEN AND ITS COMPOUNDS IN ANALYSIS

1. Tungsten Reagents

The heavy alkali metals, rubidium, cesium, and francium, form insoluble salts with tungsten-containing heteropoly acids (499,510,940). Heavy alkali metals (511) and other ions (512) are also separated by selective sorption in columns containing tungsten ferrocyanide. Cerium(IV) tungstate (965), titanium tungstate (785), and 12-tungstophosphate heteropoly ion (153), too, are used for column separation of various metal ions. Chromatographic separation of several metal ions is achieved on paper impregnated with titanium tungstate (382) or tin tungstate (787). Various radioactive elements are separated from uranium on paper impregnated with 12-tungstosilicate heteropoly ion (600).

Sodium tungstate is suggested as a titrant for the determination of iron(III) with amperometric end point detection (890). Titration of vanadium(V), chromium(VI), and other oxidized substances is possible with tungsten(V) (905). A photometric procedure is used to monitor the increase of W(V) at 770 nm after the end point. Tungsten(IV), in the form of the cyanotungstate complex, serves as an oxidation-reduction titrant for reactions with arsenic (1010) and chromium or iron (426). Heteropoly tungstates can be titrated quantitatively with $R_4N^+OH^-$ titrants in mixed methylethylketone-i-butyl alcohol solvent (656). The procedure is used for determining the heteropoly ion, either phosphorus or silicon (457). Potentiometric or conductometric end point detection is employed. The 18-tungstodiarsenophosphate ion, $[As_2W_{18}O_{62}]^{6-}$, is used as an oxidation-reduction indicator in the titration of Ti(III) with Ce(IV) (64). Some studies have been made of the possibility of a phosphate-sensitive specific ion electrode using a tungstophosphate heteropoly ion (333).

Vanadium (125,1020) and indirectly aluminum (479) are determined colorimetrically through tungsten heteropoly formation. Ultraviolet absorbance measurements of phosphorus and arsenic-containing heteropoly tungstates (410) serve for their quantitative determination. Other applications of heteropoly tungstates in analysis are reviewed (409,410,-759,847,1013).

Applications of tungstates and heteropoly tungstates to colorimetric determination of organic substances (1006), especially biological and medical substances (1012), are reviewed. Various tungstates serve as precipitating agents for the isolation of alkaloids (332,375,507,685,719) and proteins (842), especially in blood samples (58,59,837). Directions for the preparation and storage of a stable tungstic acid reagent are given (6,604). Quantitative determination of alkaloids (760,1090) and proteins (61) is possible by titration with a heteropoly tungstate and amperometric end point detection. Turbidimetric end point detection is also used (68,516).

Other applications of tungsten-containing reagents include the determination of uric acid in blood (677), the study of soaps (1073) and surface-active agents (440), the separation of sugars by paper electrophoresis (25), a spray reagent in the detection of steroids by paper chromatography (603), and the staining of samples for microscopic viewing (104,546,572,793).

2. Tungsten Electrodes

Applications of tungsten electrodes in chemical analysis are reviewed (1030). Tungsten-calomel (95) and bimetallic electrode pairs, for example, W-Sb (902), are sometimes used to monitor hydrogen ion concentration. The nature of this tungsten electrode response has been studied (213,388). The course of acid-base titrations in both aqueous (693) and acetic acid solvent (692) are indicated by potential changes from a W–Ag electrode pair. Precipitation titrations involving silver ion and halides also cause a response from W–Ag or W–Pt electrodes (418,992). Tungsten itself is determined by potentiometric precipitation titration with Pb(II) titrant and a tungsten-indicating electrode (293).

Complexometric titrations of metal ions with EDTA can be followed with a W–Pt electrode pair (424,961). The application of W–Pt electrodes to oxidation-reduction titrations has been studied (483), and manganese (945), molybdenum (110), and titanium (111) have been determined potentiometrically using a tungsten electrode.

A tungsten electrode is suggested for polarographic studies (414), especially in fused electrolytes (154,853). Amperometric titrations of weak acids have been followed with a tungsten electrode (398).

Electrodes prepared from sodium tungsten bronzes are suitable for measurement of hydrogen ion concentration and oxidation-reduction reactions (480). Use of a sodium tungsten bronze electrode is reported for complexometric titration of metal ions with EDTA (1029) and for measurement of dissolved oxygen (342,343).

Pertaining to spectrographic analyses, tungsten electrodes are sometimes used in place of graphite for the analysis of cast irons and steels (493,608).

3. Miscellaneous Applications

Tungsten trioxide is used as an additive in carbon–hydrogen determinations to aid in eliminating interference (471), and tungsten-trioxide-coated platinum wire responds to the presence of hydrogen in certain types of gas detectors (872).

III. DISSOLUTION OF TUNGSTEN, TUNGSTEN ALLOYS, AND TUNGSTEN-CONTAINING MATERIALS

A. TUNGSTEN, TUNGSTEN ALLOYS AND COMPOUNDS

Tungsten metal is resistant to attack by the common mineral acids. It does dissolve in a mixture of HF-HNO_3 (487). Alternately the powdered metal can be ignited to the oxide, which dissolves in strong alkali solution, but if trace impurities, for example, molybdenum, are to be determined their loss can occur during ignition and the HF-HNO_3 treatment is preferred (544). Tungsten is soluble in 30% hydrogen peroxide (654) and in peroxotrifluoroacetic acid (857).

Ferrotungsten dissolves in a mixture of oxalic acid–hydrogen peroxide (113). Tungsten and tungsten-containing alloys can also be put into solution by fusing with Na_2O_2 (1040), NaOH (98), or NaOH plus an oxidizing agent, for example, KNO_2 (487). Tungsten-containing steels are generally dissolved in HCl–HNO_3 until only a light colored residue remains. This residue, insoluble hydrous tungsten(VI) oxide, is then completely precipitated with cinchonine or another organic precipitating agent. See Sections VI.A.1 and IX.A for further details of this procedure. The presence of a complexing agent for tungsten aids in dissolving tungsten alloys. Hydrogen peroxide–hydrofluoric acid mixtures are used for tungsten-containing steels (449) as is the addition of phosphoric acid or tartaric acid to steel samples treated with H_2SO_4–HNO_3 (242). A procedure for colorimetric determination of tungsten in steels with thiocyanate ion following dissolving in H_3PO_4–HNO_3 is described in Section IX.C. Electrolytic dissolution of tungsten and tungsten alloys is possible (83,520). The preparation of tungsten steels prior to spectroscopic analysis is described (187). Tungsten in titanium alloys is placed in solution with sulfuric acid–fluoroboric acid (214). Various tungsten nickel alloys are dissolved in an H_2O_2–$H_2C_2O_4$ mixture (634) or by fusion with $K_2S_2O_7$ and extraction of the cooled melt with tartaric acid solution (214). A procedure for determining tungsten in niobium or tantalum alloys using HF–HNO_3 to dissolve the sample is described in Section IX.D.

Tungsten(VI) oxide, formed from treating a tungsten-containing sample with HCl–HNO_3 or by other means, dissolves in alkali hydroxide or ammonia solution forming the simple tungstate ion, WO_4^{2-} (215). Tungsten carbides can be dissolved with HF–HNO_3 (147), by fusion with potassium pyrosulfate (221), or by ignition in a high-frequency induction-type furnace (895). Tungsten borides and other refractory materials are placed in solution by fusion with Na_2CO_3–$NaNO_3$ mixtures (67) or with HF–HNO_3 (508).

B. ORES

Finely powdered scheelite ore, $CaWO_4$, is decomposed by HCl–HNO_3. Following this the tungsten present is quantitatively precipitated with cinchonine (574). Perchloric acid–hydrogen peroxide is suggested for dissolving calcium tungstate (174). It is somewhat difficult to decompose wolframite ores, $FeWO_4$–$MnWO_4$, with hydrochloric–nitric acid mixtures (574,863), and sodium carbonate fusion is an alternate means of attacking these ores (130). Caution, however, is necessary as excess alkali metal can interfere with cinchonine precipitation of tungsten(VI) oxide (574). Other flux materials for tungsten ores and minerals are $KHSO_4$ (371), NH_4NO_3–NH_4F (504), and KOH–Na_2O_2 (828). A procedure for tungsten in low-grade ores using HCl–HNO_3 is given in Section IX.E.

C. MISCELLANEOUS MATERIALS

Soil samples analyzed for tungsten are first fused with Na_2CO_3–NaCl–KNO_3 (80) or digested with HF–HNO_3 (609), while tungsten in plant material is found after destroying the organic matter with $HClO_4$–HNO_3 (662) or HF–H_2O_2 (673).

IV. SEPARATION FROM OTHER ELEMENTS

Traditionally tungsten, as acid-insoluble tungsten oxide, is separated from other ions by dissolving a tungsten-containing sample in strong acid solution. Separation is, however, incomplete, and organic coprecipitating agents, double precipitation, and correction for unprecipitation tungsten are all necessary for quantitative separation. Ion-exchange procedures offer a cleaner separation of tungsten from other ions but frequently involve working with hydrofluoric acid solutions, which necessitates the use of special inert containers. Few tungsten complexes have been employed for separation by extraction. Paper and thin-layer chromatography are, however, more successful, especially for isolation prior to qualitative identification. Separation procedures for tungsten have been reviewed (91,268,491).

A. PRECIPITATION METHODS

Tungsten remains as insoluble hydrous tungsten oxide upon treating a tungsten-containing sample with mineral acid (770). Precipitation, however, is incomplete, and an organic reagent is usually added to bring down the last traces of tungsten. A mixed reagent containing both tannin and cinchonine is more effective than either used alone in achieving quantitative results (622), although many other chemicals are also employed (119,531,563,642). Certain alcohols and other oxygen-containing reagents, however, increase tungsten solubility and should be avoided (306). The tungsten precipitate obtained by acid hydrolysis is not pure, containing several entrapped metal ions (280), and must be redissolved and reprecipitated to free trapped impurities. This is generally accomplished by basic fusion (168,844) with sodium carbonate or sodium hydroxide. Frequently the tungsten-containing sample is fused directly. This results in formation of soluble tungsten(VI) while many heavy metal ions remain as insoluble hydroxides. Acidification of the solution from the alkaline melt precipitates insoluble hydrous tungsten oxide. Addition of calcium chloride to the soluble tungstate solution precipitates calcium tungstate (917).

Trace amounts of tungsten are frequently coprecipitated with iron(III) by careful pH adjustment (640,694,754). An iron(III) salt is added to the tungstate solution, and the pH is adjusted to approximately 7 with ammonia. Further increase in basicity beyond approximately pH 8 results in decreased tungsten coprecipitation. Other hydrous oxides (753), especially the zirconium(IV) species (752), are used to coprecipitate tungsten. Coprecipitation also occurs on metastannic acid in the presence of nitric acid (683) and on the molybdophosphate heteropoly anion (44,967).

Precipitation with benzidine separates tungsten from phosphorus and arsenic (578). Precipitation with 8-quinolinol separates tungsten from tin in the presence of oxalate (416) and from vanadium(IV) in the presence of EDTA (807). Tungsten is also separated from vanadium by precipitating the latter with cupferron in the presence of HF (161). Tungsten is separated from samples containing tantalum and niobium with a mixed cinchonine–tannin reagent (850).

B.   EXTRACTION METHODS

Separation of tungsten-containing mixtures by extraction involves, generally, removal of other components while tungsten remains behind in the aqueous phase. Tungsten is, itself, extractable providing suitable complexing agents and proper pH control are maintained. Proper conditions are, for example, reported for maximum extraction of tungsten in the presence of thiocyanate ion and varying HCl concentrations using methyl isobutyl ketone (316) and other extracting solvents (982). Ternary complexes containing tungsten, thiocyanate, and an auxiliary ligand are extractable (919,1086) and have been used both for separation and spectrophotometric determination of tungsten in metal alloys (10). The extraction of tungsten with various high-molecular-weight amine oxides has been studied (209) as has the counter-current separation of tungsten from large amounts of molybdenum with methyl isobutyl ketone (431).

Table 6 lists various extracting agents for tungsten.

C.   ION EXCHANGE METHODS

Tungsten is conveniently separated from other metals by ion exchange. Both cation and anion exchange resins are used, and tungsten is either retained or eluted depending upon solution conditions that determine whether a cation or anion form of tungsten is present. In dilute acid, neutral, and basic solutions tungsten, as an anion, is retained by anion resins. Careful consideration is necessary, however, in establishing optimum conditions for quantitative ion exchange as tungsten readily forms isopoly species in solutions of increasing tungsten concentration and increasing acidity (384,887). Hydrogen peroxide and/or organic complexing agents are frequently added to tungsten solutions to prevent precipitation when carrying out separations in acid solutions of hydrogen ion concentration greater than $0.1\ M$. The effect of acetate, citrate, formate, and malonate ions on tungsten separation with both cation and anion exchangers has been investigated (889). Depending upon the concentration of complexing agent positive, neutral, or negative complexes form.

TABLE 6

Extracting Agents for Tungsten

Aqueous phase	Complexing agent	Extractant	Interference	Reference
1.5 M HCl + ascorbic acid	Aniline	Aniline	Cu, Mo, V, Zn	972
0.5 M HCl	α-Benzoin oxime	CHCl₃	Mo	730
4 M HCl	N-Benzoyl-N-phenylhydroxylamine	CHCl₃	Mo, V	678
4 M HCl	Benzohydroxamic acid	BuOH–CHCl₃ 1:1 vol./vol.	Ga, Ge, Mo	763
0.1 M HCl	3,5-Dinitropyrocatechol + diantipyrylmethane	CHCl₃	Mo, Nb, Ti, V	666
1 M HCl + 12 M LiCl	Mesityl oxide	BuCOMe	Cu, F, Hg, Mo	881
3 M HCl + 8 M LiCl	4-Methyl-2-pentanol	Benzene	Hg, Mo, U, V	723
pH 2.4, phthalate	8-Quinolinol	CHCl₃	Fe, Mo, V	937
9 M HCl	Thenoyltrifluoroacetone	BuOH–acetophenone	Mo	179
4 M HCl	KSCN	AmOH–CHCl₃, 1:1 vol./vol.	F, Fe, Mo	739
9 M HCl	SnCl₂ + KSCN + Ph₄AsCl	CHCl₃	Nb, V	10
12 M HCl + SnCl₂	Toluene-3,4-dithiol	AmOAc	Co, Mo	38
9 M HCl	Tributyl phosphate	TBP	Cu, Mo, Ti, V	180

465

TABLE 7

Ion Exchange Separations of Tungsten

Separation from	Resin	Aqueous phase	Eluent	Reference
Co, Ni	Dowex 50-X12 (H⁺)	$0.1\ M\ Li_2CO_3 + 0.3\ M$ tartaric acid	$0.5\ M$ HCl (for Li) $4\ M$ HCl (for Co, Ni)	922
Ti	SBS (H⁺)	$0.1\ M$ HCl + 1% H_2O_2	$0.1\ M$ HCl + 1% H_2O_2 (for W) 10% H_2SO_4 (for Ti)	17,827
Ag, Si	Dowex 1-X8 (SO_4^{2-})	$0.2\ M$ HF + $0.07\ M$ $H_2SO_4 + 1.2\ M\ H_2O_2$	$0.01\ M$ HF + $0.10\ M\ H_2SO_4 + 0.3\ M\ H_2O_2$ (for Ag) $0.01\ M$ HF + $0.10\ M\ H_2SO_4$ (for H_2O_2) $8\ M$ HCl (for Si)	978
Fe, Mo, U	Dowex 1-X10 (Cl⁻)	$0.5\ M$ HCl + $1\ M$ HF	$0.5\ M$ HCl + $1\ M$ HF (for Fe, U) $7\ M$ HCl + $1\ M$ HF (for W) $1\ M$ HCl (for Mo)	505
Mo, Nb, Ti, V, Zr	Dowex 1-X8 (F⁻)	$3\ M$ HF	$10.1\ M$ HF (for V) $27.5\ M$ HF (for Ti, Mo, Zr) $7\ M$ HCl + $3\ M$ HF (for W) $3\ M\ NH_4Cl + 1\ M$ HF (for Nb)	237

466

Mo, Nb, Ta, Ti, Zr	Dowex 1-X8 (Cl⁻)	0.3 M KHSO₄ + 0.5 M H₂C₂O₄	1.5 M HCl + 0.5 M H₂C₂O₄ + 0.007 M H₂O₂ (for Zr, Ti, Nb) 3 M HCl + 0.5 M H₂C₂O₄ (for Ta) 4 M HCl + 0.1 M citric acid (for W) 19 M NH₄Cl + 0.44 M NH₄ citrate (for Mo)	42
Cr, Mo, V	AGl-X8 (SO₄²⁻)	0.05 M H₂SO₄ + 0.1% H₂O₂	0.05 M H₂SO₄ + 0.1% H₂O₂ (for Cr) 0.5 M H₂SO₄ + 0.1% H₂O₂ (for V) 3 M H₂SO₄ + 0.1% H₂O₂ (for Mo)	939
Na, K	Wofatit SBW (Cl⁻)	0.1 M HCl + 0.5 M H₂C₂O₄ + 1 M H₂O₂	2 M NH₄NO₃ + 0.5 M NH₃ (for W) 0.1 M HCl + 0.5 M H₂C₂O₄ + 1 M H₂O₂ (for Na, K)	195
Cr, Mo	EDE-10P (Cl⁻)	0.5 M HF	H₂O (for Cr) 1 M HCl (for Mo) 1.0 M NH₄Cl + 5% NH₃ (for W)	707
Re	Amberlite IRA-400 (Cl⁻)	0.5 M NaOH	0.5 M NaOH + 0.5 M NaCl (for W) 8 M HCl (for Re)	698
Mo, Re	DEAE (SCN⁻)	aqueous	0.02 M NH₄SCN + OAc⁻, pH 3 (for Re) 0.1 M NH₄SCN + OAc⁻, pH 5 (for Mo) 0.1 M NaOH + 0.1 M NaCl (for W)	394
Mo, V	0.20 M Aliquot 336 on XAD-2 (Cl⁻)	0.15 M H₂SO₄ + 0.5% H₂O₂	0.15 M H₂SO₄ + 0.5% H₂O₂ (for V) 0.30 M H₂SO₄ + 0.5% H₂O₂ (for W) 4.0 M H₂SO₄ + 0.5% H₂O₂ (for Mo)	262

Ion-exchange separations involving tungsten are frequently performed in the presence of HF (232,1044). In a typical procedure (192) for separating components of a complex alloy the sample is dissolved in 48% HF plus a small amount of 16 M HNO_3, evaporated to near dryness, the residue dissolved in 1 M HF, and the sample placed on the strong anion-exchange resin, De-Acidite FF (Cl⁻ form). Eluation with 1 M HF separates Al, V(IV), Cr(III), Mn(II), Fe(III), Co(II), Ni, and Cu(II), which pass through the column. Ti(IV), Zr, W(VI), Nb(V), Mo(VI), and Ta are retained. These are eluted in order with 0.01 M HF–9 M HCl (for Ti, Zr), 3 M HF–10 M HCl (for W), 0.2 M HF–7 M HCl (for Nb), 3 M HF–3 M HCl (for Mo), and 1 M NH_4F–4 M NH_4Cl (for Ta). To avoid handling large amounts of HF an alternate separation of Ti, Zr, W, Nb, Mo, and Ta has been developed that utilizes varying mixtures of HCl, $H_2C_2O_4$, citrate, NH_4Cl, and H_2O_2 (42).

Particular interest is centered about the separation of tungsten and molybdenum, and specific resins have been developed for this purpose (769). Using commercially available resins both molybdenum and tungsten are retained on the strong-base anion resin, AV-17, from 0.5 M HF solution (707). Molybdenum is eluted first with 1.0 M HCl, followed by tungsten eluted with 1.0 M NH_4Cl–0.8 M NH_3. This same resin retains both molybdenum and tungsten from solutions of 1 M HF–0.1 M HCl, while iron, chromium, nickel, and vanadium pass through the column (944). The moderately weak anion exchange resin, EDE-10P, retains both molybdenum and tungsten from solutions of 0.1 M HCl. Molybdenum is eluted with 4 M HCl, followed by tungsten eluted with 8 M HCl (886). Dowex 1-X2 (acetate form) retains both molybdenum and tungsten from acetic acid solutions (385). Molybdenum is eluted with 2 M NH_4OAc–NH_3 adjusted to pH 5.3, followed by tungsten eluted with 2 M NH_4OAc–NH_3 adjusted to pH 9.

Conditions for the ion exchange separation of tungsten from various ions are listed in Table 7.

D. OTHER CHROMATOGRAPHIC METHODS

Tungsten(VI), molybdenum, and niobium are separated from each other by selective elution after all are retained on an alumina column (459). Molybdenum is eluted first with NaCl solution adjusted to pH 8–8.5, followed by tungsten eluted with 10% NH_3 at pH 12. Niobium remains in the column. Alumina columns are also used in separating rhenium from tungsten (473).

A Teflon support coated with methyl isobutyl ketone or other organic phase separates tungsten from rhenium (643), various refractory metals

TABLE 8

Paper and Thin Layer Chromatographic Separation of Tungsten

Chromatographic support	Developing solvent	Separation from	Reference
FN₆ paper	AmOH–HCl–30% H₂O₂, 19:6:2 vol./vol.	V, Mo, Re, Fe	292
Whatman No. 4	AmOH–85% HCOOH, 9:11 vol./vol.	Cr, V, Mo, Fe	527
Whatman No. 1	BuOH–12 M HCl-sym-collidine, 38:52:10 vol./vol.	Cr, Mo, V	1053
Whatman No. 1	BuOH–70% HNO₃, 2:3 vol./vol.	Mo, Co, V, Ni, Ti, Fe	786
S&S No. 2040a	BuOH–30% H₂O₂ sat'd with 1 ml of 1 M HNO₃, 4:1 vol./vol.	V, Mo	151
Whatman No. 1	CHCl₃–Me₂CO–AmOH–12 M HCl, 47.6:23.8:23.4:4.5 vol./vol.	Mo, U, V, Fe	62
Whatman No. 3MM	10% HOAc + 10% NaOAc, aqueous	Cr, Mo, V	755
Paper	MeOH–Me₂CO–benzyl alcohol–0.05 M Na₂H₂ EDTA, 3:7:1:1 vol./vol.	Cr, Mo	948
Silica Gel G	AmOH sat'd with 12 M HCl	Mo, Re	791
Al₂O₃	BuOH–12 M HCl, 1:1 vol./vol.	Re, Mo, V	269
Silica Gel G	t-BuOH–HOAc, 5:4 vol./vol.	U, V, Mo	420
Wako-Gel B-0 + 10% starch	MeOH–3 M HCl, 7:3 vol./vol.	Re, Mo	523

469

(261), and components in stainless steels (220). Selective elution occurs with various mineral acid–hydrofluoric acid mixtures.

Tungsten is separated from iron after first forming the heteropoly tungstophosphate. The heteropoly complex is retained on cellulose, while iron(III) is not (630).

Foam flotation separates tungsten and certain other oxyanions from multicomponent systems (138).

Paper chromatographic separation of inorganic ions, including tungsten, has been reviewed (552). Various butanol–complexing agent solvents successfully separate rhenium, vanadium, molybdenum, and tungsten (150). A 70% HNO_3–BuOH (6:4 vol./vol.) solvent separates tungsten from a variety of other ions (786). With this solvent tungsten does not migrate, while other ions advance along the paper to varying degrees. Various treated papers separate tungsten from other ions with which it is associated. Some of these and other paper chromatographic separations are summarized in Table 8.

Conditions for separation of tungsten from other ions on thin-layer plates coated with silica and alumina have been studied using various alcohol and alcohol–acid developing mixtures (270,882). These too are summarized in Table 8. Table 9 lists spray reagents used in detecting tungsten on developed paper and thin-layer chromatograms.

Vapor-phase chromatographic separation of tungsten and other metal carbonyls (766) and fluorides (423) is reported. The fluoride procedure

TABLE 9

Spray Reagents for Detection of Tungsten After Paper or Thin-Layer Chromatographic Separation

Spray reagent	Observed color	Reference
0.01% alizarine in EtOH, expose solution to NH_3 vapors before spraying	Rose	172
0.1% chlorogenic acid in acetone	Yellow	670
5% pyrogallol in EtOH	Brown	1053
1% quercetin in EtOH	Yellow	625
5% aqueous tannic acid	Brown	374
5% $SnCl_2$ in 6 M HCl	Blue	62
5% $SnCl_2$ in 6 M HCl, overspray with 20% aqueous KSCN	Yellow	755
5% $TiCl_3$ in 0.1 M HCl	Blue	135,152
0.1% toluene-3,4-dithiol in 0.25 M NaOH	Blue-green	523
0.001 M vanadyldinitrofluorone in 100 ml EtOH containing 1 ml of 1 M H_2SO_4	Brown	593

has been applied to quantitative determination of tungsten in molybdenum–tungsten alloys.

E. MISCELLANEOUS METHODS

Volatilization of tungsten chlorides and oxychlorides has been used for separating tungsten from other elements (408,843,1084); however, the procedure is not recommended for quantitative separation (1).

Tungsten(VI) migration by paper electrophoresis in the presence of oxalate (65) or lactate (66) serves in separating tungsten from molybdenum.

Tungsten is separated from arsenic (276) and other ions (1077) by electrodialysis.

V. DETECTION

Tungsten belongs in the acid-insoluble group of the classical hydrogen sulfide scheme of qualitative chemical analysis and is precipitated with silver, lead, and mercury(I) by addition of hydrochloric acid to the test solution (137,422,773,1025). Addition of other mineral acids also precipitates tungsten from soluble tungsten(VI) solutions. Upon boiling pale-yellow insoluble tungsten trioxide forms. Lead is separated from the acid-insoluble chlorides and tungsten by leaching with hot water. This is followed by addition of excess ammonia causing mercury to precipitate while tungsten and silver dissolve in the filtrate. Tungsten trioxide is soluble in basic solutions. Silver is separated from tungsten by precipitating silver iodide or, after addition of Na_2HPO_4 and acidifying to form a soluble tungstophosphate, precipitation as silver chloride. Separation of tungsten from silica, if the latter is also present, is achieved by fusing the combined oxides with potassium pyrophosphate and leaching tungstate from the cooled melt with a saturated solution of ammonium carbonate (140).

Confirmatory tests for tungsten include reduction with tin(II), or another reducing agent, in acid solution to a blue-colored oxide containing tungsten in a lower valence state (tungsten blue), formation of yellow tungsten(V)-thiocyanate complex, or formation of other suitable colored complexes (295). These tests are described in more detail in the following sections. R_f values and color-forming spray reagents used in separating and detecting tungsten by paper and thin-layer chromatography are discussed in Section IV.D.

A. PRECIPITATION REACTIONS

Samples containing soluble tungstate or freshly precipitated tungsten oxides upon acidification and addition of a suitable reducing agent, generally tin(II) chloride (918), form reduced oxide species having a characteristic blue color. The appearance of tungsten blue, as it is called, is a positive test for tungsten. Tungsten blue is insoluble, and if the concentration of tungsten is sufficient a precipitate forms. The test is positive for 5 μg or more of tungsten (233). Moderate amounts of molybdenum interfere with this test, and in the presence of molybdenum addition of KSCN along with $SnCl_2$ produces a red color due to molybdenum(V)-thiocyanate. The red color fades upon standing and upon addition of excess tin(II) chloride, which further reduces molybdenum. The tungsten blue color can then be observed. Other species giving colored reduction products also obscure the tungsten blue color, especially if the tungsten concentration is low. Alternate reducing agents used to form tungsten blue include zinc(II) (215), titanium(III) (980), sodium dithionite (638), and mercury(I) nitrate-potassium iodide (1046). Electrolytic (282) and photochemical (915) reduction are also employed.

Diphenyline, 4,2'-diaminodiphenyl, forms a precipitate with tungstate ion in slightly acid solution upon standing for at least 15 min (233). The test is positive for 6 μg or more of tungsten. Sulfate ion and small amounts of molybdenum do not interfere.

In the absence of molybdenum(VI) and vanadium(V), which give similar responses, tungsten(VI) catalyses the oxidations of thallium(I) in ammonical solution by hydrogen peroxide (234). Appearance of a brown precipitate of thallium(III) hydroxide indicates the presence of tungsten. The test is positive for 0.05 μg or more of tungsten.

Numerous insoluble tungsten compounds are known, but the precipitating species are, in general, not specific for tungsten and are not widely used for qualitative identification of tungsten especially in the presence of other ions. Some of these precipitating agents are discussed in Section VI.A.

B. COLOR REACTIONS

Yellow tungsten(V)-thiocyanate complex forms in 0.05–0.5 M NaOH solutions of tungsten(VI) upon addition of KSCN and tin(II) chloride (235,998). The test is positive for a few micrograms or more of tungsten. Molybdenum interferes by forming the red molybdenum(V)-thiocyanate complex, but this fades upon standing. The yellow complex is also observed upon the surface of strong anion-exchange resin beads (660). With this procedure a 200-fold increase in sensitivity is reported for tungsten detection compared with solution formation. Prior precipitation, ion-exchange separation, and extraction remove interfering ions.

Dithiol, toluene-3,4-dithiol, added to a solution containing tungstate in 10–11 M HCl, plus a small amount of H_3PO_4 to prevent precipitation, produces a blue-green complex with tungsten (628). The solution is heated at 70°C for 5–10 min, and the color is observed in a isoamyl acetate extract. Addition of tin(II) prior to adding dithiol minimizes interference of molybdenum and rhenium, which otherwise form green complexes with dithiol. The test is positive for a few micrograms or more of tungsten.

Rhodamine B forms a violet-colored tungsten complex in solutions containing 0.1–0.2 M HCl (208). The test is positive for a few micrograms or more of tungsten, and moderate amounts of molybdenum do not interfere.

Tungsten(VI) in 0.1 M HCl solution catalyses the decoloration of malachite green by titanium(III) chloride (833). A drop of titanium(III) solution is added to the test solution followed by a drop of malachite green solution. The color rapidly fades to a very pale violet. The test is positive for 0.1 μg or more of tungsten. Nitrate, fluoride, and ions that react with titanium(III) must be absent, but moderate amounts of molybdenum do not interfere.

C. PHYSICAL METHODS

The emission spectrum of tungsten is identified by the following characteristic spectral lines (353,533,1075):

430.211 nm
429.461
400.875
294.698
294.440

With flame emission the detection limit for tungsten in a fuel-rich oxy-acetylene flame is 4 ppm of tungsten when monitoring the 400.8 nm line (227). This value is lowered to 0.5 ppm of tungsten in a nitrous oxide–acetylene flame (744). Extensive lists of X-ray spectral lines are published for tungsten (9,51). A selection of these include:

$K\alpha_1$	0.209 Å	59.318 KeV
$K\beta_1$	0.184	67.244
$L\alpha_1$	1.476	8.398
$L\beta_1$	1.282	9.672

The infrared spectrum of solid tungsten(VI)-nitron complex exhibits peaks at 12.3 and 11.35 μ (589). These are used for qualitative identification. Molybdenum(VI) and other species also complex with nitron and exhibit similar spectra.

Microscopic examination of colorless ammonium tungstate crystals (489) and crystals of tungsten(VI)-8-quinolinol complex (1001) are used to characterize as little as 0.15 μg of tungsten.

D. DETECTION IN SPECIFIC MATERIALS

Tungsten in steels is conveniently identified by the color of its rhodamine B complex (802). The clean metal surface is treated with several drops of sulfuric acid, two drops of NH_4F solution, and one drop of nitric acid. The solution is transferred to a spot plate containing one drop of NaOAc solution, one drop of HCl, and two drops of 0.01% rhodamine B. A blue-violet color indicates tungsten. Molybdenum, present in amounts up to 20 times that of tungsten, does not interfere. Other tests for tungsten-containing alloys include formation of tungsten blue (52,222) and formation of blue-green tungsten-dithiol complex (299,628).

Tungsten in ores and minerals is identified by fusing a portion of sample with sodium peroxide (239,686). The cooled melt is leached with water, and KSCN and Sn(II) are added. A yellow color indicates presence of tungsten(V)-thiocyanate complex. Tungsten blue formation (675), dithiol reagent (160), and 8-quinolinol reagent (916) are also used to confirm tungsten in ores and minerals.

VI. DETERMINATION OF TUNGSTEN

A. GRAVIMETRIC METHODS

Tungsten is generally precipitated as the hydrous oxide from mineral acid solution in the presence of a coprecipitating agent, usually cinchonine. Various corrections are necessary to account for other metal ions that also precipitate and for small amounts of tungsten remaining in solution. Other inorganic and organic precipitating agents too suffer from lack of selectivity, and, except for 8-quinolinol, all organic precipitates of tungsten are ignited to tungsten trioxide to achieve a stoichiometric product. Summaries of conditions and reagents for tungsten precipitation are available (132,206).

1. Tungsten(VI) Oxide

Hydrous tungsten(VI) oxide is insoluble in mineral acid solution (156,764). Precipitation is, however, incomplete unless an organic coprecipitating agent is added. Cinchonine is commonly used (622), although other alkaloids (562), other organic precipitating agents (82,532), and mixtures of organic precipitating agents (524,750) are employed. The coprecipitating agent is added to the hot tungsten solution, and digestion is continued for some time to assure complete precipitation. Most accurate results are achieved only if the initial precipitate is redissolved and reprecipitated to remove entrapped impurities (215,485). An additional correction is also necessary for small amounts of tungsten remaining in the sample solution. Silica, if present with tungsten in the initial precipitate, is removed by treating the insoluble residue with sulfuric acid–hydrofluoric acid after first destroying organic matter present. Volatile silicon tetrafluoride escapes. The remaining residue may still contain iron and other heavy metal traces. It is fused with sodium carbonate and the melt taken up in dilute ammonia. Tungsten, as tungsten(VI), is soluble, while many other components, as insoluble hydrous oxides, precipitate. Sulfide precipitation from the ammonical tungstate solution also removes metal ion impurities. Phosphate, if present with tungsten, interferes (896) as do arsenic and vanadium. Their separation from tungsten using selected precipitating agents is discussed in Section IV.A. If cinchonine is used as the coprecipitating agent, the presence of alkali and ammonium salts prevent complete precipitation of tungsten. Molybdenum seriously interferes with tungsten precipitation. From one study of many possible organic coprecipitating agents gelatin is reported to cause the least amount of molybdenum interference (389), although the desirability of gelatin in place of cinchonine for quantitative tungsten precipitation has been questioned (751).

Regardless of the organic reagent selected for coprecipitation the filtered and washed residue is dried and then ignited at a temperature between 750 and 800°C to convert all forms to tungsten trioxide, WO_3. Higher ignition temperatures result in loss of volatile WO_3, while at lower temperatures conversion to anhydrous WO_3 is incomplete.

Homogeneous precipitation of insoluble tungsten oxide occurs by slow decomposition of the chloro complex in HCl or the oxalato complex in sulfuric acid (514) and of the peroxo complex in nitric acid (175).

2. Metal Tungstates

Alkaline earth ions and certain heavy metal ions form insoluble precipitates with tungsten(VI) (57,129). Barium tungstate is the least soluble alkaline earth tungstate, but unless special precautions are taken contamination by barium carbonate causes serious interference (564). Optimum pH values for precipitation of calcium, strontium, and barium tungstates are, respectively, 11.8–8.1, 12.8–8.1, and 13.4–8.1 (635).

Mercury(I) and mercury(II) both precipitate tungsten(VI) (1051). The precipitate, however, generally contains additional mercury and must be ignited to WO_3 for weighing. Homogeneous precipitation of tungsten as the silver salt results from the thermal decomposition of diamminosilver(I) in tungstate solutions near pH 7 (1000).

3.   Organic Precipitates

Benzidine (1079) and substituted benzidines (56) precipitate tungsten. Results compare favorably with cinchonine precipitation (1080) and other gravimetric procedures (170). Precipitation is from dilute acid solution, and the precipitate is ignited to WO_3 for weighing. Sodium or potassium chloride, if present, cause low results for tungsten, but sulfide and silicate do not interfere. Phosphate, molybdenum, and sulfate interfere.

8-Quinolinol precipitates tungsten(VI) from hot solutions buffered at pH 5 (249). The precipitate is stoichiometric and can be dried at 120°C and weighed as $WO_2(C_9H_6ON)_2$ (396). It can also be ignited to WO_3 at 800°C (682). Oxalic acid and EDTA added to the sample solution before 8-quinolinol masks interfering effects of many metal ions that otherwise form insoluble oxine precipitates (320,1015). If tartrate is present, however, no tungsten precipitation occurs. This has been utilized in a procedure that separates molybdenum from tungsten with 8-quinolinol (921).

β-Naphthoquinoline hydrochloride precipitates tungsten(VI) from 0.9 M HCl solution (749,874). Vanadium does not interfere if hydrogen peroxide and nitric acid are used (158). A mixed β-naphthoquinoline–tannin reagent precipitates both tungsten and molybdenum. Molybdenum alone is then determined in a separate sample by complexometric titration after masking tungsten with tartrate (879).

Tungsten is precipitated as the tri-*n*-butylammonium complex of 12-tungstophosphate from solutions 2 M in HCl (629). Iron interference is avoided by reducing iron. Molybdenum and vanadium interfere.

Various amine reagents are used for precipitation of tungsten (577,-736,1081). Some of these and other reagents are listed in Table 10.

B.   TITRIMETRIC METHODS

Numerous titrimetric procedures for tungsten appear in the chemical literature and yield satisfactory results provided suitable precautions are taken to remove other ions that undergo similar reactions.

TABLE 10
Organic Precipitating Agents For Tungsten[a]

Reagent	Condition	Interference	Reference
Acridine 0.5%, in 5% HOAc	1 M HCl, heat	Cr, Mo, V	244
4-Amino-4'-chlorobiphenyl 0.36%, in alcohol	pH 1.6–2.2, heat	Mo	565
Anti-1,5-(di-p-methoxyphenyl)-1-hydroxylamino-3-oximino-4-pentene 0.7%, in EtOH	0.2 M HCl	Mo, Sn	421
Benzidine acetate 0.1%, in 0.2 M HOAc	pH 5, HOAc	Mo, PO_4^{3-}, SO_4^{2-}	107,668
N-Benzoyl-N-phenylhydroxylamine 3%, in EtOH	0.5–1.0 M HCl	Fe, Mo, Ti, V	428
4',4'-Bis(dimethylamino)triphenyl-methane 2%, in 0.2 M HCl	1 M HCl, heat	Mo, PO_4^{3-}, Si	1082 ·
Brucine 10%, in 3 M HCl	1 M HCl, heat	Mo	329
Cinchonine 12.5%, in 6 M HCl	H^+, boric acid, heat	Mo	687
Cinchonine 5%, in 3 M HCl + tannin 0.5%, aqueous	pH 3–4, NH_4Cl, heat	Mo	850
Cinchonine 12.5%, in 6 M HCl + α-benzoin oxime 5%, in acetone–water 95:5 vol./vol.	H^+, boric acid, heat	Mo	268
o-Dianisidine 1% in 0.7 M HOAc	HOAc	Mo, V	108
β-Naphthoquinoline 2%, in 1.4 M HNO_3	0.9 M HCl, heat	Mo	157,874
Pyramidone 0.1 M, in 2 M HCl	0.5 M HCl, heat	Si	339,783
Pyridoin 1%, in EtOH	pH 2.0–4.5	Mo	48
8-Quinolinol 1%, in 2 M HCl	0.4 M HCl, heat	Mo, V	682
Rhodamine B 1%, aqueous	1.5 M HNO_3, boil	Ag, Bi, Mo	695
Tannin 1%, aqueous + methyl violet 1%, aqueous	0.2 M HCl, heat	Mo, Ti	524,1023
Totaquine 2.5%, in 5 M HOAc	H^+, heat	Mo	741
Variamine blue, in 0.01 M HOAc	pH 5, HOAc	Mo, Pb, Hg	218

[a] The tungsten is converted by igniting to WO_3 before weighing in all procedures.

1. Acid-Base Titrations

Tungsten(VI) exhibits a potentiometric break upon titration with standard sulfuric acid solution corresponding to addition of one replaceable hydrogen, provided titration is carried out in the presence of mannitol (549,897). The sample, in sodium hydroxide solution, is first neutralized to the phenolphthalein end point, mannitol is added, and titration is continued to the methyl red end point. Other components present that react with acid titrant interfere.

2. Precipitation Titrations

a. LEAD SALTS

Quantitative precipitation of tungsten(VI) occurs upon addition of a standard solution of $Pb(NO_3)_2$ from samples buffered at pH 6 with hexamethylenetetramine and heated to near boiling. The indicator, 4-(2-pyridylazo)resorcinol, PAR, present in the sample solution changes to a permanent red color when excess Pb(II) ions are present in solution (447,541). Other indicators (86,123,492) and a potentiometric end point (331) are also used. Molybdenum reacts in the same manner as tungsten, and if both are present a second titration with EDTA titrant upon a separate sample determines only molybdenum. Tungsten content is found by difference (540). Other metal ions that precipitate with lead also interfere as do large amounts of fluoride ion (543). The titration can be followed amperometrically with a rotating platinum–calomel electrode pair at -0.5 V (75).

Frequently a known excess of lead is added to a tungsten-containing solution, and the excess lead is titrated with EDTA titrant after filtering insoluble lead tungstate (712,947). End point location is either visual or by electrometric means (136,336).

b. MISCELLANEOUS METAL SALTS

Other metal ions used in precipitation titration of tungsten include silver (530), mercury (910), thallium (1072), iron(III) (169), barium (836), and calcium (87). With silver, mercury, and thallium titration is direct, while the other titrants present in known excess are themselves titrated and tungsten content found indirectly. In all instances the presence of other species that react with the titrant cause interference.

c. ORGANIC REAGENTS

The tungsten precipitate with 8-quinolinol is collected and dissolved in sodium hydroxide, and the released 8-quinolinol is titrated with potassium bromate (501,776). In a similar procedure the tungsten precipitate with benzidine is collected, placed in water, and heated, and the released hydrogen ion is titrated with standard sodium hydroxide titrant using phenolphthalein indicator (435). Direct titation of tungsten is also possible with benzidine titrant (122) and with pinacyanol (989). The tungsten-thiocyanate complex can be titrated with diantipyrylmethane (1087).

3. Oxidation-Reduction Titrations

a. PRELIMINARY REDUCTION

Reduction of tungsten(VI) prior to titration with a standard oxidizing titrant is sometimes complicated by formation of reduction products containing tungsten in more than one valence state (231,981). Quantitative reduction to a single valence state is possible, and conditions for achieving this have been studied (289,743). A silver reductor quantitatively changes tungsten(VI) to tungsten(V) in 9–12 M HCl if phosphoric acid is present (981). Tungsten(V) is also formed by reduction with 3% Bi amalgam in concentrated HCl solution under an inert atmosphere (1052). A cadmium amalgam is used to reduce tungsten to tungsten(IV) in 4–6 M HCl (661), whereas tungsten(III) is formed if the acid concentration is 12 M HCl (289). Tungsten(III) is also quantitatively formed from reduction with lead or lead amalgam in concentrated HCl especially at elevated temperatures (369,579).

b. TITRATION OF REDUCED TUNGSTEN SPECIES

After reduction to a lower valence state tungsten is titrated directly with a standard oxidizing agent. Manganese(VII) (1022), cerium(IV) (883), and other titrants (466,865) are suitable. Alternately the reduced tungsten is passed into a solution containing a known amount of Fe(III) salt. The Fe(II) formed as a result of reduction by tungsten is titrated with standard Cr(VI) titrant (148,355).

c. DIRECT TITRATION OF TUNGSTEN(VI)

Tungsten(VI) in 8 M HCl solution is reduced to tungsten(V) by Cr(II) (145,248,652). Using standard Cr(II) solution as titrant the reaction can be followed potentiometrically. Addition of oxalic acid to the titrant is recommended (211), and the reaction is frequently carried out at elevated temperature, 70–90°C (212).

C. OPTICAL METHODS

Various organic complexing agents combine with tungsten to form species suitable for colorimetric determination. Optimum structural characteristics of these reagents has been considered (490), and detection limits of a few micrograms tungsten per milliliter are common. Both absorbance and emission techniques are applied to a variety of tungsten-containing materials. X-Ray fluorescence, too, is widely used of determination of tungsten.

1. Colorimetric

Compilations of procedures for colorimetric determinations of elements, including tungsten, are available (139,574,835). For tungsten the complexes with thiocyanate and toluene-3,4-dithiol are best known, although many other reagents have been investigated (957,1085).

a. THIOCYANATE

Progress in the development of a suitable procedure for colorimetric determination of tungsten utilizing the absorbance value of the tungsten(V)-thiocyanate complex has been summarized (252). Variations in experimental procedure are many, and a firm understanding of the underlying chemistry is lacking at the present time. Tungsten(VI) is reduced with a suitable reducing agent to tungsten(V), which forms a yellow complex with thiocyanate ions. Initially the reaction was carried out using tin(II) chloride from slightly alkaline solution of tungsten(VI) (235). Reduction from acid solution is more desirable (285) with the reducing agent added before thiocyanate to avoid excessive thiocyanate reduction. In acid solution, however, a yellow-green color frequently results from reduction of tungsten and addition of thiocyanate, perhaps from the presence of both yellow tungsten(V)-thiocyanate and tungsten blue, another reduced form of tungsten (259). Reduction from strong acid solution under specified conditions yields only a tungsten species that, upon addition of thiocyanate ion, forms the yellow complex. Solutions 8 M with respect to chloride concentration and 10 M with respect to total acid concentration have been used successfully (171), although other conditions are also employed (252). Thiocyanate concentration, too, should be carefully controlled. For the 8 M chloride–10 M hydrogen ion medium cited a 0.50 M thiocyanate concentration is optimal. Reduction of tungsten with tin(II) chloride is slow at room temperature but proceeds more rapidly if the solution is boiled. Titanium(III) reduces tungsten rapidly at room temperature and is often substituted for Sn(II) chloride (238,245). A more recent procedure uses both Sn(II) and Ti(III) to reduce tungsten and achieves more precise results than with only Sn(II) alone (253). Other reducing agents have also been tried (909).

The exact composition of the tungsten(V)-thiocyanate complex is uncertain. Any of several complexes are reported to form depending upon conditions (866,993). The complex of analytical interest, formed from reduction by titanium(III) in acid solution, with an absorbance maximum at 405 nm is assigned the formula, $[W(OH)_2(SCN)_4]^{1-}$ (318). Absorbance values at 436 nm are also used for analytical determinations (321).

Greatest interference in determining tungsten as the thiocyanate complex comes from molybdenum, which forms a molybdenum(V)-thiocyanate. In the presence of a strong reducing agent, the molybdenum present is further reduced to molybdenum(III). The extent of molybdenum interference is then lessened and depends upon the tungsten to molybdenum ratio present in solution (36). Alternately, with a mild reducing agent, molybdenum is selectively reduced, while tungsten(VI) remains unaffected. Ascorbic acid (550) and thioglycolic acid (806) are used for this purpose. The molybdenum(V)-thiocyanate complex, once formed, is extracted, and the aqueous solution then is analyzed for tungsten. Simultaneous photometric determination of both molybdenum and tungsten as their thiocyanate complexes is possible after separation of Mo and W by extraction of their α-benzoin oxime complexes in chloroform to remove other interferences (730). The extracts are ashed and dissolved, and the thiocyanate complexes form. Absorbance measurements are made at 405 nm for the tungsten complex and 490 nm for the molybdenum complex after both are extracted with i-propyl ether. Other procedures for determining tungsten in the presence of molybdenum are discussed in the review cited at the beginning of this section.

Other interference in determining tungsten by the thiocyanate method comes from metals that also form thiocyanate complexes, for example, vanadium (253) and niobuim (613); metals reacting directly with thiocyanate, for example, copper (94); complexing agents that react with tungsten, for example, phosphate (253) and tartrate (835); and oxidizing agents that attack the reduced tungsten complex, for example, nitrate (680,911). Tungsten in steels is determined by the thiocyanate procedure after removing the major steel components with methyl isobutyl ketone extraction (581).

Addition of acetone to the aqueous tungsten(V)-thiocyanate solution has little effect upon increasing absorbance values (171). Extraction of the complex is sometimes employed. Di-i-propyl ether (730) and various mixed solvents (835) are used. Extraction with chloroform of the species formed between tungsten(V)-thiocyanate and tetraphenylarsonium chloride (10,251) or other ternary amine salts (610) removes many of the interfering elements commonly associated with tungsten.

Permanent color standards for visual comparison of the tungsten(V)-thiocyanate color have been prepared using ammonium chromate (241).

b. TOLUENE-3,4-DITHIOL

A blue-green complex forms with toluene-3,4-dithiol and a reduced form of tungsten. Maximum absorbance for this complex, after extraction with amyl acetate or other extracting solvent, is 640 nm. Various conditions are reported for optimum formation of the complex (215,-1017). Most investigators generally agree that formation is best in strong acid solution. A mixed sulfuric acid–hydrochloric acid medium of 3.8 M HCl and containing sufficient H_2SO_4 to produce a total hydrogen concentration of 9–10 M is satisfactory (364). Titanium(III) sulfate is used as reducing agent, and the solution is heated at 80–90°C for several minutes both before and after adding the dithiol reagent. The complex is extracted with diethyl ether, various esters, or carbon tetrachloride before absorbance measurements are made.

Molybdenum seriously interferes with this procedure by also forming a dithiol complex. It is possible, however, to selectively extract the molybdenum complex from more dilute acid solution, approximately 4 M, before reduction of tungsten(VI) (38). Addition of sufficient reducing agent to convert the molybdenum present to molybdenum(III) also removes its interfering effects upon tungsten determination (412,443).

The presence of iron salts increases the absorbance value of the tungsten-dithiol complex (18).

There are problems in preparation of a stable toluene-3,4-dithiol solution (159). For tungsten determination the reagent has been dissolved in amyl acetate (325), used as a suspension in alcohol in the form of its zinc salt (860,979), and dissolved in aqueous 2% NaOH solution containing thioglycolic acid (364).

The dithiol procedure is widely used for determining tungsten in a variety of refractory metal alloys, minerals, soils, and other media (193). Specific references to these procedures are cited in Sections VII and VIII.

c. Miscellaneous Color Forming Reagents

Tungsten(VI) forms a red complex with hydroquinone in concentrated sulfuric acid solution with an absorbance maximum at 450 nm. Measurements, however, are generally made at 530 nm to minimize interference. Optimum conditions for formation of this complex include the presence of small amounts of phosphoric acid and water (691). Tin(II) chloride is generally added to reduce iron, molybdenum, and other ions, thus decreasing their interference. Titanium seriously interferes with this procedure. Niobuim, too, forms a colored complex with hydroquinone, and both tungsten and niobuim can be determined simultaneously in binary mixtures with hydroquinone (448,614).

Heteropoly ions containing tungsten are used for spectrophotometric tungsten determination. For molybdotungstate reduction hydrazine hydrochloride is most satisfactory for heteropoly blue formation (340).

Depending upon the ratio of Mo to W either molybdotungstates or tungstomolybdates form, and both tungsten and molybdenum have been determined as their heteropoly ion (811). Mixed heteropoly complexes are also used for tungsten determination (335,904).

Various polyhydric compounds are used as color forming agents with tungsten. Some of these and other reagents are listed in Table 11.

2. Infrared

Some infrared spectra of tungsten-containing complexes have been recorded but, in general, have not been used for quantitative tungsten determination. Among these are the potassium bromide pellet scans of 8-quinolinol complexes of various metals, including tungsten (742). Intensity measurements of the characteristic tungsten-8-quinolinol peak at 10.61 μ is used for quantitative tungsten determination. Satisfactory results are obtained in the presence of either molybdenum or vanadium but not when both appear in the tungsten-containing sample (590).

3. Fluorimetric

Measurement of fluorescence emission intensity when tungsten is combined with a ligand capable of fluorescing is used for quantitative tungsten determination. Procedures specifically for tungsten utilize alizarin S with sample solutions buffered at pH 4.8–6.2 (143), carminic acid with sample solutions buffered at pH 4.6 (470), and 3-hydroxyflavone with sample solutions buffered at pH 2.5–5.5 (676). With the latter reagent determination of tungsten in steels and certain refractory alloys is possible after removing interfering ions by cation exchange (78,79). Fluorimetric studies with tungsten and other metals have been made using various other reagents including cochineal (313), 8-quinolinol (848), and rhodamine B (653).

4. Polarimetric

The optical activity of L-ascorbic acid (951), L-maleic acid (404), and D-tartaric acid (405) is altered in the presence of tungsten(VI) ion. Variation of specific rotation with varying tungsten concentration is linear over a limited concentration range and, using D-tartaric acid, is suitable for quantitative tungsten determination in ferrotungsten (655).

5. Emission

Effects on spectral line intensity of tungsten emissions with carbon electrodes and in the presence of other reagents has been studied (801).

TABLE 11
Complex Forming Agents for Photometric Tungsten Determination

Agent	Condition	Wavelength	Interference	Reference
Alazarin-3-sulfonate sodium salt 0.001 M, aqueous	pH 3.5–5.8	470	Al, Cr, Fe, Mo, U	899
5,7-Dibromo-8-hydroxyquinoline 0.004 M, in dichloroethane	1 M H$^+$, extract with dichloroethane	375	Cr, Mo, Ni, Ti	1085
cis-1,2-Dicyanoethylene dithiolate disodium salt 0.2%, aqueous	2 M HCl–acetonitrile 1:1 vol./vol.	570	Fe, Mo, Ni	131
Dihydroxychromenol 0.06%, in EtOH	pH 2, 20% EtOH, 0.04% gelatin	500	Nb, Ta	761
3,5-Dinitropyrocatechol 0.01 M, in EtOH + diantipyrrylmethane 0.25 M, in 0.5 M HCl	0.2 M HCl, extract with CHCl$_3$	400	Mo, Nb, Ti, V	666
Gossypol 0.1%, in EtOH	H$^+$, extract with 20% i-AmOH in C$_6$H$_6$	510	Mo, Nb, Ta, Ti	458
Hydrogen peroxide 3%, aqueous	7.2 M H$_2$SO$_4$, heat 40°C 30 min	262	Fe, Mo, Ti, U, V	721
Hydrogen peroxide 0.01 M, aqueous + magneson XC 0.02%, in acetone–H$_2$O 1:1 vol./vol.	1 M HCl, wait 2 hr	580	Mo, Nb, Ta, Ti, Zr	840
Hydrogen peroxide 0.01 M, aqueous + sulfonitrophenol M, 0.1% aqueous	0.1–0.5 M HCl, wait 1 hr	650	Cu, Fe, Mo	839
Hydroquinone 8%, in 18 M H$_2$SO$_4$	H$_3$PO$_4$, SnCl$_2$, H$_2$O	530	Cr, Fe, Mo, V	691
8-Mercaptoquinoline 0.25 M, in 6 M HCl–EtOH 1:1 vol./vol.	pH 2, wait 1 hr, extract with i-BuOH–CHCl$_3$ 1:1 vol./vol.	412	Cu, Fe, Mo, V	665

Reagent	Conditions	λ (nm)	Elements	Ref.
Methyl salicylate 0.2 ml/sample	2 M HCl, H$_2$O$_2$, Ac$_2$O, HOAc, heat 40°C	470	Cu, Mo, V	800
Morin 0.0002 M, in BuOH	pH 3.5, 4% DMSO	413	Mo	393
Phenylfluorone 0.02%, in acetone	pH 3.8–4.4, heat 90°C 4 min	453	Bi, Cu, Fe, Mo	100
KSCN 20% aqueous	H$^+$, SnCl$_2$, TiCl$_3$, wait 15 min	400	Cu, Mo, NO$_3^-$, V	139
KSCN 1.5 M + tetraphenylarsonium Cl 0.025 M, aqueous	8 M HCl, SnCl$_2$, heat, extract with CHCl$_3$	406	Mo, NO$_3^-$	10
KSCN 1 M + zephiramine 0.01 M, aqueous	pH 3–5, SnCl$_2$ extract with CHCl$_3$	405	Mo, V	610
Pyrocatechol 0.45 M, in 3% Na$_2$S$_2$O$_5$ + 0.4% NaOH aqueous	pH 5.2–5.5, EDTA	350, 370	Mo	112
Pyrocatechol 2 M, in 1 M HCl + diantipyrylmethane 0.1 M, in 1 M HCl	H$^+$, extract with CHCl$_3$	440	Mo, Nb, Ti	964
Pyrocatechol violet 0.02%, aqueous	pH 1.9–2.1	560	Fe, Mo, Nb, V	551,722
Pyrogallolsulfonic acid potassium salt 2%, aqueous	pH 0.5–1.2	415	Mo	373
Rhodamine B 1%, aqueous	0.15 M HCl, extract with CHCl$_3$	600	Al, Mo, Sn	109,757
Salicylfluorone 0.05%, in EtOH	0.01 M HCl, gelatin, citrate, wait 2 hr	530	Ge, Mo, Nb, Sn	762
Stilbazogall I 10^{-4} M, aqueous	pH 1.5–2.3	545	Cr, Mo, Ti, V	395
Tiron 3.1%, aqueous	pH 5.0, wait 1 hr	313	Nb, Ta, Ti, U	880
Toluene-3,4-dithiol zinc salt 0.4%, suspension in EtOH	HCl, Ti°, heat, extract with i-AmOAc	640	Mo	113

Carbides and, if silica is present, silicides of tungsten are formed along with various oxides. One procedure for spectroscopic determination of tungsten calls for addition of silica to the sample for increased intensity of tungsten emissions (201). Sulfur (3) and germanium dioxide (190) have also been added to tungsten-containing samples prior to excitation. With samples containing less than 0.01% W a longer excitation time aids in improved determination (205). The presence of various halides results in formation of more volatile tungsten species. Silver chloride (831), phosphorus pentachloride (601), and other chloride- or fluoride-containing reagents (399) are used. Hydrofluoric acid is used to remove tungsten by volatilization when determining trace constituents in the presence of large amounts of tungsten (257).

Iron emission lines near the tungsten lines of 430.21, 429.46, and 400.88 nm cause interference when determining tungsten. One procedure for tungsten first removes iron by use of a cation-exchange resin (726). A similar interference is observed from titanium emission near the tungsten 400.88 nm line. Among other emission lines measured for quantitative tungsten determination are those at 294.70 and 289.65 nm. When trace amounts of tungsten are present in a sample, concentration before excitation is desirable. Among the methods used are precipitation in the presence of a suitable collecting agent. Alumina (856), ammonium molybdophosphate (464), and various organic precipitating agents (1048) are satisfactory. Various internal standards are added to tungsten-containing samples. Some of these and the measured line pairs include W 289.60/Cu 288.29 nm (568), W 551.47/Fe 551.23 nm (569), W 304.97/Mo 305.02 nm (464), and W 294.70/Si 297.04 nm (12). Other standards are also used (671,792). Directions are given for the preparation of synthetic tungsten-containing standards in metal alloys (286) and for sample preparation prior to excitation (11). Laser excitation of samples containing a variety of elements including tungsten is successfully employed in spectrochemical analyses (41,718).

Tungsten is determined by emission spectroscopy in rocks (400,525), steels (456), various alloys containing high percentages of tungsten (437), and simultaneously with molybdenum in a variety of samples (669,953). Specific laboratory directions for tungsten have been compiled (215).

Flame excitation for tungsten can be used for samples containing moderate amounts of tungsten. With a fuel-rich oxygen–acetylene flame a detection limit of 90 μg W/ml is reported from measurements of the 400.88 nm line (228). This limit is improved if a premix burner is used (204). High-frequency plasma excitation has resulted in a lowering of the detection limit to less than 1 μg W/ml from measurements upon the

429.46 nm line (314). Other studies of high-frequency plasma excitation for tungsten are also available (650,966).

6. Atomic Absorption Spectroscopy

Atomic absorption spectroscopic determination of tungsten is made from intensity measurements upon the 255.14 or 400.88 nm line. Using a nitrous oxide–acetylene flame the detection limit is a few μg W/ml (23,744). A study of various acids upon the intensity of the 400.88 nm line indicates maximum absorption for a specified tungsten concentration in hydrofluoric acid solutions (970). Less satisfactory results occur in sulfuric or phosphoric acid solutions. Potassium ion is frequently added to the tungsten solution to suppress ionization interference. A nonflame dc arc source has been used for tungsten atomization and provides a detection limit of 1.0 μg W/ml with measurement at 255.14 nm (602).

Tungsten is determined by atomic-absorption methods in steels (380) and other highly alloyed materials (823).

7. X-Ray Spectroscopy

Intensity measurements of tungsten fluorescing lines upon bombardment of tungsten-containing samples by an X-ray source are used to construct calibration curves for quantitative determination of tungsten. The $L\alpha_1$ line at 1.476 Å and other characteristic tungsten lines are monitored (37,986). If the X-ray generating tube used for excitation contains a tungsten target, special filters are necessary to remove tungsten lines from the generating source (557). As with many fluorescent procedures sample composition affects the accuracy of results (346,450,576,632). In addition some elements have X-ray lines very close to those of tungsten. Resolution of tungsten and niobium lines, in particular, has been studied (93). Prior separation of other heavy metal ions present in complex samples is desirable, and the resulting tungsten-containing portion is frequently converted to the oxide and fused into a borax matrix before analysis (334,580). Copper (539), nickel (37), and platinum (935) are used as internal standards. Consistent background radiation, too, can serve as an internal standard (347). Gamma-ray sources are sometimes used for initiating fluorescence radiation in tungsten containing samples (216,605,778).

In addition to fluorescence techniques tungsten is determined by X-ray emission of its $L\gamma_1$ line at 1.098 Å (225). Bromine, as sodium bromide, is added as an internal standard. Absorption edge measurements also provide information for quantitative determination of tungsten (746,747). Various diffraction studies, too, are used for quantitative determination of tungsten (296) and for identification of tungsten-containing species (30,267,803,995).

X-Ray techniques serve for determination of tungsten in ores (575), steels (451,958), refractory materials (308,347), and a variety of other substances (450). The electron-beam microprobe is also utilized for analysis of tungsten-containing alloys (345,594,637,841).

D. VOLTAMMETRIC METHODS

1. Direct-Current Polarography

Polarographic studies of tungsten have centered about reduction of tungsten(VI) in the presence of strong mineral acids and mineral acid–organic acid solutions (927). Two reduction steps occur in 7–11 M HCl, the first, a irreversible process with an $E_{1/2}$ value of approximately -0.015 V versus Hg pool electrode, and the second, a reversible process with an $E_{1/2}$ value of approximately -0.400 V versus Hg pool electrode (898). The $E_{1/2}$ potential of the second wave shifts to more negative values with decreasing hydrochloric acid concentration. It is -0.56 V versus SCE in 12 M HCl and -0.64 V versus SCE in 4 M HCl with intermediate acid concentrations producing intermediate half-wave potential values (567). Various mechanisms are proposed to explain the observed polarograms, and although generally attributed to the stepwise reduction of tungsten(VI) to tungsten(V), a one-electrode process, followed by the two-electron reduction to tungsten(III) (291), reversed or anionic scanning of the reduced species indicates formation of a tungsten(IV) intermediate (528). An alternate explanation for the two waves proposes initial reduction of tungsten(VI) to tungsten(III) followed by reaction of tungsten(VI) with tungsten(III) producing tungsten(V) and hydrogen (812,925,926). With this explanation both observed waves are considered complex in nature, resulting in part from reduction of tungsten and in part from formation of hydrogen. No polarographic wave is observed for tungsten in 1.4 M HCl unless a large amount of chloride ion is present. A satisfactory wave does result in 1.4 M HCl–10.1 M LiCl with diffusion current proportional to tungsten concentration (500). Three ill-defined waves are observed for tungsten(VI) in sulfuric acid solutions (397), while in perchloric acid solution tungsten(VI) produces catalytic reduction waves (529). No polarographic waves are observed for tungsten in ammonia or strong base solutions (326).

Polarographic reduction of various isopoly tungstates has been studied in hydrochloric acid solution and other media (818), and the resulting waves have been compared to those of simple tungstate reduction (914,976). A one-electron reduction wave for tungsten(VI) results in solutions 4 M in phosphoric acid (716). In 6 M H_3PO_4 this wave has an $E_{1/2}$ value of -0.56 V versus SCE (446). Tungsten and niobium are determined simultaneously in 6 M H_3PO_4–14.5 M H_2SO_4, the tungsten wave appearing first. A single tungsten reduction wave appears with $E_{1/2}$ value of -0.49 V versus Hg pool in 18 N pyrophosphoric acid (522). The $E_{1/2}$ value shifts to more negative values as the $H_4P_2O_7$ concentration decreases. In 10 N pyrophosphoric acid tungsten is determined in the presence of titanium, their reduction waves occurring at different potentials. Tungsten is distinguished from molybdenum in pyrophosphoric acid solution as the latter produces two reduction waves (717). Polarographic studies have been conducted on various tungstophosphates and other tungsten-containing heteropoly ions primarily as an aid in understanding the composition and reduction mechanism of these species (607,767,913).

Tungsten exhibits a single reduction wave in solutions containing both hydrochloric acid and oxalic acid, while two waves appear if either citric or tartaric acid is used with HCl (517,715). In the presence of oxalic acid over the pH range, 0.75–3.0, the $E_{1/2}$ value of tungsten shifts from -0.6 to -0.8 V versus SCE as the oxalic acid concentration decreases from 0.4 to 0.01 M (950). Tungsten is determined quantitatively in the presence of tin (478) and other ions (189) by polarographic measurement of solutions containing hydrochloric acid–oxalic acid. Hydrochloric acid–citric acid solutions serve as electrolyte for determining tungsten in the presence of small amounts of molybdenum (943), and hydrochloric acid–tartaric acid solutions are used for determining tungsten in steels (1061) and rocks (808). After studying various combinations of mineral acid–organic acid solutions (69) the optimum mixture for determining from 0.1 to 20 μg W/ml was found to be 9–10 M $HClO_4$ plus 0.6–0.7 M tartaric acid, $E_{1/2}$ value -0.58 V versus Hg pool. For samples containing up to 100 μg W/ml the optimum acid mixture was either 9–10 M HCl plus 0.3 M tartaric acid plus 0.01% gelatin, $E_{1/2}$ value -0.65 V versus SCE, or 9–10 M HCl plus 0.1 M citric acid plus 0.01% gelatin, $E_{1/2}$ value -0.63 V versus SCE.

Kinetic polarographic waves, attributed to formation of peroxotungstate species, result from reduction of tungsten(VI) in the presence of acid and hydrogen peroxide (484,1066). In solutions containing 0.010 M $H_2C_2O_4$–0.080 M H_2O_2 the kinetic current maximum occurs at $+0.23$ V versus SCE and, although proportional to tungsten concentration, is not linear, thus necessitating construction of a calibration curve (704).

2. Other Polarographic Techniques

Alternating current polarographic scanning of tungsten(VI) produces two peaks in acid solution and is suitable for quantitative tungsten determination (89). Tungsten is determined in the presence of niobium and tantalum by alternating current polarography after first extracting tungsten as the dithiol complex (932). Oscillographic polarograms of tungsten are made in acid solution (70) and are used for determination of tungsten in solutions acidified with hydrochloric acid (277) and phosphoric acid (521). Vanadium, chromium, molybdenum, and tungsten have been studied by inverse polarography in HCl solutions at pH 3 and above (290), and tungsten is determined in the presence of rhenium by square-wave polarography (438).

E. RADIOCHEMICAL METHODS

Tungsten can be determined in various alloys by the isotopic dilution technique. A radioactive tungsten isotope of known activity is added to the sample, and, following chemical isolation of tungsten, the observed activity is proportionated to give the amount of tungsten that would be present had complete recovery been achieved. ^{181}W ($t_{1/2}$ 130 days) is added to solutions of tungsten–molybdenum alloys and following dithiol extraction gamma emissions are counted (612). ^{185}W ($t_{1/2}$, 75 days) serves the same function in analyzing high-alloyed steels for tungsten, emissions being counted following separation of tungsten by extraction (39) or precipitation (554).

The intensity of reflected, backscattered, beta radiation from a β-emitter placed on the surface of a sample understudy is proportional to the number of heavy nuclei in the backing material (381). The technique is used for determining tungsten and is most satisfactory for binary tungsten-containing mixtures in which there is a large difference between the atomic number of tungsten and other component present. The procedure is applied to quantitative determination of tungsten in hydration catalysts (713) and to the tungsten content of calcium tungstate precipitate employed in the separation of tungsten from tungsten-containing materials (377). A $^{90}Sr + {}^{90}Y$ beta source is used with the backscattering technique for determining tungsten carbide–titanium carbide mixtures (217). Other sources are used for finding the tungsten content of various iron alloys (417) and steels (73,923). In these procedures variations in matrix composition and surface finish affect the intensity of the reflected beta radiations. Ternary iron–molybdenum–tungsten alloys are analyzed by this technique using aluminum filters of varying thickness between the beta source and the detector (513). With this modification it is possible to determine all three components from the data obtained. Backscattering is also used for determining tungsten in ores (892).

Neutron activation of tungsten-containing samples generates several radioactive tungsten isotopes. Of these ^{187}W ($t_{1/2}$, 23.9 hr) from ^{186}W (natural abundance, 28.4%) is most often used for radiochemical determination. ^{187}W exhibits gamma and beta emissions. The gamma peaks at 0.072, 0.134, 0.480, and 0.686 MeV are among those frequently monitored, while characteristic beta emissions occur at 0.41, 0.63, and 1.33 MeV (658). Irradiation of tungsten-containing samples varies from several seconds for determination of alloying elements in aluminum (415), to several hours for determining impurities in vanadium pentoxide (7,274), to a few days for analysis of certain mineral samples (4) and alloys (47). Following irradiation the samples are allowed to stand from several hours (5) to a few days (571). This allows time for decay of other radioactive isotopes which, if present, interfere with measurement of ^{187}W. Interference from manganese (624) and various alkaline earth elements (679) is avoided in this way. Tungsten can be separated from other elements present in a sample either before or after irradiation. Precipitation (359,376), extraction (294), and ion exchange (230,419) are frequently used for this purpose.

By using short irradiations and rapid counting techniques the radioactive ^{183m}W isotope ($t_{1/2}$, 5.2 sec) formed from both ^{182}W (natural abundance, 26.4%) and ^{183}W (natural abundance, 14.4%) can be monitored for determination of tungsten (24). ^{183m}W has characteristic gamma peaks at 0.06, 0.10, and 0.16 MeV (407). Procedures utilizing this isotope include the analysis of tungsten-containing steels (639) and the analysis of molybdenum for trace tungsten contamination (559).

F.   METHODS BASED ON CATALYTIC ACTIVITY

Tungsten is one of several metal species that catalyzes the oxidation of iodide ion by hydrogen peroxide (1065). Providing other, interfering, ions are absent the rate of formation of iodine is proportional to the concentration of tungsten present. From plots of iodine concentration versus time the slope of the curve at some chosen time interval is proportional to the concentration of tungsten catalyzing the oxidation. Iodine concentration is monitored spectrophotometrically through the iodine-starch complex (1070), amperometrically (102) through enthalopy changes during the reaction (243) or by use of an iodide-specific ion electrode (19). An automated procedure utilizing this reaction is capable of determining microgram quantities of tungsten (341). Molybdenum, zirconium, iron, and other metal ions also catalyze this reaction and must be removed prior to quantitative measurement of tungsten. The mechanism of the I_2–H_2O_2 reaction in the presence of tungsten has been studied (1069). Tungsten catalyzes the reaction of hydrogen peroxide with other oxidizable materials, including indigo carmine (498,506), various sulfur species, (173,1067) and rubeanic acid (714). The latter reaction is applied to the determination of tungsten in steels; however, molybdenum and copper interfere.

Reduction of malachite green by titanium(III) in dilute acid solution is catalyzed by minute amounts of tungsten (833,884). Absorbance values for malachite green are plotted as a function of time, and from the slope of this curve, after a specified time interval, the tungsten concentration is determined. Molybdenum and other ions interfere (699). The reaction has been used for determining tungsten in steels (315). Greater sensitivity for tungsten, to 10^{-8} M concentration, is claimed if basic blue K is substituted for malachite green in this procedure (700).

Other reactions catalyzed by tungsten include the oxidation of p-phenylenediamine by iodate ion (1071) and the reaction of chlorine with tin(II) (74).

VII. ANALYSIS OF TUNGSTEN, ITS ALLOYS AND COMPOUNDS

A. HIGH-PURITY MATERIALS

1. Tungsten Metal

Small amounts of impurities markedly affect the properties of pure tungsten metal (32), and procedures describing analysis of high-purity tungsten have been reviewed (354,433,963). Spectroscopic (266,278,283), X-ray fluorescence (592,780), activation techniques (197,224), and spark-source mass spectrometry (164) are used to determine a variety of impurities in tungsten. Selected impurities are also measured by spectrophotometric (1089), atomic absorption (651), and polarographic methods (960). Procedures for gaseous impurities in tungsten have been compared (474,595), and specific methods for oxygen, nitrogen, and hydrogen are available (247,260,287,641,810). Table 12 lists procedures for specific elements that have been determined in pure tungsten.

2. Tungsten Compounds

Emission spectroscopy is used for determination of trace heavy metal impurities in tungsten(VI) oxide (178,636,738,777), while for the alkali

TABLE 12
Methods for Determination of Trace Elements in Pure Tungsten

Element	Method	Reference
Al	Spectrophotometric	198
Ar	Mass spectrometry	696
As	Titrimetric	357
Au	Polarographic	901
B	Spectrophotometric	472
C	Conductometric	535
Ca	Titrimetric	310
Co	Atomic absorption	199
Cu	Atomic absorption	379
F	Potentiometric	789
Fe	Spectrophotometric	732
Ga	Spectrophotometric	118
H	Vacuum fusion	475
Hf	Spectroscopic	727
K	Atomic absorption	674
Mn	Spectrophotometric	200
Mo	Spectrophotometric	582
N	Spectrophotometric	35
Na	Spectroscopic	50
Nb	Spectrophotometric	1058
Ni	Spectrophotometric	324
O	Vacuum fusion	1059
P	Spectrophotometric	706
Re	Spectrophotometric	196
S	Titrimetric	875
Si	Spectrophotometric	142
Sn	Titrimetric	1083
Ta	Activation analysis	165
Th	Radiometric	16
Ti	Spectrophotometric	49
Tl	Spectrophotometric	997
Zn	Atomic absorption	199
Zr	Spectrophotometric	1057

and alkaline earth contaminates flame emission is suitable after separation from the tungsten matrix (358,502). Table 13 lists procedures for determining selected elements in tungsten(VI) oxide.

Tungsten in various tungsten carbides is determined by cinchonine precipitation (265,975), while the carbon content is found by converting carbon to carbon dioxide and manometric measurement (646). Cobalt, frequently present as a binder for tungsten carbide, is determined by X-ray fluorescence (1060) or atomic absorption spectrometry (560). The oxygen and nitrogen content of carbides is found by inert gas fusion

TABLE 13
Methods for Determination of Trace Elements in Pure
Tungsten(VI) Oxide

Element	Method	Reference
Al	Spectrophotometric	284
As	Spectroscopic	1007
Cl	Turbidimetric	611
Cu	Spectrophotometric	323
Fe	Titrimetric	830
Mo	Spectrophotometric	537
Ni	Spectrophotometric	324
O	Manometric	367
P	Turbidimetric	356
Re	Polarographic	188
S	Titrimetric	536
Si	Titrimetric	876

(616). Silica, alumina, and lime are also measured in powdered carbide samples (868). Specific procedures appear for determining iron (55) and titanium (598) in carbide samples. Tungsten carbide is analyzed for several trace components by emission spectroscopy (804). Analytical procedures for other refractory tungsten compounds include determination of boron in tungsten boride (515) and measurement of free and total silicone in tungsten silicide (496,949).

Colorimetric methods are used for determining trace amounts of molybdenum in sodium tungstate (116,1049), while flame emission serves for finding the concentration of calcium in ammonium tungstate (146). Rare-earth tungstates are analyzed by titration with EDTA (309). The water content of various heteropoly tungstates is found by Karl Fisher titration (877).

The tungsten bronzes, containing nonstoichiometric amounts of metal ion (generally an alkali metal ion), tungsten, and oxygen are frequently analyzed for their metal ion, tungsten, and oxygen contents (584,788). Various procedures are used for determining lithium (1039), sodium (591,809), potassium (618), rhubidium (1031), and certain rare-earth ions (1032) including emission spectroscopy and neutron-activation analysis.

B. TUNGSTEN ALLOYS

Ferrotungsten, obtained directly from reduction of tungsten ores, serves as a starting material for the manufacture of tungsten-containing steels. Generally it contains between 70 and 80% tungsten, along with iron and small amounts of other elements. Various methods for its analysis are available (20,26). Tungsten, in ferrotungsten, is itself determined by precipitation with cinchonine (469) and by titration with lead using visual end point detection (447,486) or amperometric titration (182). Other elements present in ferrotungsten and methods for their determination are listed in Table 14.

Thorium(IV) oxide, generally 0.5–2.0%, is added to tungsten to improve the performance of tungsten filaments in lamps and electron tubes. Procedures for determining thorium in tungsten include addition of known excess EDTA and back titration of unreacted excess (319,-1043), colorimetric determination with thorin (688), and a radiochemical procedure based upon [232]Th (15).

Tungsten forms a continuous series of alloys with molybdenum, and alloys containing tungsten as the major constituent are analyzed for their molybdenum content by colorimetry (597) and controlled potential coulometry (814). Tungsten itself is obtained by radiochemical measurement of the [181]W isotope (612). Trace impurities in molybdenum–tungsten alloys are identified by emission spectroscopy (556) and electron microprobe (1063) technique.

Tungsten alloyed with rhenium exhibits improved low-temperature ductility and high-temperature strength. X-Ray spectroscopy (60) and colorimetric determination of rhenium with α-furildioxime (167) are used to find the rhenium content of these alloys over a wide range of compositions. Rhenium in alloys containing approximately 20% Re is determined by amperometric titration with titanium(III) (271). Thiocyanate ion (258) and dimethylglyoxime (383) are used for colorimetric determination of rhenium in Re–W alloys containing up to 10% Re. Rhenium is precipitated from a Re–Ta–W alloy with tetraphenylarsonium chloride after separating tantalum and tungsten by cinchonine precipitations (805).

TABLE 14
Methods for Determination of Trace Elements in Ferrotungsten

Element	Method	Reference
Al	Gravimetric	900
As	Spectrophotometric	1064
Bi	Spectrophotometric	588
Cu	Spectroscopic	954
Mn	Titrimetric	906
P	Titrimetric	1019
Pb	Spectroscopic	988
Sb	Spectrophotometric	317
Si	Gravimetric	1062
Sn	Spectrophotometric	829

Niobium in binary niobium–tungsten alloys is determined by precipitations from ammonical solution (878). For most accurate results the dried precipitate is fused with $K_2S_2O_7$, and the small amount of coprecipitated tungsten is determined in the melt. Following niobium precipitations tungsten in the sample filtrate is itself precipitated with β-naphthoquinoline (305). Trace amounts of niobium in a tungsten matrix are found colorimetrically using 1-(2-pyridylazo)resorcinol (210).

A similar reagent, 1-(2-pyridylazo)-2-naphthol, is used for colorimetric determination of iron in iron–tungsten alloys (781), while tungsten in iron–tungsten alloys is found by EDTA titration of known excess lead(II) required to precipitate lead tungstate (772). Up to several percent each of copper, nickel, and iron are determined in a tungsten-based alloy by selective EDTA titration achieved through pH control and addition of complexing agents (542), while selective spectrophotometric reagents are used to determine copper, nickel, iron, and cobalt in another tungsten-based alloy (690). Zinc is determined in a Ni–Cu–W alloy by polarography (300), and various gravimetric and titrimetric procedures are used for finding both major and trace components in a Cr–V–W alloy (503). Silicon is also measured in various tungsten alloys (731).

Procedures for tungsten in alloys in which tungsten is not the major constituent are described in Section VIII.A.2.

VIII.   DETERMINATION IN SPECIFIC MATERIALS

A.   METALLURGICAL SAMPLES

1.   Iron and Steel

Both pig iron (962) and cast iron (1042) have been analyzed specifically for tungsten using colorimetric procedures with thiocyanate ion.

In steels tungsten is present from a few tenths of a percent to about 20% depending upon the intended use of the particular steel. Those steels containing the higher percentages of tungsten are well known for their hardness and are widely used for tools, drills, and other applications where strength, especially high-temperature strength, is needed. Tungsten in steels has traditionally been determined by precipitation of hydrous tungsten(VI) oxide from strong acid solution in the presence of cinchonine or other organic reagent (485). A procedure for this determination is given in Section IX.A. Colorimetric determination of tungsten as the thiocyanate complex (71,251,281) is, generally, more rapid than precipitation, and details of this procedure are given in Section IX.C. From 0.02 to 0.8% tungsten in steel is determined colorimetrically with dithiol (38,585). Molybdenum, if present in the steel, is first extracted as the dithiol complex, additional strong acid is added, and the tungsten complex is extracted. Other colorimetric reagents for determining tungsten in steels include hydroquinone (386) and phenylfluorone (13).

Emission spectroscopy has long been used for steel analysis, and specific procedures for tungsten in steels are available (246,279,495,946). Atomic absorption spectrometry is an alternate approach for emission analysis, and a procedure is reported for determination of nine common alloying elements in steels, including from 1 to 5% tungsten, by atomic absorption (476). Another atomic-absorption procedure analyzes steels containing up to 19% tungsten (823). X-Ray fluorescence (538) is capable of determining tungsten in steels over a wide concentration range. From 2 to 10% tungsten in steels is determined by conventional dc polarography (758), while lesser amounts are found by oscillopolarographic techniques (596).

Radiochemical procedures utilizing tungsten are applied to the analysis of steels. Isotopic dilution using ^{185}W (554) and backscattering of reflected beta radiation incident upon a sample surface (250,312,1078) are used for steels containing up to 20% tungsten. Neutron activation analysis, too, is applied for the determination of tungsten in steels (47,553,971).

Carbides in alloy steels are analyzed by first isolating the carbide phase, generally by electrolysis, separating the residue and analyzing by appropriate means (771,790). Tungsten in the carbide phase is found by a dithiol colorimetric procedure (728), activation analysis (701), or other means (838).

Tungsten in slags, resulting from the production of steels, is determined by emission spectroscopy (92,987).

2. Nonferrous Alloys

In addition to its use as an alloying element in steels tungsten is frequently combined with other metals to produce many useful alloys. Cobalt-based alloys, significant for their high-temperature strength, frequently contain about 10% tungsten. Tungsten in a cobalt matrix is determined by precipitation of tungsten(VI) oxide (708), polarography (99), X-ray fluorescence (973), and atomic absorption (1036). Tungsten in heat-resistant nickel-based alloys is found by a colorimetric method with thiocyanate (387) or atomic absorption (1035) and in silver–nickel–tungsten alloys by precipitation of tungsten(VI) oxide (697).

Tungsten–molybdenum alloys are well known, and tungsten content is found by a variety of methods including precipitation of tungsten with β-

naphthoquinoline (648), colorimetrically with thiocyanate (115), and emission spectroscopy (462). β-Naphthoquinoline precipitation of tungsten is also used with tungsten–rhodium alloys following separation of rhodium using a mercury cathode electrolysis (304). Mercury cathode deposition of palladium, too, achieves its separation from tungsten in tungsten–palladium alloys. Tungsten content is then found by colorimetric measurement with thiocyanate (303). Tungsten can also be precipitated as the hydrous oxide from tungsten–palladium alloys (518).

Tungsten in tungsten–niobium alloys is first separated from niobium by anion exchange (615) or adsorption chromatography (460). With the latter procedure niobium is retained on an alumina column from basic solution. Tungsten passes through the column and is determined by the thiocyanate colorimetric procedure. It is possible to quantitatively precipitate tungsten(VI) oxide from niobium-based alloys (236). Trace amounts of tungsten in ferroniobium are found by extraction and photometric measurement of the ternary tetraphenylarsonium chloride-thiocyanate-tungsten complex (429).

Small amounts of tungsten in zirconium-based alloys are found by colorimetric measurement of the thiocyanate complex (144) and by X-ray fluorescence (659). Titanium alloys containing tungsten are analyzed for their tungsten content by colorimetry using dithiol or thiocyanate (1055) and by spark-source mass spectrometry (1011). Tungsten is determined in uranium-based alloys by precipitation (127) or colorimetrically with hydroquinone (90), in vanadium-based alloys using polarography (219), and in copper–bismuth alloys colorimetrically with phenylfluorone (101).

B. ELEMENTS AND COMPOUNDS

Tungsten, as an impurity in other pure elements, has been most thoroughly studied in relation to those elements with which it is generally associated. In molybdenum, for example, microgram amounts of tungsten are determined colorimetrically with thiocyanate ion (1014) or dithiol (431), by emission spectroscopy (463), and by neutron-activation analysis (559). Colorimetric procedures with thiocyanate ion and dithiol are also used for finding trace amounts of tungsten in niobium (364,519), tantalum (348), and titanium (689,891), while emission spectroscopy serves to measure tungsten in niobium (229), titanium (1018), and zirconium (184).

Table 15 lists procedures for determining trace amounts of tungsten in high-purity elements.

TABLE 15

Methods for Determination of Trace Tungsten in Various High Purity Elements

Element	Method	Reference
Al	Spectrophotometric, salicylfluorone	667
B	Spectroscopic	968
Be	Spectrophotometric, dithiol	77
Cu	Spectrophotometric, dithiol	1088
Fe	X-Ray fluorescence	587
Hf	Spectroscopic	88
Li	Spectroscopic	705
Mg	Spectroscopic	186
Mo	Spectrophotometric, thiocyanate	1014
Nb	Spectrophotometric, dithiol	432
Ni	Spectrophotometric, hydroquinone	955
Pu	Spectroscopic	46
Se	Neutron activation	294
Si	Spectroscopic	1005
Ta	Spectrophotometric, thiocyanate	14
Te	Neutron activation	867
Th	Spectroscopic	185
Ti	Spectrophotometric, thiocyanate	689
U	Spectroscopic	756
Zr	Polarographic	1056

Ammonium molybdate is analyzed for small amounts of tungsten colorimetrically with thiocyanate ion (873), and the catalytic effect of tungsten upon the iodide–hydrogen peroxide reaction is used for finding trace amounts of tungsten in single crystals of CdS and LiF (103).

Methods for determining tungsten as an impurity in various metal oxides are summarized in Table 16.

TABLE 16

Methods for Determination of Trace Tungsten in Various Metal Oxides

Oxide	Method	Reference
MoO_3	Spectroscopic	464
Nb_2O_5	Spectroscopic	929
SiO_2	Neutron activation	430
Ta_2O_5	Spectrophotometric, thiocyanate	14
TiO_2	Spectroscopic	256
U_3O_8	Spectroscopic	733
V_2O_5	Spectroscopic	1076

C. ROCKS AND MINERALS

Tungsten, generally in the range of a few micrograms per gram of sample, is determined by emission spectroscopy in a variety of rock types including silicates, granites, and others (403,526,920,1048). Standard rock samples, containing less than 1 μg W/gram sample, are analyzed for tungsten by neutron-activation analysis and monitoring of the ^{187}W beta emissions (351). Neutron activation is also employed for tungsten in other rock samples (273,419). Thiocyanate ion (834,1004,-1054) and dithiol (445) colorimetric procedures are also applied for determining tungsten in various rock samples. A polarographic procedure for tungsten is also reported (808).

Tungsten is determined in meteorites by neutron-activation analysis (31,350).

The principle tungsten ores are ferberite (FeWO$_4$), hübnerite (MnWO$_4$), wolframite (a mixed FeWO$_4$–MnWO$_4$), and scheelite (CaWO$_4$). Tungsten is also found in a variety of other minerals (561). Steps for the complete analysis of wolframite are available by both wet chemical (930) and spectroscopic (825) means. Scheelite, too, is analyzed by wet chemical (606) and spectroscopic (826) procedures. These procedures are extended to the analysis of other tungsten-containing ores (155,849).

Tungsten, itself, is determined in tungsten-containing ores by a variety of methods. Traditionally determination is by precipitation of the acid-insoluble hydrous tungsten(VI) oxide and ignition to WO$_3$ (619,703). Details of this procedure using cinchonine to aid in removing the last traces of tungsten are found in Section IX.A of this report. Precipitation of insoluble tungsten-8-quinolinol complex gives satisfactory results for tungsten in the analysis of a variety of tungsten-containing ores (207). Precipitation of insoluble lead tungstate with a known excess of lead(II) ion followed by complexation titration of the remaining lead with EDTA is used for analysis of tungsten ores (711). Tungsten is found in ores by a colorimetric procedure using thiocyanate ion (681). Emission (134) and atomic absorption (784,798) spectroscopic methods are also used for finding tungsten in ores. Other procedures for tungsten include X-ray fluorescence (128), radiochemical procedures, (436) and neutron-activation analysis (748).

Precipitation of tungsten in low-grade ores using cinchonine and other organic precipitating agents is compared (737). Colorimetric procedures for low-grade tungsten ores are also commonly employed with thiocyanate ion (240,330,885), dithiol (934), and other color-forming agents (757). Emission spectroscopy is also utilized for determining tungsten in low-grade ores (442,725). A polarographic procedure for determining tungsten in low grade ores is given in Section IX.E of this report.

Specific procedures for finding selected trace elements in tungsten ores, concentrates and minerals are listed in Table 17.

Tungsten is frequently associated with other mineral species. In the tin ore, cassiterite, for example, precipitation of tungsten(VI) with mercury(II) (1034), colorimetry with thiocyanate ion (226), and neutron-activation analysis (81) are all used to find the tungsten content. Trace amounts of tungsten in molybdenite are found by neutron-activation analysis (794) and in niobium–tantalum concentrate by colorimetry (974) and polarographic measurement (311). Procedures specifically for tungsten are available for sulfide (465), silicate (453), and antimony (2) ores.

D. SOILS AND PLANTS

Tungsten in soils is generally rendered soluble by basic fusion with the residue taken up in water. The soluble tungstate is determined colorimetrically with thiocyanate ion (609,1024) or dithiol (80,684,931). The colored complex is usually extracted with an appropriate immisible solvent (e.g., i-butanol or amyl acetate) prior to absorbance measurement. Reported tungsten content ranges from a few parts per million to a few hundred parts per million.

The effect of tungsten on plants (372,845) and the uptake by plants of tungsten from soil (858,1047) have been studied. Determination of tungsten in plant material, following sample digestion with strong

TABLE 17
Methods for Determination of Trace Elements in Tungsten
Ores, Concentrates, and Minerals

Element	Method	Reference
As	Spectroscopic	352
Bi	Spectrophotometric	908
Ca	X-Ray fluorescence	121
Ga	Spectrophotometric	959
Mo	Spectrophotometric	413
P	Spectrophotometric	1050
Pb	Gravimetric	871
Nb	Spectrophotometric	724
Sb	Spectrophotometric	337
Sc	Spectrophotometric	301
Si	Gravimetric	621
Sn	Titrimetric	710
Ta	Spectroscopic	402

mineral acid, is by thiocyanate ion and colorimetric measurement (322) or polarographically (662). Tungsten in plants is also determined colorimetrically with dithiol following extraction (411) or cation-exchange separation (1026). Neutron-activation analysis is also used to find tungsten content of plants (673).

E.　ANIMAL MATTER

Tungsten is one of several transition metal ions determined in marine organisms by neutron-activation analysis (264). Tungsten specifically is identified in blood (745) and urine (33) samples. From a neutron-activation procedure nanograms per milliliter of tungsten is reported in human blood (97). Tungsten, as one of many constituents, is detected in human heart tissue by a radiochemical procedure (1041) and in bone (912) and tooth samples (832) using X-ray spectrometry.

F.　WATER AND AIR

Greatest concern in determining tungsten in natural waters, found generally at the nanogram per liter concentration level (133,390), is in concentrating the tungsten from a large water sample. Iron(III) hydroxide (391), calcium carbonate (627), and cadmium sulfide (183) are used as coprecipitating agents. Alternately ion exchange (444,815) or solvent extraction (494) are employed. Following concentration, determination of tungsten is by colorimetry using dithiol (392,488) or thiocyanate ion (765). Emission spectroscopy (626) and neutron-activation analysis (5) are also applied for analysis of tungsten content in waters.

Tungsten in airborne particulates is found in the nanogram per cubic meter range from air samples collected through a filter during a 24 hr period (176). Various methods are used for determining tungsten after separation from the collection filter by treatment with NaOH or HCl–H_2O_2 (1008). Sodium carbonate fusion is also used to render tungsten soluble (1037). Color development with thiocyanate ion (1008), emission spectroscopy (568), and neutron-activation analysis (599) are suitable for tungsten. Special attention has been given to the amount of tungsten in air samples from metallurgical plants (996).

G.　MISCELLANEOUS MATERIALS

The effects of tungsten in glass has been reported (1016), and tungsten has been determined by X-ray spectroscopy (344). Tungsten catalysts for conversion of alkanes to alkenes are analyzed, and the tungsten is determined by precipitation of hydrous tungsten(VI) oxide (734). Other constituents are found by appropriate means. Procedures are available for determining tungsten in drug preparations (63) and in explosives (861).

IX. RECOMMENDED PROCEDURES

A. DETERMINATION OF TUNGSTEN USING CINCHONINE (215,363,485)

Although insoluble in acid solution hydrous tungsten oxide is not quantitatively precipitated unless an organic precipitating agent is also present. Cinchonine is generally used to remove the last traces of tungsten and gravimetric determination of tungsten by precipitation from acid solution in the presence of cinchonine is widely used for determination of tungsten in a variety of materials. With proper choice of sample size, the procedure is applied both to samples containing only a few percent tungsten and to samples in which tungsten is the major component. Because the precipitate contains impurities, it is generally rendered soluble by sodium carbonate fusion and the impurities removed as insoluble hydrous oxides. The original filtrate from the initial precipitation may, itself, contain traces of tungsten even after addition of cinchonine. A correction for this is also possible and is described in this procedure.

The precipitation of tungsten from steels and ores is described.

Reagents

Nitric acid, 16 M.
Nitric acid, 8 M.
Hydrochloric acid, 12 M.
Hydrochloric acid, 6 M.
Hydrochloric acid, 1 M.
Hydrofluoric acid, 48%.
Sulfuric acid, 9 M.
Ammonia, 15 M.
Ammonia, 6 M.
Ammonia wash solution, 10% NH_3 (vol./vol.) + 1% NH_4Cl (wt./vol.). Dilute 100 ml of 15 M NH_3 with 900 ml of distilled water. Dissolve into this solution 10 grams NH_4Cl.

Cinchonine, 12% (wt/vol.). Dissolve 120 grams of cinchonine in 1000 ml of 6 M HCl.

Cinchonine wash solution. Add to 25 ml of 12% cinchonine solution 30 ml of 12 M HCl, and dilute the mixture with distilled water to a volume of 1 liter.

Sodium carbonate, anhydrous. Solid.

Procedure

For steel samples place an accurately weighed 2 gram sample in a 400 ml beaker, add 50 ml 6 M HCl, and heat gently. When only a small insoluble residue remains, carefully add 10 ml 8 M HNO$_3$. Digest the sample at near-boiling temperature until no dark residue remains. Only insoluble yellow hydrous tungsten(VI) oxide should be visible. Dilute the solution to approximately 100 ml with distilled water. The next step is addition of cinchonine reagent.

For ore samples, place an accurately weighed 1 gram sample of finely ground ore in a 400 ml beaker. Add 5 ml of water, swirl to thoroughly wet the sample, and then add 100 ml of 12 M HCl. Cover the beaker, and heat at 50–60°C for 1 hour with occasional stirring. Rinse the beaker cover, and carefully evaporate the sample to about 50 ml. Carefully add 40 ml 12 M HCl and 15 ml 16 M HNO$_3$. Swirl to thoroughly wet the residue, and boil gently until the volume is reduced to approximately 50 ml. Add 5 ml more of 16 M HNO$_3$, and stir and evaporate at boiling until only 5–10 ml of solution remains. Dilute the sample to approximately 150 ml wth distilled water. The next step is addition of cinchonine reagent.

Heat the sample solution to near boiling. Add 5 ml 12% cinchonine solution, and digest the sample at 80–90°C for 30–45 min. Filter the hot solution using Whatman No. 540 filter paper. Add some ashless filter paper pulp to the funnel to aid in retaining the precipitate. Thoroughly wash the precipitate with hot cinchonine wash solution.

The filtrate and washings from this initial tungsten precipitation may be set aside, 5 ml of additional cinchonine solution added, and, if after several hours, additional precipitate forms it is collected and added to the main tungsten precipitate. For most accurate results, it may be desirable to transfer the initial tungsten precipitate back to its original beaker, using not more than 25 ml water in the transfer step, redissolve, and reprecipitate the tungsten. To redissolve the precipitate, add 6 ml 15 M ammonia solution, cover the beaker, and warm gently. Rinse the beaker walls with warm ammonia wash solution, and, using additional wash solution, filter the sample through the same paper used initially to collect the tungsten precipitate. Thoroughly rinse the beaker and residue with additional warm ammonia wash solution. The residue, containing iron and other insoluble hydroxides (residue A), is retained for correction of possible coprecipitated tungsten. Directions for this correction are described further on in this procedure. The ammonical filtrate from this second precipitation is boiled to expel ammonia, and then 20 ml 12 M HCl and 10 ml 16 M HNO$_3$ are added. Boil the acid solution and evaporate until only 5–10 ml of solution remains. Dilute the sample to approximately 150 ml with distilled water. Heat the sample to near boiling, and add 10 ml cinchonine solution. Stir thoroughly, and digest at 80–90°C for 30–45 min. Cool, add a small amount of filter paper pulp, and filter through a second sheet of Whatman No. 540 filter paper. Thoroughly wash the precipitate with hot cinchonine wash solution. The filtrate and washings from this second precipitation may also be set aside, and, if after several hours additional tungsten precipitate forms, it can be combined with the main tungsten precipitate.

If it is deemed necessary, residue A, from the procedure described in the preceding paragraph, can be redissolved with warm 1 M HCl, and traces of tungsten can be precipitated by repeating the previously described procedure beginning with addition of 6 ml 15 M ammonia. Any additional tungsten-cinchonine precipitate collected is combined with the bulk of the tungsten precipitate.

The tungsten precipitate, or collected tungsten precipitates if any or all of the additional purification steps are employed, is transferred to a large weighed platinum crucible. Moisture is removed from the paper by drying over a low flame. The paper is then charred, and eventually all the carbon is burned away. This is done at the lowest possible temperature to avoid loss of tungsten(VI) oxide. Cool the crucible, add one to two drops of 9 M H_2SO_4 and 2–3 ml 48% HF. Evaporate to dryness, expelling silica. Finally ignite the dry residue to constant weight at 750–800°C in a muffle furnace. Lower temperatures result in incomplete dehydration of tungsten(VI) oxide, and higher temperatures, especially above 850°C, result in volatilization of tungsten(VI) oxide. The weight of impure WO_3 is found by difference.

The oxide obtained by ignition contains several impurities notably iron(III) oxide. To correct for these impurities, place the residue in a crucible, add 5 grams anhydrous sodium carbonate, and fuse, maintaining the molten condition for 10 min. Cool and extract the melt with 100 ml of hot water. Filter the insoluble residue through a Whatman No. 40 paper, and thoroughly wash the residue with hot distilled water. Place the filter paper containing the insoluble hydrous oxide impurities in a platinum crucible, char over a low heat, burn away the carbon, and finally ignite at 750–800°C. Repeat the fusion this time, adding 0.5–1 gram Na_2CO_3 to the ignited residue. Take up the cooled melt in 25 ml of hot distilled water; filter; after thoroughly washing with hot water again transfer the residue, this time to a weighed platinum crucible; char; expel carbon; and finally ignite the residue. The weight of cooled residue is subtracted from the weight of tungsten(VI) oxide.

The filtrates and washings from the carbonate fusions may still contain appreciable amounts of foreign ions and, in addition, traces of tungsten.

Collect the various filtrates into a single container. Acidify the solution with 6 M HCl. Boil to expel carbon dioxide and, at the same time, reduce the volume to 40–50 ml. Add an excess of 6 M NH$_3$ making the solution basic. If a precipitate forms it should be collected, dried, and carried through the sodium carbonate fusion described in the preceding paragraph. The ammonical solution may still contain molybdenum, chromium, vanadium, and phosphate ions. These are best determined individually by appropriate colorimetric procedures. The reader is referred to the original references or other parts of the treatise for procedures used to determine these elements.

Interference

As previously stated to obtain the last traces of tungsten precipitate, it may be necessary to let the sample solution sit for several hours after the bulk of the tungsten has been removed. The presence of alkali metal and/or ammonium salts tends to enhance retention of tungsten. Fluoride ion and organic acids too, through complexation, contribute to incomplete tungsten precipitation.

The tungsten precipitate itself contains various impurities. Among these impurities are silica, tin, niobium, and tantalum. Molybdenum, vanadium, and phosphorus, too, are generally carried down with the tungsten precipitate as are iron, chromium and other similar metal ions. Basic fusion renders tungsten soluble, while the heavy metal hydroxides remain insoluble, however, because of the affinity of tungsten for adhering to these insoluble hydroxides a double fusion is recommended. Molybdenum, vanadium, chromium(VI), and phosphorus are not separated from tungsten by basic fusion, and individual corrections for these ions must be made. The reader is referred to Section IV and the references therein for details of separation procedures for specific contaminates.

B. DETERMINATION OF TUNGSTEN WITH 8-QUINOLINOL (1015)

Quantitative precipitation of tungsten(VI) using 8-quinolinol occurs without interference from a number of other ions provided suitable precautionary measures are taken (775,776). Proper acid adjustment, pH 4.7–5.0; the presence of oxalate buffer rather than acetate buffer, favoring formation of a stoichiometric compound; addition of EDTA to mask interfering ions; and precipitation from a mixed solvent solution to enhance crystalline formation all contribute to the desirability of this method. The procedure given applies to Ni–Cr–W alloys containing approximately 51% Ni, 26% Cr, 16% W, 4% Fe, and minor amounts of other metals. As proposed the final precipitate is weighed. Alternately it can be redissolved and the 8-quinolinol content found by bromometric titration (775).

Reagents

Nitric acid, 16 M.

Hydrochloric acid, 12 M.

Hydrochloric acid, 4 M.

Hydrochloric acid, 2 M.

Hydrofluoric acid, 48%.

Sulfuric acid, 18 M.

Sodium hydroxide, 4 M.

Sodium hydroxide, 1% (wt/vol.), aqueous.

Oxalic acid, 10% (wt/vol.), aqueous.

Disodium dihydrogen EDTA, 10% (wt/vol.), aqueous.

8-Quinolinol, 3% (wt/vol.). Dissolve 3 grams of 8-quinolinol in acetone. Dilute to 100 ml with additional acetone. Prepare fresh as needed.

Potassium bromide, 20% (wt/vol.), aqueous.

Potassium bromate, 0.1000 N. Dissolve 2.784 grams of $KBrO_3$, previously dried at 150°C for 1 hr, in distilled water, and dilute with additional distilled water to 1 liter in a volumetric flask.

Potassium iodide, 10% (wt/vol.), aqueous.

Sodium thiosulfate, 0.1 N. Dissolve approximately 25 grams of $Na_2S_2O_3 \cdot 5H_2O$ in distilled water that has been freshly boiled and cooled. Add approximately 0.1 gram Na_2CO_3. Dilute to 1 liter with additional, freshly boiled, distilled water. Store the solution in the dark. Standardize against a suitable primary standard, for example, iodine.

Methyl red, 0.1% (wt/vol.). Dissolve 0.1 gram of methyl red in 30 ml of ethanol, and dilute to 100 ml with distilled water.

Starch. Triturate 2 grams of soluble starch with 10 mg of HgI_2 and about 30 ml of distilled water. Pour this into 1 liter of boiling distilled water, and continue heating until the solution is clear. Cool and store in a glass-stoppered bottle.

Acetone.

Procedure

Dissolve approximately 0.1 gram of sample (containing 15–20 mg W) in 10 ml of 16 M HNO_3. Cover the sample, and heat gently. Add 5 ml of 12 M HCl, and continue heating. If any residue remains carefully add dropwise 48% HF until all has dissolved. Add 2–3 ml of 18 M H_2SO_4, and carefully evaporate the sample until thick fumes are evolved. Dissolve the residue in a minimum amount of 1% NaOH, and transfer the sample to a 200 ml beaker rinsing with additional 1% NaOH. Add 10% oxalic acid solution until the sample is acidic, and then add 5 ml excess oxalic acid solution. Heat the sample to boiling, and add 10–25 ml of 10% EDTA solution (the larger amount is necessary, if, from prior knowledge of the sample, the concentration of ions to be masked is high). Continue boiling, and add sufficient 1% NaOH until the pH of the solution (indicator paper) is about 4.7–5.0. Cool and dilute with water to a volume of about 70 ml. Add 50 ml of acetone. Follow this with 10–15 ml of 3% 8-quinolinol solution. Stir, and place in a constant–temperature bath at 55–60°C. Precipitate formation begins in about 30–40 min and is complete after 2–3 hr of heating. Filter through a previously weighed number 3 or 4 (fine) sintered-glass crucible. Wash the precipitate two to three times with cold 2 M HCl and then with 50–60 ml of hot water added in small portions. The precipitate is dried at 120°C for 1 hr, cooled, and weighed. The precipitate is $WO_2(C_9H_6ON)_2$, molecular weight 504.15, and contains 36.48% W.

Alternate to determining tungsten by weighing, the precipitate can be dissolved and the 8-quinolinol titrated by the usual bromination reaction (485,501). This procedure is, in fact, recommended by those who question the stoichiometry of the tungsten oxinate precipitate (162).

Immerse the crucible containing the tungsten precipitate in 30–40 ml of 4 M NaOH. After the precipitate has dissolved, remove the crucible, and thoroughly rinse it with distilled water. Add the rinsings to the tungsten solution. Add 20–30 ml of 10% oxalic acid solution (to prevent tungsten precipitation), and neutralize with 4 M HCl to a pH of about 5 (methyl red). Add 20 ml excess 10% oxalic acid solution. Add 5 ml of 20% KBr solution and two to four drops of methyl red indicator. Titrate slowly with standard 0.1000 N $KBrO_3$ solution, mixing thoroughly after the addition of each increment of titrant. Continue the titration until the indicator changes from red to yellow, signaling the presence of a slight excess of free bromine. Stopper the flask, and allow it to stand for 2 min. Then add 10 ml of 10% KI solution, and mix thoroughly to assure complete replacement of I_2 for Br_2. Titrate the liberated iodine with standard sodium thiosulfate solution. As the end point approaches add 5 ml of starch indicator, and continue the titration to the disappearance of the blue iodine-starch complex color. The amount of bromine reacted with 8-quinolinol is found by subtracting the milliequivalents of excess bromine (iodine replacement) from the total milliequivalents of bromine added ($BrO_3^- $–$Br^-$ reaction). Four atoms of bromine are consumed for each molecule of 8-quinolinol present; therefore, 1 mmol of WO_2^{2+} requires 8 mequiv. of bromine.

Interference

The alkali metal ions (except NH_4^+), nitrate, chloride, sulfate, and acetate do not interfere. Moderate amounts of SiO_2, Sn(IV), Zr(IV), and PO_4^{3-} do not interfere. Studies with solutions containing 20 mg W and each of these four ions separately resulted in quantitative tungsten recovery in the presence of 30 mg SiO_2, 100 mg Sn(IV), 100 mg Zr(IV), and 150 mg PO_4^{3-}. Uranyl oxinate precipitates along with tungsten under the experimental conditions employed; however, the uranyl precipitate is soluble in 2 M HCl and is removed in the washing procedure. Four-hundred milligrams of uranium do not interfere in determining 20 mg of W by this method. Fifty milligrams each of Co(II), Cr(III), Mn(II), Ni(II), and V(IV) do not interfere with determination of 20 mg W if EDTA is present in the sample solution. Vanadium(V), if present, is reduced to vanadium(IV) by oxalate, and chromium(III) is masked with EDTA upon heating. Twenty milligrams of iron(III) present in a sample containing 20 mg W gave satisfactory results for tungsten, but precipitation in the presence of 50 mg of iron(III) gave slightly high results for tungsten. Other authors suggest precipitation of tungsten oxinate in the presence of iron at pH 3.6 rather than 4.7–5.0 to minimize iron interference (775). Ammonium ion interferes with this procedure and should be avoided. More seriously, titanium(IV) interferes. Results for tungsten by this procedure are low if titanium is present in the sample solution.

C. COLORIMETRIC DETERMINATION WITH THIOCYANATE ION IN STEELS (251,253)

The tungsten content of steels varies over a wide range, typically from less than 0.2 to 20%. By proper choice of sample size and variations in the dissolving procedure, this wide range can be accommodated in a colorimetric procedure for determining tungsten with thiocyanate ion. Greatest interference with this procedure comes from molybdenum, frequently present with tungsten in steel samples. The authors found that with their procedure samples containing more than 2.5% W are not affected even when molybdenum concentrations were as high as 97.5%. For steels containing less tungsten the allowable tungsten range and maximum molybdenum content were, respectively; 1.25–2.5% W/60% Mo, 0.5–1.25% W/12% Mo, 0.25–0.5% W/3.0% Mo, and less than 0.25%

W/0.3% Mo. It is in the latter case, with tungsten content less than 0.25%, that molybdenum interference is most likely to be encountered. To overcome this, the authors modified their original procedure (253) to include an additional extraction step (251). By doing this, samples containing less than 0.25% W could be analyzed satisfactorily even in the presence of 25% Mo.

Representative results using the thiocyanate procedure gave, respectively, observed and known tungsten content for standard steels of 0.16% compared to 0.17%, 1.24% compared to 1.30%, and 19.3% compared to 19.4% W. A low-tungsten steel containing 2.6% Mo resulted in a tungsten value of 0.053% when the additional separation was used. This compares to 0.055% W by a dithiol colorimetric procedure for the same sample. Twelve replicate determinations on a standard sodium tungstate solution of 4.42×10^{-5} M produced a mean absorbance value of 0.650 with a standard deviation of 0.5%.

Use 0.5 gram samples for steels containing less than 0.5% W. Use 0.2 gram samples for steels containing between 0.5 and 1.25% W. Use 0.1 gram samples for steels containing greater than 1.25% W.

Reagents

Nitric acid, 16 M.
Hydrochloric acid, 12 M.
Hydrochloric acid, 6 M.
Phosphoric acid, 15 M.
Tin(II) chloride, 10% (wt/vol.). Dissolve 20 grams $SnCl_2 \cdot 2H_2O$ in 12 M HCl. Dilute to 200 ml with additional 12 M HCl. Prepare fresh every several days.
Titanium(III) chloride, 15% (wt/vol.) Commercially available solution.
Sodium thiocyanate, 2 M. Dissolve 81.1 grams NaSCN in distilled water, and dilute to 500 ml with additional distilled water. Discard the solution if a pink color develops.
Tetraphenylarsonium chloride, 0.025 M. Dissolve 1.05 grams of $(C_6H_5)_4AsCl$ in distilled water, and dilute to 100 ml with additional distilled water. Prepare fresh every 2 weeks.
Chloroform, containing 0.08% hydroquinone. Add 20 ml of 1% $C_6H_4(OH)_2$ in ethanol to 250 ml chloroform. The hydroquinone solution is stable for several days in a dark bottle. The chloroform mixture is prepared fresh daily.
Ammonium hydrogen bifluoride, solid.
Methyl isobutyl ketone.
Tungsten standard, 25.0 μg W/ml. Dissolve 2.2432 grams reagent-grade $Na_2WO_4 \cdot 2H_2O$ in distilled water, and dilute to 500 ml in a

volumetric flask. Five milliliters of this solution is transferred to a second 500 ml volumetric flask and diluted to exactly 500 ml with 12 M HCl. One ml of this diluted tungsten solution contains 25.0 μg W.

Procedure

If molybdenum interference is not a problem, dissolve the appropriate weight of sample, based upon the expected tungsten contents cited in the introductory remarks, in 5 ml 15 M H_3PO_4 plus 25 ml 16 M HNO_3. Warm gently until dissolution is complete. Heat to boiling, and boil several minutes to expel nitrogen oxides. Cool, and transfer the sample to a 50 ml volumetric flask. With samples containing more than 2.5% W use a 100 ml volumetric flask. In either case dilute the sample to the specified volume with 12 M HCl. For steels containing less than 0.25% W where molybdenum interference is expected dissolve an accurately weighed 0.5 gram sample in 25 ml 12 M HCl plus sufficient 16 M HNO_3, five to ten drops, to achieve dissolution. Warm gently until the sample is dissolved; then boil several minutes to expel nitrogen oxides. Cool and transfer the sample solution to a 50 ml volumetric flask. Dilute the sample to exactly 50 ml with 12 M HCl. If a small residue remains after boiling, filter it and dissolve it in a minimum amount of 2:1 (vol./vol.) H_3PO_4–HNO_3. Combine this with the main sample after the extraction step to block molybdenum interference. Transfer 10.0 ml of sample solution to a 100 ml separatory funnel containing 10 ml water and 20 ml methyl isobutyl ketone. Mix vigorously to extract iron from the sample. Molybdenum interference is eliminated if iron is taken from the sample. An alternate approach is to remove molybdenum itself (251). Allow the layers to separate, and drain the aqueous tungsten-containing layer into a 100 ml conical flask. Wash the organic layer twice with 4 ml portions of 6 M HCl. Add the washings to the tungsten solution. Discard the methyl isobutyl ketone layer. If it was necessary to dissolve a portion of the sample on H_3PO_4, that portion should now be combined with the extracted tungsten sample. Evaporate the aqueous layer to near dryness, and immediately proceed with the next step.

To determine tungsten take the condensed sample from which molybdenum interference has been eliminated or take an appropriate aliquot of the tungsten sample for which molybdenum interference is not a problem and transfer it to a 100 ml conical flask. For samples containing between 0.25 and 2.5% W a 5 ml aliquot of the 50 ml sample solution is satisfactory. For samples diluted to 100 ml use a 10 ml aliquot if the steel has between 2.5 and 5.0% W, a 5 ml aliquot if the steel has between 5.0 and 12.5% W, or a 1 ml aliquot for steels containing more than 12.5% W. Add to the sample sufficient 12 M HCl to give a final volume of 16 ml.

Add 4 ml SnCl$_2$ solution, and mix thoroughly. The solution should be nearly colorless. If it is greenish, indicating incomplete removal of nitrogen oxides, boil the sample until the green color is gone. Now add additional SnCl$_2$ until the solution becomes colorless. Add 0.2 ml TiCl$_3$ solution, and boil the sample gently for 5 min. Cool the solution quickly in an ice bath.

Transfer the sample to a 100 ml separatory funnel with a Teflon stopcock. Use two 10 ml portions of 6 M HCl to wash the sample into the separatory funnel. Add 1 ml of tetraphenylarsonium chloride solution. Swirl the solution, and then add 3 ml NaSCN solution. Mix again, and then extract the tetraphenylarsonium-tungsten-thiocyanate complex with 9, 8, and 7 ml of hydroquinone-treated chloroform, respectively. Add ten drops of tetraphenylarsonium chloride solution before each of the last two extractions. Transfer the chloroform layers to a second 100 ml separatory funnel with Teflon stopcock. Add 10 ml distilled water and 1.5–2.0 grams NH$_4$HF. Mix vigorously, allow the layers to separate, and pass the chloroform layer through a Whatman No. 1 filter paper directly into a 25 ml calibrated flask. Rinse the aqueous layer with 1 ml additional chloroform, and pass this through the filter into the flask. Wash the filter paper with a small amount of chloroform, and combine this with the sample solution. Dilute the sample to exactly 25 ml with additional chloroform. Mix thoroughly, and transfer a portion of this sample to a 1 cm cell. Measure the absorbance at 402 nm against a reagent blank. The blank is obtained starting with 16 ml 12 M HCl, adding Sn(II) and so forth as outlined earlier in this paragraph.

A calibration curve is prepared by transferring between 1.00 and 10.0 ml standard tungsten solution to separate 100 ml conical flasks and adding sufficient 12M HCl to each to give a final volume of 16 ml. Continue with the procedure given in the preceding paragraph starting with the addition of Sn(II) solution.

Interference

This procedure was tested by its authors on a variety of standard steels with a wide range of tungsten content. The interference from excessive amounts of molybdenum is described in the procedure presented here. Vanadium, up to fiftyfold excess, does not interfere, being eliminated when tungsten complex is extracted into chloroform. The back extraction of the chloroform-containing tungsten layer removes niobium. Other studies (835) report that milligram quantities of Al, Sb, Bi, Pb, Mn, Ti, and Zr do not interfere with the thiocyanate procedure for tungsten. Large amounts of copper precipitate copper(I) thiocyanate. Fluoride and nitrate ions interfere.

Care must be taken in extracting iron from the tungsten sample in HCl solution. If the sample is allowed to stand for more than 5 min, precipitation of hydrous tungsten oxide begins and the sample is lost. Removal of iron to eliminate molybdenum interference is satisfactory, and the interference has been shown to depend upon the molybdenum to iron ratio (251). For samples of low tungsten content where removal of iron is necessary to prevent molybdenum interference, phosphoric acid cannot be used to dissolve the sample. Phosphate complexes with the iron present preventing its extraction and hence not eliminating molybdenum interference.

Addition of hydroquinone to the chloroform extracting solvent is necessary to remove trace amounts of oxidizing impurity present in chloroform. Finally, use of Ti(III) in conjunction with Sn(II) results in more precise data especially with samples containing high percentages of tungsten. An excess of Ti(III), however, should be avoided, and the authors omit this reagent when analyzing samples containing less than 0.25% W.

D. COLORIMETRIC DETERMINATION IN NIOBIUM AND TANTALUM WITH TOLUENE-3,4-DITHIOL (364,432)

Although originally developed for determination of 0.005–0.1% W in niobium (364) this colorimetric procedure was modified (432) for application to samples containing 0.00005–0.008% W in either niobium or tantalum. The change was necessary because present day refining techniques frequently result in samples containing less than 0.005% W. The alteration consists of increased use of hydrofluoric acid to keep the Nb or Ta in solution when large size samples, up to 10 grams, are taken for analysis. For samples greater than 5 grams an additional extraction, to remove the bulk of Nb or Ta, is suggested before proceeding with the W determination. Molybdenum, in amounts similar to tungsten, is also determined with dithiol. The molybdenum dithiol complex is separated first, employing solution conditions that do not affect extraction of tungsten. This is necessary as the presence of molybdenum interferes with subsequent tungsten determination. Using prepared samples the authors report 24.7 μg and 25.2 μg W by this procedure for, respectively, 10 and 5 gram Nb samples to which was added 25 μg W. They likewise report 26.3 μg and 13.3 μg W for, respectively, 10 and 5 gram Ta samples known to contain 25 μg and 12.5 μg W.

Reagents

Nitric acid, 16 M.
Hydrochloric acid, 12 M.

Hydrofluoric acid, 48%.

Sulfuric acid, 18 M.

Acid mixture, 4 M HCl + 5 M HF.

Hydroxylamine hydrochloride, solid.

Sodium sulfate, anhydrous, solid.

Methyl isobutyl ketone. Saturate with acid mixture.

Carbon tetrachloride.

Methyl orange, 0.1% (wt/vol.), aqueous.

Toluene-3,4-dithiol, 0.5% (wt/vol.). Break a 5 gram vial of dithiol under the surface of 500 ml 1 M NaOH. Stir until dissolved, heating at 35°C if necessary. Add 10 grams thioglycollic acid. Dissolve this, and dilute the entire sample to 1 liter with distilled water. Store in 4 oz. polyethylene bottles in a refrigerator.

Titanium(III) sulfate, 0.4% (wt/vol.). Dissolve 1 gram Ti sponge in 175 ml water plus 25 ml 18 M H_2SO_4. If necessary heat in a boiling water bath to aid in dissolving. Transfer the solution to a 250 ml volumetric flask, and dilute to 250 ml with distilled water.

Tungsten standard, 10.0 μg W/ml. Dissolve 0.1794 gram reagent-grade $Na_2WO_4·2 H_2O$ in distilled water, and dilute to exactly 1000 ml in a 1 liter volumetric flask. Transfer 50.0 ml of this solution to a 500 ml volumetric flask, and dilute this to 500 ml with distilled water. One milliliter of this second solution contains 10.0 μg W.

Procedure

Place an accurately weighed 10 gram sample (containing between 5 and 75 μg W) in a 200 ml Teflon beaker. Carefully add 25 ml 48% HF, followed by sufficient dropwise addition of 16 M HNO_3 to achieve complete solution. Evaporate the sample solution with steam or a heat lamp to a syrupy consistency. Add 10 ml 48% HF and 5 ml 12 M HCl. Stir and again evaporate to a syrupy consistency.

If less than 5 grams of sample is required to achieve the proper amount of tungsten this next step, an extraction to remove Nb or Ta, can be omitted. For samples containing over 5 grams of Nb or Ta, add slowly the minimum volume of acid mixture (4 M HCl + 5 M HF) to produce a clear solution after warming. For a 10 gram sample this will be approximately 120 ml. Cool the sample, and transfer it to a 200 ml polyethylene separatory funnel containing 200 ml of acid-saturated methyl isobutyl ketone. A few drops of methyl orange added at this point will aid in distinguishing between solvent layers. Shake the funnel for 1 min, allow the layers to separate, and return the aqueous layer to the Teflon sample beaker. Wash the organic layer with 25 ml of acid mixture. Repeat the washing with a fresh 25 ml of acid mixture.

Washings are added to the main sample. Evaporate the entire acid solution to a syrupy consistency. Discard the methyl isobutyl ketone solution containing Nb or Ta.

To the syrupy sample solution from either of the preceding steps add 10 ml 18 M H_2SO_4 and 75 ml acid mixture. Cover the beaker with a sheet of polyethylene, and secure it to the beaker with a rubber band. Heat the mixture on a steam bath until a clear uniform solution is obtained. Remove the cover, add 0.1 gram $NH_2OH \cdot HCl$, and 10 ml dithiol reagent. Heat the sample for an additional 25 min, and then allow the beaker and its contents to cool to room temperature.

To remove molybdenum from the sample transfer the contents of the beaker to a glass separatory funnel containing 10 ml CCl_4. Shake the mixture for 1 min, allow the layers to separate, and remove the CCl_4 layer into a suitable calibrated flask for subsequent molybdenum determination. Wash the tungsten-containing aqueous layer twice with separate 7 ml portions of CCl_4, and combine all the nonaqueous extracts. Drain the aqueous phase into the original Teflon sample beaker.

To determine tungsten first add to the tungsten solution 25 ml 0.4% Ti(III) solution. Follow this with 70 ml 12 M HCl and 10 ml dithiol reagent. Heat the sample for 30 min on a steam bath, and cool to room temperature. Place the sample in a glass separatory funnel containing 10 ml CCl_4. Shake the mixture for 1 min, and allow the layers to separate. Transfer the nonaqueous layer to a 25 ml volumetric flask. Wash the aqueous layer twice with separate 7 ml portions of CCl_4, and combine all CCl_4 extracts in the volumetric flask. Add a few crystals of anhydrous Na_2SO_4 to the flask to remove possible moisture, and dilute to exactly 25 ml with additional CCl_4. Transfer a portion of this solution to a 2 cm photometer cell, and measure the absorbance of the tungsten complex at 640 nm. Use as a reference sample a reagent blank beginning with 10 ml 18 M H_2SO_4 and 75 ml acid mixture, treating this solution in the same sequence of steps as the tungsten-containing sample.

A calibration curve is prepared by placing from 0.5 to 8.0 ml of standard tungsten solution, equivalent to 5–80 μg W in separate Teflon beakers. To each sample add 10 ml 18 M H_2SO_4 and 75 ml acid mixture. Treat these samples in the same sequence of steps as the tungsten-containing unknown sample. The amount of tungsten in the unknown sample is read from a calibration curve prepared from measurements made upon the standard solutions.

Interference

Samples containing 50 μg W showed no interference when 1 mg of each of the following ions were present: Fe, Ni, Cr, V, Al, and Co. The interference of molybdenum is removed by extracting molybdenum dithiol complex prior to formation of the tungsten dithiol complex. Up to 20% each of Sn, Mn, and V do not interfere with the tungsten dithiol procedure (214). Copper, bismuth, mercury, and silver, if present, can be removed by dithizone extraction from weakly acid solution before determining tungsten, and up to 2 mg each of Zr and U and 1 mg Ti do not affect determination of 20 μg W (835). Antimony and less than 1 mg of lead do not interfere (113).

The color-forming reagent dithiol is, itself, unstable and should be prepared fresh prior to use. The high total acid concentration, 9–10 M, is necessary for proper complex formation, and the presence of sulfuric acid is necessary for determining W in Nb and Ta. The high proportion of hydrofluoric acid is also necessary with Nb and Ta to prevent hydrolysis, especially with Ta. Other acid ratios, totaling approximately 9–10 M, are used for W determination in Nb containing 0.005–0.1% W with 0.5 gram samples (364). Initial steps in this procedure are carried out in Teflon and polyethylene containers. The final tungsten extraction is performed in a glass separatory funnel with no serious attack from HF provided the extracting phase, CCl_4, is added before the HF-containing tungsten solution. The use of HF requires the usual precautions and careful handling skills associated with this corrosive and dangerous substance.

E. POLAROGRAPHIC DETERMINATION OF TUNGSTEN IN LOW-GRADE ORES (573)

Tungsten(VI) in hydrochloric acid solution exhibits a characteristic polarographic plateau suitable for quantitative determination. The procedure described here utilizes this wave to determine tungsten in low-grade ores containing from 0.001 to 0.100% WO_3. Several possible interfering ions are first leached from the finely ground sample by boiling it in concentrated nitric acid. After filtering, the tungsten is removed from other possible interfering ions by leaching with concentrated hydrochloric acid. Following thorough washing of the insoluble residue the tungsten-containing hydrochloric acid solution is diluted to a fixed volume with constant boiling hydrochloric acid, which serves as supporting electrolyte, and the polarogram is recorded. Values obtained for WO_3 content by this method tested on actual ore samples agree favorably with WO_3 values obtained from a gravimetric procedure on the same ores using cinchinone precipitation followed by precipitation and weighing of lead tungstate. To test the precision of this method its author analyzed eighteen different ores, with tungsten oxide concentrations ranging from 0.004 to 0.060% WO_3, for a total of 49 determinations.

He calculated for each determination the ratio of plateau diffusion current, in microamperes, to WO_3 concentration, in millimoles/liter. The presumably constant value for the 49 determinations resulted in a mean value of 4.15 with a range from 3.54 to 5.25 and a standard deviation of ± 0.4. The mean value of i_d/C for 22 standard sodium tungstate solutions of similar concentration in constant boiling HCl was 4.19 with no standard deviation given. Analysis time from start to finish is approximately 6 hr.

Reagents

Nitric acid, 16 M.
Hydrochloric acid, 12 M.
Hydrochloric acid, 6 M.
Hydrochloric acid, constant boiling, approximately 6 M (570). Add 1000 ml of 12 M HCl to 850 ml of distilled water. Boil for 0.5 hr, cool, and use without further preparation.
Tungsten standard, 0.4 mg WO_3/ml. Dissolve 0.569 gram of reagent-grade $Na_2WO_4 \cdot 2H_2O$ in distilled water, and dilute to exactly 1000 ml in a volumetric flask. Standardize this solution gravimetrically for exact tungsten concentration.

Procedure

Ten grams of finely ground sample (containing between 0.004 and 0.040% WO_3) is accurately weighed and placed in a 600 ml beaker. Add 50 ml of 16 M HNO_3, cover the beaker, and boil gently while stirring for 1 hr. Filter the mixture with suction through a fine-pore sintered glass crucible. Wash the residue thoroughly five times with hot distilled water. Transfer the residue back to the 600 ml beaker, washing the crucible thoroughly with 12 M HCl. Add 200 ml additional 12 M HCl, and heat at 80°C with stirring for 2 hr. Filter the mixture with suction through the same sintered glass crucible. Wash the residue thoroughly five times with 6 M HCl. The filtrate and washings are combined and condensed by gentle boiling to approximately 15 ml total volume. Transfer the sample, using constant boiling HCl, to a calibrated 30.0 ml flask. Dilute the sample to exactly 30 ml with additional constant boiling HCl. The sample is now transferred to a polarograph cell, deaerated for 10 min with oxygen-free nitrogen, and a polarogram is obtained scanning from -0.4 to -0.8 V versus SCE. The diffusion current value is read from the plateau of this curve. The diffusion current is related to actual tungsten concentration by either preparing a standard series calibration curve using appropriate amounts of standard sodium tungstate solution (containing 0.40–4.00 mg WO_3/30-ml sample) dissolved in constant boiling HCl, or, as the author proposes, by measuring the value of i_d/C for a series of known solutions and using this value along with the i_d reading from the unknown solution. To illustrate the author obtained 4.15 as the value of i_d/C. Then,

$$\% \ WO_3 = \frac{i_d \cdot 0.232 \cdot 0.030 \cdot 100}{4.15 \cdot \text{sample wt in grams}} = \frac{0.168 \ i_d}{\text{sample wt in grams}}$$

Interference

Small quantities of each of the following ions in their various oxidation states were added separately to constant boiling HCl, and polarograms were recorded in the region of the tungsten wave. Elements that gave no wave in the region of interest were Al, Ag, Au, B, Ba, Bi, Ca, Cd, Co, Cr, Fe, Hg, Mg, Mn, P, Pt, Rh, Sb, Sr, Ti, U, Zn, and Zr. Although possessing interfering polarographic waves the following ions are removed in the leaching step with nitric acid: Ni, V, Pb, Cu, As, and Mo. Tin also possesses an interfering polarographic wave, but is not extracted with tungsten in concentrated HCl. Rather it remains with the insoluble residue. To test the effect of the leaching steps 0.1 gram amounts of each of the interfering ions, Ni through Sn, in their various oxidation states, were added to standard sodium tungstate samples. The described procedure was carried out and resulted in a i_d/C value of 4.09. This lies within one standard deviation of the i_d/C value for analysis of various tungsten ores and is considered by the author to indicate negligible interference.

Paper filtering medium for the separation steps is not satisfactory. It was shown that traces of carbohydrate material from paper, specifically 5-hydroxymethylfurfural, gave a polarographic wave that overlaps that of tungsten.

Ores containing amounts of WO_3 greater than 0.040% require smaller samples; otherwise, when the HCl solution of tungsten is evaporated to 15 ml a precipitate forms that does not redissolve.

Low results are then obtained for tungsten. The procedure is not recommended for ores containing over 10% WO_3.

F. DETERMINATION BY ATOMIC ABSORPTION IN HIGHLY ALLOYED MATERIALS (823)

Various steels and nickel-based alloys containing from 1 to 20% tungsten are successfully analyzed for tungsten by atomic-absorption spectrometry using a fuel-rich nitrous oxide–acetylene flame. This procedure has been tested on steels containing up to 20% W, on a nickel-based alloy containing 30% Mo and 4% W, and on a nickel-based alloy containing 25% Cr, 6% Fe, 6% Cu, 6% Mo, 3% Si, and 1% W. Because of interfering effects on tungsten absorbance from the other alloying elements present, it is necessary that the calibration standard matrix approximate closely the actual sample composition. If this cannot be accomplished, the method of standard addition is recommended. The authors report for the Ni–Cr–Fe–Cu–Mo–Si–W alloy 1.21% W by the proposed method with a coefficient of variation of 2.3% relative. A gravimetric tungsten procedure for this same alloy gave 1.22% W. For samples containing up to 5% W, 1 gram samples are recommended. For samples containing between 5 and 10% W 0.5 gram samples are used. Smaller samples can be used for alloys containing more than 10% W, but with less precise results. Samples containing less than 0.2% W are not recommended for analysis by this technique. The use of hydrofluoric acid for sample dissolution requires the usual precautions and careful handling skills associated with this corrosive and dangerous substance.

Reagents

Hydrochloric acid, 12 *M*.
Nitric acid, 16 *M*.
Hydrofluoric acid, 48%.
Sodium chloride, 2.5% (wt/vol.), aqueous.
Tungsten standard, 10.0 mg W/ml. Fuse 1.262 grams of freshly ignited tungsten(VI) oxide with approximately 5 grams of NaOH in a silver crucible. Dissolve the melt with water, transfer to a 100 ml volumetric flask, and dilute to 100 ml with additional distilled water. One milliliter of this solution contains 10.0 mg W.

Operating Conditions

Wavelength	255.1 nm
Band pass	0.15 nm
Burner height	Adjust for optimum reading
Nitrous oxide	9.9 liter/min
Acetylene	6.0 liter/min

Procedure

Weigh accurately duplicate 1 gram samples of finely ground material (containing approximately 1–5% W) into separate Teflon beakers. Treat each sample separately. Add 10 ml HCl followed by 5 ml HNO_3 and 5 ml HF. The order of acid addition is not critical, and if appreciable amounts of the sample dissolve in, for example, HNO_3 this acid should be added first and the reaction allowed to proceed before adding the other acids. Careful and gentle warming may aid in dissolving the sample. Because the sample solution is often opaque, it may be that minute sample particles remain after the bulk of the sample solution has been transferred to a 100 ml polypropylene volumetric flask. If this occurs, add an additional 5 ml of the 2:1:1 mixture of $HCl-HNO_3-HF$. Dissolve the residue, and combine it with the bulk of the sample solution in the volumetric flask. Add to the sample 10 ml of 2.5% NaCl solution. Dilute one of the samples to exactly 100 ml with distilled water, and set aside the other sample until the appropriate amount of standard tungsten solution is ascertained for determination by the standard addition technique.

A series of standards is prepared by placing appropriate amounts, from 1 to 10 ml, of standard tungsten solution in 100 ml polypropylene volumetric flasks, adding amounts of each acid ($HCl-HNO_3-HF$) equivalent to the amount used in dissolving the sample, adding 10 ml 2.5% NaCl, and diluting to exactly 100 ml with distilled water. The samples and standard solutions are now ready to be aspirated into the fuel-rich nitrous oxide–acetylene flame.

Prepare a calibration curve of absorbance, 100-%T, versus the amount of tungsten from the data obtained with the standards. The authors found nonlinearity in this plot for absorbance values greater than 0.5 and hence adjusted their samples appropriately to stay within the linear portion of their curve. Aspirate the unknown sample that has been diluted to exactly 100 ml, and from the absorbance observed obtain an approximate idea of the tungsten content.

Sufficient standard tungsten solution is now added to the identical, undiluted sample to produce an absorbance reading approximately twice that of the unknown solution. The unknown plus standard is now diluted to exactly 100 ml with distilled water, and its actual absorbance value is observed. Should this reading of unknown plus standard fall outside the linear portion of the calibration curve appropriate dilution is necessary before the determination can proceed.

With absorbance values for both unknown sample and unknown sample plus known amount of added standard, the amount of tungsten in the unknown can be calculated from the following relationship:

$$\frac{\text{absorbance (unknown)}}{\text{absorbance (unknown + standard)}} = \frac{\text{amount (unknown)}}{\text{amount (unknown + standard)}}$$

Interferences

Synthetic 1 gram samples containing 50 mg W and separately each of the other elements present in the alloys under study were analyzed for tungsten using this procedure. In all cases significant differences in absorbance values for tungsten were observed. Not only each element singly but combinations of the alloying elements produced variations in the absorbance values for tungsten. It is therefore essential to closely approximate the sample matrix if a standard series calibration curve is to be used. The standard addition procedure described avoids this difficulty to some extent.

Because of the corrosive nature of the acid solvent concern should be given to attack upon the burner heads as the sample is aspirated into the flame. The authors used a burner constructed of stainless steel with no serious difficulty, however, they caution that a titanium burner may suffer from attack by the acid solvent.

ACKNOWLEDGMENT

Assistance from Ms. Nancy E. Parker and Ms. Carol J. Ionson for collecting reference material and Ms. Sandra K. Flick for typing the manuscript is gratefully acknowledged.

REFERENCES

1. Aaremae, A., and G. Assarsson, *Z. Anal. Chem.*, **144,** 412 (1955).
2. Abashidze, N. F., *Tr. Kavkazsk. Inst. Mineral. Syr'ya*, **1,** 137 (1960); through *Chem. Abstr.*, **58,** 3876b (1963).
3. Abashidze, N. F., *Tr. Metrol. Inst. SSSR*, No. 96, 165 (1968); through *Chem. Abstr.*, **70,** 34036b (1969).
4. Abdullaev, A. A., E. M. Lobanov, A. P. Noviko, and A. A. Khaidarov, *Prib. Ilya Geofiz. Issled. Skvazhin Radioackt. Metadami* (Kiev: Akad. Nauk Ukr. SSR) *Sb.*, **1962,** 23; through *Chem. Abstr.*, **60,** 9054a (1964).
5. Abdullaev, A. A., A. Sh. Zakhidov, and P. Kh. Nishanov, *Izv. Akad. Nauk Uzb. SSR, Ser. Fiz.-Mat. Nauk*, **12,** 60 (1968); through *Chem. Abstr.*, **70,** 99462q (1969).
6. Abrahamson, E. M., *Amer. J. Clin. Pathol., Tech. Suppl.*, **4,** 75 (1940).
7. Adams, F., and J. Horste, *Acta Chim. Acad. Sci. Hung.*, **52,** 115 (1967).
8. Adams, J. G., US Dept. Commerce, AD 274,588 (1962).
9. Adler, I., in L. Meites, Ed., *Handbook of Analytical Chemistry*, McGraw-Hill, New York, 1963.
10. Affsprung, H. E., and J. W. Murphy, *Anal. Chim. Acta*, **30,** 501 (1964).
11. Agatonova, V. A., L. D. Bednaya, I. I. Bockkareva, V. G. Vites, N. M. Gegechkori, O. A. Dyatlova, and Z. A. Efimova, *Fiz. Sb. L'vov Univ.*, **4,** 44 (1958); through *Chem. Abstr.*, **54,** 20624d (1960).

12. Ahrens, L. H., *J.S. Afr. Chem. Inst.*, **23**, 21 (1943).

13. Akhmedova, Kh. A., O. A. Tataev, and Kh. G. Buganov, *Sovrem. Metody Khim. Tekhnol. Kontrol Proizvod.*, **1968**, 122; through *Chem. Abstr.*, **72**, 18221u (1970).

14. Akiyama, K., and Y. Kobayashi, *Bunseki Kagaku*, **14**, 292 (1965).

15. Aleksandrov, A. N., Uch. *Zap. Mordovsk Univ.*, **1961**, 15; through *Chem. Abstr.*, **59**, 12371b (1963).

16. Aleksandrov, L. N., *Zavod. Lab.*, **26**, 975 (1960).

17. Alimarin, I. P., and A. M. Medvedeva, *Tr. Moskov. Inst. Tonkoi Khim. Tekhnol. M.V. Lomonosova*, No. 6, 3 (1956); through *Chem. Abstr.*, **53**, 13885i (1959).

18. Allen, S. H., and M. B. Hamilton, *Anal. Chim. Acta*, **7**, 483 (1952).

19. Altinata, A., and B. Pekin, *Anal. Lett.*, **6**, 667 (1973).

20. American Society for Testing and Materials, *1973 Annual Book of ASTM Standards*, Ferrous Castings Ferroalloys, Part 2, Designation A144-66, American Society for Testing and Materials, Philadelphia, Pa., 1973.

21. Amiruddin, A., and W. D. Ehmann, *Geochim. Cosmochim. Acta*, **26**, 1011 (1962).

22. Ammar, I. A., and R. Salim, *Corros. Sci.*, **11**, 591 (1971).

23. Amos, M. D., and J. B. Willis, *Spectrochim. Acta*, **22**, 1325 (1966).

24. Anders, O. U., *Anal. Chem.*, **33**, 1706 (1961).

25. Angus, H. J. F., E. J. Bourne, F. Searle, and H. Weigel, *Tetrahedron Lett.*, **1**, 55 (1964).

26. Anonymous, *Ferrotungsten Methods of Chemical Analysis*, Standartov, Moscow, USSR, 1969; through *Chem. Abstr.*, **75**, 143275c (1971).

27. Argent, B. B., and G. J. C. Milne, *J. Less Common Metals*, **2**, 154 (1960).

28. Armstrong, R. D., K. Edmondson, and R. E. Firman, *J. Electroanal. Chem.*, **40**, 19 (1972).

29. Asamoto, R. R., and P. E. Novak, *Rev. Sci. Instrum.*, **38**, 1047 (1967).

30. Asztalos, I., *Hiradastech. Ipari Kut. Intez. Kozlem.*, **7**, 63 (1967); through *Chem. Abstr.*, **69**, 40993g (1968).

31. Atkins, D. H. F., and A. A. Smales, *Anal. Chim. Acta*, **22**, 462 (1960).

32. Atkinson, R. H., G. H. Keith, and R. C. Koo, ''Tungsten and Tungsten-based Alloys,'' in M. Semchyshem and J. J. Harwood, Eds., *Refractory Metals and Alloys*, Wiley-Interscience, New York, 1961.

33. Aull, J. C., and F. W. Kinard, *J. Biol. Chem.*, **135**, 119 (1940).

34. Aveston, J., *Inorg. Chem.*, **3**, 981 (1964).

35. Awashti, S. P., S. Sahasranaman, and M. Sundaresan, *Analyst (London)*, **92**, 650 (1967).

36. Babko, A., and O. F. Drako, *Zh. Anal. Khim.*, **12**, 342 (1957).

37. Babusci, D., *Anal. Chim. Acta*, **32**, 175 (1965).

38. Bagshawe, B., and R. J. Truman, *Analyst (London)*, **72**, 189 (1947).

39. Baishya, N. K., and R. B. Heslop, *Anal. Chim. Acta*, **53**, 87 (1971).

40. Ballou, J. E., US Atomic Energy Commission, HW-64112 (1960).

41. Ban, B., S. Leach, G. Taieb, and M. Velghe, *J. Chim. Phys.*, **64**, 397 (1967).

42. Bandi, W. R., E. G. Buyok, L. L. Lewis, and L. M. Melnick, *Anal. Chem.*, **33**, 1275 (1961).

43. Barborik, M., *Pract. Lek.*, **18**, 241 (1966); through *Chem. Abstr.*, **65**, 17593e (1966).
44. Bartecki, A., and M. Cieslak, *Rocz. Chem.*, **45**, 949 (1971).
45. Barth, V. D., NASA Accession Rep., No. 65-25319 (1965).
46. Barton, H. N., *Appl. Spectrosc.*, **19**, 159 (1965).
47. Barwinski, A., A. Buczek, L. Gorski, J. Janczyszyn, S. Kwiecinski, L. Loska, and M. Geisler, *Isotopenpraxis*, **4**, 15 (1968); through *Chem. Abstr.*, **70**, 120832q (1969).
48. Batt, A. N., and B. D. Jain, *Z. Anal. Chem.*, **195**, 424 (1963).
49. Bausova, N. V., *Tr. Inst. Khim.*, *Akad. Nauk SSSR, Ural. Filial.*, **10**, 97 (1966); through *Chem. Abstr.*, **66**, 34586y (1967).
50. Bausova, N. V., and G. A. Reshetnikova, *Tr. Inst. Khim.*, *Akad. Nauk SSSR, Ural. Filial.*, **10**, 103 (1966); through *Chem. Abstr.*, **66**, 34578x (1967).
51. Bearden, J. A., *Rev. Mod. Phys.*, **39**, 78 (1967).
52. Beermann, C., and H. Hartmann, *Arch. Eisenhuettenw.*, **22**, 159 (1951).
53. Beiles, R. G., and E. M. Beiles, *Zh. Neorg. Khim.*, **12**, 1399 (1967).
54. Beiles, R. G., R. A. Satina, and E. M. Beiles, *Zh. Neorg. Khim.*, **6**, 1612 (1961).
55. Belcher, C. B., *Anal. Chim. Acta*, **29**, 340 (1963).
56. Belcher, R., and A. J. Nutten, *J. Chem. Soc.*, **1951**, 1516 (1951).
57. Berkem, A. R., *Rev. Fac. Sci. Univ. Istanbul*, **8**, 332 (1943); through *Chem. Abstr.*, **40**, 2717[4] (1946).
58. Berkman, S., R. J. Henry, O. J. Golub, and M. Segalove, *J. Biol. Chem.*, **206**, 937 (1954).
59. Berry, E. R., L. Rosenfeld, and A. Chanutin, *Arch. Biochem. Biophys.*, **62**, 318 (1956).
60. Bertin, E. P., *Anal. Chem.*, **36**, 826 (1964).
61. Betso, S. R., and P. W. Carr, *Anal. Chim. Acta*, **69**, 161 (1972).
62. Bhatnagar, R. P., and N. S. Poonia, *Anal. Chim. Acta*, **30**, 211 (1964).
63. Bickford, C. F., W. S. Jones, and J. S. Keene, *J. Amer. Pharm. Assoc., Sci. Ed.*, **37**, 255 (1948).
64. Blanchet, Mme, and L. Malaprade, *Chim. Anal. (Paris)*, **42**, 603 (1960).
65. Blasius, E., and A. Czekay, *Z. Anal. Chem.*, **156**, 81 (1957).
66. Blum, L., *Rev. Chim. (Bucharest)*, **9**, 28 (1958).
67. Blumenthal, H., and W. Fall, *Powder Metall. Bull.*, **6**, 48 (1951).
68. Bobtelsky, M., and I. Barzily, *Anal. Chim. Acta*, **35**, 520 (1966).
69. Bock, R., and B. Bockholt, *Z. Anal. Chem.*, **216**, 21 (1966).
70. Bodor, E., and M. Maleczki-Szeness, *Chem. Zvesti*, **16**, 280 (1962).
71. Bogdanchenko, A. G., and A. D. Sapir, *Zavod. Lab.*, **15**, 11 (1949).
72. Bogdanov, G. A., T. M. Kurokhtina, N. A. Korotchenko, and G. P. Aleeve, *Zh. Fiz. Khim.*, **44**, 1212 (1970).
73. Bogdanov, N. A., V. L. Reitblat, V. F. Funke, and A. A. Zhukhovitskii, *Pr. Radioaktiv. Izotopov. Met. Sb.*, **34**, 283 (1955); through *Chem. Abstr.*, **52**, 2648i (1958).
74. Bognar, J., and O. Jellinek, *Mikrochim. Acta*, **1967**, 193 (1967).
75. Bogovina, V. I., V. P. Novak, and V. F. Mal'tsev, *Zh. Anal. Khim.*, **20**, 951 (1965).

76. Bonds, W. D., *Diss. Abstr. Int. B,* **31,** 5235 (1971).

77. Booth, E., and A. Parb, Atomic Energy Research Establishment (Great Britain), AM 10 (1959).

78. Bottei, R. S., and A. Trusk, *Anal. Chem.,* **35,** 1910 (1963).

79. Bottei, R. S., and A. Trusk, *Anal. Chim. Acta,* **41,** 374 (1968).

80. Bowden, P., *Analyst (London),* **89,** 771 (1964).

81. Bowen, H. J. M., *Radiochem. Radioanal. Lett.,* **7,** 75 (1971).

82. Box, F. W., *Analyst (London),* **69,** 272 (1944).

83. Box, W. D., *Nucl. Appl.,* **2,** 299 (1966).

84. Boyer, M., *J. Electroanal. Chem.,* **31,** 441 (1971).

85. Boyer, M., P. Souchay, and F. Gracian, *Rev. Chim. Mineral,* **8,** 591 (1971).

86. Brantner, H., *Mikrochim. Acta,* **1962,** 125 (1962).

87. Braun, T., *Acta Chem. Acad. Sci. Hung.,* **41,** 199 (1964).

88. Brayer, R. C., R. F. O'Connell, A. S. Powell, and R. H. Gale, *Appl. Spectrosc.,* **15,** 10 (1961).

89. Breyer, B., and S. Hacobian, *Anal. Chim. Acta,* **16,** 497 (1957).

90. Bricker, C. E., and G. R. Waterbury, *Anal. Chem.,* **29,** 1093 (1957).

91. Brinkman, U. A. Th., G. de Vries, and R. Kuroda, *J. Chromatogr.,* **85,** 216 (1973).

92. Brintzinger, H., and R. Titzmann, *Z. Anal. Chem.,* **128,** 486 (1948).

93. Brissey, R. M., *Anal. Chem.,* **24,** 1034 (1952).

94. British Iron and Steel Research Association Methods of Analysis Committee, *J. Iron Steel Inst.,* **178,** 267 (1954).

95. Britton, H. T. S., and E. N. Dodd, *J. Chem. Soc.,* **1931,** 829 (1931).

96. Brown, D. H., and D. Forsyth, *J. Chem. Soc.,* **1962,** 1837 (1962).

97. Brown, H. J. M., *Biochem. J.,* **77,** 79 (1960).

98. Brunck, O., and R. Hoeltje, *Angew. Chem.,* **45,** 331 (1932).

99. Bucklow, I. A., and T. P. Hoar, *Metallurgia,* **48,** 317 (1953).

100. Buganov, Kh. G., K. N. Bagdasorov, and O. A. Tataev, *Sb. Statei Molodykh Uch. Dagestan Fil. Akad. Nauk SSSR,* **1969,** 198; through *Chem. Abstr.,* **75,** 104784y (1971).

101. Buganov, Kh. G., K. N. Bagdasarov, S. A. Tataev, and Kh. A. Akhmedova, *Sovrem. Metody Khim. Tekhnol. Kontr. Proizvod.,* **1968,** 84 (1968); through *Chem. Abstr.,* **72,** 8942r (1970).

102. Bulgakova, A. M., and A. P. Mirnaya, *Metody Anal. Khim. Reaktivov Prep.,* No. 13, 143 (1966); through *Chem. Abstr.,* **69,** 92664v (1968).

103. Bulgakova, A. M., and N. P. Zalyubovskaya, *Zh. Anal. Khim.,* **18,** 1475 (1963).

104. Bulmer, D., *Quart. J. Microsc. Sci.,* **103,** 311 (1962).

105. Bundzhe, V. G., V. M. Gorbacheva, Yu. D. Dunaev, and G. Z. Kir'yakov, *Tr. Inst. Khim. Nauk, Akad. Nauk Kaz. SSR,* **15,** 15 (1967); through *Chem. Abstr.,* **67,** 60309a (1967).

106. Burr, A. F., *Advan. X-Ray Anal.,* **11,** 241 (1967).

107. Buscarons Ubeda, F., and A. Herrera de la Sota, *An. Real Soc. Espan. Fis. Quim., Ser. B,* **45,** 403 (1949).

108. Buscarons Ubeda, F., and E. Loriente Gonzales, *An. Fiz. Quim. (Madrid)*, **41**, 249 (1945).

109. Busev, A. I., V. M. Byrko, N. P. Kovtun, and L. G. Karalashvili, *Zh. Anal. Khim.*, **25**, 237 (1970).

110. Busev, A. I., and G. Li, *Zh. Anal. Khim.*, **13**, 519 (1958).

111. Busev, A. I., and G. Li, *Zavod. Lab.*, **25**, 30 (1959).

112. Busev, A. I., and T. A. Sokolova, *Zh. Anal. Khim.*, **23**, 1348 (1968).

113. Busev, A. I., V. G. Tiptsova, and V. M. Ivanov, *Handbook of the Analytical Chemistry of Rare Elements*, Ann Arbor Science, Ann Arbor, Mich., 1970.

114. Busev, A. I., V. G. Tiptsova, and A. D. Khlystova, *Zavod. Lab.*, **28**, 1414 (1962).

115. Bush, G. H., and D. G. Higgs, *Analyst (London)*, **80**, 536 (1955).

116. Buss, H., H. W. Kohlschuetter, and L. Walter, *Z. Anal. Chem.*, **191**, 273 (1962).

117. Butsik, L. A., M. M. Derbaremdiker, and M. M. Korolev, *Vop. Geol. Sev.-Zap. Sekt. Tikhookean. Poyasa, Vladivostok*, **1966**, 166; through *Chem. Abstr.*, **66**, 97531x (1967).

118. Buxbaum, P., and K. G. Vadasdi, *Chem. Anal. (Warsaw)*, **14**, 429 (1969).

119. Bykovskaya, Yu. I., *Tr. Inst. Mit. A.A. Baikova*, No. 10, 239 (1962); through *Chem. Abstr.*, **57**, 16124c (1962).

120. Caillet, P., and P. Saumagne, *J. Mol. Struct.*, **4**, 351 (1969).

121. Campbell, W. J., and J. W. Thatcher, US Bureau of Mines, Report of Investigations, No. 5416 (1958).

122. Campo, A. del, and F. Sierra, *An. Real Soc. Espan. Fis. Quim.*, **31**, 356 (1933).

123. Campo, A. del, and F. Sierra, *An. Real Soc. Espan. Fis. Quim.*, **33**, 364 (1935).

124. Capilna, S., L. Ababei, E. Ghizari, and M. Stefan, *Fiziol. Norm. Patol.*, **10**, 445 (1964); through *Chem. Abstr.*, **63**, 6231c (1965).

125. Cardenas, A. L., US Atomic Energy Commission, GATT-725 (1960).

126. Carey, M. A., B. A. Rady, and C. V. Banks, *Anal. Chem.*, **36**, 1166 (1964).

127. Carpenter, R. L., R. D. Gardner, W. H. Ashley, and A. L. Henicksman, US Atomic Energy Commission, LA-2270 (1959).

128. Carr-Brion, K. G., and K. W. Payne, *Analyst (London)*, **93**, 441 (1968).

129. Carriere, E., and R. Berkem, *Bull. Soc. Chim. Fr.*, **4**, 1907 (1937).

130. Casado, F. L., *Rev. R. Acad. Cienc. Exactas, Fis. Nat. Madrid*, **39**, 489 (1945); through *Chem. Abstr.*, **45**, 2380c (1951).

131. Chakrabarti, A. K., and S. P. Bag, *Anal. Chim. Acta*, **59**, 225 (1972).

132. Chalmers, R. A., in C. L. Wilson and D. W. Wilson, Eds., *Comprehensive Analytical Chemistry*, Vol. Ic, Elsevier, Amsterdam, 1962, p. 598.

133. Chan, K. M., and J. P. Riley, *Anal. Chim. Acta*, **39**, 103 (1967).

134. Chandola, L. C., *India, At. Energy Comm., Bhabha At. Res. Cent. [Rep.]*, 1970, B.A.R.C.-468; through *Chem. Abstr.*, **74**, 82787b (1971).

135. Chang, C. C., and H. H. Yang, *Hua Hsueh Tung Pao*, **1965**, 45; through *Chem. Abstr.*, **63**, 7634d (1965).

136. Chang, T. H., F. Y. Tsao, and S. M. Shu, *Hua Hsueh Hsueh Pao*, **30**, 230 (1964); through *Chem. Abstr.*, **61**, 6392b (1964).

137. Chao, T. P., and J. T. Yang, *J. Chem. Educ.*, **25**, 388 (1948).

138. Charewicz, W., and R. B. Grieves, *Anal. Lett.*, **7**, 233 (1974).

139. Charlot, G., *Colorimetric Determination of Elements*, Elsevier, Amsterdam, 1964.

140. Charlot, G., and D. Bezier, *Quantitative Chemical Analysis*, Methuen, London, 1957.

141. Charlot, G., A. Collumeau, and M. J. C. Marchon, *Selected Constants; Oxidation-Reduction Potentials of Inorganic Substances in Aqueous Solution*, Butterworths, London, 1971.

142. Chelnokova, M. N., and L. N. Zvyagina, *Nauch. Tr. Perm. Politekh. Inst.*, No. 52, 6 (1969); through *Chem. Abstr.*, **74**, 9335y (1971).

143. Ch'en, K. C., and C. T. Chen, *Hsia Men Ta Hsueh Pao She Hui K'o Hsueh*, No. 1, 121 (1957); through *Chem. Abstr.*, **56**, 2885i (1962).

144. Cheng, K. L., G. W. Goward, and B. B. Wilson, US Atomic Energy Commission, WAPD-CTA(GLA)-180 (1957).

145. Chernikhov, Yu. A., and V. G. Goryushina, *Zavod. Lab.*, **12**, 397 (1946).

146. Chiaki, E., Y. Tomita, and M. Ezawa, *Bunseki Kagaku*, **15**, 976 (1966).

147. Chmelar, Z., *Czech Patent*, 107,839; through *Chem. Abstr.*, **60**, 5106h (1964).

148. Choi, Q. W., and K. R. Min, *J. Korean Chem. Soc.*, **7**, 186 (1963).

149. Chojnacha, J., *J. Inorg. Nucl. Chem.*, **33**, 1345 (1971).

150. Chou, S. F., H. H. Ch'ien, *Hua Hsueh Hsueh Pao*, **27**, 14 (1961); through *Chem. Abstr.*, **59**, 12148f (1963).

151. Chou, S. F., and S. C. Liang, *Sci. Sinica (Peking)*, **8**, 196 (1959).

152. Chou, S. F., and S. C. Liang, *Sci. Sinica (Peking)*, **11**, 207 (1962).

153. Choudhuri, D., and S. K. Mukherjee, *J. Inorg. Nucl. Chem.*, **32**, 1023 (1970).

154. Chovnyk, N. G., *Zh. Fiz. Khim.*, **30**, 277 (1956).

155. Chuenko, L. I., *Tr. Vses. Nauchn.-Issled. Geol. Inst.*, **125**, 66 (1966); through *Chem. Abstr.*, **67**, 70254t (1967).

156. Claeys, A., *Anal. Chim. Acta*, **17**, 360 (1957).

157. Claeys, A., *Anal. Chim. Acta*, **19**, 114 (1958).

158. Claeys, A., *Anal. Chim. Acta*, **24**, 493 (1961).

159. Clark, R. E. D., and R. G. Neville, *J. Chem. Educ.*, **36**, 390 (1959).

160. Clark, R. E. D., and C. E. Tamale-Ssali, *Analyst (London)*, **84**, 16 (1959).

161. Clarke, S. G., *Analyst (London)*, **52**, 527 (1927).

162. Clerg, M. de, and C. Duval, *Anal. Chim. Acta*, **5**, 40 (1951).

163. Connor, J. A., and E. A. V. Ebsworth, "Peroxy Compounds of Transition Metals," in H. J. Emeleus and A. G. Sharpe, Eds., *Advances in Inorganic Chemistry and Radiochemistry*, Vol. 6, Academic, New York, 1964.

164. Cornu, A., R. Bourguillot, and R. Stefani, *Energ. Nucl.*, **9**, 386 (1967).

165. Corth, R., *Anal. Chem.*, **34**, 1607 (1962).

166. Cotton, F. A., and G. Wilkinson, *Advanced Inorganic Chemistry*, 3rd edit., Wiley-Interscience, New York, 1972.

167. Cotton, T. M., and A. A. Woolf, *Anal. Chem.*, **36**, 248 (1964).

168. Cremer, E., and B. Fetkenheuer, *Wiss. Veroffentlich. Siemens-Komzern*, **5**, 199 (1927); through *Chem. Abstr.*, **22**, 1744[6] (1928).

169. Crisan, I. A., and D. C. Pricop, *Stud. Univ. Babes-Bolayi, Ser. Chem.*, **13,** 27 (1968); through *Chem. Abstr.*, **71,** 18586w (1968).

170. Crossland, B., and T. R. F. W. Fennell, *Analyst (London)*, **94,** 989 (1969).

171. Crouthamel, C. E., and C. E. Johnson, *Anal. Chem.*, **26,** 1284 (1954).

172. Crouzoulon, P., *Chim. Anal. (Paris)*, **50,** 483 (1968).

173. Csanyi, L., and J. Batyai, *Magy. Kem. Foly.*, **69,** 103 (1963).

174. Curry, R. H., and J. R. Carter, *Anal. Chem.*, **36,** 926 (1964).

175. Dams, R., and J. Hoste, *Talanta*, **8,** 664 (1961).

176. Dams, R., J. A. Robbins, K. A. Rahn, and J. W. Winchester, *Anal. Chem.*, **42,** 861 (1970).

177. Dantin Gallego, J., and M. Lopez de Azcona, *Int. Congr. Occupational Health, 14th, Madrid, 1963,* 504 (1964).

178. Day, G. T., *Can. Spectrosc.*, **9,** 118 (1964).

179. De, A. K., and M. S. Rahaman, *Anal. Chem.*, **36,** 685 (1964).

180. De, A. K., and M. S. Rahaman, *Talanta*, **11,** 601 (1964).

181. Dedman, A. J., T. J. Lewis, and D. H. Richards, *J. Chem. Soc.*, **1963,** 5020.

182. Degterev, N. M., *Zavod. Lab.*, **22,** 167 (1956).

183. Degtyarenko, A. P., R. I. Libina, and A. D. Miller, *Gikrokhim. Mater.*, **29,** 264 (1959); through *Chem. Abstr.*, **54,** 20024e (1960).

184. Degtyareva, O. F., and L. G. Sinitsyna, *Zh. Anal. Khim.*, **20,** 603 (1965).

185. Degtyareva, O. F., and L. G. Sinitsyna, *Zh. Anal. Khim.*, **22,** 1500 (1967).

186. Degtyareva, O. F., L. G. Sinitsyna, and A. E. Prokuryakova, *Zh. Anal. Khim.*, **17,** 926 (1962).

187. De Leo, E., *Metall. Ital.*, **58,** 299 (1966).

188. Demkin, A. M., and S. I. Sinyakova, *Zavod. Lab.*, **35,** 773 (1969).

189. Deshmukh, G. S., and J. P. Srivastava, *Zh. Anal. Khim.*, **15,** 601 (1960).

190. Diaz-Guerra, J. P., *An. Quim.*, **67,** 135 (1971).

191. Dickens, P. G., and M. S. Whittingham, *Quart. Rev. (London)*, **22,** 30 (1968).

192. Dixon, E. J., and J. B. Headridge, *Analyst (London)*, **89,** 185 (1964).

193. Dobkina, B. M., and G. B. Sazikova, *Zavod. Lab.*, **34,** 32 (1968).

194. Dobson, G. R., *Organometel. Chem. Rev., Sect. B*, **6,** 1035 (1970).

195. Doege, H. G., *Anal. Chim. Acta*, **38,** 207 (1967).

196. Doege, H. G., and H. Gross-Ruyken, *Mikrochim. Acta*, **1967,** 98 (1967).

197. Doege, H. G., and H. Grosse-Ruyken, *Isotopenpraxis*, **4,** 262 (1968); through *Chem. Abstr.*, **70,** 111383s (1969).

198. Donaldson, E. M., *Talanta*, **18,** 905 (1971).

199. Donaldson, E. M., D. J. Charette, and V. H. E. Rolko, *Talanta*, **16,** 1305 (1969).

200. Donaldson, E. M., and W. R. Inman, *Talanta*, **13,** 489 (1966).

201. Donati, A., *Ann. Chim. Appl.*, **17,** 14 (1927); through *Chem. Abstr.*, **21,** 1421[9] (1927).

202. Dontsov, G. I., *Sov. Med.*, **29,** 14 (1966); through *Chem. Abstr.*, **66,** 53816g (1967).

203. Dontsov, G. I., *Vop Pitan.*, **25,** 66 (1966); through *Chem. Abstr.*, **65,** 7892h (1966).

204. D'Silva, A. P., R. N. Kniseley, and V. A. Fassel, *Anal. Chem.*, **36**, 1287 (1964).

205. Dubov, R. I., and E. M. Kvyatkovskii, *Izv. Vyssh. Ucheb. Zaved., Geol. Razuedka*, No. 2, 102 (1959); through *Chem. Abstr.*, **53**, 21421h (1959).

206. Duval, C., *Anal. Chim. Acta*, **5**, 401 (1951).

207. Easton, A. J., and A. A. Moss, *Mineral Mag.*, **35**, 995 (1966).

208. Eegriwe, E., *Z. Anal. Chem.*, **70**, 400 (1927).

209. Ejaz, M., *Anal. Chim. Acta*, **71**, 383 (1974).

210. Elinson, S. V., L. I. Pobedina, and A. T. Rezova, *Zh. Anal. Khim.*, **20**, 676 (1965).

211. El-Shamy, H. K., and M. F. Barakat, *Egypt. J. Chem.*, **2**, 191 (1959).

212. El Wakkad, S. E. S., and H. A. M. Rize, *Analyst (London)*, **77**, 161 (1952).

213. El Wakkad, S. E. S., H. A. Rizk, and I. G. Ebaid, *J. Phys. Chem.*, **59**, 1004 (1955).

214. Elwell, W. T., and D. F. Wood, *Analysis of the New Metals*, Pergamon, Oxford, 1966.

215. Elwell, W. T., and D. F. Wood, *Analytical Chemistry of Molybdenum and Tungsten*, Pergamon, Oxford, 1971.

216. Enomoto, S., Commission Energic At. (French), Rapp. CEA-R-3369, 1968.

217. Enomoto, S., T. Furuta, and C. Mori, *Nagoya Kogyo Gijutsu Shikensho Hokoku*, **12**, 263 (1963); through *Chem. Abstr.*, **61**, 13e (1964).

218. Erdey, L., I. Buzas, and K. Vigh, *Talanta*, **14**, 515 (1967).

219. Eristavi, D. I., V. D. Eristavi, and A. G. Daneliya, *Zh. Anal. Khim.*, **26**, 183 (1971).

220. Espanol, C. E., and A. M. Marafuschi, *J. Chromatogr.*, **29**, 311 (1967).

221. Evans, B. S., and F. W. Box, *Analyst (London)*, **68**, 203 (1943).

222. Evans, B. S., and D. H. Higgs, *Analyst (London)*, **70**, 75 (1945).

223. Evans, H. T., "Heteropoly and Isopoly Complexes of the Transition Elements of Groups 5 and 6," in J. D. Dunitz and J. A. Ibers, Eds., *Perspectives in Structural Chemistry*, Vol. 4, Wiley, New York, 1971.

224. Eychenne, M., P. Bayle, D. Blanc, J. Leverlochere, and J. LeStrat, *Chim. Anal. (Paris)*, **49**, 355 (1967).

225. Fagel, J. E., H. A. Liebhafsky, and P. D. Zemany, *Anal. Chem.*, **30**, 1918 (1958).

226. Faleev, P. V., *Zavod. Lab.*, **8**, 1174 (1939).

227. Fassel, V. A., and D. W. Golightly, *Anal. Chem.*, **39**, 466 (1967).

228. Fassel, V. A., R. B. Myers, and R. N. Kniseley, *Spectrochim. Acta*, **19**, 1187 (1963).

229. Featheringham, J. A., C. F. Lentz, and R. M. Jacobs, US Atomic Energy Commission, WAPD-CTA(GLA)-631-5 (1958).

230. Fedoroff, M., *Ann. Chim. (Paris)*, **6**, 159 (1971).

231. Fedorov, A. A., *Sb. Tr. Tsentr. Nauchn.-Issled. Inst. Chern. Met.*, No. 37, 33 (1964); through *Chem. Abstr.*, **62**, 15424c (1965).

232. Fedorova, N. L., V. I. Kurbatova, V. L. Zolotavin, M. I. Isupova, and V. K. Zharikova, *Tr. Vses. Nauchn.-Issled. Inst. Standartn. Obraztosov Spektr. Etalonov*, **3**, 82 (1967); through *Chem. Abstr.*, **70**, 16781n (1969).

233. Feigl, F., *Qualitative Analysis by Spot Tests*, Elsevier, New York, 1946.

234. Feigl, F., and C. Costa Neto, *Monatsh. Chem.*, **86**, 336 (1955).

235. Feigl, F., and P. Krumholz, *Angew. Chem.*, **45**, 674 (1932).

236. Ferraro, T. A., *Talanta*, **15**, 923 (1968).

237. Ferraro, T. A., *Talanta*, **16**, 669 (1969).

238. Fer'yanchich, F. A., *Zavod. Lab.*, **3**, 301 (1934).

239. Fer'yanchich, F. A., *Zavod. Lab.*, **12**, 630 (1946).

240. Fer'yanchich, F. A., *Zavod. Lab.*, **13**, 668 (1947).

241. Fer'yanchich, F. A., and D. N. Iordanskii, *Zavod. Lab.*, **7**, 866 (1938).

242. Fettweis, F., *Stahl Eisen*, **45**, 1109 (1925).

243. Feys, R., J. Devynck, and B. Tremillon, *Talanta*, **22**, 17 (1975).

244. Fidler, J., *Collect. Czech. Chem. Commun.*, **14**, 645 (1949).

245. Finkel'shtein, D. N., *Zavod. Lab.*, **22**, 911 (1956).

246. Fischer, P., R. Spiers, and P. Lisan, *Ind. Eng. Chem. Anal. Ed.*, **16**, 607 (1944).

247. Fischer, W., and R. Mehlhorn, *Reinstoffe Wiss. Tech. Intern. Symp., 1, Dresden*, **1961**, 525 (1961); through *Chem. Abstr.*, **60**, 15132f (1964).

248. Flatt, R., and F. Sommer, *Helv. Chim. Acta*, **27**, 1518 (1944).

249. Fleck, H. R., *Analyst (London)*, **62**, 378 (1937).

250. Fodor, J., and C. Varga, *Proc. UN Intern.-Conf. Peaceful Uses Atomic Energy, 2nd Geneva*, **19**, 215 (1959).

251. Fogg, A. G., T. J. Jarvis, D. R. Marriott, and D. T. Burns, *Analyst (London)*, **96**, 475 (1971).

252. Fogg, A. G., D. R. Marriott, and D. T. Burns, *Analyst (London)*, **95**, 848 (1970).

253. Fogg, A. G., D. R. Marriott, and D. T. Burns, *Analyst (London)*, **95**, 854 (1970).

254. Forsythe, W. E., and E. Q. Adams, *J. Opt. Soc. Amer.*, **35**, 108 (1945).

255. Foster, A. R., *Precision Metal Molding*, **20**, 39 (1962).

256. Fratkin, Z. G., *Zavod. Lab.*, **30**, 170 (1964).

257. Fratkin, Z. G., and V. S. Shebunin, *Tr. Komis. Anal. Khim. Akad. Nauk SSSR, Geokhim. Anal. Khim.*, **15**, 127 (1965); through *Chem. Abstr.*, **64**, 2731a (1966).

258. Freedman, M. L., *Anal. Chem.*, **34**, 865 (1962).

259. Freund, H., M. L. Wright, and R. K. Brookshier, *Anal. Chem.*, **23**, 781 (1951).

260. Friedrich, K., J. Barthel, and J. Kunze, *J. Less Common Metals*, **14**, 55 (1968).

261. Fritz, J. S., and L. H. Dahner, *Anal. Chem.*, **40**, 20 (1968).

262. Fritz, J. S., and J. J. Topping, *Talanta*, **18**, 865 (1971).

263. Fujieda, S., *Bussei, 10,* 486 (1969); through *Chem. Abstr.*, **72**, 8747f (1970).

264. Fukai, R., and W. W. Meinke, *Limnol. Oceanogr.*, **7**, 186 (1962).

265. Furey, J. J., and T. R. Cunningham, *Anal. Chem.*, **20**, 563 (1948).

266. Gabler, R. C., and M. J. Peterson, US Bureau of Mines Report on Investigations, No. 6632 (1965).

267. Gado, P., *Magy. Kem. Foly.*, **67**, 189 (1961).

268. Gahler, A. R., in N. H. Furman, Ed., *Standard Methods of Chemical Analysis*, Vol. 1, Van Nostrand, Princeton, N.J., 1962.

269. Gaibakyan, D. S., *Arm. Khim. Zh.*, **22**, 13 (1969).

270. Gaibakyan, D. S., *Arm. Khim. Zh.*, **22**, 219 (1969).

271. Gallai, Z. A., and T. Ya. Rubinskaya, *Zh. Anal. Khim.*, **21**, 961 (1966).

272. Ganguly, P., and D. N. Misra, *J. Mol. Biol.*, **12**, 385 (1965).

273. Ganiev, A. G., P. Kh. Nishanov, and D. V. Karimkulov, *Aktiv. Anal. Elem. Sostava Geol. Ob'ektov,* **1967**, 134; through *Chem. Abstr.*, **69**, 24282u (1968).

274. Ganiev, A. G., M. K. Nazmitdinov, and S. Sabirov, *Izv. Akad. Nauk Uzb SSR, Ser. Fiz.-Mat. Nauk,* **12**, 40 (1968); through *Chem. Abstr.*, **69**, 15710e (1968).

275. Garner, C. D., in B. F. G. Johnson, senior reporter, *Inorganic Chemistry of the Transition Elements,* Vol. 1, The Chemical Society, London, 1972.

276. Gartner, K., L. Ebert, and E. Taubig, *Acta Chem. Acad. Sci. Hung.,* **7**, 215 (1961).

277. Gavrilko, Yu. M., P. N. Kovalenko, and K. N. Bagdasarov, *Ukr. Khim. Zh.,* **32**, 514 (1966).

278. Gavrilov, F. F., I. A. Voronezhskaya, and M. I. Fedorovskaya, *Tr. Ural. Politekh. Inst.,* No. 121, 95 (1962); through *Chem. Abstr.*, **59**, 8117e (1963).

279. Gazzi, V., *Gazz Chim. Ital.,* **64**, 102 (1934); through *Chem. Abstr.*, **28**, 5004[7] (1934).

280. Gebauhr, W., *Z. Anal. Chem.,* **197**, 212 (1963).

281. Geld, I., and J. Carroll, *Anal. Chem.,* **21**, 1098 (1949).

282. Gelfayan, G. T., *Izv. Akad. Nauk Armyan SSR, Fiz.-Mat., Estestven. Tekh. Nauk,* **3**, 523 (1950); through *Chem. Abstr.*, **46**, 2951d (1952).

283. Gentry, C. H. R., and G. P. Mitchell, *Metallurgia,* **46**, 47 (1952).

284. Gentry, C. H. R., and L. G. Sherrington, *Analyst (London),* **71**, 432 (1946).

285. Gentry, C. H. R., and L. G. Sherrington, *Analyst (London),* **73**, 57 (1948).

286. Gerber, W. O., *Appl. Spectrosc.,* **6**, 5 (1952).

287. Geyer, R., and K. Friedrich, *Z. Anal. Chem.,* **213**, 259 (1965).

288. Geyer, R., and G. Henze, *Z. Anal. Chem.,* **177**, 185 (1960).

289. Geyer, R., and G. Henze, *Wiss. Z. Tech. Hochsch. Chem. Leuna-Merseburg,* **3**, 99 (1960/1961); through *Chem. Abstr.*, **56**, 3110i (1962).

290. Geyer, R., G. Henze, J. Henze and E. G. Mueller, *Proc. Conf. Appl. Phys.-Chem. Methods Chem. Anal., Budapest,* **1**, 211 (1966).

291. Geyer, R., G. Henze and F. Sobeck, *Wiss. Z. Tech. Hochsch. Chem. Leuna-Merseburg,* **6**, 247 (1964); through *Chem. Abstr.*, **63**, 15862c (1965).

292. Geyer, R., W. Herrmann, and B. Bogatzki, *Wiss. Z. Tech. Hochsch. Chem. Leuna-Merseburg,* **9**, 1 (1967); through *Chem. Abstr.*, **67**, 121950b (1967).

293. Geyer, R., and M. Neumann, *Acta Chim. (Budapest),* **60**, 349 (1969).

294. Gil'bert, E. N., and V. A. Pronin, *Izv. Sib. Otd. Akad. Nauk SSSR, Ser. Khim. Nauk,* **1969**, 68; through *Chem. Abstr.*, **71**, 77034n (1969).

295. Gillis, J., *Mikrochemie Ver. Mikrochim. Acta,* **31**, 58 (1943).

296. Glazer, W., *Chem. Anal. (Warsaw),* **3**, 567 (1958).

297. Glemser, O., and W. Holznagel, *Angew. Chem.,* **72**, 918 (1960).

298. Glemser, O., J. Weidelt, and F. Freund, *Z. Anorg. Allg. Chem.,* **332**, 299 (1964).

299. Gleu, K., and R. Schwab, *Chem.-Ztg.,* **74**, 301 (1950).

300. Godovannaya, I. N., T. N. Nazarchuk, and O. I. Popova, *Zh. Anal. Khim.,* **21**, 1020 (1966).

301. Gokhale, Y. W., and T. R. Bhat., *Talanta,* **14**, 435 (1967).

302. Goldstein, G., C. M. Wolff, and J. P. Schwing, *Bull. Soc. Chim. Fr.,* **1971**, 1201.

303. Golubeva, I. A., *Zh. Anal. Khim.,* **24**, 889 (1969).

304. Golubeva, I. A., *Zh. Anal. Khim.*, **25**, 2433 (1970).

305. Golubtsova, R. B., *Zh. Anal. Khim.*, **6**, 357 (1951).

306. Golubtsova, Z. G., L. I. Lebedeva, and T. P. Likhareva, *Vestn. Leningrad. Univ.*, *Ser. Fiz. Khim.*, **23**, 129 (1968).

307. Golyakova, L. P., *Gig. Tr. Prof. Zabol.*, **15**, 4 (1971); through *Chem. Abstr.*, **75**, 24867n (1971).

308. Gonzales, R., E. Muratori, P. Frere, and R. Durand, *Mem. Sci. Rev. Met.*, **64**, 403 (1967); through *Chem. Abstr.*, **68**, 18301n (1968).

309. Gorbenko, F. P., V. S. Smirnaya, E. I. Vel'shtein, and E. S. Golyakova, *Tr. Uses. Nauchn.-Issled. Inst. Khim. Reaktivov Osobo Chist. Khim. Veshchestv*, No. 31, 264 (1969); through *Chem. Abstr.*, **75**, 136747e (1971).

310. Gorbenko, F. P., E. U. Volodko, and E. M. Nemirovskaya, *Zavod. Lab.*, **36**, 277 (1970).

311. Gordon, B. E., and R. M. Tanklevskaya, *Ukr. Khim. Zh.*, **29**, 1310 (1963).

312. Gorski, L., and A. Lubecki, *Chem. Anal. (Warsaw)*, **10**, 191 (1965).

313. Goto, H., *J. Chem. Soc. Jap.*, **60**, 937 (1939).

314. Goto, H., H. Hirokawa, and M. Suzuki, *Z. Anal. Chem.*, **225**, 130 (1967).

315. Goto, H., and S. Ikeda, *J. Chem. Soc. Jap.*, **73**, 654 (1952).

316. Goto, H., Y. Kakita, and M. Namiki, *Nippon Kagaku Zasshi*, **82**, 580 (1961).

317. Goto, H., Y. Kakita, and M. Sase, *Nippon Kinzoku Gakkaishi*, **21**, 385 (1957); through *Chem. Abstr.*, **56**, 9397f (1962).

318. Gottschalk, G., *Z. Anal. Chem.*, **187**, 164 (1962).

319. Gottschalk, G., *Tech.-Wiss. Abh. Osram-Ges.* **8**, 201 (1963); through *Chem. Abstr.*, **61**, 4956d (1964).

320. Gottschalk, G., *Talanta*, **14**, 61 (1967).

321. Gottschalk, G., *Tech.-Wiss. Abh. Osram-Ges.*, **9**, 251 (1967); through *Chem. Abstr.*, **69**, 113252b (1968).

322. Gran, G., *Svensk Papperstidn*, **54**, 764 (1951); through *Chem. Abstr.*, **46**, 6831d (1952).

323. Green, T. E., NASA Accession Report, No. N63-19,800 (1963).

324. Green, T. E., *Anal. Chem.*, **37**, 1595 (1965).

325. Greenberg, P., *Anal. Chem.*, **29**, 896 (1957).

326. Grenier, J. W., and L. Meites, *Anal. Chim. Acta*, **14**, 482 (1956).

327. Griffith, W. P., *J. Chem. Soc.*, **1963**, 5345 (1963).

328. Griffith, W. P., and P. J. B. Lesniak, *J. Chem. Soc. A*, **1969**, 1066 (1969).

329. Grimaldi, F. S., and N. Davidson, *US Geol. Survey Bull.*, No. 950, 135 (1946).

330. Grimaldi, F. S., and V. North, *Ind. Eng. Chem. Anal. Ed.*, **15**, 652 (1943).

331. Grubitsch, H., N. Ozbil, and K. Kluge, *Z. Anal. Chem.*, **166**, 114 (1959).

332. Guglialmelli, L., *An. Soc. Quim. Argent.*, **6**, 57 (1919); through *Chem. Abstr.*, **13**, 215[4] (1919).

333. Guibault, G. G., and P. J. Brignac, *Anal. Chim. Acta*, **56**, 139 (1971).

334. Guillon, A., and M. Lebrun, Commission Energic At. (French), Rapp., CEA-R-3474 (1968).

335. Gullstrom, D. K., and M. G. Mellon, *Anal. Chem.*, **25**, 1809 (1953).

336. Gupta, C. M., *J. Proc. Inst. Chem. (India)*, **38**, 211 (1966).

337. Gur'ev, S. D., and N. F. Saraeva, *Sb. Nauch. Tr. Gos. Nauchn.-Issled. Inst. Tsvet. Metal.*, No. 18, 37 (1961); through *Chem. Abstr.*, **60**, 2315g (1964).

338. Gurtsiev, O. N., *Sb. Nauch. Tr., Sev.-Oset. Gos. Med. Inst.*, No. 18, 169 (1967); through *Chem. Abstr.*, **70**, 90567n (1969).

339. Gusev, S. I., and V. I. Kumov, *Zh. Anal. Khim.*, **3**, 373 (1948).

340. Guyon, J. C., and J. Y. Marks, *Anal. Chem.*, **40**, 837 (1968).

341. Hadjiioannou, T. P., and C. G. Valkana, *Chim. Chron. A*, **32**, 89 (1967).

342. Hahn, P. B., D. C. Johnson, M. A. Wechter, and A. F. Voigt, *Anal. Chem.*, **46**, 555 (1974).

343. Hahn, P. B., M. A. Wechter, D. C. Johnson, and A. F. Voigt, *Anal. Chem.*, **45**, 1016 (1973).

344. Hahn-Weinheimer, P., and H. Johanning, *Glastech. Ber.*, **36**, 183 (1963).

345. Hakkila, E. A., H. L. Barker, G. R. Waterbury, and C. F. Metz, US Atomic Energy Commission, LA-3157, (1964).

346. Hakkila, E. A., S. N. Deming, and R. G. Hurley, US Atomic Energy Commission, LA-4007, (1968).

347. Hakkila, E. A., and G. R. Waterbury, *Talanta*, **6**, 46 (1960).

348. Hakkila, E. A., G. R. Waterbury, and G. B. Nelson, US Atomic Energy Commission, TID-7629, 55 (1961).

349. Halmekoski, J., *Ann. Acad. Sci. Fenn., Ser. A2*, No. 96 (1959); through *Chem. Abstr.*, **54**, 1151h (1960).

350. Hamaguchi, H., *Nippon Kagaku Zasshi*, **82**, 1493 (1961).

351. Hamaguchi, H., R. Kuroda, T. Shimizu, I. Tsukahara, and R. Yamamoto, *Geochim. Cosmochim. Acta*, **26**, 503 (1962).

352. Hanna, Z. G., *Monatsh. Chem.*, **94**, 950 (1963).

353. Harrison, G. R., *Massachusetts Institute of Technology Wavelength Tables*, MIT, Cambridge, Mass., 1969.

354. Hay, D. R., R. K. Skogerboe, and E. Scala, *J. Less Common Metals*, **15**, 121 (1968).

355. Headridge, J. B., and M. S. Taylor, *Analyst (London)*, **88**, 590 (1963).

356. Hegedus, A. J., and M. Dvorszky, *Magy. Tudomanyos Akad. Kem. Tudomanyok Osztalyanak Kozlemenyei*, **11**, 405 (1959); through *Chem. Abstr.*, **54**, 4263g (1960).

357. Hegedus, A. J., and M. Dvorszky, *Mikrochim. Acta*, **1961**, 169 (1961).

358. Hegedus, A. J., J. Neugebauer, and M. Dvorszky, *Magy. Kem. Foly.*, **65**, 159 (1959).

359. Herold, C., *Reinstaffe Wiss. Tech. Intern. Symp. I. Dresden*, **1961**, 465; through *Chem. Abstr.*, **61**, 6380h (1964).

360. Herve, G., and P. Souchay, *C.R.H. Acad. Sci., Ser. C*, **265**, 805 (1967).

361. Heumann, Th., and N. Stolica, *Electrochim. Acta*, **16**, 643 (1971).

362. Heumann, Th., and N. Stolica, *Electrochim. Acta*, **16**, 1635 (1971).

363. Hillebrand, W. F., G. E. F. Lundell, H. A. Bright, and J. I. Hoffman, *Applied Inorganic Analysis*, Wiley, New York, 1953.

364. Hobart, E. W., and E. P. Hurley, *Anal. Chim. Acta*, **27**, 144 (1962).

365. Hodara, I., and I. Balouka, *Electrochim. Acta*, **15**, 283 (1970).

366. Hodara, I., and A. Glasner, *Electrochim. Acta*, **15**, 923 (1970).

367. Hoekstra, H. R., and J. J. Katz, *Anal. Chem.*, **25**, 1608 (1953).

368. Holden, N. E., and F. W. Walker, *Chart of the Nuclides*, 10th edit., US Atomic Energy Commission, Washington, D.C., 1969.

369. Holt, M. L., and A. G. Gray, *Ind. Eng. Chem. Anal. Ed.*, **12**, 144 (1940).

370. Holt, M. L., and L. Kahlenberg, *Quart. Rev. Amer. Electroplat. Soc.*, **19**, 41 (1933).

371. Hope, R. P., *Proc. Aust. Inst. Min. Metall.*, **1958**, 51 (1958).

372. Hopkins, D. P., *Mfg. Chemist*, **27**, 520 (1956).

373. Horak, J., and A. Okac, *Collect. Czech. Chem. Commun.*, **29**, 188 (1964).

374. Hu, Z. T., and S. C. Shi, *Hua Hsueh Tung Pao*, **1963**, 312; through *Chem. Abstr.*, **60**, 6d (1964).

375. Huang, M. C., and C. L. Chou, *Yao Hsueh Hsueh Pao*, **12**, 301 (1965); through *Chem. Abstr.*, **63**, 13686a (1965).

376. Huaringa Ricci, M., *Nature*, **177**, 85 (1956).

377. Hudec, I., *Jad. Energ.*, **13**, 23 (1967); through *Chem. Abstr.*, **66**, 72191y (1967).

378. Hurd, D. T., Tungsten, in T. Lyman, Ed., *Metals Handbook*, 8th edit., American Society for Metals, Metals Park, Ohio, 1961.

379. Husler, J. W., *At. Absorption Newslett.*, **8**, 1 (1969).

380. Husler, J. W., *At. Absorption Newslett.*, **10**, 60 (1971).

381. Husain, S. A., *Proc. UN Intern. Conf. Peaceful Uses At. Energy, 2nd Geneva*, **19**, 213 (1958).

382. Husain, S. W., *Analysis*, **1**, 314 (1972).

383. Huseya, M., *Bunsekl Kagaku*, **13**, 122 (1964).

384. Iguchi, A., *Sci. Papers Coll. Educ., Univ. Tokyo*, **6**, 49 (1956); through *Chem. Abstr.*, **51**, 2355d (1957).

385. Iguchi, A., *Sci. Papers Coll. Educ., Univ. Tokyo*, **6**, 153 (1956); through *Chem. Abstr.*, **51**, 9255e (1957).

386. Ikenberry, L., J. L. Martin, and W. J. Boyer, *Anal. Chem.*, **25**, 1340 (1953).

387. Il'ina, L. I., *Fiz.-Khim. Metody Anal. Metal. Splavov*, **1969**, 120; through *Chem. Abstr.*, **74**, 119694x (1971).

388. Iofa, Z. A., and B. I. Petrov, *Zavod. Lab.*, **3**, 728 (1934).

389. Isaeva, A. B., *Tr. Inst. Okeanol., Akad. Nauk SSSR*, **47**, 159 (1961); through *Chem. Abstr.*, **56**, 10908f (1962).

390. Ishibashi, M., *Rec. Oceanog. Works Japan* [N.S.], No. 1, 88 (1953); through *Chem. Abstr.*, **48**, 6175b (1954).

391. Ishibashi, M., T. Fujinaga, T. Kuwamoto, M. Koyama, and S. Sugibayashi, *Nippon Kagaku Zasshi*, **81**, 392 (1960).

392. Ishibashi, M., T. Shigematsu, and Y. Nakagawa, *Bull. Inst. Chem. Res. Kyoto Univ.*, **32**, 199 (1954).

393. Ishibashi, M., and H. Kohara, *Bunseki Kagaku*, **13**, 239 (1964).

394. Ishida, K., and R. Kuroda, *Anal. Chem.*, **39**, 212 (1967).

395. Ishii, H., and H. Einaga, *Nippon Kagaku Zasshi*, **88**, 183 (1967).

396. Ishimaru, S., *J. Chem. Soc. Japan,* **55,** 201 (1934).

397. Issa, R. M., B. A. Abd-el-Nabey and A. M. Hindawey, *Z. Anal. Chem.,* **240,** 9 (1968).

398. Ito, M., and S. Musha, *Bunseki Kagaku,* **12,** 439 (1963).

399. Ivanova, G. F., *Zh. Anal. Khim.,* **20,** 82 (1965).

400. Ivanova, G. F., *Geokhimiya,* **1966,** 1052 (1966).

401. Ivanova, G. F., in V. F. Barabanov, Ed., *Mineral Geokhim. Vol'framitovykh Mestorozhd, Tr. Vses. Soveshch., 1st,* Leningrad University Izd., Leningrad, 85 (1967); through *Chem. Abstr.,* **70,** 89594n (1969).

402. Ivanova, L. B., in V. F. Barabanov, Ed., *Mineral Geokhim. Vol'framitovykh Mestorozhd, Tr. Vses. Soveshch., 1st,* Leningrad University Izd., Leningrad, 235 (1967); through *Chem. Abstr.,* **70,** 120817p (1969).

403. Izyumova, L. G., *Spektral, Anal. Geol. Geokhim., Mater. Sib. Soveshch. Spektrosk., 2nd [Itkutsk], USSR,* **1963,** 175; through *Chem. Abstr.,* **68,** 119203n (1968).

404. Jacobsohn, K., and M. Deodata de Azevedo, *Rev. Port. Quim.,* **5,** 99 (1963).

405. Jacobsohn, K., and M. Deodata de Azevedo, *Z. Anal. Chem.,* **202,** 417 (1964).

406. Jahr, K. F., and J. Fuchs, *Angew. Chem.,* **78,** 725 (1966).

407. Janczyszyn, J., L. Loska, and L. Gorski, *Radiochem. Radioanal. Lett.,* **8,** 363 (1971).

408. Jander, G., and D. Mojert, *Z. Anorg. Allgem. Chem.,* **175,** 270 (1928).

409. Jean, M., *Ann. Chim. (Paris),* **3,** 470 (1948).

410. Jean, M., *Anal. Chim. Acta,* **42,** 545 (1968).

411. Jeffery, P. G., *Rec. Geol. Survey Uganda,* **1953,** 75 (1955); through *Chem. Abstr.,* **49,** 10121g (1955).

412. Jeffery, P. G., *Analyst (London),* **81,** 104 (1956).

413. Jeffery, P. G., *Analyst (London),* **82,** 588 (1957).

414. Jensovsky, L., *Chem. Tech. (Berlin),* **7,** 159 (1955).

415. Jervis, R. E., and W. D. Mackintosh, *Proc. UN Intern. Conf. Peaceful Uses At Energy, 2nd Geneva,* **28,** 470 (1958).

416. Jilek, A., and A. Rysanek, *Collect. Czech. Chem. Commun.,* **8,** 246 (1936).

417. Jirkovsky, B., *Radiochim. Conf. Abstr. Pap., Bratislava,* **1966,** 51; through *Chem. Abstr.,* **68,** 56206t (1968).

418. Jirkovsky, R., *Chem. Listy,* **42,** 100 (1948).

419. Johansen, O., and E. Steinnes, *Talanta,* **17,** 407 (1970).

420. Johri, K. N., and H. C. Mehra, *Mikrochim. Acta,* **1971,** 317 (1971).

421. Jones, A. L., and J. H. Yoe, *Virginia J. Sci.,* **3,** 301 (1943).

422. Jones, W. F., *Mikrochim. Acta,* **1967,** 1019 (1967).

423. Juvet, R. S., and R. L. Fisher, *Anal. Chem.,* **38,** 1860 (1966).

424. Kabanov, B. N., and L. Ya. Polyak, *Zh. Anal. Khim.,* **11,** 678 (1956).

425. Kabanov, V. Ya., and V. F. Chuvaev, *Zh. Fiz. Khim.,* **38,** 1317 (1964).

426. Kabir-ud-Din, A. A. Khan, and M. A. Beg, *Experientia,* **24,** 1093 (1968).

427. Kabir-ud-Din, A. A. Khan, and M. A. Beg, *J. Electroanal. Chem.,* **20,** 239 (1969).

428. Kaimal, V. R. M., and S. C. Shome, *Anal. Chim. Acta,* **31,** 268 (1964).

429. Kajiyama, R., K. Ichihashi, and K. Ichikawa, *Bunseki Kagaku*, **18**, 1500 (1969).

430. Kalinin, A. I., R. A. Kuznetsov, and V. V. Moiseev, *Radiokhimiya*, **4**, 578 (1962).

431. Kallmann, S., E. W. Hobart, and H. K. Oberthin, *Anal. Chim. Acta*, **41**, 29 (1968).

432. Kallmann, S., E. W. Hobart, and H. K. Oberthin, *Talanta*, **15**, 982 (1968).

433. Kallmann, S., R. Liu, H. K. Oberthin, R. Stevenson, and F. Lux, US Department of Commerce, AD 622275 (1965).

434. Kammori, O., K. Sato, and F. Kurosawa, *Bunseki Kagaku*, **17**, 1270 (1968).

435. Kanchev, V. K., *J. Russ. Phys. Chem. Soc.*, **46**, 729 (1914); through *Chem. Abstr.*, **9**, 1881[7] (1915).

436. Kanter, I. A., *Tsvet. Metal.*, **41**, 24 (1968).

437. Kantor, T., E. Kocsis, and M. T. Vandorffy, *Acta Chim. Acad. Sci. Hung.*, **48**, 209 (1965).

438. Kaplan, B. Ya., and I. A. Sorokovskaya, *Zavod. Lab.*, **29**, 391 (1963).

439. Karantassis, T., *Bull. Sci. Pharmacol.*, **31**, 561 (1924).

440. Kasai, Y., W. Yano, and W. Kimura, *Kogyo Kagaku Zasshi*, **74**, 668 (1971); through *Chem. Abstr.*, **75**, 22946p (1971).

441. Kauffman, G. B., and P. F. Vartanian, *J. Chem. Educ.*, **47**, 212 (1970).

442. Kaufman, D., and S. K. Derderian, *Anal. Chem.*, **21**, 613 (1949).

443. Kawabuchi, K., *Bunseki Kagaku*, **14**, 52 (1965).

444. Kawabuchi, K., and R. Kuroda, *Anal. Chim. Acta*, **46**, 23 (1969).

445. Kawabuchi, K., and R. Kuroda, *Talanta*, **17**, 67 (1970).

446. Kawahata, M., H. Mochizuki, and R. Kajuyama, *Bunseki Kagaku*, **8**, 25 (1959).

447. Kawahata, M., H. Mochizuki, R. Kajiyama, and K. Ichihashi, *Bunseki Kagaku*, **12**, 659 (1963).

448. Kawahata, M., H. Mochizaki, and T. Misaki, *Bunseki Kagaku*, **11**, 188 (1962).

449. Kawamura, K., *Bunseki Kagaku*, **2**, 417 (1953).

450. Kawashima, I., *Bunko Kenkyu*, **17**, 22 (1968); through *Chem. Abstr.*, **70**, 43760k (1969).

451. Kawashima, I., and K. Tokiwa, *Nippon Kinzoku Gakkaishi*, **29**, 1201 (1965); through *Chem. Abstr.*, **67**, 70260s (1967).

452. Kaye, S. V., *Health Phys.*, **15**, 399 (1968).

453. Keller, E., and L. Michael, *At. Absorption Newslett.*, **9**, 92 (1970).

454. Kepert, D. L., "Isopolytungstates," in F. A. Cotton, Ed., *Progress in Inorganic Chemistry*, Vol. 4, Wiley-Interscience, New York, 1962.

455. Kergoat, R., and J. E. Guerchais, *C.R.H. Acad. Sci., Ser. C*, **268**, 2304 (1969).

456. Kethelyi, J., *Chem. Anal. (Warsaw)*, **7**, 135 (1962).

457. Khachaturyan, O. B., L. V. Myshlyaeva, I. V. Sedova, and G. C. Mikulenok, *Zh. Prikl. Khim. (Leningrad)*, **43**, 1617 (1970).

458. Khadeeva, L. A., and Sh. T. Talipov, *Nauch. Tr. Tashkent Gos. Univ.*, No. 264, 119 (1964); through *Chem. Abstr.*, **65**, 14419h (1966).

459. Kharlamov, I. P., N. P. Krivenkova, and V. M. Bry'ko, *Zavod. Lab.*, **36**, 1428 (1970).

460. Kharlamov, I. P., P. Ya. Yakovlev, and M. I. Lykova, *Zavod. Lab.*, **26**, 786 (1960).

461. Khlystova, A. D., *Zh. Anal. Khim.*, **23,** 211 (1968).

462. Khlystova, A. D., and E. T. Kabesheva, *Vestn. Mosk. Univ., Khim.*, **12,** 98 (1971).

463. Khlystova, A. D., and N. I. Tarasevich, *Zavod. Lab.*, **34,** 1327 (1968).

464. Khlystova, A. D., and N. I. Tarasevich, *Zh. Anal. Khim.*, **25,** 515 (1970).

465. Khukhiya, V. L., and T. G. Matsaberidze, *Sb. Nauchn.-Tekh. Inform. Minister-stva Geol. Okhrany Nedr,* No. 1, 126 (1955); through *Chem. Abstr.,* **53,** 4005h (1959).

466. Kiboku, M., *Bunseki Kagaku,* **6,** 356 (1957).

467. Kim, T. K., R. W. Mooney, and V. Chiola, *Separ. Sci.,* **3,** 467 (1968).

468. Kipling, M.D., *Int. Congr. Occupational Health, 14th, Madrid,* **1963,** 680 (1964).

469. Kiriyama, S., and S. Nishida, *Sumitomo Metals,* **4,** 310 (1952); through *Chem. Abstr.,* **48,** 6909c (1954).

470. Kirkbright, G. F., T. S. West, and C. Woodward, *Talanta,* **13,** 1637 (1966).

471. Kissa, E., and M. Seepere-Yllo, *Mikrochim. Acta,* **1967,** 287 (1967).

472. Klitina, V. I., *Opred. Mikroprimesei,* No. 2, 21 (1968); through *Chem. Abstr.,* **71,** 119259x (1969).

473. Klofutar, C., F. Krasovec, and A. Kodre, *J. Radioanal. Chem.,* **5,** 3 (1970).

474. Klyachko, Yu. A., T. A. Izmanov, and E. M. Chistyakova, *Sb. Tr. Tsentr. Nauchn.-Issled. Inst. Chern. Met.,* **31,** 133 (1963); through *Chem. Abstr.,* **59,** 5768e (1963).

475. Klyachko, Yu. A., T. A. Izmanova, and E. M. Chistyakova, *Zavod. Lab.,* **29,** 1425 (1963).

476. Knight, D. M., and M. K. Pyzyna, *At. Absorption Newslett.,* **8,** 129 (1969).

477. Koerner, W. E., *Met. Chem. Eng.,* **15,** 522 (1916).

478. Kogan, F. I., P. N. Kovalenko, and Z. I. Ivanova, *Zh. Anal. Khim.,* **20,** 329 (1965).

479. Kokorin, A. I., *Nauch. Dokl. Vyssh. Shkoly, Khim. Khim. Tekhnol.,* No. 3, 475 (1958); through *Chem. Abstr.,* **53,** 2915a (1959).

480. Kobsharov, A. G., and V. F. Ust-Kachkintsev, *Uch. Zap. Perm. Gos. Univ.,* No. 111, 63 (1964); through *Chem. Abstr.,* **64,** 1617f (1966).

481. Kolthoff, I. M., and I. Hodara, *J. Electroanal. Chem.,* **5,** 165 (1963).

482. Kolthoff, I. M., and J. J. Lingane, *Polarography,* Vol. II, Wiley-Interscience, New York, 1952.

483. Kolthoff, I. M., and E. R. Nightingale, *Anal. Chim. Acta,* **19,** 593 (1958).

484. Kolthoff, I. M., and E. P. Perry, *J. Amer. Chem. Soc.,* **73,** 5315 (1951).

485. Kolthoff, I. M., E. B. Sandell, E. J. Meehan, and S. Bruckenstein, *Quantitative Chemical Analysis,* Macmillan, London, 1969.

486. Konkin, V. D., and V. I. Zhikhareva, *Sb. Tr. Ukr. Nauchn.-Issled. Inst. Metal.,* No. 8, 337 (1962); through *Chem. Abstr.,* **58,** 11955d (1963).

487. Kopelman, B., and J. S. Smith, "Wolfram and Wolfram Alloys," in E. P. Dukes, Ed., *Kirk-Othmer Encyclopedia of Chemical Technology,* Vol. 22, Wiley-Interscience, New York, 1970.

488. Kopylova, M. M., and A. V. Kharlamova, *Novye Methody Anal. Khim. Sostava Podzemn. Vod.,* **1967,** 30; through *Chem. Abstr.,* **68,** 62531m (1968).

489. Korenman, I. M., *J. Appl. Chem. USSR,* **18,** 571 (1945).

490. Korenman, I. M., and E. I. Levina, *Zh. Anal. Khim.*, **9**, 170 (1954).

491. Korkisch, J., *Modern Methods for the Separation of Rarer Metal Ions*, Pergamon, Oxford, 1969.

492. Korobov, N. N., *Trans. Inst. Chem. Ivanovo, USSR*, No. 2, 33 (1939); through *Chem. Abstr.*, **33**, 9189[1] (1939).

493. Korotkov, V. F., *Sb. Tr. Tsentr. Nauchn.-Issled. Inst. Chern. Met.*, No. 37, 105 (1964); through *Chem. Abstr.*, **63**, 9042h (1965).

494. Korrey, J. S., and P. D. Goulden, *At. Absorption Newslett.*, **14**, 33 (1975).

495. Kosheleva, E. D., and P. F. Lokhov, *Sb. Nauchn.-Tekhn. Tr. Nauchn.-Issled. Inst. Met. Chelyab. Sovnarkhoza*, No. 4, 178 (1961); through *Chem. Abstr.*, **58**, 6188a (1963).

496. Kosolapova, T. Ya., L. N. Kugai, K. D. Modyleyskaya, S. V. Radzikovskaya, and O. G. Seraya, *Tr. Seminar Zharostoikim Mater., Akad. Nauk Ukr. SSR, Inst. Metal. Spets. Splavov, Kiev*, No. 6, 69 (1960); through *Chem. Abstr.*, **56**, 8005i (1962).

497. Kotkowski, S., and A. Lassocinska, *Rocz. Chem.*, **42**, 527 (1968).

498. Kotkowski, S., and J. Matenko, *Pozman. Towarz. Przyjaciol Nauk, Wydzial Mat.-Przyrod Prace Komisji Mat.-Przyrod.*, **10**, 51 (1962); through *Chem. Abstr.*, **61**, 15399h (1964).

499. Kourin, V., A. K. Lavrukhina, and S. S. Rodin, *J. Inorg. Nucl. Chem.*, **21**, 375 (1964).

500. Kovalenko, P. N., and G. P. Protsenko, *Elektrokhim. Optechn. Methody Anal. Sb.*, **1963**, 120; through *Chem. Abstr.*, **61**, 32a (1964).

501. Kozyreva, L. S., and A. F. Kuteinikov, *Knostrukts. Uglegrafit Mat.* (Moscow.-Met.) *Sb.*, **1964**, 325; through *Chem. Abstr.*, **62**, 5889h (1965).

502. Krainer, H., H. M. Ortner, K. Muller, and H. Spitzy, *Talanta*, **21**, 933 (1974).

503. Kral, S., *Hutn. Listy*, **22**, 699 (1967); through *Chem. Abstr.*, **69**, 49015x (1968).

504. Krasil'nikova, L. N., *Sb. Tr. Vses. Nauchn.-Issled. Gorno-Met. Inst. Tsvet. Met.*, No. 5, 20 (1959); through *Chem. Abstr.*, **55**, 26845b (1961).

505. Kraus, K. A., and F. Nelson, *J. Amer. Chem. Soc.*, **77**, 3972 (1955).

506. Krause, A., P. Meteniowski, and L. Wachowski, *Monatsh. Chem.*, **102**, 494 (1971).

507. Krepelka, J., and O. Stieber, *Collect. Czech. Chem. Commun.*, **11**, 540 (1939).

508. Kriege, O. H., US Atomic Energy Commission, LA-2306 (1959).

509. Krishaiah, K. S. R., *Proc. Indian Acad. Sci., Sect. A*, **67**, 222 (1968).

510. Krivy, I., and J. Krtil, *Collect. Czech. Chem. Commun.*, **29**, 587 (1964).

511. Krtil, J., *J. Inorg. Nucl. Chem.*, **27**, 233 (1965).

512. Krtil, J., *J. Inorg. Nucl. Chem.*, **27**, 1862 (1965).

513. Kryukov, S. N., B. S. Bokshtein, T. I. Degal'tseva, and A. A. Zhurkhovitskii, *Zavod. Lab.*, **24**, 1305 (1958).

514. Ku, Y. T., and S. T. Wang, *J. Chin. Chem. Soc. (Taipei)*, **17**, 289 (1950).

515. Kugai, L. N., *Tr. Seminara Zharostoikim Mater. Akad. Nauk Ukr. SSR, Inst. Metal. Spets. Splavov, Kiev*, No. 6, 45 (1960); through *Chem. Abstr.*, **56**, 12300a (1962).

516. Kuleshova, M. I., *Nek. Vop. Lekarstvoved.*, **1959**, 21; through *Chem. Abstr.*, **54**, 21638i (1960).

517. Kulev, I. I., and E. F. Speranskaya, *Zh. Anal. Khim.*, **22**, 1371 (1967).

518. Kunenkova, E. N., *Tr. Inst. Met. A.A. Baikova,* No. 11, 227 (1962); through *Chem. Abstr.*, **59**, 22g (1963).

519. Kunenkova, E. N., and T. Kh. Bobrova, *Probl. Bol'shoi Met. Fiz. Khim. Novykh Splavov, Akad. Nauk SSSR, Inst. Met.*, **1965**, 315; through *Chem. Abstr.*, **63**, 17143g (1965).

520. Kunin, L. L., and I. V. Tulepova, *Sb. Tr. Tsentr. Nauchn.-Issled. Inst. Chern. Met.,* No. 24, 39 (1962); through *Chem. Abstr.*, **58**, 11022e (1963).

521. Kurbatov, D. I., and G. A. Nikitina, *Tr. Inst. Khim., Akad. Nauk SSSR, Ural. Filial*, **10**, 51 (1966); through *Chem. Abstr.*, **66**, 34548n (1967).

522. Kurbatov, D. I., and I. S. Skoryina, *Zh. Anal. Khim.*, **17**, 711 (1962).

523. Kuroda, R., K. Kawabuchi, and T. Ito, *Talanta*, **15**, 1486 (1968).

524. Kuznetsov, V. I., V. N. Obozhin, and E. S. Pal'shin, *Zh. Anal. Khim.*, **10**, 32 (1955).

525. Kuznetsova, A. I., and Ya. D. Raikhbaum, *Zavod. Lab.*, **33**, 1076 (1967).

526. Kuznetsova, E. F., V. L. Marzuvanov, and O. I. Mozhaeva, in M. I., Korsunskii, Ed., *Vop. Obshch. Prikl. Fiz., Tr. Respub. Konf., 1st.,* Izd. "Nauka" Kaz. SSR, Alma-Ata, USSR, **1967**, 221 (1969); through *Chem. Abstr.*, **72**, 38590s (1970).

527. Lacourt, A., *Ind. Chim. Belge,* **20**, 267 (1955).

528. Laitinen, H. A., K. B. Oldham, and W. A. Ziegler, *J. Amer. Chem. Soc.*, **75**, 3048 (1953).

529. Laitinen, H. A., and W. A. Ziegler, *J. Amer. Chem. Soc.*, **75**, 3045 (1953).

530. Lal, S., *Z. Anal. Chem.*, **255**, 210 (1971).

531. Lambie, D. A., *Analyst (London),* **68**, 74 (1943).

532. Lambie, D. A., *Analyst (London),* **70**, 124 (1945).

533. Lange, N. A., Ed., *Handbook of Chemistry,* 10th edit., McGraw-Hill, New York, 1961.

534. Larrabee, R. D., Massachusetts Institute of Technology Research Laboratory of Electronics, Technical Report, No. 328 (1957).

535. Lassner, E., *Mikrochim. Acta,* **1970**, 820 (1970).

536. Lassner, E., and R. Pueschel, *Monatsh. Chem.*, **95**, 812 (1964).

537. Lassner, E., R. Pueschel, and K. Katzengruber, *Mikrochim. Acta,* **1969**, 527 (1969).

538. Lassner, E., R. Pueschel, and H. Schedle, *Talanta*, **12**, 871 (1965).

539. Lassner, E., R. Pueschel, and H. Schedle, *Metall (Berlin)*, **20**, 724 (1966).

540. Lassner, E., and R. Scharf, *Chem.-Anal.*, **49**, 68 (1960).

541. Lassner, E., R. Scharf, and R. Pueschel, *Z. Anal. Chem.*, **165**, 29 (1959).

542. Lassner, E., R. Scharf, and P. L. Reiser, *Z. Anal. Chem.*, **165**, 88 (1959).

543. Lassner, E., and H. Schedle, *Talanta*, **13**, 326 (1966).

544. Lassner, E., and H. Schedle, *Talanta*, **19**, 1670 (1971).

545. Latimer, W. M., *The Oxidation States of the Elements and Their Potentials in Aqueous Solutions,* Prentice-Hall, New York, 1952.

546. Laubie, H., and G. Tempere, *Bull. Soc. Pharm. Bordeaux,* **90**, 15 (1952).

547. Laun, D. D., *J. Res. Nat. Bur. Stand., A,* **68**, 207 (1964).

548. Launay, J. P., *C.R.H. Acad. Sci., Ser. C*, **269,** 941 (1969).

549. Lazarev. A. I., *Zavod. Lab.*, **26,** 935 (1960).

550. Lazarev, A. I., and V. I. Lazareva, *Zavod. Lab.*, **24,** 798 (1958).

551. Lebedeva, L. I., Z. G. Golubtosva, and N. G. Yanklovich, *Zh. Anal. Khim.*, **26,** 1962 (1971).

552. Lederer, M., and C. Majani, *Chromatogr. Rev.*, **12,** 239 (1970).

553. Leliaert, G., J. Hoste, and Z. Eeckhaut, *Talanta*, **2,** 115 (1959).

554. Leliaert, G., J. Hoste, and Z. Eeckhaut, *Rev. Trav. Chim.*, **79,** 557 (1960).

555. Le Meur, B., and F. Chauveau, *Bull. Soc. Chim. Fr.*, **1970,** 3834 (1970).

556. Lenskaya, K. K., O. F. Tikhomirova, V. N. Golubeva, N. N. Sorokina, and L. M. Suchelenkova, *Sb. Tr. Tsentr. Nauchn.-Issled. Inst. Chern. Met.*, No. 49, 48 (1966); through *Chem. Abstr.*, **66,** 25846f (1967).

557. Leroux, J., and M. Mahmud, *Can. J. Spectrosc.*, **13,** 19 (1968).

558. Lesne, J. P., and P. Caillet, *Can. J. Spectrosc.*, **18,** 69 (1973).

559. Leushkina, G. V., E. M. Lobanov, A. G. Dutov, and N. P. Matveeva, *Zavod. Lab.*, **35,** 715 (1969).

560. Levine, S. L., *At. Absorption Newslett.*, **8,** 58 (1969).

561. Li, K. C., and C. Y. Wang, *Tungsten*, Reinhold, New York, 1955.

562. Liang, S. C., and K. N. Chang, *Sci. Record*, **2,** 295 (1949).

563. Liang, S. C., and K. N. Chang, *J. Chin. Chem. Soc.*, **18,** 25 (1951).

564. Liang, S. C., and P. Y. Hsu, *Acta Chin. Sinica*, **22,** 93 (1956).

565. Liang, S. C., and S. J. Wang, *Hua Hsueh Hsueh Pao*, **24,** 117 (1958); through *Chem. Abstr.*, **53,** 970c (1959).

566. Lingane, J. J., *Electroanalytical Chemistry*, Wiley-Interscience, New York, 1958.

567. Lingane, J. J., and L. A. Small, *J. Amer. Chem. Soc.*, **71,** 973 (1949).

568. Liplavk, I. L., and K. G. Chernousova, *Sb. Nauch. Rabot. Inst. Okhrany Truda, Vses. Tsentr. Sovet Profsoyuzov*, **1962,** 97; through *Chem. Abstr.*, **60,** 11375e (1964).

569. Lishanskii, G. Ya., *Sovrem. Metody Khim. Spektral. Anal. Mater.*, **1967,** 297; through *Chem. Abstr.*, **67,** 104892s (1967).

570. Little, K., and J. D. Brooks, *Anal. Chem.*, **46,** 1343 (1974).

571. Lobanov, E. M., L. E. Krasivina, B. P. Zverev, and M. M. Usmanova, *Akt. Anal. Christ. Mater.*, 1968, 16; through *Chem. Abstr.*, **71,** 56351y (1969).

572. Locke, M., and N. Krishnan, *J. Cell Biol.*, **50,** 550 (1971).

573. Love, D. L., *Anal. Chem.* **27,** 1918 (1955).

574. Lowenheim, F. A., "Tungsten," in F. D. Snell and L. S. Ettre, Eds., *Encyclopedia of Industrial Chemical Analysis*, Vol. 19, Wiley-Interscience, New York, 1974.

575. Lubecki, A., and H. Vogg, US Atomic Energy Commission, KFK-1728 (1973).

576. Lucas-Tooth, J., and C. Pyne, *Advan. X-Ray Anal.*, **7,** 523 (1964).

577. Lucena-Conde, F., and J. Zato, *An. Real. Soc. Espan. Fis. Quim., Ser. B*, **52,** 353 (1956); through *Chem. Abstr.*, **52,** 13520f (1958).

578. Lukas, J., and A. Jilek, *Chem. Listy*, **24,** 320 (1930).

579. Luke, C. L., *Anal. Chem.*, **33,** 1365 (1961).

580. Luke, C. L., *Anal. Chem.*, **35,** 1551 (1963).

581. Luke, C. L., *Anal. Chem.*, **36**, 1327 (1964).

582. Luke, C. L., *Anal. Chim. Acta*, **34**, 302 (1966).

583. Lundgren, K. D., and A. Swensson, *Acta Med. Scand.*, **145**, 20 (1953).

584. Lutz, C. W., and L. E. Conroy, *Anal. Chem.*, **38**, 139 (1966).

585. Machlan, L. A., and J. L. Hague, *J. Res. Nat. Bur. Stand.*, **59**, 415 (1957).

586. MacInnis, M., "Wolfram Compounds," in E. P. Dukes, Ed., *Kirk-Othmer Encyclopedia of Chemical Technology*, Vol. 22, Wiley-Interscience, New York, 1970.

587. Maeda, F., and T. Hayasaka, *Bunseki Kagaku*, **19**, 1036 (1970).

588. Maekawa, S., Y. Yoneyama, and E. Fujimori, *Busneki Kagaku*, **10**, 345 (1961).

589. Magee, R. J., S. A. F. Shahine, and C. L. Wilson, *Mikrochim. Ichnoanal. Acta*, **1964**, 479 (1964).

590. Magee, R. J., and A. S. Witwit, *Anal. Chim. Acta*, **29**, 517 (1963).

591. Magneli, A., *Ark. Kem.*, **1**, 273 (1949).

592. Major, M., and P. Gado, *Magy. Kem. Foly.*, **74**, 196 (1968).

593. Maleczki, E., *Veszpremi Vegyipari Egyetem Tudomanyas Ulesszakanak Eloadasai*, **1957**, 19. through *Chem. Abstr.*, **55**, 12155d (1961).

594. Malissa, H., and H. H. Arlt, *Radex Rundschau*, **1964**, 204; through *Chem. Abstr.*, **62**, 5874c (1965).

595. Mallett, M. W., *Talanta*, **9**, 133 (1962).

596. Mal'tsev, V. F., A. P. Martynov, V. P. Novak, and V. I. Bogovina, *Proizvod. Trub*, No. 22, 259 (1969); through *Chem. Abstr.*, **74**, 150872c (1971).

597. Malyutina, T. M., B. M. Dobkina, and V. A. Pisareva, *Zavod. Lab.*, **31**, 648 (1965).

598. Mamedov, I. A., and A. M. Mukimov, *Izv. Vyssh. Ucheb. Zaved. Khim. Khim. Tekhnol.*, **5**, 889 (1962); through *Chem. Abstr.*, **58**, 11954h (1963).

599. Mamuro, T., Y. Matsuda, A. Mizohata, T. Takeuchi, and A. Fujita, *Radioisotopes*, **20**, 117 (1971).

600. Marcu, G., *Stud. Univ. Babes-Bolyai, Ser. Chem.*, **12**, 139 (1967); through *Chem. Abstr.*, **68**, 44997p (1968).

601. Margolin, L. S., and N. A. Nikitina, *Nov. Metod. Tekh. Geol. Rabot, Leningrad*, **2**, 123 (1959); through *Chem. Abstr.*, **55**, 20757d (1961).

602. Marinkovic, M., and T. J. Vickers, *Appl. Spectrosc.*, **25**, 319 (1971).

603. Martin, R. P., *Biochim. Biophys. Acta*, **25**, 408 (1957).

604. Martinek, R. G., *Chem.-Anal.*, **53**, 108 (1964).

605. Martinelli, P., and P. Blanquet,*Radiochem. Methods Anal. Proc. Symp., Salzburg, Austria*, **2**, 451 (1964).

606. Martinez, H. O., *An. Dir. Nacl. Quim. (Buenos Aires)*, **5**, 68 (1952); through *Chem. Abstr.*, **48**, 5021b (1954).

607. Massart, R., and G. Herve, *Rev. Chim. Miner.*, **5**, 501 (1968).

608. Mathien, V., M. Lacomble, and L. Charlet, *Colloq. Spectros. Intern., 8th, Lucerne, Switz.*, **1959**, 195; through *Chem. Abstr.*, **60**, 7437g (1964).

609. Matsaberidze, T. G., *Ezhegodnik Kaukazsk. Inst. Mineral Syr'ya* (Moscow, Gos. Izd. Geol. Lit.), **1957**, 51 (1959); through *Chem. Abstr.*, **57**, 7901h (1962).

610. Matsuo, H., S. Chaki, and H. Shigeki, *Bunseki Kagaku*, **17**, 752 (1968).

611. Matviak, M., *Chem.-Anal.*, **32**, 83 (1943).

612. McClendon, L. T., and J. R. DeVoe, *Anal. Chem.*, **41**, 1454 (1969).

613. McDuffie, B., W. R. Bandi, and L. M. Melnick, *Anal. Chem.*, **31**, 1311 (1959).

614. McKaveney, J. P., *Anal. Chem.*, **33**, 744 (1961).

615. McKaveney, J. P., *Anal. Chem.*, **35**, 2139 (1963).

616. Mead, A. S., *UK At. Energy Authority, Res. Group Rep.* AERE-R 6537 (1970).

617. Meites, L., Ed., *Handbook of Analytical Chemistry*, McGraw-Hill, New York, 1963.

618. Menapace, L. M., and A. F. Voigt, *J. Radioanal. Chem.*, **6**, 219 (1970).

619. Mendonca Pinto, C., *An. Assoc. Quim. Brasil*, **7**, 65 (1948); through *Chem. Abstr.*, **43**, 2893c (1949).

620. Mennicke, H., *Molybdäns, Vanadiums, und Wolframs*, Verlag M. Krayn, Berlin, 1913.

621. Merkova, M. V., *Vestn. Leningr. Univ.*, **14**, No. 6, *Ser. Geol. Geogr.*, No. 1, 157 (1959); through *Chem. Abstr.*, **53**, 17753i (1959).

622. Merz, E., *Z. Anal. Chem.*, **191**, 416 (1962).

623. Mezentseva, N. V., *Poroshk. Metall.*, **4**, 112 (1964); through *Chem. Abstr.*, **61**, 3604c (1964).

624. Mezhiborskaya, Kh. B., and M. I. Krasikova, *Zh. Anal. Khim.*, **25**, 581 (1970).

625. Michal, J., *Chem. Listy*, **50**, 77 (1956).

626. Mihalka, S., and N. W. Ghelberg, *Hidroteh. Gospodarirea Apelor Meterol. (Bucharest)*, **12**, 610 (1967); through *Chem. Abstr.*, **68**, 98534d (1968).

627. Miller, A. D., and R. I. Libina, *Zh. Prikl. Khim.*, **32**, 2624 (1959).

628. Miller, C. C., *Analyst (London)*, **69**, 109 (1944).

629. Miller, C. C., and D. H. Thow, *Talanta*, **5**, 129 (1960).

630. Miller, C. C., and D. H. Thow, *Talanta*, **8**, 43 (1961).

631. Mimoun, H., I. S. De Roch, and L. Sajus, *Bull. Soc. Chim. Fr.* **1969**, 1481 (1969).

632. Mitchell, B. J., *Anal. Chem.*, **32**, 1652 (1960).

633. Mo, C. S., *Acta Geol. Sinica*, **37**, 181 (1957); through *Chem. Abstr.*, **52**, 18107h (1958).

634. Modylevskaya, K. D., *Zavod. Lab.*, **33**, 948 (1967).

635. Mokhosoev, M. V., and L. R. Tokareva, *Zh. Neorg. Khim.*, **16**, 2159 (1971).

636. Molenda, K., *Pr. Inst. Mech. Precyz.*, **18**, 46 (1970); through *Chem. Abstr.*, **74**, 71191b (1971).

637. Moll, S. H., *Intern. Conf. Electron Ion Beam Sci. Technol., 1st Toronto*, **1964**, 825 (1965); through *Chem. Abstr.*, **64**, 13362f (1966).

638. Monastyrskii, D. N., *Tr. Leningrad. Politekh. Inst. M. I. Kalinina*, No. 201, 1718 (1959); through *Chem. Abstr.*, **54**, 15086b (1960).

639. Monnier, D., R. Daniel, and W. Haerdi, *Chimica*, **20**, 428 (1966).

640. Morachevskii, Yu. V., L. G. Shipunova, and L. D. Novozhilova, *Uch. Zap. Leningrad. Gos. Univ. A.A. Zhdanova, Ser. Khim. Nauk*, No. 19, 58 (1961); through *Chem. Abstr.*, **55**, 24209e (1961).

641. Moseev, L. I., V. I. Blokhin, A. I. Lastov, and A. P. Smirnov-Averin, *Metody Opred. Issled. Sostovaniya Gasov Metal.*, **1968**, 108; through *Chem. Abstr.*, **70**, 92918b (1969).

642. Moser, L., and W. Blaustein, *Monatsh. Chem.*, **52**, 351 (1929).

643. Moskvin, L. N., L. I. Shur, L. G. Tsaritsyna, and G. T. Martysh, *Radiokhimiya*, **9**, 377 (1967).

644. Moss, J. R., and B. L. Shaw, *J. Chem. Soc., A*, **1970**, 595 (1970).

645. Motorkina, R. K., *Zh. Neorg. Khim.*, **2**, 92 (1957).

646. Mrozinski, J., *Rudy Met. Niezelaz*, **12**, 25 (1967); through *Chem. Abstr.*, **67**, 39913h (1967).

647. Mueller, J., I. Malatova, M. Houskova, and S. Furgyik, *Prac. Lek.*, **21**, 185 (1969); through *Chem. Abstr.*, **72**, 75391y (1970).

648. Mukhina, Z. S., L. I. Il'ina, and N. S. Kondukova, *Khim. Svoistva Metody Anal. Tugoplavkikh Soedin.*, **1969**, 108; through *Chem. Abstr.*, **72**, 50661y (1970).

649. Mullins, W. T., and G. W. Leddicotte, US Atomic Energy Commission, NAS-NS 3042 (1961).

650. Muntz, J. H., *Appl. Spectrosc.*, **21**, 300 (1967).

651. Muntz, J. H., and L. L. Roush, US Atomic Energy Commission, AED-CONF-66-172-9 (1966).

652. Muraki, I., *J. Chem. Soc. Japan, Pure Chem. Sect.*, **76**, 193 (1955).

653. Murata, A., and F. Yamauchi, *Nippon Kagaku Zasshi*, **77**, 1259 (1956).

654. Murau, P. C., *Anal. Chem.*, **33**, 1125 (1961).

655. Musil, A., and H. Faber, *Z. Anal. Chem.*, **202**, 412 (1964).

656. Myshlyaeva, L. V., and I. V. Sedova, *Zh. Neorg. Khim.*, **16**, 2206 (1971).

657. Nadeenko, V. G., *Gig. Sanit.*, **31**, 10 (1966); through *Chem. Abstr.*, **65**, 15968a (1966).

658. Nadkarni, R. A., and B. C. Halder, *J. Radioanal. Chem.*, **8**, 45 (1971).

659. Nakajima, T., Y. Ouchi, and K. Kato, *Bunseki Kagaku*, **20**, 330 (1971).

660. Nakatsukasa, Y., and M. Fujimoto, *Microchem. J.*, **9**, 465 (1965).

661. Nakazono, T., *J. Chem. Soc. Japan*, **48**, 76 (1927).

662. Nangniot, P., *Collect. Czech. Chem. Commun.*, **30**, 4070 (1965).

663. Navratil, O., and J. Fikr, *Omagiu Raluca Ripan*, **1966**, 383; through *Chem. Abstr.*, **67**, 68131g (1967).

664. Nazarenko, V. A., and E. N. Poluektova, *Zh. Neorg. Khim.*, **14**, 204 (1969).

665. Nazarenko, V. A., and E. N. Poluektova, *Zh. Anal. Khim.*, **26**, 1331 (1971).

666. Nazarenko, V. A., E. N. Poluektova, and G. G. Shitareva, *Zh. Anal. Khim.*, **28**, 101 (1973).

667. Nazarenko, V. A., M. B. Shustova, G. Ya. Yagnyatinskaya, E. N. Poluektova, and E. I. Shelikhina, *Tr. Khim. Khim. Tekhnol.*, **1969**, 129; through *Chem. Abstr.*, **75**, 114596b (1971).

668. Nazarenko, V. A., and L. E. Shvartsburd, *Zavod. Lab.*, **16**, 357 (1950).

669. Nazarova, L. V., and V. A. Dagaev, *Uch. Zap. Kishinev. Gos. Univ.*, **56**, 39 (1960); through *Chem. Abstr.*, **56**, 1992i (1962).

670. Neales, T. F., *J. Chromatogr.*, **16**, 262 (1964).

671. Nedler, V. V., *Bull. Acad. Sci. URSS, Ser. Phy.*, **4**, 142 (1940); through *Chem. Abstr.*, **35**, 1725[5] (1941).

672. Nelson, D. L., R. R. Reeves, and H. H. Richtol, *J. Chem. Educ.*, **50**, 810 (1973).

673. Neuburger, M., and A. Fourey, *J. Radioanal. Chem.*, **1**, 289 (1968).

674. Neumann, G. M., *Talanta*, **18**, 1047 (1971).

675. Neumhoeffer, O., *Z. Anorg. Allg. Chem.*, **296**, 208 (1958).

676. Nevskaya, E. M., and V. A. Nazarenko, *Zh. Anal. Khim.*, **27**, 1699 (1972).

677. Newton, E. B., *J. Biol. Chem.*, **120**, 315 (1937).

678. Ni, C. M., C. F. Chu, and S. C. Liang, *Hua Hsueh Hsueh Pao*, **29**, 247 (1963); through *Chem. Abstr.*, **62**, 5872a (1965).

679. Nishanov, P. Kh., and N. V. Mamadaliev, *Atkiv. Anal. Blagorod. Metal*, **1970**, 122; through *Chem. Abstr.*, **76**, 121162g (1972).

680. Nishida, H., *Bunseki Kagaku*, **3**, 25 (1954).

681. Nishida, H., *Bunseki Kagaku*, **6**, 299 (1957).

682. Nishida, H., *Bunseki Kagaku*, **13**, 263 (1964).

683. Nishida, H., *Bunseki Kagaku*, **13**, 760 (1964).

684. North, A. A., *Analyst (London)*, **81**, 660 (1956).

685. North, E. O., and G. D. Beal, *J. Amer. Pharm. Ass.*, **13**, 889 (1924).

686. North, V., and F. S. Grimaldi, *US Geol. Survey Bull.*, No. 950, 129 (1946).

687. Norwitz, G., *Anal. Chem.*, **33**, 1253 (1962).

688. Norwitz, G., *Metallurgia*, **66**, 297 (1962).

689. Norwitz, G., and M. Codell, *Anal. Chim. Acta*, **11**, 359 (1954).

690. Norwitz, G., and H. Gordon, *Anal. Chem.*, **37**, 417 (1965).

691. Norwitz, G., and H. Gordon, *Anal. Chim. Acta*, **69**, 59 (1974).

692. Novak, V., *Chem. Listy*, **49**, 848 (1955).

693. Novak, V., *Chem. Listy*, **49**, 934 (1955).

694. Novikov, A. I., and G. K. Ryazanova, *Radiokhimiya*, **10**, 369 (1968).

695. Oats, J. T., *Eng. Mining J.*, **144**, 72 (1943).

696. Oblas, D. W., and H. Hoda, *J. Appl. Phys.*, **39**, 6106 (1968).

697. Obolonchik, V. A., and K. D. Modglevskaya, *Zavod. Lab.*, **23**, 912 (1957).

698. Okuna, H., M. Honda, and T. Ishimori, *Bunseki Kagaku*, **4**, 386 (1955).

699. Omarova, E. S., E. F. Speranskaya, and M. T. Kozlovskii, *Izv. Akad. Nauk Kaz SSR, Ser. Khim.*, **18**, 32 (1968); through *Chem. Abstr.*, **69**, 32763s (1968).

700. Omarova, E. S., E. F. Speranskaya, and M. T. Kozlovskii, *Zh. Anal. Khim.*, **23**, 1826 (1968).

701. Opravil, O., B. Zitnansky, and I. Sebastian, *Chem. Listy*, **57**, 1294 (1963).

702. Ortner, H. M., *Anal. Chem.*, **47**, 162 (1975).

703. Osenjo, H. O., *Rev. Minera, Geol. Mineral.*, **29**, 49 (1970); through *Chem. Abstr.*, **74**, 94082f (1971).

704. O'Shea, T. A., and G. A. Parker, *Anal. Chem.*, **44**, 184 (1972).

705. Owens, L. E., and J. Y. Ellenbury, *Anal. Chem.*, **23**, 1823 (1951).

706. Pakalns, P., *Anal. Chim. Acta*, **50**, 103 (1970).

707. Pakholkov, V. S., and V. E. Panikarovskihk, *Izv. Vyssh. Uch. Zaved. Khim. Khim. Tekhnol.*, **10**, 168 (1967); through *Chem. Abstr.*, **67**, 50044w (1967).

708. Pajuste, O., *Sb. Nauchn.-Tekh. Statei, Proekt.-Tekhnol. Nauchn.-Issled. Inst. Tallin*, No. 8, 177 (1968); through *Chem. Abstr.*, **73**, 62315x (1970).

709. Pan, K., and T. M. Hseu, *Bull. Chem. Soc. Japan,* **26,** 126 (1953).

710. P'an, Y. L., *Hua Hsueh Shih Chieh,* **1959,** 532; through *Chem. Abstr.,* **54,** 12887a (1960).

711. P'an, Y. L., *Yu Se Chin Shu,* No. 20, 29 (1959); through *Chem. Abstr.,* **58,** 10726a (1963).

712. Pang, Y. L., *Hua Hsueh Hsueh Pao,* **28,** 341 (1963); through *Chem. Abstr.,* **59,** 12173f (1963).

713. Panidi, I. S., Ya. M. Paushkin, L. P. Starchik, and I. V. Yakovlev, *Neftepererab. Neftekhim. (Moscow),* **1969,** 25; through *Chem. Abstr.,* **71,** 77090c (1969).

714. Pantaler, R. P., *Zh. Anal. Khim.,* **18,** 603 (1963).

715. Pantani, F., *Ric. Sci.,* **3,** 873 (1963).

716. Pantani, F., and P. Desideri, *Ric. Sci.,* **1,** 249 (1961).

717. Pantani, F., and M. Migliorini, *Ric. Sci.,* **3,** 1085 (1963).

718. Panteleev, V. V., and A. A. Yankovskii, *Zh. Prikl. Spektrosk., Akad. Nauk Belorussk. SSR,* **3,** 350 (1965); through *Chem. Abstr.,* **64,** 11841b (1966).

719. Pardo Arguedas, A., *An. Fac. Farm. Bioquim., Univ. Nac. Mayor San Marcos,* **8,** 421 (1957); through *Chem. Abstr.,* **53,** 22735f (1959).

720. Parish, R. V., "The Inorganic Chemistry of Tungsten," in H. J. Emeleus and A. G. Sharpe, Eds., *Advances in Inorganic Chemistry and Radiochemistry,* Vol. 9, Academic, New York, 1966.

721. Parker, G. A., and D. F. Boltz, *Anal. Lett.,* **1,** 679 (1968).

722. Pashchenko, E. N., and V. F. Mal'tsev, *Zavod. Lab.,* **34,** 12 (1968).

723. Patil, S. P., and V. M. Shinde, *Anal. Lett.,* **6,** 709 (1973).

724. Patrovsky, V., *Sb. Geol. Ved., Tekhnol. Geokhim.,* No. 7, 155 (1966); through *Chem. Abstr.,* **65,** 14420d (1966).

725. Pavlenko, L. I., *Zh. Anal. Khim.,* **15,** 463 (1960).

726. Pavlenko, L. I., *Zh. Anal. Khim.,* **15,** 716 (1960).

727. Peck, E. S., *Anal. Chem.,* **40,** 324 (1968).

728. Pemberton, R., *Analyst (London),* **77,** 287 (1952).

729. Penchev, N. P., and G. S. Nikolov, *God. Sofii, Univ., Khim. Fak.,* **57,** 57 (1962/ 1963); through *Chem. Abstr.,* **63,** 7882g (1965).

730. Peng, P. Y., and E. B. Sandell, *Anal. Chim. Acta,* **29,** 325 (1963).

731. Pen'kova, E. F., *Zavod. Lab.,* **15,** 475 (1949).

732. Penner, E. M., and W. R. Imman, *Talanta,* **9,** 1027 (1962).

733. Pepper, C. E., and G. R. Blank, US Atomic Energy Commission, NLCO-1105 (1974).

734. Perchik, F. I., and O. I. Khotsyanivs'kii, *Vesn. Politekh Inst. Ser. Khim. Mashinobuduv Tekhnol.,* No. 3, 150 (1966); through *Chem. Abstr.,* **68,** 56323d (1968).

735. Pervozkin, V. K., and A. N. Baraboshkin, *Tr. Inst. Elektrokhim., Akad. Nauk SSSR, Ural. Fil.,* No. 11, 35 (1968); through *Chem. Abstr.,* **70,** 16603e (1970).

736. Pesis, A. S., *Tr. Permsk. Med. Inst.,* No. 31, 157 (1960); through *Chem. Abstr.,* **56,** 6657h (1962).

737. Peterson, H. E., and W. L. Anderson, US Bureau of Mines, Reports on Investigations, No. 3709 (1943).

738. Pevtsov, G. A., L. K. Raginskaya, T. G. Manova, and L. A. Pavlova, *Methody Anal. Khim. Reaktivov Prep.*, No. 15, 61 (1968); through *Chem. Abstr.*, **69**, 92703g (1968).

739. Pfeifer, V., *Mikrochim. Acta*, **1960**, 518 (1960).

740. Pham-Huu-Chanh, and S. Chanvattey, *Agressologie*, **8**, 433 (1967); through *Chem. Abstr.*, **68**, 76620b (1968).

741. Philipp, P., *An. Ass. Quim. Brasil*, **6**, 161 (1947); through *Chem. Abstr.*, **42**, 6268h (1948).

742. Phillips, J. P., and J. F. Deye, *Anal. Chim. Acta*, **17**, 231 (1957).

743. Pickering, W. F., *Crit. Rev. Anal. Chem.*, **3**, 271 (1973).

744. Pickett, E. E., and S. R. Koirtyohann, *Spectrochim. Acta*, **23B**, 235 (1968).

745. Pincussen, L., and B. Minz, *Biochem. Z.*, **234**, 19 (1931).

746. Plaksin, I. N., E. V. Anchevskii, E. I. Russkaya, and L. P. Starchik, *Tr. 1-go [Pervogo] Vses. Koordinats. Soveshch. Aktivatsionnoma Anal., Inst. Yadern. Fiz. Akad. Nauk Uz. SSR, Tashkent*, **1962**, 172; through *Chem. Abstr.*, **63**, 3618h (1965).

747. Plaksin, I. N., G. I. Mikhailov, L. P. Starchik, and M. J. Silenko, *Tsvet. Metal.*, **40**, 27 (1967).

748. Plaksin, I. N., I. F. Slepchenko, and L. P. Starchik, *Dokl. Akad. Nauk SSSR*, **137**, 880 (1961).

749. Platunov, B. A., *Vestn. Leningrad. Univ., Ser. Mat. Fiz. Khim.*, **7**, 137 (1952); through *Chem. Abstr.*, **48**, 4356c (1954).

750. Platunov, B. A., and A. E. Deitch, *Vestn. Leningrad. Univ. Ser. Mat. Fiz. Khim.*, **6**, 45 (1950); through *Chem. Abstr.*, **47**, 12117a (1953).

751. Platunov, B. A., and M. I. Guseva, *Uch. Zap. Leningrad. Gos. Univ. A.A. Zhdanova, Ser. Khim. Nauk*, No. 11, 66 (1952); through *Chem. Abstr.*, **49**, 15619d (1955).

752. Plotnikov, V. I., and V. L. Kochetkov, *Zh. Anal. Khim.*, **21**, 1260 (1966).

753. Plotnikov, V. I., and V. L. Kochetkov, *Zh. Neorg. Khim.*, **13**, 203 (1968).

754. Plotnikov, V. I., V. L. Kochetkov, and V. P. Chimaeva, *Izv. Akad. Nauk Kaz. SSR, Ser. Khim.*, **18**, 15 (1968); through *Chem. Abstr.*, **69**, 80893f (1968).

755. Pluchet, E., and M. Lederer, *J. Chromatogr.*, **3**, 290 (1960).

756. Podobnik, B., and L. Lorenzini, Comite Nazionale Energia Nucleari, RT/CHI-63(6) (1963); through *Chem. Abstr.*, **59**, 12174b (1963).

757. Pollock, J. B., *Analyst (London)*, **83**, 516 (1958).

758. Polotbnova, N. A., and L. M. Danilina, *Zavod. Lab.*, **36**, 261 (1970).

759. Polotebnova, N. A., L. M. Damilina, L. V. Derkach, A. A. Kozlenko, S. V. Krachun, Ya. L. Neimark, K. K. Radul, E. F. Tkach, and L. A. Furtune, *Sb. Nauch. Statei, Kishinev. Gos. Univ., Estestv. Mat. Nauki*, 1969, 130; through *Chem. Abstr.*, **74**, 93979s (1971).

760. Polotebnova, N. A., J. Krtil, F. Kh. Ravitskaya, and S. V. Krachun, *Uch. Zap. Kishinev. Gos. Univ.*, **68**, 71 (1963); through *Chem. Abstr.*, **63**, 8122e (1965).

761. Poluektova, E. N., *Zh. Anal. Khim.*, **21**, 187 (1966).

762. Poluektova, E. N., and V. A. Nazarenko, *Zh. Anal. Khim.*, **19**, 856 (1964).

763. Poluektova, E. N., and V. A. Nazarenko, *Zh. Anal. Khim.*, **22**, 746 (1967).

764. Polyakov, V. I., *Dokl. Akad. Nauk Uzb. SSR*, **27**, 31 (1970).

765. Polykovskaya, N. A., *Gig. Sanit.*, **32**, 50 (1967); through *Chem. Abstr.*, **66**, 88479j (1967).

766. Pommier, C., and G. Guiochon, *J. Chromatogr. Sci.*, **8**, 486 (1970).

767. Pope, M. T., and E. Papaconstantinou, *Inorg. Chem.*, **6**, 1147 (1967).

768. Pope, M. T., and G. M. Varga, *Inorg. Chem.*, **5**, 1249 (1966).

769. Popov, I. F., V. F. Matrenkin, and O. B. Klebanov, *Sin. Svoistova Ionoobmen. Mater.*, 1968, 69; through *Chem. Abstr.*, **71**, 82225e (1969).

770. Popov, M. A., *Zavod. Lab.*, **13**, 379 (1947).

771. Popova, N. M., and A. F. Platonova, *Zavod. Lab.*, **15**, 267 (1949).

772. Popova, O. I., and O. G. Seraya, *Zh. Anal. Khim.*, **23**, 791 (1968).

773. Porter, L. E., *Ind. Eng. Chem., Anal. Ed.*, **6**, 138 (1934).

774. Pregosin, P. S., and L. M. Venanzi, *J. Coord. Chem.*, **3**, 145 (1973).

775. Pribil, R., *Analytical Applications of EDTA and Related Compounds*, Pergamon, Oxford, 1972.

776. Pribil, R., and V. Sedlar, *Collect. Czech. Chem. Commun.*, **16**, 69 (1951).

777. Prochazkova, V., and V. Jara, *Z. Anal. Chem.*, **161**, 251 (1958).

778. Przyborski, W., J. Chwaszczewska, W. Czarnacki, and M. Szymczak, US Atomic Energy Commission, INR-820(II)PS (1968).

779. Pueschel, R., and E. Lassner, "Chelates and Chelating Agents in the Analytical Chemistry of Molybdenum and Tungsten," in H. A. Flaschka and A. J. Barnard, Eds., *Chelates in Analytical Chemistry*, Vol. 1, Marcel Dekker, New York, 1967.

780. Pueschel, R., and E. Lassner, *J. Less Common Metals*, **17**, 313 (1969).

781. Pueschel, R., E. Lassner, and K. Katzengruber, *Chem.-Anal.*, **56**, 63 (1967).

782. Pyatnitskii, I. V., and L. F. Kravtsova, *Ukr. Khim. Zh.*, **34**, 706 (1968).

783. Pyatnitskii, I. V., and N. N. Silich, *Ukr. Khim. Zh.*, **33**, 834 (1967).

784. Quin, B. F., and R. R. Brooks, *Anal. Chim. Acta*, **65**, 206 (1972).

785. Qureshi, M., J. P. Gupta, and V. Sharma, *Talanta*, **21**, 102 (1974).

786. Qureshi, M., and F. Khan, *J. Chromatogr.*, **34**, 222 (1968).

787. Qureshi, M., K. N. Mathur, and A. H. Issaili, *Talanta*, **16**, 503 (1969).

788. Raby, B. A., and C. V. Banks, *Anal. Chem.*, **36**, 1106 (1964).

789. Raby, B. A., and W. E. Sunderland, *Anal. Chem.*, **39**, 1304 (1967).

790. Radwan, M., B. Rawienska-Kosciukowa, and E. Zarzecka, *Hutn.*, **33**, 299 (1966); through *Chem. Abstr.*, **66**, 34587z (1967).

791. Rai, J., and V. P. Kukreja, *Chromatographia*, **1969**, 404 (1969).

792. Raikhbaum, Ya. D., *Zavod. Lab.*, **8**, 601 (1939).

793. Rambourg, A., *C.R.H. Acad. Sci., Ser. D*, **265**, 1426 (1967).

794. Randa, Z., J. Benada, J. Kuncir, and M. Vobecky, *Radiochem. Radioanal. Lett.*, **3**, 227 (1970).

795. Randin, J. P., *J. Chem. Educ.*, **51**, 32 (1974).

796. Rani, S., and S. K. Banerju, *Indian J. Appl. Chem.*, **33**, 263 (1970).

797. Rao, C. N. R., *Solid State Chemistry*, Marcel Dekker, New York, 1974.

798. Rao, P. D., *At. Absorption Newslett.*, **9**, 131 (1970).

799. Rasmussen, P. G., *J. Chem. Educ.*, **44**, 277 (1967).

800. Raspi, W. G., and G. Ciantelli, *Chim. Ind. (Milan)*, **45**, 1515 (1963).

801. Ratschke, R., and C. Dowe, *Acta Chim. Acad. Sci. Hung.*, **80**, 147 (1974).

802. Reboul, P., *Chim. Ind. (Paris)*, **64**, 574 (1950).

803. Redmond, J. C., *Anal. Chem.*, **19**, 773 (1947).

804. Redmond, J. C., *Steel*, **122**, 86 (1948).

805. Reed, J. F., *Anal. Chem.*, **33**, 1337 (1961).

806. Reef, B., and H. G. Doege, *Talanta*, **14**, 967 (1967).

807. Rehak, B., and M. Malinek, *Z. Anal. Chem.*, **153**, 166 (1956).

808. Reichen, L. E., *Anal. Chem.*, **26**, 1302 (1954).

809. Reuland, R. J., and A. F. Voigt, *Anal. Chem.*, **35**, 1263 (1963).

810. Revel, G., and P. Albert, *C.R.H. Acad. Sci., Ser. C*, **265**, 1443 (1967).

811. Reznik, B. E., G. M. Ganzburg, and G. V. Mal'tseva, *Zh. Anal. Khim.*, **23**, 1848 (1968).

812. Reznik, L. B., and P. N. Kovalenko, *Ukr. Khim. Zh.*, **30**, 28 (1964).

813. Rieck, G. D., *Tungsten*, Pergamon, Oxford, 1967.

814. Rigdon, L. P., and J. E. Harrar, *Anal. Chem.*, **40**, 1641 (1968).

815. Riley, J. P., and D. Taylor, *Anal. Chim. Acta*, **40**, 479 (1968).

816. Ringbom, A., *Complexation in Analytical Chemistry*, Wiley-Interscience, New York, 1963.

817. Ripan, R., and C. Calu, *Talanta*, **14**, 887 (1967).

818. Ripan, R., A. Duca, N. Calu, *Stud. Cercet. Chim.*, **11**, 7 (1960); through *Chem. Abstr.*, **55**, 10148e (1961).

819. Ripan, R., and G. Marcu, *Acad. Rep. Populare Romine, Studii Cercet. Chim.*, **10**, 201 (1959); through *Chem. Abstr.*, **54**, 24068b (1960).

820. Ripan, R., and G. Marcu, *An. Stiint. Univ. "Al. I. Cuza," Iasi, Sect. I*, **6**, 869 (1960); through *Chem. Abstr.*, **59**, 3521c (1963).

821. Robin, M. B., and P. Day, Mixed Valence Chemistry, in H. J. Emeleus and A. G. Sharpe, Eds., *Advances in Inorganic Chemistry and Radiochemistry*, Vol. 10, Academic, New York, 1967.

822. Rollinson, C. L., "Tungsten," in J. C. Bailar, Senior Editor, *Comprehensive Inorganic Chemistry*, Vol. 3, Pergamon, Oxford, 1973.

823. Rooney, R. C., and C. G. Pratt, *Analyst (London)*, **97**, 400 (1972).

824. Ross, R. B., *Metallic Materials*, Chapman and Hall, London, 1968.

825. Rossmanith, K., and Z. G. Hanna, *Monatsh. Chem.*, **90**, 76 (1959).

826. Rossmanith, K., and Z. G. Hanna, *Monatsh. Chem.*, **91**, 238 (1960).

827. Ryabchikov, D. I., and V. E. Bukhtiarov, *Zh. Anal. Khim.*, **15**, 242 (1960).

828. Sajo, I., *Acta Chim. Acad. Sci. Hung.*, **6**, 233 (1955).

829. Saki, T., *Nippon Kinzoku Gakkzishi*, **30**, 473 (1966); through *Chem. Abstr.*, **65**, 9731g (1966).

830. Sakuraba, S., and Y. Suzuki, *J. Electrochem. Soc. Japan*, **18**, 325 (1950).

831. Salcheva, M., B. Belchev, and R. Tencheva, *Rudodobiv Metal. (Sofia)*, **23**, 15 (1968); through *Chem. Abstr.*, **69**, 56802e (1968).

832. Samsahl, K., and R. Soremark, *Proc. Intern. Conf. Modern Trends Activation Anal.*, **1961**, 149 (1961).

833. Sandell, E. B., *Ind. Eng. Chem. Anal. Ed.*, **10**, 667 (1938).

834. Sandell, E. B., *Ind. Eng. Chem. Anal. Ed.*, **18**, 163 (1946).

835. Sandell, E. B., *Colorimetric Determination of Traces of Metals*, Wiley-Interscience, New York, 1958.

836. Santos Romero, M., *An. Fiz. Quim. (Madrid)*, **42**, 985 (1946); through *Chem. Abstr.*, **41**, 5052e (1947).

837. Sanyal, A. B., and P. Ganguly, *Biochim. Biophys. Acta*, **133**, 535 (1967).

838. Sato, T., T. Nishizawa, and K. Murai, *Tetsu Hagane*, **45**, 409 (1959); through *Chem. Abstr.*, **54**, 3124e (1960).

839. Savvin, S. B., E. G. Namvrina, and L. A. Okhanova, *Zh. Anal. Khim.*, **28**, 1119 (1973).

840. Savvin, S. B., E. G. Namvrina, and R. S. Tramm, *Zh. Anal. Khim.*, **27**, 108 (1972).

841. Sawatani, T., and S. Murota, *Nippon Kinzoku Gakkaishi*, **32**, 1167 (1968); through *Chem. Abstr.*, **70**, 53674z (1969).

842. Scandrett, F. J., *Nature*, **186**, 558 (1960).

843. Schaefer, H., C. Pietruck, and U. Grozinger, *Z. Anal. Chem.*, **141**, 24 (1954).

844. Schapiro, W., and A. Schapiro, *Ann. Chim. Anal.*, **17**, 323 (1913).

845. Scharrer, K., and W. Schropp, *Z. Pflanzenernahr. Dung Bolenk*, **34A**, 317 (1934); through *Chem. Abstr.*, **29**, 1136[1] (1935).

846. Schmidt, F. F., and H. R. Ogden, US Department of Commerce, AD 425,547 (1963).

847. Schmitt, T. M., *Diss. Abstr. Int. B*, **31**, 3890 (1971).

848. Schneider, H., and M. E. Roselli, *An. Asoc. Quim. Argent.*, **57**, 39 (1969).

849. Schoeller, W. R., *Sand, Clays and Minerals*, **2**, 67 (1934).

850. Schoeller, W. R., and C. Jahn, *Analyst (London)*, **52**, 504 (1927).

851. Schroeder, H. A., *Air Qual. Monogr.*, No. 7015, 1 (1970); through *Chem. Abstr.*, **74**, 115475d (1971).

852. Schufle, J. A., *J. Chem. Educ.*, **52**, 325 (1975).

853. Schwabe, K., and R. Ross, *Z. Anorg. Allg. Chem.*, **325**, 181 (1963).

854. Scharzenbach, G., G. Geier, and J. Littler, *Helv. Chem. Acta*, **45**, 2601 (1962).

855. Schwarzkoff, P., and R. Kieffer, *Refractory Hard Metals*, Macmillan, New York, 1953.

856. Scobie, A. G., *Ind. Eng. Chem. Anal. Ed.*, **15**, 79 (1943).

857. Scott, A. F., and J. G. Shell, *J. Amer. Chem. Soc.*, **81**, 2278 (1959).

858. Seidov, I. M., *Gig. Sanit.*, **28**, 93 (1963); through *Chem. Abstr.*, **60**, 16449g (1964).

859. Seidov, I. M., *Vop. Pitan.*, **23**, 73 (1964); through *Chem. Abstr.*, **60**, 14897c (1964).

860. Seki, J., and T. Tanaka, *Tokyo Gakugei Daigaku Kiyo, Dai-4-Bu*, **24**, 150 (1972); through *Chem. Abstr.*, **78**, 23558n (1973).

861. Selig, W., US Atomic Energy Commission, URCL-7873, 41 (1964).

862. Sell, H. G., G. H. Keith, R. C. Koo, R. H. Schnitzel, and R. Corth, US Department of Commerce, AD 266,300 (1961).

863. Senderova, V. M., *Tr. Mineral. Muz. Akad. Nauk SSSR,* No. 8, 108 (1957); through *Chem. Abstr.,* **53,** 5010d (1959).

864. Senilova, M. N., *Mikroelem. Sib.,* No. 5, 55 (1967); through *Chem. Abstr.,* **70,** 103699j (1969).

865. Sevcik, J., and J. Cihalik, *Collect. Czech. Chem. Commun.,* **31,** 3140 (1966).

866. Shakhov, A. S., *Zavod. Lab.,* **10,** 470 (1941).

867. Shamaev, V. I., *Radiokhimiya* **2,** 624 (1960).

868. Shanahan, C. E. A., *Analyst (London),* **70,** 421 (1945).

869. Shanks, H. R., P. H. Sidles, and G. C. Danielson, "Electrical Properties of the Tungsten Bronzes," in R. F. Gould, Ed., *Nonstoichiometric Compounds,* Advances in Chemistry Series 39, American Chemical Society, Washington, D.C., 1963.

870. Shapiro, K. Ya., Z. G. Karov, and V. V. Kulakova, *Zh. Neorg. Khim.,* **13,** 487 (1968).

871. Sharples, H., *Chem.-Anal.,* **36,** 40 (1947).

872. Shaver, P. J., *Appl. Phys. Lett.,* **11,** 255 (1967).

873. Shcherbakov, V. G., *Sb. Tr. Vses. Nauchn.-Issled. Inst. Tverd. Splavov,* **1964,** 274; through *Chem. Abstr.,* **62,** 1077g (1965).

874. Shcherbakov, V. G., and Z. K. Stegendo, *Sb. Tr. Vses. Nauchn.-Issled. Proekt. Inst. Tugoplavk. Met. Tverd. Splavov,* No. 10, 182 (1970); through *Chem. Abstr.,* **75,** 157888s (1971).

875. Shcherbakov, V. G., and R. M. Veitsman, *Sb. Tr. Vses Nauchn.-Issled. Inst. Tverd. Splavov,* **1960,** 23; through *Chem. Abstr.,* **57,** 38f (1962).

876. Shcherbakov, V. G., R. M. Veitsman, and R. A. Antonova, *Sb. Tr. Vses. Nauchn.-Issled. Proekt. Inst. Tugoplavk. Met. Tverd. Splavov,* No. 10, 192 (1970); through *Chem. Abstr.,* **75,** 83874k (1971).

877. Sherman, F. B., and V. A. Klimova, *Izv. Akad. Nauk SSSR, Ser. Khim.,* **1971,** 120 (1971).

878. Sheskolskaya, A. Ya., *Zh. Anal. Khim.,* **20,** 1250 (1965).

879. Sheskolskaya, A. Ya., *Zh. Anal. Khim.,* **22,** 812 (1967).

880. Shimizu, T., K. Kato, S. Oyama, and K. Hosohara, *Bunseki Kagaku,* **15,** 120 (1966).

881. Shinde, V. M., and S. M. Khopkar, *Talanta,* **16,** 525 (1969).

882. Shiobara, Y., *Bunseki Kagaku,* **19,** 243 (1970).

883. Shiokawa, T., *J. Chem. Soc. Japan,* **67,** 53 (1946).

884. Shiokawa, T., *Sci. Rept. Res. Inst., Tohoku Univ. Ser. A,* **2,** 287 (1950).

885. Shishkov, D. A., *Godishnik Minnv.-Geol. Inst. Sofiya,* **2,** 147 (1954–1955); through *Chem. Abstr.,* **54,** 11850i (1960).

886. Shishkov, D. A., *Godishnik Minnv.-Geol. Inst. Sofiya,* **8,** 447 (1961–1962); through *Chem. Abstr.,* **62,** 3382a (1965).

887. Shishkov, D. A., *C.R. Acad. Bulg. Sci.,* **20,** 935 (1967).

888. Shishkov, D. A., *Dokl. Bolg. Akad. Nauk,* **24,** 769 (1971); through *Chem. Abstr.,* **75,** 144394c (1971).

889. Shishkov, D. A., and B. Velcheva, *C.R. Acad. Bulg. Sci.,* **18,** 231 (1965).

890. Shivahare, G. C., and R. C. Verma, *Trans. Soc. Adv. Electrochem. Sci. Technol.*, **4**, 81 (1969); through *Chem. Abstr.*, **72**, 106743u (1970).

891. Short, H. G., *Analyst (London)*, **76**, 710 (1951).

892. Shumilovskii, N. N., L. V. Mel'ttser, A. A. Kalmakov, and E. S. Kokumov, *Radioizotopnye Metody Avtomat. Kontrolya Akad. Nauk Kirg. SSR, Tr. Rasshiren Soveshch. Vses. Seminara Po Primeneniya Radioaktiun. Izotopov Izmeritel'n. Tekhn. Priborostr.*, **1**, 86 (1961); through *Chem. Abstr.*, **60**, 232c (1964).

893. Sillen, L. G., and A. E. Martell, *Stability Constants of Metal-Ion Complexes*, Special Publication No. 17, The Chemical Society, London, 1964.

894. Sillen, L. G., and A. E. Martell, *Stability Constants of Metal-Ion Complexes*, *Supplement No. 1*, Special Publication No. 25, The Chemical Society, London, 1971.

895. Simons, E. L., J. E. Fagel, and E. W. Balis, *Anal. Chem.*, **27**, 1123 (1955).

896. Simpson, S. G., W. C. Schumb, and M. A. Sieminski, *Ind. Eng. Chem. Anal. Ed.*, **10**, 243 (1938).

897. Sinclair, A. G., *Talanta*, **16**, 459 (1969).

898. Singh, S. S., *Z. Phys. Chem. (Frankfurt am Main)*, **56**, 249 (1967).

899. Sinha, S. N., and A. K. Dey, *Z. Anal. Chem.*, **183**, 182 (1961).

900. Sinkai, S., and T. Nagata, *J. Soc. Chem. Ind., Japan*, **42**, 397 (1939).

901. Sinyakova, S. I., and L. S. Chulkina, *Zh. Anal. Khim.*, **23**, 841 (1968).

902. Slizys, R., and I. Bubelis, *Tr. 3-ei [Tret'ei] Stud. Nauchn.-Tekhnol. Konf. Pribaltiki Belorus SSR, Latv. Gos. Univ. Petra Studhki, Riga*, **1958**, 29; through *Chem. Abstr.*, **55**, 9115d (1961).

903. Sljukic, M., N. Vuletic, B. Kojic-Prodic, and B. Matkovic, *Croat. Chem. Acta*, **43**, 133 (1971).

904. Smith, D. P., and M. T. Pope,*Anal. Chem.*, **40**, 1906 (1968).

905. Smith, E. J., and J. M. Fitzgerald, *Anal. Chim. Acta*, **60**, 367 (1972).

906. Smith, G. F., J. A. McHard, and K. L. Olson, *Ind. Eng. Chem. Anal. Ed.*, **8**, 350 (1936).

907. Smithells, C. J., *Tungsten*, Chapman-Hall, London, 1952.

908. Sollenberger, C. L., and J. Smith, *Anal. Chem.*, **23**, 1490 (1951).

909. Songina, O. A., and P. A. Karpova, *Zavod. Lab.*, **13**, 38 (1947).

910. Songina, O. A., V. Z. Kotlyarskaya, and A. P. Voiloshnikova, *Izv. Akad. Nauk Kaz. SSR, Ser. Khim.*, No. 8, 77 (1955); through *Chem. Abstr.*, **49**, 13022g (1955).

911. Songina, O. A., and M. T. Kozlovskii, *Zavod. Lab.*, **13**, 677 (1947).

912. Soremark, R., and B. Bergman, *Acta Isotopica*, **2**, 5 (1962).

913. Souchay, P., *Ann. Chim. (Paris)*, **19**, 102 (1944).

914. Souchay, P., and J. P. Launay, *C.R.H. Acad. Sci., Ser. C*, **268**, 1354 (1969).

915. Sousa, A. de, *Anal. Chim. Acta*, **7**, 24 (1952).

916. Sousa, A. de, *Mikrochem. Ver. Mikrochim. Acta*, **40**, 104 (1952).

917. Sousa, A. de, *Anal. Chim. Acta*, **10**, 29 (1954).

918. Souza Castelo, M. del P., and F. Bermejo Martinez, *Michrochem. J.*, **16**, 94 (1971).

919. Spaccamela Marchetti, E., and M. T. Cereti Mazza, *Ann. Chim. (Rome)*, **59**, 902 (1969).

920. Spackov, A., *Collect. Czech. Chem. Commun.*, **30**, 1255 (1965).

921. Spacu, P., C. Gheorghiu, and I. Paralescu, *Acad. Rep. Populare Romine Studii Cercet. Chim.*, **10**, 157 (1962); through *Chem. Abstr.*, **58**, 3881f (1963).

922. Spano, E. F., T. E. Green, and W. J. Campbell, US Bureau of Mines, Report of Investigations, No. 6308 (1963).

923. Spauszus, S., and H. Spanier, *Wiss. Z. Tech. Hochsch. Chem. Leuna-Merseburg*, **6**, 128 (1964); through *Chem. Abstr.*, **62**, 7095f (1965).

924. Speranskaya, E. F., *Izv. Vyssh. Uch. Zaved. Khim. Khim. Tekhnol.*, **6**, 195 (1963); through *Chem. Abstr.*, **59**, 12174a (1963).

925. Speranskaya, E. F., *Zh. Anal. Khim.*, **18**, 9 (1963).

926. Speranskaya, E. F., and D. B. Manbeeva, *Zh. Fiz. Khim.*, **39**, 1837 (1965).

927. Speranskaya, E. F., V. E. Mertsalova, and I. I. Kulev, *Usp. Khim.*, **35**, 2129 (1966).

928. Spiridonova, V. S., *Nov. Dannye Toksikol. Redk. Metal. Ikh Soedin.*, 1967, 77; through *Chem. Abstr.*, **70**, 40489t (1969).

929. Spitz, J., J. J. Chazee, and T. V. Danh, *Method. Phys. Anal.*, **4**, 375 (1968).

930. Stahl, W., *Chem.-Ztg.*, **56**, 175 (1932).

931. Stanton, R. E., *Aust. Inst. Mining Met., Proc.*, No. 236, 59 (1970).

932. Stashkova, N. V., V. V. Feofanova, and V. I. Kurbatova, *Tr. Vses. Nauchn.-Issled. Inst. Standartn. Obraztsov Spektral. Etalonov*, **3**, 47 (1967); through *Chem. Abstr.*, **69**, 64401p (1968).

933. Stepan, J., *Cesk. Farm.*, **15**, 43 (1966); through *Chem. Abstr.*, **65**, 1188f (1966).

934. Stepanova, N. A., and G. A. Yakunina, *Zh. Anal. Khim.*, **17**, 858 (1962).

935. Stever, K. R., J. L. Johnson, and H. H. Heady, *Advan. X-Ray Anal.*, **4**, 474 (1961).

936. Stolbova, E. D., and N. V. Sharshukova, *Sb. Nauch. Tr., Ryazansk. Med. Inst.*, **15**, 129 (1962); through *Chem. Abstr.*, **61**, 15252a (1964).

937. Stolyarov, K. P., *Vestn. Leningrad. Univ., Ser. Fiz. Khim.*, No. 4, 140 (1963); through *Chem. Abstr.*, **60**, 8638h (1964).

938. Stonehart, P., J. G. Koren, and J. S. Brinen, *Anal. Chim. Acta*, **40**, 65 (1968).

939. Strelow, F. W. E., and C. J. C. Bothma, *Anal. Chem.*, **39**, 595 (1967).

940. Strohal, P., D. Letic, and J. Tuta, *Croat. Chem. Acta*, **38**, 205 (1966).

941. Strohmeier, W., *Angew. Chem.*, **75**, 1024 (1963).

942. Studenikova, Z. V., M. I. Glinkina, and K. I. Kornilova, *Mezhdunarod. Geol. Kongress, 21-ya [Dvadtsat Pervaya] Sessiya Doklady Sovet. Geologov, Problema*, **1**, 178 (1960); through *Chem. Abstr.*, **55**, 20827a (1961).

943. Studenskaya, L. S., and G. N. Emasheva, *Tr. Vses. Nauchn.-Issled. Inst. Standartn. Obraztsov Spektr. Etalonov*, **1**, 18 (1964); through *Chem. Abstr.*, **64**, 14959f (1966).

944. Studenskaya, T. S., N. D. Fedorova, V. V. Stepin and V. L. Zolotavin, *Tr. Vses. Nauchn.-Issled. Inst. Standartn. Obraztsov Spektr. Elalonov*, **1**, 22 (1964); through *Chem. Abstr.*, **64**, 14937h (1965).

945. Suarez Acosta, R., *Inst. Hierro Acero, Madrid [Publ.]*, **11**, 177 (1958); through *Chem. Abstr.*, **53**, 1993g (1959).

946. Subbotina, G. A., and I. P. Manzhosov, *Mater. Ural. Soveshch. Spektroskopii, 4th, Sverdlousk*, **1963**, 128 (1965); through *Chem. Abstr.*, **64**, 18403e (1966).

947. Sucha, L., *Sb. Vys. Sk. Chem.-Technol. Praze, Anal. Chem.*, **3**, 113 (1968); through *Chem. Abstr.*, **72**, 106823x (1970).

948. Suchy, K., and E. Buchar, *Sb. Vys. Pedagog. (Prague), Prir. Vedy*, **1**, 111 (1957); through *Chem. Abstr.*, **52**, 14422b (1958).

949. Sugawara, K. F., *Anal. Chim. Acta*, **35**, 127 (1966).

950. Susic, M. V., *Bull. Boris Kidrich Inst. Nucl. Sci.*, **13**, 9 (1962); through *Chem. Abstr.*, **59**, 1285b (1963).

951. Susic, M. V., D. S. Veselinovic, and D. Z. Suznjevic, *Glasnik Hem. Drustva Beograd.*, **29**, 121 (1964); through *Chem. Abstr.*, **64**, 15364h (1966).

952. Suvorov, S. V., *Gig. Sanit.*, **28**, 24 (1963); through *Chem. Abstr.*, **59**, 4468d (1963).

953. Suzuki, M., *Bunseki Kagaku*, **19**, 207 (1970).

954. Svehla, A., and O. Kvopkova, *Hutn. Listy*, **16**, 588 (1961).

955. Sverak, J., *Z. Anal. Chem.*, **201**, 12 (1964).

956. Syre, R., *Niobium, Molybdenum, Tantalum, Tungsten—A Summary of Their Properties with Recommendations for Research and Development*, North Atlantic Treaty Organization, Advisory Group for Aeronautical Research and Development, Paris, 1961.

957. Sysoev, V. A., *Tr. Moskovskogo Tekhnol. Inst. Legkoi Prom. L.M. Kaganovicha*, No. 3, 169 (1941); through *Chem. Abstr.*, **40**, 3069[9] (1946).

958. Szeiman, S., *Kohasz Lapok*, **98**, 134 (1965); through *Chem. Abstr.*, **63**, 4930h (1965).

959. Szucs, A., and O. Klug, *Magy. Kem. Lapja*, **21**, 328 (1966); through *Chem. Abstr.*, **65**, 113341 (1966).

960. Szucs, A., and O. Klug, *Magy. Kem. Lapja*, **24**, 323 (1969); through *Chem. Abstr.*, **71**, 66996b (1969).

961. Tachikawa, T., *Bunseki Kagaku*, **14**, 697 (1965).

962. Takahashi, S., and M. Matsushima, *Bunseki Kagaku*, **15**, 511 (1966).

963. Talalaevskii, M., *The Use of Colorimetric Methods for Determination of Impurities and Admixters in Tungsten and Molybdenum Wires*, Saransk, Mordovsk. Kn. Izd., 1965; through *Chem. Abstr.*, **64**, 14969f (1966).

964. Tananaiko, M. M., O. N. Godovana, and N. I. Verblyudova, *Visn. Kiiv. Univ. Ser. Khim.*, No. 10, 7 (1969); through *Chem. Abstr.*, **74**, 19094h (1971).

965. Tandon, S. N., and J. S. Gill, *Talanta*, **20**, 585 (1973).

966. Tappe, W., and J. van Calker, *Z. Anal. Chem.*, **198**, 13 (1963).

967. Tarasevich, N. I., A. D. Khlystova, and K. A. Semenenko, *Tr. Kom. Anal. Khim., Akad. Nauk SSSR, Inst. Geokhim. Anal. Khim.*, **15**, 263 (1965); through *Chem. Abstr.*, **63**, 6310f (1965).

968. Tarasevich, N. I., and A. A. Zheleznova, *Zh. Anal. Khim.*, **18**, 1345 (1963).

969. Termendzhyan, Z. Z., and D. S. Gaibakyan, *Arm. Khim. Zh.*, **23**, 230 (1970).

970. Thomas, P. E., and W. F. Pickering, *Talanta*, **18**, 123 (1971).

971. Thompson, B. A., and P. D. LaFleur, *Anal. Chem.*, **41**, 852 (1969).

972. Titov, P. D., *Izv. Sib. Otd. Akad. Nauk SSSR, Ser. Khim. Nauk*, **1964**, 94; through *Chem. Abstr.*, **62**, 2227g (1965).

973. Tomkins, M. L., G. A. Borun, and W. A. Fahlbusch, *Anal. Chem.*, **34**, 1260 (1962).

974. Tong, Y. L., *Malaysia Min. Lands Mines, Ann. Rept. Geol. Surv.*, **1968**, 109 (1970); through *Chem. Abstr.*, **75**, 83800h (1971).

975. Touhey, W. O., and J. C. Redmond, *Anal. Chem.*, **20**, 202 (1948).

976. Tourne, C., *Bull. Soc. Chim. Fr.*, **1967**, 3199 (1967).

977. Tourne, C., *C.R.H. Acad. Sci., Ser. C*, **266**, 702 (1968).

978. Toy, C. H., and R. T. Van Santen, *Anal. Chem.*, **36**, 151 (1964).

979. Tramm, R. S., and E. G. Namvrina, *Khim. Metody Anal. Niobiya Splavov Ego Osn.*, **1970**, 29; through *Chem. Abstr.*, **76**, 41548x (1972).

980. Travers, A., *C.R.H. Acad. Sci.*, **116**, 416 (1918).

981. Treadwell, W. D., and R. Nieriker, *Helv. Chim. Acta*, **24**, 1098 (1941).

982. Troitskii, K. V., *Pr. Mech. At. Anal. Khim. Akad, Nauk SSSR, Inst. Geokhim. Anal. Khim.*, **1955**, 133; through *Chem. Abstr.*, **50**, 3940d (1956).

983. Tsigdinos, G. A., *Heteropoly Compounds of Molybdenum and Tungsten*, Bulletin Cdb-12a, Climax Molybdenum, Ann Arbor, Mich., 1969.

984. Tsigdinos, G. A., and C. T. Hallada, *Isopoly Compounds of Molybdenum, Tungsten and Vanadium*, Bulletin Cdb-14, Climax Molybdenum, Ann Arbor, Mich., 1969.

985. Tsigdinos, G. A., and F. W. Moore, in K. Niedenzu and H. Zimmer, Eds., *Annual Reports in Inorganic and General Syntheses-1973*, Academic, New York, 1974.

986. Tsukiyama, H., and I. Iwamoto, *Bunko Kenkyu*, **17**, 17 (1968); through *Chem. Abstr.*, **70**, 64009n (1969).

987. Tumanov, A. K., *Zavod. Lab.*, **26**, 1366 (1961).

988. Tumanova, T. G., *Sb. Tr. Chelyabinsk. Elektromet. Komb.*, No. 1, 185 (1968); through *Chem. Abstr.*, **71**, 56266z (1969).

989. Turov, P. P., and D. S. Turova, *Zh. Anal. Khim.*, **25**, 934 (1970).

990. Tytko, K. H., and O. Glemser, *Z. Naturforsch., B*, **25**, 429 (1970).

991. Tytko, K. H., and O. Glemser, *Z. Naturforsch., B*, **26**, 658 (1971).

992. Ueno, K., and T. Tachikawa, *Bunseki Kagaku*, **7**, 757 (1958).

993. Ul'ko, N. V., and M. L. Parubocha, *Visn. Kiiv. Univ., Ser. Khim.*, No. 10, 18 (1969); through *Chem. Abstr.*, **73**, 105040y (1970).

994. Ul'ko, N. V., and R. A. Savchenko, *Zh. Neorg. Khim.*, **12**, 328 (1967).

995. Umanskii, Ya. S, and S. S. Khidekel, *Zavod. Lab.*, **8**, 49 (1939).

996. Urusova, T. M., *Gig. Sanit.*, **34**, 72 (1969); through *Chem. Abstr.*, **71**, 63826k (1969).

997. Vadasdi, K., P. Buxbaum, and A. Salamon, *Anal. Chem.*, **43**, 318 (1971).

998. Vanossi, R., *An. Asoc., Quim. Argent.*, **40**, 176 (1952); through *Chem. Abstr.*, **47**, 3179h (1953).

999. Varga, G. M., E. Papaconstantinou, and M. T. Pope, *Inorg. Chem.*, **9**, 662 (1970).

1000. Varughese, K., and K. S. Rao, *Anal. Chim. Acta*, **57**, 219 (1971).

1001. Vasilenko, V. D., B. E. Reznik, and L. A. Nakonechnaya, *Tr. Kom. Anal. Khim. Akad. Nauk SSSR, Otd. Khim. Nauk*, **5**, 117 (1954); through *Chem. Abstr.*, **49**, 12181f (1955).

1002. Vasil'ev, R. I., and A. V. Yakovleva, *Opt. Spektrosk.*, **5**, 620 (1958).

1003. Vas'ko, A. T., *Electrochemistry of Tungsten*, Tekhnika, Kiev, 1969; through *Chem. Abstr.*, **73**, 41236g (1970).

1004. Vas'kova, A. G., V. B. Avilov, O. A. Kiseleva, and V. N. Kaminskaya, *Zavod. Lab.*, **35**, 795 (1969).

1005. Vecsernyes, L., *Z. Anal. Chem.*, **182**, 429 (1961).

1006. Vejdelek, Z. J., and B. Kakai, *Farbreaktionen in der spektrophotometrische Analyse organischen Verbindungun*, Band II, Veb. Gustav Fischer Verlag, Jena, 1973.

1007. Veleker, T. J., *Anal. Chem.*, **32**, 1181 (1960).

1008. Vengerskaya, Kh. Ya., *Metody Opred. Vredn. Veshchestv. Vozdukhe, Moscow Sb.*, **1961**, 32; through *Chem. Abstr.*, **58**, 8404d (1963).

1009. Vengerskaya, Kh. Ya., and S. S. Salikhodzheav, *Gig. Tr. Prof. Zabol.*, **6**, 27 (1962); through *Chem. Abstr.*, **57**, 8843c (1962).

1010. Vicente-perez, S., I. Martin del Molino, and F. Lucena-Conde, *Inform. Quim. Anal. (Madrid)*, **20**, 151 (1966).

1011. Vidal, G., P. Galmard, and P. Lanusse, *Rech. Aerosp.*, No. 130, 27 (1969).

1012. Vignoli, L., B. Cristau, and A. Pfister, *Bull. Soc. Pharm. Marseille*, **4**, 71 (1955).

1013. Vignoli, L., B. Cristau, and A. Pfister, *Chim. Anal. (Paris)*, **38**, 392 (1956).

1014. Vinogradov, A. V., and M. I. Dronova, *Zh. Anal. Khim.*, **20**, 343 (1965).

1015. Vinogradov, A. V., and M. I. Dronova, *Zh. Anal. Khim.*, **23**, 696 (1968).

1016. Volf, M. B., *Veda Vyzkum Prumyslu Sklarskem*, **7**, 25 (1961); through *Chem. Abstr.*, **57**, 14719i (1962).

1017. Vonderbrink, S. A., *Diss. Abstr.*, **27B**, 1764 (1966).

1018. Vorsatz, B., and S. Haltenberger, *Magy. Tudomanyos Akad. Koz. Fiz. Kutato Intezetenek Kozlemenyei*, **3**, 5 (1955); through *Chem. Abstr.*, **52**, 16123h (1958).

1019. Vovsi, A. M., M. M. Grinblat, E. R. Borshchevskaya, *Tr. Leningrad. Metal. Zavod.*, **1962**, 264; through *Chem. Abstr.*, **60**, 4789h (1964).

1020. Wallace, G. W., and M. G. Mellon, *Anal. Chem.*, **32**, 204 (1961).

1021. Walton, R. A., *Chem. Commun.*, **1968**, 1385 (1968).

1022. Wang, C. H., *Hua Hsueh Shih Chieh*, **14**, 298 (1959); through *Chem. Abstr.*, **54**, 20663b (1960).

1023. Wang, M. L., and K. J. Li, *Hua Hsueh Shih Chieh*, **14**, 119 (1959); through *Chem. Abstr.*, **53**, 21380e (1959).

1024. Ward, F. N., *US Geol. Survey, Circ.*, **119**, 1 (1951).

1025. Waters, K. L., C. J. Brockman, and W. H. Waggoner, *J. Chem. Educ.*, **34**, 137 (1957).

1026. Watkinson, J. H., *New Zealand J. Sci.*, **1**, 201 (1958).

1027. Weakley, T. J. R., in Dunitz, J. D., P. Hemmerich, J. A. Ibers, C. K. Jorgensen, J. B. Neilands, R. S. Nyholm, D. Reinen, and R. J. P. Williams, Eds., *Structure and Bonding*, Vol. 18, Springer-Verlag, Berlin, 1974.

1028. Weakley, T. J. R., and R. D. Peacock, *J. Chem. Soc. A*, **1971**, 1836 (1971).

1029. Wechter, M. A., P. B. Hahn, G. M. Ebert, P. R. Montoya, and A. F. Voigt, *Anal. Chem.*, **45**, 1267 (1973).

1030. Wechter, M. A., H. R. Shanks, G. Carter, G. M. Ebert, G. Guglielmino, and A. F. Voigt, *Anal. Chem.*, **44**, 850 (1972).

1031. Wechter, M. A., and A. F. Voigt, *Anal. Chem.*, **38,** 1681 (1966).

1032. Wechter, M. A., and A. F. Voigt, *Anal. Chim. Acta,* **41,** 181 (1968).

1033. Weeks, M. E., *Discovery of the Elements,* 7th edit., Chemical Education, Easton, Pa., 1968.

1034. Weiss, I., *Rev. Brasil Quim.,* **20,** 337 (1945).

1035. Welcher, G. G., and O. H. Kriege, *At. Absorption Newslett.,* **8,** 97 (1969).

1036. Welcher, G. G., and O. H. Kriege, *At. Absorption Newslett.,* **9,** 61 (1970).

1037. Welford, G. A., W. R. Collins, R. S. Mose, and D. Sutton, *Talanta,* **5,** 168 (1960).

1038. Wendling, E., *Rev. Chim. Miner.,* **4,** 425 (1967).

1039. Wendt, R. H., and V. A. Fasse, *Spectrochim. Acta,* **21,** 1691 (1965).

1040. Wenger, P., and E. Rogovine, *Helv. Chim. Acta,* **10,** 242 (1927).

1041. Wester, P. O., D. Brune, and K. Samsahl, *Intern. J. Appl. Radition Isotopes,* **15,** 59 (1964).

1042. Westwood, W., and A. Meyer, *Analyst (London),* **72,** 464 (1947).

1043. Wilkins, D. H., *Anal. Chim. Acta,* **19,** 440 (1958).

1044. Wilkins, D. H., *Talanta,* **2,** 355 (1959).

1045. Wilkinson, W. D., "Tungsten," in *Encyclopedia of Science and Technology,* Vol. 14, McGraw-Hill, New York, 1971.

1046. Willard, H. H., and H. Diehl, *Advanced Quantitative Analysis,* Van Nostrand, Princeton, N.J., 1943.

1047. Wilson, D. O., and J. F. Cline, *Nature,* **209,** 941 (1966).

1048. Wilson, S. H., and M. Fields, *Analyst (London),* **69,** 12 (1944).

1049. Winterstein, C., *Z. Erzbergbau Metall.,* **10,** 549 (1957); through *Chem. Abstr.,* **52,** 4404d (1958).

1050. Winterstein, C., *Z. Erzbergbau Metall.,* **13,** 546 (1960); through *Chem. Abstr.,* **58,** 4242c (1961).

1051. Wirtz, H., *Metall. Enz.,* **41,** 84 (1944).

1052. Witwit, A. S., and R. J. Magee, *Anal. Chim. Acta,* **27,** 366 (1962).

1053. Witwit, A. S., R. J. Magee, and C. L. Wilson, *Talanta,* **9,** 495 (1962).

1054. Wohlmann, E., *Geologie (Berlin),* **13,** 845 (1964).

1055. Wood, D. F., and R. T. Clark, *Analyst (London),* **83,** 326 (1958).

1056. Wood, D. F., and R. T. Clark, *Analyst (London),* **87,** 342 (1962).

1057. Wood, D. F., and J. T. Jones, *Analyst (London),* **90,** 125 (1965).

1058. Wood, D. F., and J. T. Jones, *Analyst (London),* **93,** 131 (1968).

1059. Wood, D. F., and G. Wolfenden, *Anal. Chim. Acta,* **38,** 385 (1967).

1060. Wybenga, F. T., *Appl. Spectrosc.,* **19,** 193 (1965).

1061. Yakovlev, P. Ya., R. D. Malinina, G. P. Razumova, and M. S. Dymova, *Teoriya Praktika Polyarograf. Anal. Akad. Nauk Moldausk. SSR, Materialy Pervogo Vses. Soveshch.,* **1962,** 198; through *Chem. Abstr.,* **59,** 2154h (1963).

1062. Yakovleva, E. F., *Sb. Tr. Tsentr. Nauchn.-Issled. Inst. Chern. Met.,* No. 19, 57 (1960); through *Chem. Abstr.,* **56,** 9414e (1962).

1063. Yakowitz, H., R. E. Michaelis, and D. L. Vieth, *Advan. X-Ray Anal.,* **12,** 418 (1969).

1064. Yamaguchi, N., A. Hata, and M. Hasegawa, *Bunseki Kagaku*, **17**, 1118 (1968).

1065. Yatsimirskii, K. B., *Catalysis Chem. Kinet.*, **1964**, 201 (1964).

1066. Yatsimirskii, K. B., and L. I. Budarin, *Zh. Neorg. Khim.*, **7**, 1824 (1962).

1067. Yatsimirskii, K. B., and E. F. Naryshkina, *Zh. Neorg. Khim.*, **3**, 346 (1958).

1068. Yatsimirskii, K. B., and K. E. Prik, *Zh. Neorg. Khim.*, **9**, 178 (1964).

1069. Yatsimirskii, K. B., and K. E. Prik, *Zh. Neorg. Khim.*, **9**, 1838 (1964).

1070. Yatsimirskii, K. B., and V. I. Rigin, *Zh. Anal. Khim.*, **13**, 112 (1958).

1071. Yatsimirskii, K. B., and V. F. Romanov, *Zh. Neorg. Khim.*, **9**, 1578 (1964).

1072. Yoshimura, C., S. Uno, and H. Noguchi, *Bunseki Kagaku*, **12**, 42 (1963).

1073. Yoshimura, K., and M. Morita, *Bull. Nat. Hyg. Lab., Tokyo*, No. 73, 141 (1955); through *Chem. Abstr.*, **50**, 7483d (1956).

1074. Youinou, M. T., and I. E. Guerchais, *Bull. Soc. Chim. Fr.*, **1968**, 40 (1968).

1075. Zaidel', A. N., V. K. Profof'ev, S. M. Raiskii, V. A. Slavnyi, and E. Ya. Shreider, *Tables of Spectral Lines*, IFI/Plenum, New York, 1970.

1076. Zakhariya, N. F., and A. I. Staikov, *Zh. Anal. Khim.*, **26**, 1348 (1971).

1077. Zarinskii, V. A., M. M. Farafonov, and V. V. Zateeva, *Zh. Anal. Khim.*, **12**, 677 (1957).

1078. Zarzecka, E., *Nukleonika*, **9**, 759 (1964); through *Chem. Abstr.*, **62**, 5889h (1965).

1079. Zato, J., *An. Real. Soc. Espan. Fis. Quim., Ser. B*, **50**, 977 (1954).

1080. Zato, J., *Inform. Quim. Anal. (Madrid)*, **9**, 192 (1955).

1081. Zato, J., *Inform. Quim. Anal. (Madrid)*, **10**, 161 (1956).

1082. Zato, J., *Inform. Quim. Anal. (Madrid)*, **12**, 68 (1958).

1083. Zelenina, T. P., K. F. Gladysheva, and L. D. Zinov'eva, *Sb. Nauch. Tr. Vses. Nauchn.-Issled. Gorno-Met. Inst. Tsvent. Metal.*, No. 9, 124 (1965); through *Chem. Abstr.*, **63**, 14039h (1965).

1084. Zelikman, A. N., O. E. Krein, L. A. Nisel'son, and Z. I. Ivanova, *Zh. Prikl. Khim.*, **35**, 1467 (1962).

1085. Zharovskii, F. G., and D. O. Gorina, *Zh. Anal. Khim.*, **26**, 766 (1971).

1086. Zhivopistsev, V. P., I. N. Ponosov, and E. A. Selezneva, *Zh. Anal. Khim.*, **18**, 1432 (1963).

1087. Zhivopistsev, V. P., and G. A. Sadakov, *Uch. Zap. Perm. Gos. Univ.*, No. 14, 174 (1966); through *Chem. Abstr.*, **68**, 56321b (1968).

1088. Zopatti, L. P., and E. N. Pollock, *Anal. Chim. Acta*, **32**, 178 (1965).

1089. Zvyagina, L. N., M. Ya. Grankina, and G. V. Fominykh, *Opred. Mikroprimesei*, No. 2, 78 (1968); through *Chem. Abstr.*, **71**, 108750z (1969).

1090. Zyka, J., *Pharmazie*, **10**, 170 (1955).

INDEX